AF616613

BOSTON STUDIES IN THE PHILOSOPHY OF SCIENCE

VOLUME VIII

SYNTHESE LIBRARY

MONOGRAPHS ON EPISTEMOLOGY,

LOGIC, METHODOLOGY, PHILOSOPHY OF SCIENCE,

SOCIOLOGY OF SCIENCE AND OF KNOWLEDGE,

AND ON THE MATHEMATICAL METHODS OF

SOCIAL AND BEHAVIORAL SCIENCES

BOSTON STUDIES IN THE PHILOSOPHY OF SCIENCE

VOLUME VIII

PSA 1970

IN MEMORY OF RUDOLF CARNAP

PROCEEDINGS OF THE 1970 BIENNIAL MEETING
PHILOSOPHY OF SCIENCE ASSOCIATION

EDITED BY ROGER C. BUCK AND ROBERT S. COHEN

D. REIDEL PUBLISHING COMPANY / DORDRECHT-HOLLAND

Library of Congress Catalog Card Number 73–20858

ISBN 90 277 0187 3

Printed in The Netherlands by D. Reidel, Dordrecht

PREFACE

This book contains the papers presented at the second biennial meeting of the Philosophy of Science Association, held in Boston in Fall, 1970. We have added the paper by Jaakko Hintikka which he was unable to present due to illness, and we have unfortunately not received the paper of Michael Scriven. Otherwise, these proceedings are complete so far as formal presentations.

The meeting itself was dedicated to the memory of Rudolf Carnap. This great man and distinguished philosopher had died shortly before. The five talks from the session devoted to recollections of Professor Carnap are printed at the beginning of this book, and they are followed by eight other tributes and memories. We are particularly grateful to Wolfgang Stegmüller for permitting us to include a translation of his *éloge* which was broadcast in Germany. The photographs were kindly contributed by Hannah Thost-Carnap.

ROGER C. BUCK

Department of History and Philosophy of Science,
Indiana University

ROBERT S. COHEN

Boston Center for the Philosophy of Science,
Boston University

Photograph by Francis Schmidt, 1935

Photograph by Adya, 1962

TABLE OF CONTENTS

CONTRIBUTED PAPERS

I. Observation

II. Philosophical Problems of Biology

III. Equivalence, Analyticity, and In-Principle Confirmability

IV. Probability, Statistics and Acceptance

V. Problems in Quantum Physics; Genetic Epistemology

VI. Theoretical Pluralism; Understanding; Methodological Agreement

VII. Induction and Reduction

VIII. Scientific Theories: Comparison and Change

HOMAGE TO RUDOLF CARNAP

At the Boston meeting, October 23rd, 1970, Adolf Grünbaum, President of the Philosophy of Science Association, introduced the Carnap Memorial meeting.

> We are assembled here to pay tribute to Rudolf Carnap because his work made an immense contribution to 20th century philosophy of science. But we also honor his memory because his life as a human being enriched the lives of those who had personal contact with him professionally or otherwise. A few of these colleagues, friends or former students of his will address you now. Herbert Feigl, Richard Jeffrey, Willard Quine and Abner Shimony will do so in person themselves. Carl Hempel is still recovering from eye surgery and has asked me to read his tribute to you on his behalf.

1

Our great master and good friend, Rudolf Carnap, passed away on September 14, 1970. His death came rather swiftly (at the age of 79), after a brief but severe illness. The world has lost one of its truly great thinkers. After Bertrand Russell who, perhaps along with Gottlob Frege, was his prime 'Father figure', Carnap was in the opinion (not only) of his disciples one of the greatest logicians and philosophers of science of our century. Since it is impossible to summarize in a few minutes his tremendously important contributions, I shall take the liberty of reminiscing a bit about the forty-four years of his friendship I was privileged to enjoy.

There is a great deal about Carnap's life and work that can be learned from the volume *The Philosophy of Rudolf Carnap*, that magnificent book in the *Library of Living Philosophers* (Vol. II, Open Court, 1963) edited by Paul Arthur Schilpp. There is a highly interesting autobiographical account of Carnap's intellectual development, and there are also his many thoroughgoing replies, remarks and rebuttals in response to the tweny-six articles by disciples as well as critics. Carnap devoted a great deal of thought and energy to these parts written by him for the Schilpp volume.

Even before Carnap came for his first visit to the Vienna Circle (1925), some of us – especially Schlick, Hahn, Waismann and I, had read some of his early work (*Der Raum*, *Die Aufgabe der Physik*, and the voluminous typescript which was later to become – in somewhat abridged form – his first major contribution to epistemology, published under the somewhat romantic title (dreamed up by Schlick) *Der Logische Aufbau der Welt.* I was tremendously impressed by this exact logical reconstruction of the concepts of empirical knowledge. Once I came to know Carnap better personally, i.e., after he had moved to Vienna, I kept pestering him with my objections to his phenomenalism (i.e., the more exact and much more fully worked out development of the epistemologies of Mach, Avenarius and the early B. Russell). Although I pleaded for a critically realist view (essentially along the lines of, e.g. O. Külpe and the early Schlick), Carnap was firm in what he took to be a 'metaphysically neutral' stand – rejecting as cognitively meaningless both subjective idealism and critical realism. But in contradistinction to the positivism of Mach and Ostwald he always emphasized his 'empirical realism' in regard to the 'external world', the entities of microphysics, and of scientific theories in general. Carnap retained, though in a semantically refined and elaborated form, his metaphysical neutrality. This was explicitly stated in his by now justly famous and 'classical' article, 'Empiricism, Semantics, and Ontology' (1951). Ontologically oriented metaphysicians have often been surprised over Carnap's indifference to what they consider matters of profound truth or falsity. Given Carnap's distinction between questions concerning the categorial frame of knowledge, and questions pertaining to issues within a given frame, one can understand (even if one does not fully agree) that he regarded 'ontological' questions as issues about the choice of a language.

I remember vividly Carnap's first lecture (1925) to the Vienna Circle. He presented his Space-Time topology (the 'C-T System', cf. *Introduction to Symbolic Logic*, Chapter G, p. 197 ff.) in the manner an engineer might explain the structure of a machine he had just invented. To the non-logicians Carnap indeed seemed to be no philosopher at all. Some of this sort of misunderstanding was Carnap's fate throughout most of his long life. But this did not seem to disturb him at all.

As I recall it, Moritz Schlick and Hans Hahn were to decide on an appointment of a *Privatdozent* in Philosophy at the University of Vienna (in 1925–26). They found it difficult to choose between Carnap and Hans Reichenbach who by then were good friends of each other, as well as of Schlick's. I am sure it was Hans Hahn, the great Vienna mathematician and admirer of Bertrand Russell who finally swayed the committee in favor of Carnap. Hahn felt that Carnap was in the process of fulfilling the program that Russell had only informally sketched in his *Our Knowledge of the External World.* This, in fact, was Carnap's achievement in the *Logische Aufbau.* (Impressive developments, such as Nelson Goodman's *The Structure of Appearance* were to follow).

Beginning with a (second) reading and discussion of Wittgenstein's *Tractatus,* Carnap clearly became the leading spirit of the Vienna Circle. His task-oriented, unemotional, (hardly ever ego-involved), penetrating and thorough ways of work, thought and expression impressed most of us much more than the intuitive (really basically artistic, though ingenious) manner of philosophizing of the elusive and hermitic Wittgenstein. (Only Schlick and Waismann succumbed to the almost hypnotic spell of Wittgenstein.)

Carnap characterized himself as 'schizoid' (rather than 'cyclothymic'), as 'introvert' and 'cerebrotonic'.

I remember him always as extremely rational, calm, well-balanced, eager to absorb new information, never reluctant to change his mind even in vitally important matters. While his great sense of humor was not immediately obvious even to his close friends, it did manifest itself in his occasional irony (and even in 'pulling someone's leg'). The chapter (12) on 'The Magic View of Language' in his *Philosophical Foundations of Physics* (Basic Books, New York, London, 1966) will surprise readers who know Carnap only from reading his austerely 'exact' books.

Carnap, originally very much like Schlick, was reluctant to propagandize and proselytize in behalf of the Vienna Circle's Logical Positivism (or Logical Empiricism). His was a quiet, sober way of teaching and discussing – with colleagues as well as with students. But the enthusiasm and the enormous energy of his friend Otto Neurath proved infectious. Thus he collaborated wholeheartedly in the composition of our 'declaration of independence' (from traditional philosophy) in the famous pamphlet: *Wissenschaftliche Weltauffassung der Wiener Kreis* (1929). For the ten following years he was an active co-editor, with Hans Reichenbach, of the periodical *Erkenntnis*. Many articles of the greatest significance for logical theory, foundations of mathematics, epistemology, and the philosophy of the empirical sciences, etc., appeared in that memorable publication. There were important contributions by Carnap, Reichenbach, M. Schlick, F. Waismann, O. Neurath, P. Frank, E. Schrödinger, P. Jordan, C. G. Hempel, W. Köhler, P. Hertz, M. Strauss, K. Ajdukiewicz, K. Grelling, and P. Oppenheim, A. J. Ayer, E. Nagel, and many others.

Carnap was also a key figure in the Unity of Science Congresses (e.g., Prague 1929, Königsberg 1930, Paris 1935, etc.). I also managed to persuade him to speak in a special session (with long discussion, eminent discussants) of the International Congress of Philosophy in Mexico City, 1963.

On several occasions I was privileged to watch (and even in a very minor way, to catalyze) some of the major forward steps of Carnap's thought. I remember a long walk and discussion in the beautiful Türkenschanzpark of Vienna. It was then (perhaps in 1928) that Carnap developed a first sketch of the idea of a logical syntax of language. Quick with labels for new conceptions (rather than carrying them out in exact detail) I told Carnap that the syntax (he then still called it 'semantics', a term which he applied – following Tarski after 1934 or 1935 in the later well-established sense) he was thinking of as formulated in a metalanguage, amounted to a 'Hilbertization' of *Principia Mathematica*. He accepted, smilingly, my designation as essentially correct.

I believe Carnap was the first to understand and to endorse Kurt Gödel's epochmaking discovery (made, I think, in 1930 – first published in 1931) regarding the essential incompleteness of a large class of mathematical systems. Gödel, in turn, had utilized Carnap's (metalinguistic, i.e., syntactical techniques) in his famous proof.

Throughout the many years of our intimate friendship I visited Carnap in Chicago, at Harvard (Ernest Nagel was with us there in 1936), a summer in Maine, Princeton, Sante Fe and Los Angeles, usually once a year, sometimes for several weeks, sometimes just for a few days. (In the summers of 1930, 1931, 1932, and 1935 we spent wonderful vacations together in the Austrian Alps.) It was during the summer vacation of 1932 (in Burgstein, Tyrol) that I introduced Karl Popper to Carnap. Several weeks of exciting discussions in that beautiful mountain spot ensued. Popper reported briefly about that episode in his contribution to the Carnap-Schilpp Volume. As is well known, Carnap and Popper diverged sharply later, especially in matters of inductive logic. Other, to me unforgettable, occasions were our reunion (for the autumn quarter) of 1940 at Harvard – with a vertitable 'galaxy' of stars: B. Russell, A. Tarski, R. von Mises, E. G. Boring, S. S. Stevens, W. V. O. Quine, I. A. Richards, and others; and the small conference in St. Louis, 1946 with intensive discussions between Carnap, H. Reichenbach, Norbert Wiener *et al.* on issues of probability and induction. Carnap and I conducted a seminar together with Paul Feyerabend and F. von Hayek at Alpbach, Tyrol in 1964; and colloquia at the University of Hawaii, as well as a most pleasant vacation in the Hawaiian Islands in 1967. We exchanged letters frequently, and had occasional long distance telephone conversations between Los Angeles and Minneapolis during the last few years. Physicialism, Philosophy of Physics, Unity of Science, Semantics, Psychology and Logical Behaviorism, and Inductive Logic were the major topics of our many conversations. Carnap was always a patient and attentive listener – and a most constructive critic of my (often deviant) ideas.

In 1938 we stayed on for two or three days after the meeting of the American Philosophical Association at Urbana, Illinois and I urged Carnap to apply his enormous analytic powers to the problems of induction and probability (this had been one of *my* hobbies ever since 1922 – but Carnap had paid only scant attention to these matters, leaving them largely to R. von Mises and Reichenbach, later also to Ernest Nagel). Carnap immediately began sketching in many hours of intensive discussion of what later became his great and influential work in Inductive Logic. Even if his (still highly controversial) contributions to this field of thought will eventually be displaced by different approaches (some already in existence!), there can be no question as to their great importance. If Carnap failed in this ambitious, (and I don't say he did) enterprise (of more than 30 years) he failed as nobly as did (if I may mention a very distant analogue) Einstein with his Unified Field Theory. Carnap's work, as well as his socialist-pacifist-world federation ideals (here he always acknowledged the incisive influence of his late friend Otto Neurath) will live on. His work will long continue to be influential, and his great personality will be remembered and cherished by his friends as long as they live!

Minnesota Center for the Philosophy of Science, HERBERT FEIGL
University of Minnesota

REFERENCES

For more detailed accounts of Carnap's role in the Vienna Circle and in recent philosophy generally, see his autobiography in the Schilpp Volume; also V. Kraft, *The Vienna Circle* (A. Pap, transl.) Philosophical Library, New York, 1953; J. Jorgensen's *The Development of Logical Empiricism* in *International Encyclopedia of Unified Science*, Vol. II, No. 9, 1951; A. J. Ayer (ed.) *Logical Positivism*, The Free Press, New York, 1959; and H. Feigl, 'The Wiener Kreis in America' in D. Fleming and B. Bailyn (eds.) *The Intellectual Migration, Europe and America 1930–1960*, Harvard University Press, 1969.

2

At this gathering in memory of Rudolf Carnap, it may be fit for me to record a few of the impressions of Carnap the man and Carnap the philosopher that I formed as his student and later also as his friend.

I first met Carnap in 1929, at the Prague conference on the epistemology of the exact sciences. I had gone there from Berlin to meet the author of two books, published the year before, which had struck me as singularly powerful and illuminating: *Der Logische Aufbau der Welt* and *Scheinprobleme in der Philosophie*. Living in a philosophical climate where loose speculative thinking and convoluted writing were often thought marks of profundity, I was captivated by the extraordinary lucidity, rigor, and solidity of Carnap's analyses and reconstructions, as well as by the boldness and ingenuity of the logical imagination from which they had sprung.

With his neat distinction between cognitively meaningful and meaningless components in philosophical problems and doctrines, Carnap appeared to me to hold the key to the solution or dissolution of any philosophical perplexity; his mode of analysis seemed to put all philosophical issues into a hard, clear light, leaving nothing in a mysterious penumbra, nothing in a frustrating fog.

On meeting him in Prague, I raised the question of coming to Vienna for the following semester to study with him, and Carnap readily agreed to let me attend his seminar in addition to his lecture course. In subsequent years, I visited him for shorter periods in Vienna and then in Prague; and it was Carnap who made it possible for me in 1937 to come to this country when he secured a year's assistantship for me at the University of Chicago, where he was then teaching.

Carnap was a dedicated teacher. His lectures were carefully planned and organized and were delivered in a calm and sober manner, without any trace of histrionics. His style of lecturing and of discussing bespoke a deep and inspiring concern with the exploration of philosophical issues and an abiding dedication, so evident also in his writings, to the highest standards of clarity, precision, and cogency of reasoning.

His students must have appreciated his open-minded readiness to listen to their questions and to given serious attention to their objections – an attitude that stood in marked contrast to the traditional aloofness of most professors at continental universities. He was fair and even-tempered; and he never put down a student by ridicule or by display of professorial superiority. Pretentiousness and self-importance were totally and naturally alien to him.

His openness to objections, whether in class discussion or in professional exchanges, was genuine: he treated a debate not as a duel aimed at scoring points against an opponent, but as a joint effort to advance knowledge and understanding; and he was prepared, therefore, to change his views in response to telling arguments.

As is well known, he did change his views on various subjects: he broadened the earlier testability requirement for empirically significant sentences into a requirement of confirmability, and later liberalized it further in an effort to give an adequate account of the logical status of theoretical terms. He similarly liberalized, step by step, the idea that scientific terms should be definable by means of an observational vocabulary; he broadened his conception of philosophy as the logical syntax of the language of science by introducing semantic and pragmatic considerations as well. In these and other cases, Carnap characteristically did not just abandon an earlier theoretical conception which had shown itself to be inadequate: he replaced it by a carefully elaborated new, more flexible and inclusive conceptual system.

Carnap's readiness to make such changes has sometimes been referred to an ironical overtones, almost as though it reflected a weakness of moral fibre, akin to a betrayal of one's professed moral principles in times of trial. But surely, readiness to change one's beliefs in response to pertinent reasons is a prerequisite for all objective theoretical inquiry; and in adhering to this standard, Carnap was implementing his dedication to the conception of philosophical exploration and discussion as objective theoretical pursuits.

Carnap had a wide-ranging and intense intellectual curiosity that extended far beyond the borders of his philosophical work; he had advanced training in mathematics and physics; he read and thought a good deal about psychological, social, and political issues; and he wondered about future developments in these fields. He was intrigued by science fiction; and in a conversation about suspended animation, Carnap once remarked that if it were feasible, he might exchange whatever years remained in his present life for a change of being put into that quiescent state and resuscitated at intervals, to witness different stages of the future.

He kept steadily at work, even when a painful back obliged him to remain stretched out on his couch; and in the waking hours during the days of his final illness, he spoke of his plans for further work.

To observers who did not know him closely, Carnap may well have appeared as intellect and rationality incarnate, as a powerful thinker moving in a theoretical realm far distant from the domain of human passions and hopes and fears; but this picture does not do him justice. Carnap formed very strong human bonds; he was a warm-hearted and loyal friend, a most interested and sympathetic listener, and a broad-minded observer of the human scene who sought earnestly and with imagination to understand – sometimes in psychoanalytic terms – personalities of a case quite different from his own.

His deep social and political commitments and his human compassion speak from a memorandum that he must have written not long before his death. It is a report, sent to a number of friends and colleagues, about a visit he paid in January 1970, when he was nearly 79, to two Mexican philosophers in a prison in Mexico City, where they had been detained as political prisoners since September 1968. Carnap describes how, in the presence of police guards, he talked with the two men, de Gortari and Molina, about their interests, their work, and their prospects for release; and he ends with a passage from which I would like to quote:

> Mrs. de Gortari (who had accompanied Carnap) handed me two blank cards and asked me to write a few words for each of the two philosophers .... I wrote for each of them some words of admiration for their fortitude, tenacity, and stoic equanimity with which they bear their hard fate, devoting their time to positive, fruitful work; and I also expressed the hope that the day of liberation would not be too much delayed .... Both men read the cards and were visibly moved; they said that they would keep and cherish the cards forever.
>
> Suddenly trumpets and drums gave the signal for the end of the visiting period. I

took a very cordial farewell with an abrazo from de Gortari and then from Molina. They expressed their thanks very warmly; they said that this had been their best day since September 1968. Then I walked with Mrs. de Gortari the long way between the rows of barracks. Several times I looked back; I saw the two men standing in the doorway and I waved my hand to them.

The words we are saying tonight are words that we would have like to write on little cards for Carnap: alas, it is too late to give them to him.

Princeton University CARL G. HEMPEL

3

While we mourn Carnap's death, let us not omit to celebrate his life·

I knew him during his last twenty-five years – the period which he devoted primarily to inductive logic – so I will talk about that, and say a bit about how he seemed, through my eyes, at various stages.

I first heard of Carnap in a metaphysics seminar at Boston University, when I was seventeen. Professor Brightman had mentioned that the Logical Positivists (who were atheists) held metaphysics to be meaningless, and that seemed right. I elected to write a paper on them, and became captivated by a little book – a 'Psyche Miniature' – by Rudolf Carnap: *Philosophy and Logical Syntax.* I never looked back. As soon as I could, after the war, I went to Chicago and took all of Carnap's courses. (During the war, stationed for a while in Chicago, I had managed to visit one of his classes: I sat at the back of the room, full of reverence and misery and longing, and never spoke to him.)

After five years, having exhausted my government subsidy, I took my M. A. and went out into the Great World, which I found wanting. So eventually I returned to Academia – to Princeton – and wrote a doctoral dissertation with Hempel and Putnam on Carnapian inductive logic. Following an implausible suggestion of Hempel's, I sent a copy to Carnap, *and he read it*! (I did not know, then, that he read everything people sent him, and took notes, and wrote long, careful replies. Of course, he had read my M. A. thesis in Chicago, but that was his job!) In this reply, he pointed out that unless he had missed something, there was a serious error in the last chapter: A construction which I had assumed would yield probability measures would not do so. John Kemeny had proved as much, and his proof was reproduced in some dittographed lecture notes of

Carnap's, which he sent me. (The notes also reported work by my friend and fellow-student at Chicago, Abner Shimony.) I consoled myself with the reflection that it was too late for Princeton to withdraw my doctorate; that the rest of the dissertation survived; and that the final section of Carnap's monograph, *The Continuum of Inductive Methods*, was tarred with the same brush.

That was in 1957. I was in England at the time. Carnap suggested that when I returned to the United States, I visit him and talk about probability and induction. He wrote to me as to a kindred spirit whose opinion he valued. I was thrilled. I went.

It turned out that his system of inductive logic was a work in progress, and that much had changed since Chicago – since the massive tome, *Logical Foundations of Probability* (1950). I had attended his lectures on inductive logic, in the late forties, but had heard them rather passively. I had had some further thoughts about the matter at Princeton where, in Hempel's seminar, I had turned away from Carnap's conception of probability, toward the subjectivistic or personalistic view of Ramsey and DeFinetti and Savage; but it had never occurred to me that Carnap himself had changed. But in 1958 I learned that he had never stopped changing.

The *Logical Foundations of Probability* had been projected as the first of two volumes under the general title, 'Probability and Induction'. *The Continuum of Inductive Methods* (1952) was to have been a preliminary study for the second volume. But *The Continuum* was no sooner published than vitiated, in Carnap's eyes: None of the *c*-functions in the continuum were adequately responsive to analogy influence. In 1952–3, at the Institute for Advanced Study, Carnap and Kemeny developed a generalization of the continuum, within which they hoped that an adequate inductive method might be found. (Carnap had already worked out a more limited generalization, which was eventually published in 1959, as *Anhang B*, Section VIII of Carnap and Stegmüller, *Induktive Logik und Wahrscheinlichkeit*.)

My wife and I spent the summer of 1959 in Los Angeles, where I worked with Carnap nearly every day. (He worked *every* day, especially Sundays, on grounds of atheism.) Kemeny came out for a week or so, and got an enormous amount done, as did Haim Gaifman, who was then Carnap's assistant. We would talk for three hours in the morning or the afternoon, usually sitting in the garden. At issue might be anything from a bit of notation to the very core of Carnap's view of induction: Carnap would bring to it the same calm reluctance to abandon debate until all participants were of one mind, for he had no doubt of the universality of reason, at least in philosophical matters. Your reasons for judgement must have equal authority for me, if they are genuinely reasons; and if they are not, best suspend judgement. Mere conviction was of no interest to Carnap.

Memories of Carnap are inseparable from memories of ina, I find. (Ina was Mrs. Carnap. She invented her name, and insisted on the lower case 'i'.) She was in the background throughout the philosophical talks, and was a major participant (although not at great length) in the talks about about everything else at mealtimes. It was her idea, to replace Volume II of *Probability and Induction* by a nonperiodical journal, to be called 'Studies in Induction and Probability'. Some quick, cheap method of publication would be used, so that the era of dittographed *arcana*, distributed within the circle of the initiates, would end. The volumes would be generally available; cheap enough for poverty-stricken students to buy; and fresh enough to prevent fiascos like the one in my dissertation. The first volume was to contain an article by Carnap which would play the foundational role of the outdated 1950 book. There would be an article by me, explaining measure theory and DeFinetti's representation theorem to readers of Carnap's earlier work. (He had decided to recast his exposition in the measure-theoretic idiom current in mathematical statistics and probability theory.) And then there would be articles by Kemeny and Gaifman, applying DeFinetti's representation theorem to Carnap's system. Again, the system itself would be a generalization of its predecessors.

That was to be the first volume, which the four of us would prepare for early publication – say, toward the end of 1961. Other volumes were projected: I have a list of twenty titles of articles which the four of us proposed to write, and it was hoped that after the first volume, other collaborators would appear. But Carnap's foundational article grew and grew. As the years passed, it developed into a sizeable monograph, of which the sections, dittographed, were circulated among the initiates as they appeared. And then they were revised in turn, and the revisions recirculated.

When I last saw Carnap, over a year ago, before going off to England for a year, he suggested that Volume 1 of the *Studies* should contain only the first half of his monograph, since that part was then in final form. While the book was being printed and bound, he would revise the second half for Volume 2. I agreed to look after the editorial busywork, in England. Well, that took longer than we had expected, but now it is done: Volume 1 of *Studies in Inductive Logic and Probability* is now at the printer's. It will be published in 1971, by the University of California Press.

After ina died, in 1964, Carnap's daughter, Hanna Thost, came to live with him and look after him. She stayed until the end, and after. On the day following her father's death, she received a packet from his typist. It was the final revision of the penultimate sections of part 2 of Carnap's monograph, 'A Basic System of Inductive Logic'. He must have sent it off just before going into the hospital. He had all but finished it. The rest is now being revised by two of his associates over the years, Lary Kuhns and Gordon Matthews, and the whole will appear in Volume 2 of the *Studies*, in 1972. Wolfgang Stegmüller, who heard from him just before his death, tells me that Carnap intended to write some further material for the 'Basic System', but no draft of this exists. He was full of ideas and plans, to the end. It is unlike him, to rest.

But although we shall never hear all that he had to say, we have not yet heard the last from our old master, our teacher and friend. He was a great, and good man, and I am proud to have known him.

University of Pennsylvania RICHARD C. JEFFREY

4

Carnap is a towering figure. I see him as the dominant figure in philosophy from the 1930's onward, as Russell had been in the decades before. Russell's well-earned glory went on mounting afterward, as the evidence of his historical importance continued to pile up; but the leader of the continuing developments was Carnap. Some philosophers would assign this role rather to Wittgenstein; but many see the scene as I do.

Russell had talked of deriving the world from experience by logical construction. Carnap, in his *Aufbau*, undertook the task in earnest. It was a grand project, and yet a self-effacing one, when so few philosophers understood technical logic. Much ingenuity went into the constructions, much philosophical imagination, much understanding of psychology and physics. If the book did not achieve its exalted purpose, it did afford for the first time an example of what a scientific philosopher might aspire to in the way of rigor and explicitness. If afforded detailed glimpses also, and philosophically exciting ones, of how our knowledge of the external world could in considerable part turn out to be, in Eddington's phrase, a put-up job. And it provided techniques of construction that continue to be useful.

In his *Logical Syntax* Carnap again vigorously exploited the resources of modern logic for philosophical ends. The book is a mine of proof and opinion on the philosophy of logic and the logic of philosophy. During a critical decade it was the main inspiration of young scientific philosophers. It was the definitive work at the center, from which the waves of tracts and popularizations issued in ever widening circles. Carnap more than anyone else was the embodiment of logical positivism, logical empiricism, the Vienna Circle.

Ultimately Carnap saw limitations in his thesis of syntax. Thus came his third phase: books and papers on semantics, which have given Carnap a central place in the controversies over modal logic.

Meanwhile Carnap's *Logical Foundations of Probability* continued to develop, a monument to his unwavering concern with the logic of science. Two months ago I had a lively letter from him about some supplementary work that he was doing on this subject. Also he sent me a sheaf of material from the new work in progress.

Carnap was my greatest teacher. I got to him in Prague 38 years ago, just a few months after I had finished my formal studies and received my Ph.D. I was very much his disciple for six years. In later years his views went on evolving and so did mine, in divergent ways. But even where we disagreed he was still setting the theme; the line of my thought was largely determined by problems that I felt his position presented.

I first heard about Carnap and his *Aufbau* from John Cooley in 1931, when we were graduate students at Harvard. Herbert Feigl was then at Harvard as an International Rockefeller Fellow. He encouraged me to go to Vienna and to Carnap the following year if I got a traveling fellowship.

Carnap moved from Vienna to Prague that year, and I followed him. I attended his lectures and read his *Logische Syntax* page by page as it issued from Ina Carnap's typewriter. Carnap and Ina were a happy pair.

He was 41, she even younger. Along with their intense productivity there was an almost gay informality. If you combine strong intellectual stimulation, easy laughter, and warm friendliness, you have an unbeatable recipe for good company; and such were the Carnaps. On a day when Carnap didn't have to come into the city to lecture, my wife and I would ride the trolley to the end of the line and walk the remaining few blocks to their little house in a suburb called Pod Homolkou. As the name implies, the place is at the foot of something; and Carnap and Ina would have just come in, likely as not, from an hour on skis on that very slope. Carnap and I would discuss logic and philosophy by the hour. My wife and I would stay to lunch, or maybe dinner; but, if dinner, that was the end of philosophy and logic until another meeting. Carnap's habits were already austere: no science after dinner, on pain of a sleepless night. No alcohol ever. No coffee.

I was then an unknown young foreigner of 23, with thirteen inconsequential pages in print and sixteen at press. It was extraordinary of anyone, and characteristic of Carnap, to have been so generous of his time and energy. It was a handsome gift. It was my first experience of sustained intellectual engagement with anyone of an older generation, let alone a great man. It was my first really considerable experience of being intellectually fired by a living teacher rather than by a dead book. I had not been aware of the lack. One goes on listening respectfully to one's elders, learning things, hearing things with varying degrees of approval, and expecting as a matter of course to have to fall back on one's own resources and those of the library for the main motive power. One recognizes that his professor has his own work to do, and that the problems and the approaches that appeal to him need not coincide in any very fruitful way with those that are exercising oneself. I could see myself in the professor's place, and I sought nothing different. I suppose most of us go through life with no brighter view than this of the groves of Academe. So might I have done, but for the graciousness of Carnap.

At Harvard the following year, I lectured on Carnap's philosophy. Our correspondence was voluminous. He would write in English, practicing up for a visit to America, and I in German; and we would enclose copies for correction. By Christmas 1935 he was with us in our Cambridge flat. Four of us drove with him from Cambridge to the Philosophical Association meeting in Baltimore. The others were David Prall, Mason

Gross, and Nelson Goodman. We moved with Carnap as henchmen through the metaphysicians' camp. We beamed with partisan pride when he countered a diatribe of Arthur Lovejoy's in his characteristically reasonable way, explaining that if Lovejoy means *A* then *p*, and if he means *B* then *q*. I had yet to learn how unsatisfying this way of Carnap's could sometimes be.

Soon Carnap settled at Chicago. Two years later I took him to task for flirting with modal logic. His answer was characteristic:

> I do not indulge in this vice generally and thoroughly .... Although we do not like to apply intensional languages, nevertheless I think we cannot help analyzing them. What would you think of an entomologist who refuses to investigate fleas and lice because he dislikes them?

In 1939 Carnap came to Harvard as visiting professor. Those were historic months: Russell, Carnap, and Tarski were here together. Then it was that Tarski and I argued long with Carnap against his idea of analyticity.

Because of distances our later meetings were regrettably few. In 1949 my new and present wife and I spent some memorable days at the Carnap's in New Mexico. In 1951 he and I held a symposium on ontology in Chicago. In 1958 a reunion in California was prevented by an illness of mine. Finally in 1965. to my delight, I saw him at Popper's colloquium in London. He looked well and was vigorous and alert. When Popper confronted him on induction his defense was masterly. It carried me back to his confrontation of Lovejoy thirty years before. It was the same old Carnap. His tragic death, while still at the height of his powers, marks a sad date in the history of philosophy.

Harvard University W. V. QUINE

5

No one can expect to give an adequate impression of Rudolf Carnap to those who did not have the privilege of knowing him. However, we can hope that reports from people who had diverse relationships to him will combine to convey a partially accurate conception of him.

Of all the speakers tonight I was probably the farthest from Carnap in philosophical outlook. This was surely the case when I was actually his student at the University of Chicago, and it was probably the case thereafter, in spite of some convergence of views. For I began to study with him as an intoxicated metaphysician and in the course of many years became a sober one. His generosity in discussing problems with students, analyzing their papers, and encouraging them was not in the least contingent upon their discipleship. He did not in any way encourage discipleship, which showed how deeply his opposition to authoritarianism in the political domain was grounded in his character. Likewise, I think, his famous Principle of Tolerance in *The Logical Syntax of Language* was not merely a device to deal with *Scheinprobleme*, but an expression of his character.

Carnap was a great systematic philosopher, who had a coherent approach to nearly the whole range of classical philosophical problems. The massiveness and lucidity of his philosophical thought became increasingly apparent to a student who was challenging him and searching for weak points in its structure. It was an exercise of great value to lose in such a venture. The systematic character of Carnap's work was the expression of a most remarkable feature of his intellect: the complete mastery and recall of all that he had thought through in the past. If he had once reached a tentative solution to a problem – even if it had been several decades previously – one never saw him groping to remember the analysis. The territory he had once conquered he thoroughly occupied. This was a major component in the massiveness of this thought.

I do not want to suggest, however, in pointing to the systematic character of Carnap's thought, that he was inflexible. The other speakers tonight have sketched some of the major changes in his general philosophical position and in this theory of induction, showing his openness to criticism and to suggestions for improvements. I want to add that he took particular delight in technical advances which permitted him to widen the scope of his investigations without loss of precision. Perhaps the most important such advance for him was Tarski's work on the concept of truth, which enabled him to extend his investigations of language from syntax to semantics. In *Testability and Meaning* and *Foundations of Logic and Mathematics* he himself did the main technical work which permitted the relaxation and extension of the empiricist criterion of meaningfulness. A third and more recent example was Kemeny's work on models, which permitted him to define *c*-functions on languages which could not be readily analyzed semantically in terms of state-descriptions. The general drift of his philosophical development towards greater flexibility, openness, and richness indicates that his early program for the elimination of metaphysics was never an attempt to contract the scale of the world or to view it anthropocentrically, as some critics have claimed. Rather, it was the result of his intense desire to understand things clearly, and my personal impression was that he experienced much joy in finding that clarity is compatible with greater and greater scope.

Finally, one cannot adequately suggest how Carnap appeared to those who knew him by talking only about his intellectual qualities. To be sure, his thorough-going fair-mindedness towards others and himself was very intimately connected with his logical lucidity. But his generosity, spontaneous kindness, concern for the oppressed, and courage all went far beyond fair-mindedness. The combination of his intellectual power with these moral characteristics made him a great man.

Boston University A. SHIMONY

6

I met Carnap in person for the first time in the autumn of 1950. Until then we had only corresponded, the correspondence consisting mostly in my asking various questions concerning his publications, most of which I had read many times in full and consulted thousands of times for clarification of my ideas (I must have doubtless spent more time by an order of magnitude on the study of Carnap's writings than on those of any other author), and in his patient and detailed replies.

When I finally arrived in that autumn of 1950 with my family in Chicago on a fellowship from the Hebrew University, we were lucky in finding accommodations in walking distance from the university as well as from Carnap's apartment – we, of course, could not afford a car at that time.

In addition to my attending Carnap's lectures on Analytic Philosophy, carefully prepared and marvelously delivered (I hope that I am not betraying my trade as a university teacher when I say that I have made in later years much use of my lecture notes in my own courses on this subject), he was so kind as to set apart, during a period of five months, on each Tuesday afternoon around to hours for discussions, in which we were joined, after a few weeks, by Arthur Burks.

Carnap, who kept constantly complaining about his weak memory, always had before him a sheet of paper on which he had noted down the topics he wanted to discuss with us; we were, of course, always welcome to raise questions of our own, an opportunity of which I tried to make maximum use. This was the year in which his *Logical Foundations of Probability* had been published, and I spent most of my free time in studying this book. Carnap, who had only relatively late gotten interested in this subject, was by then utterly absorbed in it and working on its planned sequels. As for myself, to tell the truth, I always thought, and still continue to think, of Philosophy of Language as the field to which Carnap had made the most significant contribution and should have been in a position to make further decisive contributions, if he only had left himself sufficient time for this endeavor. My attempts to draw him back to this field, through leading questions and other ruses, met, though, with only very partial success. After 1950, he published only two short but highly important papers in Philosophy of Language, in addition to the *Replies* in the Schilpp volume.

I shall never forget the discussion between Carnap and Quine on the occasion of the latter's visit to Chicago, in December 1950 or January 1951, in which he gave a masterful presentation of his 'Two Dogmas', though not in the form published shortly afterwards. There were too not many of us attending that lecture, and I presume that we all preferred to watch these two giants join in battle rather than interrupt with our inconsequential questions and remarks. Needless to say that nobody 'won' – until this very day.

Until 1967, there was never a time when I would not make the greatest efforts to visit Carnap, whenever I was in the States. There was never a meeting from which I would not come away with the feeling that I had learned something new and important. Till today, whenever I tackle a philosophical problem, the first question that comes to my mind is what did Carnap say on that problem or what would he have said, had it been put before him. I am convinced, and shall probably remain so to the end of my life, that this is the best way of approaching these problems.

The Hebrew University of Jerusalem YEHOSHUA BAR-HILLEL

7

I. FIRST IMPRESSIONS: THE LOGIC CLASS

I was, candidly, set to encounter a Great Man as I waited for Symbolic Logic 320, fall term, 1941, to begin. I was not disappointed. In spite of the inevitable slight shock at finding him no larger than a moderately tall lifesize, with a small strand of hair standing up in back, there was an immediate impression of weight. He was not especially overweight physically. Rather there was a slowness, even a ponderousness, as of a tank in preliminary maneuver. The impression of controlled mass was strengthened by the slow, cello-like voice. Though German-accented, with the long syllables dwelt upon, it flowed with total assurance, without hesitation for words or grammar. The flow was not unstoppable, I found, on raising a question; and the Great Man impression grew as the answer came with courteous formality, tempered with an almost fatherly gentleness. Something of this impression was shared, I think, by the other students, even without benefit of preestablished aura. But some seemed not to find the flow as easily stopped. It was not just momentum. There was also a feeling of careful scheduling of topics for each lecture, a timetable not to be upset. And there was an impression of intense concentration underlying the steadily articulate delivery, a concentration which manifested itself in his gaze being averted from the class to something closer, often his fingernails, which he would scrutinize as if through a microscope. It contributed to a sense of remoteness which discouraged questions. This remoteness, to be sure, did not seem inappropriate to a great man; or even, perhaps, to a great teacher, in the European tradition. Indeed, Carnap often seemed to exemplify that tradition at its best, within its limitations: he was kindly but impersonal, vigorously dutiful in supplying outlines, exercises, and class drills – but all somewhat unilaterally, with little dependence on feedback. Nothing, for example, could swerve him from spending the first third of each period in a methodical review of the preceding period, or could persuade him that the level of student understanding had gone beyond that called for by his carefully considered timetable. Replies to extrapolative or speculative questions were consistently minimal. On the other hand, the material he had chosen to convey was testably, indelibly conveyed.

There was, of course, more to Carnap's method as a teacher than such a description would indicate, as I came to realize. It may be suggestive to recall the word-pictures Carnap gives us in his intellectual autobiography (pp. 4–6 in Schilpp) of his own most memorable teachers, Hermann Nohl and Frege. Nohl is praised for taking "a personal interest in the lives and thoughts of his students" and because "in his seminars and private talks he tried to give us a deeper understanding of philosophers on the basis of their attitude toward life ('Lebensgefühl') and their cultural background." Of Frege he says "Ordinarily we saw only his back, while he drew the strange diagrams of his symbolism on the blackboard, and explained them. Never did a student ask a question or make a remark, whether during the lecture or afterwards. The possibility of discussion seemed to be out of the question." Without insisting that Carnap made a conscious choice between being a lovable Nohl and an unapproachable Frege, it is worth noting that his praise of Nohl is immediately followed (with apparently unintended irony) by the remark that "On the whole, I think I learned much more in the field of philosophy by reading and by private conversations than by attending lectures and seminars." The imprint of Frege, on the other hand, is evident throughout the remainder of the autobiography.

To be sure, much of Carnap's teaching manner was a reflection of personality. The natural set of his mind was obviously closer to Frege's than to Nohl's. Yet within such bounds, there was a conscious choice of teaching styles. His style was, in fact, quite different in his 'Seminar on the Principles of Empiricism, 440' which I attended concurrently.

II. CONTRASTING IMPRESSIONS: THE SEMINAR

For the seminar, there were no notes. Readings were handled only by a reference to Ayer's writings as possible background. Exposition proceeded by a general formulation of a thesis, followed by what, for want of a better term, I shall refer to as parables or fables. These were not just related. They were acted out. Many students must still recall the animated dialogue between the idealist geographer and the realist geographer concerning the mountain they had just surveyed (and about which they disagreed only as to its ultimate ontological nature). Or the revelatory saga of the blind student who arrived, through systematic coaching, definitions, and thought, at a theoretical understanding of color words, not only as applied to physics but as used in connection with perception in others.

In the acting out of these parables, there were amusing imitations of indignant dispute, with gestures, simulated perplexities, discoveries, and the like. Yet these interludes did not seem hard to reconcile with the manner of the logic class. It seemed only a deliberate change of pace for a differently-textured area of inquiry. Dignity was relaxed but not lost. Nor was there lost, it may be conceded, a certain essence of European professor. The effect was not unpleasant. An occasional overconscious drollery, is after all, well-intended. Indeed, the naked implausibility and artificiality of the plot 'premiss', which would emerge as a parable began, had a professorial quaintness that won an amiably tolerant audience. And when the dramatist would earnestly supply circumstantial detail to add 'realism' to such a fable, the audience, amused and charmed, would visualize as cooperatively as possible. The 'concrete' details, however, did sometimes go to somewhat surprising lengths. Not only did the geographers, between ontological sallies, compare their findings in terms drawn from the appropriate technical disciplines, but, for the blind student case, teachers and ingenious pedagogical methods were detailed, as were special instruments of perception, some suggested by current technologies, others more speculative. But such methods and devices often proved to have their own epistemological points to make, so that the impression of unnecessary stage-setting would undergo repeated revisions. Indeed, such revisions would often continue after the apparent close of a parable. For example, the methods introduced for the blind student would reappear in new elaborations in a succeeding fable, this time about a far more restricted being. The needed elaborations for this case would turn out to be made possible by some principle or device mentioned, apparently only for science-fiction effect in the earlier parable.

Typically, such a cumulative linking of fables would build to some especially significant extremity. The sequence above, for example, continued on through increasingly grave but illuminating cases of perceptual deprivation, to a climax in which a nonhuman superbrain whose only perceptual ability was that of distinguishing dark from light was led to a theoretical understanding of the universe, – an understanding as comprehensive, *cognitively*, as that of the society of scientists (by now rather complex) which had educated it.

With the stage thus set, Carnap would then be able to sum up, quite simply, but with focused clarity and indelible impact, the central themes that had emerged – in this case, about differences between experience and knowledge, about relations between the communicability of information and the causal connectedness of the universe, about the empirical intelligibility of a hypothesis that distinct, causally isolated universes exist....

The impact of this unusual thesis-parable-summary cycle was, as I think back, oddly enhanced by the very remoteness that seemed a barrier in the logic class. While at first it was a mere part of the professorial quaintness that enlisted amused attention as an 'example' (parable) would begin, the psychological distance would seem to grow – and change in nature – as unexpected turns of plot appeared. As a story would shift to a future society, to other-dimensioned creatures, or to interplanetary communication, the professorial remoteness would, momentarily, become that of a more genuinely alien, but kindly, mind – with whom communication was suddenly as vital as if it had just arrived by time-machine. The effect was, for others, I think, as well as myself, that the philosophical points, when perceived, would seem to stand out with a naked independence and universality, without restriction to era or galaxy.

III. THE LOGIC CLASS RECONSIDERED

The contrasting style of the seminar may dispel an impression of unreflective rigidity in Carnap's approach to teaching. But it also prompts a reconsideration of the style of the logic class. Why should it have had to give such an impression in the first place?

To be sure, the two styles revealed certain affinities not apparent at the outset. The seemingly rambling parables showed, in retrospect, at least as much planning as the logic course. And their seeming lavishness of image and detail proved more economically deployed than it first appeared. They did, after all, maneuver the seminar to some rather special vantage points – for philosophical panoramas not ordinarily easy to encompass.

But such reflections only underline the contrasts. The unswerving exposition, unleavened by drama or speculation, must now be regarded as deliberate, the examples as consciously prosaic. Replies to extrapolative questions, as already remarked, were consistently minimal. Indeed, the minimality was so consistent and calculated that one finally felt it as a principle at work, a sort of least-action principle – with respect to commitment in general. But it is important to spell this out a bit.

There was never a suggestion of brusqueness. Every question brought a kindly and patient response. But if the question suffered from any slight confusion in the use of technical terms, Carnap's reply would not do more than point that out. Often the tone would be reluctant, sympathetic: "But in this system one could not *say*... because –." There would seem even to be an apology for the meagerness of the system's resources. But there would also be a gentle finality to such a response. The lecture would be resumed. There would seem no impulse to provoke reformulation of the question, let alone guide it, or even to imagine that there was any remaining question in the student's mind. If a reformulated question was nevertheless put, the lecture would again come to a graceful halt and the reformulation would be patiently considered as if *de novo* (no matter how quickly the reformulation had been interjected.) To be sure, if reformulations persisted, Carnap might, with an air of consciously sympathetic attentiveness, offer a trial interpretation of the student's remarks in a form beginning: 'Would you then mean to say ...?' But such suggestions would typically be obvious *reductios*, made only more frustrating by Carnap's unchallengeable pretense of earnest misunderstanding. No air of 'conscious drollery' ever relieved the problem of dealing with such a ploy. Indeed, at the time, such misunderstandings would seem discouragingly genuine. After all, it seemed not too surprising in a logic teacher, especially one teaching in a language not his own. I began to withhold more exciting questions for discussion at office hours, (there to be carefully reformulated in 'his terms'.)

Procedure at office hours, however, was just as in class, except for the greater freedom to pursue questions. With remarkable patience, as I now see it, Carnap would take up each question on my mercilessly long list-for-the-day. But, as in class, when a question was poorly formulated, he would be of little help beyond pointing out what violation of type-restrictions or mismatch of universes of discourse had occurred. Nor would any sequence of not quite correct reformulations tempt him to guess at, and reformulate, my question for me. Instead, he would finally suggest, with a polite sadness, that we pass on to the next question on my list. Even when a correct formulation was finally arrived at, the answer would still be disappointingly minimal, always at the least deep-going level appropriate to the question – often no more than an indication that we had arrived at a truism, or a contradiction. And, more often than not, the meat of my question would seem to have been lost among the reformulations. Either way was frustrating, since it always seemed to me that I had merely been trying to follow out consequences of some extraordinarily sweeping formulation of his. At such times there would come an impression, of playing a chess-like game with a disguised adversary – beneath the apparent gentleness and pedantic literalness – whose aim was to prevent any glimpse of a hidden truth.

Very slowly, however, over many sessions, enlightenments were attained. Most often, they took the form of a collapsing system of misunderstandings. Each reformulation, often undertaken with ill-concealed impatience, would require small mental readjustments that somehow cleared the way, so that when the terminal collapse came, it had the liberating shock which some ascribe to satori. On rarer occasions, a question would finally firm up, after shucking off layers of confused formulation, and Carnap would say something, in a matter-of-fact way, that would suddenly open a door to a whole new realm of concepts, or would turn some large part of my mental universe upside down.

Yet Carnap never seized seized such moments to drive a point home or rouse further enthusiasm by opening deeper doors. The response never went beyond the question at hand. I would have an impression of waiting layers of interpretations each guarded by a door with a separate key. If this was tutelage, it was of a uniquely austere and passive kind. A Zen master, at least, poses questions to provoke satori. Carnap waited for questions and then criticized them.

Nevertheless I did come to view this procedure as just such an austere teaching policy. I felt I perceived a similar policy of literalness and minimal commitment at work say, in a Supreme Court decision to settle a case on the basis of a technicality, reserving opinion on deeper issues until they are more properly formulated. Such a policy has, certainly, its deep rationale. Both law and logic rest on intricacies of definition and formulation, and one sure way of driving them home is to withhold all response beyond what is literally called for. That such a mechanical, non-directive approach can indeed teach is amply attested by the experience of computer programmers. But it is very demanding, as programmers will also acknowledge.

The benefits of this austere minimality policy were, however, not limited to the few who pursued matters at office hours. There was something in the carefully timed classroom exposition alone, in the underplayed preparation of subtopics that tended to culminate in flashes of insight as vivid as those provoked by the parables of the seminar.

One of the moments of illumination I recall came in the latter half of the second term. By that time an undeniably impressive framework of concepts had been assembled and was continuing upward, step by firmly planted step. What had seemed plodding, now seemed inexorable advance – purposive and architectonic. After an exploration of the logically definable properties of relations and their interrelationships, with the help of many careful blackboard diagrams, Carnap developed the notions of isomorphism and structure. Isomorphism was not defined for classes first, then for dyadic relations, and so on, as normal pedagogical might dictate, but directly, as a schema for the general n-adic case. Conceptualization capacities were strained and struggling. Carnap went on to define the structural properties of an n-adic relation as those hereditary with respect to isomorphism. I ventured a bewildered confession – an inability to visualize very many structural properties for the case $n=1$, i.e., for classes. Indeed, I could not see more than one. "And what is that?" was Carnap's patient question. I hopelessly replied that I could not really see how two isomorphic classes could have much in common beyond the same number of members. There was then, my memory is quite clear, a dramatic pause. Carnap smiled. Earlier in the course, after defining cardinals 1, 2, ... Carnap had announced that the concept of cardinal number in general could only be reached at a later, more advanced point in the course. The pause allowed the realization – that we had now arrived at that point – to take its own slow, many-inferenced course. My mental eye seemed to travel over a synoptic vision of the great structure of ideas that underlay this hardwon vantage point – and there was time also for visions of previously unglimpsed realms – realms that were opened up by coming upon the intuitively familiar concept of cardinal number as a special case of a new, more general notion of *relation number*, with its promise of many strange arithmetics; strange yet already illumined by that small candle of familiarity. Also illumined, then, more dimly, was a yet vaster world: that of all structural properties.

This vision was quickly followed by another, consolidating one. Carnap went on to say, after supplying the needed details of definition, that the importance and power of logic stem from the fact that all inferences about any subject matter depended only on the relational structure, and structural properties, of that subject matter. Completely corresponding results were assured for quite different subject-matters provided only that there be a structural similarity. And – my thoughts echoed – all structures and structural properties are definable by logic alone. We seemed in possession, then, of all possible forms of all possible knowledge!

Such carefully prepared moments seemed to have similar impact on at least some of the other students. I recall attending a talk Carnap gave, for a student mathematics club, in which several of us from the logic class were amazed to see Carnap retrace the whole course of our two terms, right up through the cardinality concept, using only the essential definitions, but with easy cogency. At the end, after a burst of exhiliarated remarks, we fell silent briefly. It seemed impossible, both for our great system, and for our inexorable Carnap, that he should have traversed it so quickly. One classmate finally wondered aloud:

"Why couldn't he have gone through it like that with us?" After a pause, in which we reflected upon the long two terms, there were two further remarks: "Well, the audience didn't get to understand it the way we do." "With the short treatment, we wouldn't have either."

There was, I am suggesting, a rather special sense of timing and preparation that went into the understandings achieved during those two long terms. In part it might be viewed as a sense of the dramatic expressing itself in a different way. But it went deeper. It seemed to stem from an insight into the workings of insight. The well-planned thought-explosion required the careful assembly of combustibles. Each bit of the needed conceptual structures had to be introduced and absorbed on its own terms, without hint of further exciting implications – even when interconnecting fuses were set in place. And when the flare began, carefully estimated time had to be allowed to ensure that all the remoter secondary targets were reached. The flare, of course, had to be set off by the student, in accord with his own timing, if it were to have effect. That is, while routine conceptual preparation may be set out straight-forwardly, the really difficult insights, once within range, must be attained by the student on his own. Insights, like jokes, lose impact if the point is made explicit.

Such attitudes, I am aware, may sound unCarnapian. It was, after all,

Carnap's dissatisfaction with Wittgenstein's remarks on things that could only be shown, not said, that partly motivated his *Logical Syntax*. But the attitude here ascribed to Carnap is not inconsistent. Carnap's belief in the sayability of anything concerned 'cognitive' language. His withholding of literal explicitness is a matter of pragmatics. While his writings may sometimes suggest that his acknowledgement of a realm of 'pragmatics' was only to dispose of matters he thought little about, my own experience has led me to a different estimate.

Perhaps another episode in that experience will illustrate some of this, including a revealing interplay between the literal-cognitive and the inexplicit-pragmatic in Carnap's use of language.

The word 'property' had bothered me throughout the two logic terms. And I had bothered Carnap about it a bit. He seemed to treat questions about this apparently central notion with an amiable casualness that suggested I could not be serious about such a question. A typical reply might be: "anything that can be said about anything" followed by a few examples, though examples were hardly my need. In a later more advanced course, I asked about it a last time, going on at length about features, traits, tertiary properties, relational properties .... At least that is the sort of harangue it was. In any case, it apparently convinced Carnap that I was serious in my confusion and that the time had come for a move. The genial informality did not appear. Instead Carnap became totally serious. After a short silence in which the impression of philosophical majesty grew, he announced "I shall define 'property'," and rose and walked slowly, at his greatest weight, to the board. He wrote "$P(F) \equiv F = F$". 'P' served as a symbolization for 'property'. 'F' was a monadic predicate variable in the type-theory logic we were using at that time. I stared. It defined F to be a property just on condition that it was self-identical. I said the obvious: It makes everything a property. "Everything of that type" Carnap amended. A silence began. I was not to be helped, sink or swim. I was aware of things Carnap had said about 'universal terms' in his *Syntax* and their borderline closeness to 'meaningless' questions. This and other considerations showed well enough, I thought, the unpalatable point Carnap intended to make: that in this language 'property' simply corresponds to a grammatical type. No synthetic sentence about the 'nature' of properties can be formulated, even when the language is fully interpreted. I felt cheated again. My real question seemed to remain at one remove, inaccessible. I finally blurted something of this out. Instead of replying 'cognitively', he formed a few beginnings of sentences about properties as if groping for something 'significant' to say about them. Each seemed to violate type restrictions or shift universes of discourse. Then by some quite slight grimace, he managed to convey a despairing inarticulateness. Somehow the philosophical finality got through to me. My 'real' question, if answered, had to be answered in a *language*, and that language would prompt 'real' questions with respect to *its* universal terms, and so on. In the ensuing discussion, the distinction at stake (between internal and external questions, as Carnap later put it in a famous article) became more articulate for me, but it concerned something I had already 'seen', something that I had been *shown*, not told about – something shown, moreover, by contemplation of a totally cognitive sentence (an unbendingly literal, hopelessly uninformative sentence) – and a grimace.

In this episode one can glimpse the dramatist of the seminar, usually invisible in the logic class, showing himself in conscious command of the deliberately minimal materials of the more difficult genre. There was, one can see, always a commonality of origin and intent underlying the two styles, and a careful readiness for interaction.

In the years that followed these early experiences, I came to see the two contrasting, but subtly interaction, styles – of the logic class and the seminar – as involving more than pedagogical technique. They were not confined to the classroom but appeared as basic communicational *modes*. I could recognize and trace them in his writings and in his discussions with colleagues. (The minimality of the logic class may be more evident in his writings but the elaborate parable of the seminar will remain at least as well remembered by close philosophical friends and associates. I still cherish a photograph I snapped of Carnap in mid-parable at the Institute of Advanced Studies (circa 1952) at a kind of small 'summit' meeting arranged by Herbert Feigl for his Minnesota Center for Philosophy of Science. In the picture, Carnap's eyes are raised to his right hand whose symbolic role has taken it above his head. The index finger points rigidly downward. Perhaps it is a field vector, an instrumental sighting, or a step in a rationally reconstructed discovery.)

Still further, I came to see the two modes of communication as related to characteristic modes of thought in Carnap, even to features of what might be called his 'philosophical personality'. They were modes of thinking and communicating which, whatever the native tendencies, were developed over years in accord with an increasingly articulated philosophy of (among other things) thinking and communicating.

To better dissociate them from classroom context, I can, for the logic class mode, speak rather of minimality of commitment, uninstantiated generality, literalness (often formalized), non-extra-polativeness, or the like. For the seminar mode I must reach a bit further below the pedagogical surface. Underlying the graphic, acted-out example, as used by Carnap, I see his preoccupation with the development of simplified models; e.g. his treatments of many problems via a 'rational reconstruction' of science as a language, a store of sentences, or as a sentence-processing system with protocol-sentence-input. They are often couched in language that seems to refer to procedures and decisions of working scientists. This small echo of 'parable' is harmless enough to scientists, who are accustomed both to idealizations and to informal communication about them, but it has led some philosophers of science to reject them as unrealistic. Others, attempting to relate Carnap's use of simplified models to their use in, say, physics, tend to object that the parable is not spelled out even further: in sufficient detail so that one could compare the idealized science with actual science, and ascertain measures of approximation, check predictions from the idealization by observation, and the like. But any student of a Carnap seminar, who has seen the parable aspects vigorously developed in varied science fictions is well aware that it is not fact but concept that is under study. Unlike the physical scientist's approach to complex fact by simplified model, Carnap's simplified structures are approaches to areas of conceptual complexity. Scientists, especially mathematicians, are familiar with this kind of theory (and simplification) too. The mathematical theory of games, for example, and its variety of simplified subtheories, is not judged by a measured degree of approximation to actual games. Nor are its suggestive concepts of strategy and payoff to be viewed as *models*, in any measurable sense, of preexisting concepts of strategy or payoff. Similarly, Carnap's 'explications', 'rational reconstructions', and the like, were never just analyses of given concepts. He sought to forge systems of concepts to *replace* the confusion areas. (Ironically, the fact that Carnap's simplest inductive logic systems are for systems of fixed vocabulary brought complaints that Carnap took no account of the role of conceptual innovation in the development of science, that he portrayed science as consisting of no more than the confirmation and disconfirmation of sentences. But, conceptual innovation was Carnap's *business*.)

Mention of game theory should not suggest that Carnap's constructions be judged in terms of theorem interest. Their simplicity was designed to postpone mathematically interesting complexity where possible. What theorems there were were simply to show that the concepts defined did exhibit the proper interrelationships. Similar concerns are shown in Tarski's famous study of 'The Concept of Truth in Formalized Languages', in which Tarski proves the law of excluded middle – not as a contribution to mathematics, but to show that his definition of truth for a given system works. This was the sort of theorem Carnap's structures also proved. It is almost coincidence that Gödel's theorem appeared in time to permit Tarski to cap his great study with the 'interesting' theorem that no language can define its own truth concept.

The final Tarskian theorem does, however, point to a pervasive feature of formal systems, possibly of all representational systems, that in a sense provides an intrinsic rationale for the contrast between Carnap's two thought modes. The Tarskian theorem, and the Gödel incompletability theorem, with which it is closely related, belong to a family of limitative results which, among other things, seems to put a final bar on reaching total generality in any definition of such concepts as truth, analyticity, consequence. Whatever concepts humans can form of these 'ineffables' seems therefore left to analogy with less comprehensive concepts defined for specified systems. In other words, the seeker after total generality is *forced* to approach such topics by example. In other words, the austere, uninstantiated generality of the logic class *must* be supplemented by the graphic example of the seminar, since what they are examples *of* can not always be stated, even in principle. Carnap's explicit struggle to come to terms with this necessary dualism in approach is writ large throughout his *Logical Syntax of Language* (which was strongly influenced by his conversations with Gödel during the period Gödel was working on his famous theorem.) And it was evidenced in many conversations as a central theme in his thinking. That it was so central is a measure of his enormous drive toward complete generality, toward finality, solution

The *Syntax* was, in a way, an apex of that drive. The *Aufbau* program (construction of the world!), which was the first clear exemplification of that drive, paused after its 'first volume', to consolidate and rework, in the *Syntax*, its logical underpinnings, borrowed from *Principia Mathematica.* More positively, it aimed at a maximally powerful logic system for all future such construction – a project both stimulated by the possibilities inherent in the generality of Gödel's arithmetization idea and, inevitably, circumscribed by Gödel's result. The upshot was Carnap's famous *Language II*, complete with an infinity of 'positions', and 'infinite induction' rule, and other powerful equipages, which he then, and later, proclaimed as permanently adequate for empirical science. Yet it was not, and could not, be the strongest possible, nor could the corresponding logical terms be totally general. It was, therefore, a special case, an example, a paradigm – material for a *parable*. Carnap's dissatisfaction is amply evidenced in the chapters following those dealing with LII, namely, those of Part IV on *general* syntax.

For a philosopher to concentrate very hard on any one thing, even if that thing be generality, is, for some, a sign of narrowness, the mark of the specialist. A. J. Ayer once divided all philosophers into pontiffs and journeymen, with Carnap as chief example of the latter. And Richard McKeon sorted them into holoscopic and meroscopic types, with Carnap as the chief example of the part-peerer. Ayer's division was more kindly to Carnap than McKeon's but both missed Carnap's scope. Perhaps it was because they and similar critics had not my opportunity for wide-ranging conversations with Carnap over years. But perhaps they had not adequately reflected on how they would expect a *very* holoscopic, global mind to act. Perhaps the only way I can convey my counterimpression is by a parable of my own.

Picture a *very* holoscopic mind. Suppose it is a very powerful one. After a survey of the whole scene it would, of course, form plans. The plans would require deeper study in certain areas. This deeper study would reveal broader promises and puzzles. Interrelationships would be perceived. Plans and studies, by interaction, would quickly become global. Science would have to be unified, language systematized, the foundations of reasoning and experience scrutinized. Many specialized, meroscopic jobs would have to be done. Some could be done best by the mind itself – like constructing the needed overall conceptual framework – but time is limited. Minds must organize. A journal must be started. A manifesto of the new plan must be issued, congresses scheduled, an encyclopedia planned. Delays must be expected, of course. Wars. The interaction of minds is uncertain; the interaction of groups is unguided. In the meantime, the mind applies itself to those very special but very basic jobs which few but it can perceive as essential and as promising. This brings the mind fame as a great specialist.

At some point, of course, the whole plan is seen as unlikely to progress beyond its most initial phases in the time the mind sees as available to it. But then it was never unaware of probabilities. As a holoscopic mind, it understands its predicament perfectly. It still likes the plan. It proceeds in its painstaking work as if it had milennia.

My parable, unlike those of Carnap, has no literal point – only an impression of a remarkable mind.

Michigan State University HERBERT G. BOHNERT

8

For most of his life, Carnap was a socialist humanist as well as a scientific philosopher. In the buoyant decade of socialist Vienna, logical clarification was thought to be a means of liberation from simple prejudice and sophisticated mystification, just as empirical science was seen as a tool for construction and a source of intellectual pleasure. But such a spirit of enlightenment did not prevail, and darker stronger powers came to dominate. Carnap lived through the years of fascism, imperialism, wars, these decades of barbarism, and they entered profoundly into his life. Ever the man of reason, of sceptical and probabilistic judgment as a guide for living, he was not a naive Utopian about the chance for humane relations among men and women in any class-divided or race-divided or creed-divided society, or in a world of unlimited patriotisms. A socialist and an internationalist, Carnap nevertheless lived through situations which demanded defense, retreat, self-criticism, stubborn decency, maximal intelligence about minimal possibilities – not a life of individual creativity in a communal society but of individual work and hope within a context that elicited resistance. In his passionate careful way, he was a man of the resistance. His first major work, *Der logische Aufbau der Welt* of 1928, was written, as he said, in the confident spirit which "acknowledges the bonds that tie men together, but at the same time strives for free development of the individual." His last effort was an attempt to help imprisoned Mexican philosophers whom Carnap visited in their jail just a few months ago before he died. Critical of his own life and work, Carnap was critical of other men and of their works and institutions; but always with reason, never with hysteria or fantasy. A Stoic *and* an Epicurean, an individualist *and* a socialist, a happy man *and* a somber man – the best of this troubled age.

Boston University ROBERT S. COHEN

9

When Carnap was invited to come to the Department of Philosophy at The University of Chicago I was its secretary (the nearest thing to a chairman the department then had). The initiative came from Morris, but I strongly supported it. Since the administration was hostile to the proposal, this support was doubtless necessary if the project was to succeed. I well knew how widely Carnap's views and methods diverged from mine, but then I had learned at Harvard and Freiburg in the twenties that philosophers never agree – well, hardly ever, And I thought it would be exciting to have such a man. And so it was. There was no one else in the department with whom I liked better to talk.

Carnap did not impress me intellectually quite so much as a few philosophers – Lewis, Russell, Whitehead, Lorenzen – have done, but still he was impressive. He was also likable, a very decent human being. My problem was to convince him that my ideas were clear enough to be worth discussing. If I showed him a manuscript, almost all the marginal comments were 'n.c.' for 'not clear'. Thus, when I tried to show him that 'external' and 'internal' relations could be clearly distinguished, he said that this attempt, like G. E. Moore's, was unsuccessful. He did concede a point to the contention that relations between universals and their instances were external to the universals and internal to the instances; but he balked at applying a similar asymmetrical distinction to temporal relations of earlier and later, or remembered and remembering experiences. (In short he stood by what I view as Hume's greatest mistake, as Russell,

Ayer, Von Wright, and many others have done.) I still wonder whether (a) the Carnapian criteria of clarity are too severe to make it possible to deal with this and other important philosophical questions, or (b) whether I simply lacked the necessary skill and ingenuity. I incline to (a).

I agreed with Carnap that metaphysical mistakes are lapses from legitimate to illegitimate (confused or inconsistent) uses of words; the disagreement was on Carnap's view that the alternative to metaphysical mistakes was other such mistakes – or else no metaphysics. The issue came to a focus concerning the relations of truth to time, he holding with many or most logicians – that all truths obtain timelessly and I holding that many new truths become true as "the indeterminate future becomes the irrevocable past" (Peirce). This view really annoyed Carnap! He had to admit that the Tarski elucidation of 'truth' was 'neutral' to the issue, and I never got any stronger arguments on his side than the contention that common usage and common sense take truths to be timeless (whereas I think they are undecided on the question) and the contention (which he stressed much more) that it is convenient for many purposes in logic and science to take them to be so, which I grant but regard as compatible with my position. Here I accused Carnap of trying to get away with an ontological proof of a metaphysical entity, allegedly timeless truth, timeless by a convenient definition. (Of course very abstract truths, purely logical or metaphysical, are timeless. The dispute concerned truths about particular events or classes of events.)

Charner Perry who (to my relief) followed me as officer of the department, used to tease Carnap as follows. When a highly historical dissertation, and they were mostly that, was the topic of an oral examination, Perry would solemnly inquire of Carnap if he would like to attend, only to be met once more with, "I am not much interested in the history of philosophy."

In departmental seminars Carnap took part loyally and made genuine efforts to see what speakers were driving at – in glaring contrast to another philosopher of somewhat similar views (briefly a member of the department) who invariably took a single sentence the speaker had uttered and attacked it, totally ignoring the context into which the utterance had painstakingly been placed. I was amused once when, after a fellow ex-German temporarily with us had used the expression 'the eternal human',

I asked this man, did he mean the expression literally or was it only a piece of poetry (whatever may be eternal, I do not see that humanity can be so), whereupon Carnap, who had been silent until then, burst out, "Everything he says is poetry!"

Carnap was mildly interested in, and gave a bit of technical assistance to, my effort to exhibit a contradiction in the Thomistic idea of omniscience. His question was, will not a metaphysician always find a way of wriggling out of any contradictions which his confused employment of words appears to present? The answer, I think, is, yes and no: yes, he will probably satisfy himself that he has escaped; no, he will (in many cases) not have escaped without paying a price, the price of shifting from relatively clear but inconsistent meanings to meanings whose degree of clarity is so low that the impressiveness of the position is diminished. On this answer we would, perhaps, almost have agreed.

When Carnap left Chicago I felt it as a great loss.

One can never forget Ina Carnap, her charm, or her devotion to her husband's work.

The University of Texas at Austin CHARLES HARTSHORNE

10

When Carnap arrived at UCLA in 1954, he was completing work on his brief but fundamental paper 'The Methodological Character of Theoretical Concepts'. In a series of previous articles, beginning in the Vienna Circle days, Carnap had presented successive refinements of his basic view that even for such complex theoretical structures as those characteristic of modern physics and contemporary theology it must be possible to winnow the empirically significant from the metaphysically speculative. In his typically scientific way, Carnap had backed his arguments by offering precise definitions applicable to formalized theories. 'The Methodological Character of Theoretical Concepts' contained his most recent, and by far most sophisticated, such definition.

As a graduate student, in 1958, I studied Church's review of a similar attempt by Ayer in *Language Truth and Logic*. Church had shown that Ayer's definition had unacceptable consequences, and I noticed that a result slightly stronger than Church's could be obtained by an alternative proof strategy. Richard Montague found my argument interesting and suggested that I attempt to apply it to the much more complex definition given in 'The Methodological Character of Theoretical Concepts'. To my own astonishment, a version of my criticism worked against Carnap and together with a second, related, criticism seemed to me to throw serious doubt on the feasibility of the program itself. But my argument against Carnap's definition was less perfect than Church's argument against Ayer. Carnap had stated the explicit condition that all defined notation be eliminated before the definition of empirical significance is applied, and my argument violated this condition.

At Carnap's request I prepared a brief written account of my results, and we made an appointment to discuss the matter at his home later in the week.

Climbing the garden path to Carnap's secluded home in the Santa Monica foothills, I felt some trepidation about my forthcoming interview. I knew Carnap to be a kind and even gentle man, but he was also masterful. In seminars he welcomed questions, and he plainly loved philosophical discussion. However, he always seemed to be perfectly familiar with whatever questions or objections were raised. His responses would typically sharpen the question, place it in a larger context, and then convincingly dispose of it. One slowly came to see that Carnap had a very general and powerful framework which he used to focus and clarify issues preparatory to resolving them. The distinction, for individual sentences, between those which are empirically significant and those which are not, like the distinction between those that are analytic and those that are not, played a very important, if not fundamental, role in that framework. Thus on the issues I had raised, I simply expected to be set straight. And in the general way one feels when about to confront a great power, I hoped I had not been too audacious in framing my objections.

Carnap waved at me through his study window and rose to greet me as I entered the house. He was tall and slender, probably only a few inches taller than I, though he seemed to tower over me. "Ho, Dave" he boomed in his typical hearty way, which in this instance I found somewhat unnerving. He seated me in his study and then took my report in his hands and waved it at me.

"Well", he said, "this is really wonderful." Then, with evident enthusiasm, he reflected that he had been quite wrong for about 30 years, and his critics who had been arguing that theories must be accepted or rejected as a whole (he mentioned at least Quine and Hempel) were very likely correct. He asked a few quick and pointed questions about some variations on his definition, brushed aside my hesitant mention of the fact that my argument violated his condition about primitive notation, and then turned to discussion of how Hilbert's ε-operator, applied to an entire theory, might be used to characterize the empirical content of theoretical terms. I was still somewhat stunned by his whole-hearted acceptance of my criticism.

At the end of our talk, Carnap congratulated me again on my result and urged me to communicate it immediately to a number of his co-workers, especially those who had been critical of his own attempts. Though we had frequent discussions over the last thirteen years, no subsequent piece of work of mine was ever so critical of him, nor pleased him so much.

It took me some time to understand Carnap's response. After reading my report, Carnap had improved his understanding of a subject which he had studied for many years. It was an advance in Philosophy, and whether it was initiated by a student's criticism of Carnap or by Carnap's own work made as little difference to Carnap as it did to Philosophy. His enthusiasm for the subject and his drive to understand the phenomena he studied, completely submerged any concern with his own role in the process.

Through the years I have observed many other instances of Carnap's selflessness. But the emotional impact upon a second year graduate student of seeing *Rudolf Carnap* respond to a student's argument with an enthusiasm completely unmitigated by his own 30-year investment on the other side has stayed strongly with me. It was a rare and cherished experience; Carnap taught much more than logic.

Perhaps only in a person of Carnap's enormous intellectual power and accomplishment can the love of wisdom be unadulterated by love of self. But since 1958 I have tried to be like that.

University of California, Los Angeles DAVID KAPLAN

11

My correspondence with Rudolf Carnap began in 1933. At his invitation I visited him in Prague for some memorable weeks in the summer of 1934. In 1935 I suggested to the department of philosophy at the University of Chicago that Carnap be invited as visiting professor for the winter months of 1936. The invitation was extended to him, and accepted. During this period I proposed that both Carnap and Reichenbach be invited to become permanent members of the philosophy department. This double proposal was made to the Administration, but was successful only with respect to Carnap. He began his regular teaching in October 1936, and remained at the University of Chicago until his move to the University of California at Los Angeles in 1952.

I saw much of Carnap during those years, partly because of our joint editorship (with Otto Neurath) of the *International Encyclopedia of Unified Science*. I visited him only once in California, but our correspondence and our work on this encyclopedia continued until his death in 1970. I am happy that he saw in print the final bibliography and index monograph of this large enterprise, the first monograph of which was published in 1938.

Rudolf Carnap was one of the finest human beings I have ever known. His generosity to those in need, the depth and constancy of his personal friendships, his close companionship with his dog, the place of novels and music in his range of sympathy, the complete absence of pettiness and petulance, the clarity and kindness of his smile, the combination of serenity and concern which he always manifested – these attest the largeness of the man. I treasure the remembrances of these many years.

University of Chicago and University of Florida CHARLES MORRIS

12

Nobody contests Carnap's greatness as a philosopher and as a man. He was a pioneer, an untiring worker, and a sower of fruitful ideas in modern scientific philosophy. To me he was also a generous friend and mentor, and to my husband a close associate and collaborator. I should like to convey some of my feelings and impressions regarding these relationships in the hope to contribute a little to rounding out Carnap's portrait.

Although Carnap's and Hans Reichenbach's friendship dated back to 1923, I had not known Carnap before he came to the University of California at Los Angeles in 1954 as my late husband's successor in the Philosophy Department.

I had heard wonderful things about Carnap from Hans who did not only admire and esteem him as a philosopher but also loved him as a person. He had told me how charming, interesting, warm, good, and solid Carnap was. And so I found him to be. I remember our first meeting as if it had been yesterday. It was as if I met an old friend. Both he and Ina were completely informal, offered me the familiar 'Du' immediately (but explained that 'Carnap' functioned as a first name because they disliked 'Rudolf'). I had brought some flowers as a welcoming gift. He acknowledged them by asking "Are they from your garden? I hope you did not spend any money on them!".

I saw them often and became almost a member of the family; when relatives or visitors from out of town came, I would be asked to join them, or the Carnaps brought them to my house. They were immensely sensitive and to a large extent I owe them my ultimate recovery from the shock I had experienced. Occasionally when I lost my composure because I had not quite mastered yet the loss of my husband, Carnap would put his arm around me, offer his shoulder and say "Come here. Cry! Don't have any moral scruples". And I just borrowed a little of his strength.

He was always available, always ready to help selflessly, whether I wanted to discuss a philosophical, an editorial, a pedagogical, or a personal problem with him; whether I asked him to look over a manuscript, or to write introductory comments to works by Hans that I was publishing posthumously.

We took enumerable walks together in the neighborhood of his or my house. He loved nature, the mountains, the ocean, the trees, and the flowers. He was an 'optical' person and was very excited when he could distinguish colors again after his cataract operation. He disliked talking long on the telephone because he wanted to see the other person's face.

Whenever I took a trip, he wanted to be shown exactly on the map where I had been and then would reminisce about his own travels and hikes, or his visits to the archeological excavations in Greece led by his famous uncle Wilhelm Dörpfeld. He always came to look at the slides I had taken.

Carnap had a tremendous capacity to recall earlier experiences and to communicate them. He would remember the little streets and paths that he had traveled in his youth, often on foot, particularly in Austria, Switzerland and Greece. He had walked over the mountains to Delphi. He told about his ski trips and the time he had broken a leg and almost did not make it back because he could hardly walk and was all by himself.

Carnap loved music. Once when I played a Bach recording for him he told me that the unaccompanied cello sonatas by Bach were among the favorite pieces he had played himself on the cello. "Bach is like mathematics" he said.

Carnap's interests ranged widely over areas in addition to philosophy and the sciences: languages, literature, psychology, education, politics, and people, and I marveled at his instant recall of so many diverse facts.

He did not subscribe to any daily newspaper but read carefully magazines such as *The Nation*, *Monthly Review*, *The New York Review of Books*, *I.F. Stone's Weekly*, and kept abreast by looking at the news on television in the evening.

He was fabulously organized. He had a file of cards summarizing briefly in shorthand the content of and his reaction to every book he read, and in a conversation about a particular work he would frequently pick out the relevant card and read it to me. He underlined everything he thought important, even passages on the picture postcards I sent him from abroad. His days were structured. There was work in the morning, a walk before lunch, a nap, a snack, more work, dinner and reading. He kept up this regime until the end.

During the time Carnap was teaching at UCLA, I attended his various seminars. I would pick him up in my car from his house in West Los Angeles. Since I knew that he did not like to talk on the drive to the university because he was already concentrating on the material he was going to present, I kept quiet. His seminars were really graduate lecture courses, never presentations of papers by students. They were attended by many of his colleagues at UCLA from various departments, by faculty members of other universities and colleges in the area, and by graduate students. He always prepared the content most conscientiously and had a folder with notes in shorthand in front of him. But he spoke freely and consulted them rarely. He encouraged questions and comments. Sometimes Ina came along. One semester she taped the whole seminar on 'The Philosophical

Foundations of Physics' which was later edited and published as a book. On the way home Carnap would always talk animatedly, often discuss special aspects of the respective seminar. After we came back to his house, he would rest until dinner while I talked to Ina. As often as not the subject matter of our discussion would be Carnap.

Carnap was frugal. His house was minimally furnished, there were no pictures on the walls of his rooms or office, only book shelves. He did not drink and did not smoke. I knew that he liked marzipan and I used to give him some for his birthday. It was a sight to watch him cutting off a microscopically thin slice after lunch. He made it last almost until his next birthday. He seemed to know his own worth, though, because when I drew his attention to the fact that he and Bertrand Russell were both born on May 18, he said with a straight face "Oh, so another great man was born on that date".

Carnap and my husband saw each other on many occasions in Europe and the United States. Carnap speaks about this relationship in the Schilpp volume. They carried on a steady correspondence. Much can be learned from these letters about their concerns. They dealt with mutual critiques of manuscripts and books, with long discussions of special philosophical problems, with the possibilities of finding positions for younger gifted men they had spotted, and, during Hitler's time, with the urgent necessity of rescuing mutual friends who had to leave Europe. Many a well known name is among them, and Carnap and Hans shared their happiness and relief when they had been successful in placing somebody.

Carnap's letters contain practical advice and *Lebensweisheit*. For instance, he suggested to send a manuscript by railway express because that would be the cheapest way. Even though their mutual critiques were very frank, he advised Hans not to publish too many replies to negative criticisms of his books and articles by others, but rather to correct eventual misunderstandings within the framework of other positive contributions. And I know that on the whole Hans took this advice to heart. Another time Carnap wrote "I had indeed misunderstood your position. If one has a different opinion from another, then he easily misunderstands the explanations of the other man. I see that in your letter you also misinterpret my position ...."

In one of his letters Hans offered Carnap a Flint professorship, a specially endowed chair in the Philosophy Department at UCLA for outstanding visiting professors. Carnap declined because of the loyalty oath affair. My husband regretted but fully concurred in Carnap's decision and asked him for permission to quote publicly from his letter because the matter was still being fought at UCLA.

Carnap had a kind of innocent purity and rare integrity. Ina's nickname for Carnap, 'Angelface', testifies to that. If I suggested a little white lie as a convenient way out of an uncomfortable situation, he would say with a smile "You have a criminal mind". He really lived his principle of tolerance and took Kant's Imperative seriously: to treat human beings, whether in one's own person or in that of any other, always as an end, never as a means only. Carnap would defend anybody whom I criticized by trying to understand and explain rationally other people's irrational feelings and behavior. He was a good model and when I have a problem I still catch myself thinking: I must ask Carnap.

Los Angeles City College MARIA REICHENBACH

13

Philosophers, scientific theorists and research workers in the foundations of science all over the world received the news of Rudolf Carnap's death in September 1970 with sorrow and distress. The following is an attempt to give a brief appreciation of Carnap's philosophical achievements.

For the last 35 years, Carnap lived in the United States and thus spent the greatest part of his creative life there. The course of his scientific activity began in Vienna, however, where he published one of the first German language textbooks in mathematical logic as well as his first major philosophical work, *Der Logische Aufbau der Welt (The Logical Construction of the World)*. This book represents a high point of empiricist philosophy. For centuries, empiricist philosophers had always held that our concepts and our knowledge rest upon experience. However, in principle, they never got beyond dogmatic assurances and programmatic proclamations. Carnap was the first to attempt to carry this program into action, in his *Logische Aufbau.* With his mastery of modern logic he constructed a system of concepts in which all the concepts of the factual sciences were derived from a single empirical basic concept, by chains of definitions. There are two principal grounds upon which he later gave up this constitutive theory of concepts: first, he recognized that certain concepts, namely the so-called *disposition concepts* in principle cannot be introduced by definitions. Second, he believed he could demonstrate that the *positivist sense-datum language* which he introduced in this book should be replaced by a *physicalist language*. Nevertheless, this first work exhibits an altogether extraordinary constructive and analytical talent. One of the leading contemporary logicians, Quine, once remarked that it remained a complete mystery to him how Carnap in this work could have attempted to introduce by means of definitions such concepts that no one would have thought to be definable except in a dream.

Carnap's second major work was the *Logische Syntax der Sprache (Logical Syntax of Language)*, which appeared in 1934. One may rightly say that, with this work as well, Carnap established a milestone in the history of recent philosophy. In this work, the so-called *linguistic tendency*, already hinted at in the ongoing work in analytic philosophy and methodology of science, is fully realized. Here, Carnap was oriented by the insight that the task of a scientific philosophy consists in the construction of *formal artificial languages*, adequate to the precise reconstruction of scientific propositions. By contrast to natural languages, these artificial languages are constructed according to strict rules, which bear only upon the external (abstract) form of expressions, and therefore may be characterized as *syntactical rules*.

There are three outstanding features which may be culled from the *embarras de richesses* which this work presents. Carnap was the first philosopher to show the necessity to sharply distinguish between two levels of language in the logical analysis of language: The *Object language*, in which one represents the objects of inquiry, and which is construed by analysis as an uninterpreted calculus; and the *metalanguage*, which is an interpreted language and which is used to talk about the object language. Carnap was able to show that the confusion of these two levels of language, a failure to distinguish between object-linguistic and metalinguistic concepts, is responsible for the introduction of certain contradictions, and that this confusion led even such distinguished logicians and mathematicians as Bertrand Russell and David Hilbert into serious errors.

Another important feature of this work is the distinction which Carnap introduces between *material* and *formal modes of speech*. Carnap shows that many philosophical errors arise from the confusion of those statements which are meant to denote things and their properties with those statements which in fact denote linguistic expressions. So for example, the materialist asserts "All corporeal things are constituted by subatomic elements" whereas the phenomenalist asserts "All corporeal things are constituted by sensations". There appears to be an irremediable contradiction between these two assertions. However, if both of these theses are translated into the correct formal mode of speech, the first thesis asserts that all scientific propositions about things are translatable into physicalist statements of the sort given, whereas the second thesis asserts that all propositions about things are translatable into sense-datum statements. Therefore, it is no longer an issue of contradictory metaphysical assertions, but rather one concerning two compatible and rationally discussable linguistic theses.

However, Carnap's most important contribution in this work is, without a doubt, in that he established that all logical rules are to be understood as *syntactical* rules. The earlier formulation may be stated briefly thus: Syntactical rules are rules of formation, which tell us how to construct new statements from other statements, as for example, from the two statements *A* and *B* one constructs the statement "If *A* then *B*". In this earlier formulation, the syntactical rules still contain something of content, namely, they constitute judgments, insofar as they tell us how one may derive certain judgments as consequences from other judgments. Against this view, Carnap was able to show that even purely logical relations between statements can always be derived from syntactical structures alone. As against the formation rules of grammar, the logical rules of inference can be seen as *transformation rules*, which, like all syntactical rules, take only the external or abstract forms of statements into account.

Unfortunately, Carnap's work remained unknown to linguists when it appeared, mainly because he explicitly excluded natural languages from his considerations. Thereby, linguistic theory remained unaware of the existence of grammatical transformation rules (for natural languages). The transformational grammar of Noam Chomsky, therefore, was greeted as a sensation when it appeared in 1957. As the linguist and logician Bar-Hillel once remarked, Carnap had anticipated the essential insights of this new linguistic theory 25 years earlier.

The *Logical Syntax of Language* is Carnap's most difficult work. When it appeared, only a few Polish logicians, who were working along somewhat similar lines, were able to understand it properly. In fact, it was the Polish philosopher and mathematician, *Tarski*, who had the greatest influence on Carnap, in the works which followed the *Logical Syntax*. In 1935, in an important paper on the concept of truth in formalized languages, Tarski showed how the concept of truth, as well as many other semantical concepts, could be introduced in a precise way for formal artificial languages. It was characteristic of Carnap that he not only grasped Tarski's ideas fully, but immediately set himself the task of constructing a systematic *theory* of semantics. Shortly after his resettlement in the United States, he began to publish a series of works on semantics. The conception which he presented in his *Logical Syntax* was not abandoned, but, in a significant sense, elaborated. Whereas he there conceived of languages only as uninterpreted calculi, he now rounded out this treatment of formal languages with a *theory of interpretation* of such languages. What was distinctively new here was that the idea of interpretation was not based on a vague psychological concept of meaning, but was rather introduced as an intersubjectively controllable and mathematically rigorous concept. *Truth* and *logical truth* are the two fundamental concepts of semantics.

Only when these concepts were made precise was it possible to make any clear sense of the demand for a proper foundation and for the completeness of a logical system. A logical calculus is complete only when it can be shown that there is a matching relationship between the syntactical concept of provability, defined for the calculus, and the semantic concept of logical truth. This view of the necessity of semantic completeness for a logical system is generally accepted today.

The third volume of Carnap's writings on semantics, *Meaning and Necessity*, contains some completely new ideas, namely the basis for a theory of *intensional semantics*. The earlier works on semantics, almost exclusively on a foundation of *extensional semantics*, dealt with the basic concepts of individual, class, truth-value. In the course of his investigations, Carnap recognized that this conceptual apparatus remained incomplete, and needed to be extended by a system of intensional semantics. Among the basic concepts of such a theory are such concepts as the *sense* or *meaning* of names and sentences, the *synonymity* or *meaning-equivalence* of expressions, and the *analyticity* of propositions. Above all this fully developed conceptual apparatus requires a *logic of modalities*, which bears on the concepts of possibility and necessity, as well as an *epistemic logic*, i.e. a logic of knowledge and belief. Carnap ties his analysis to that of Frege, in the main, but at the same time achieving a unified systematization as well as an essential simplification of Frege's ideas. The discussion of the nature and problems of intensional concepts, initiated at the publication of *Meaning and Necessity*, goes on until the present day. In all the subsequent investigations and in the critical confrontations in this field, Carnap's book remains *the* standard work.

In addition to his studies in semantics, Carnap continued to dedicate himself to *problems in the foundations of knowledge in the natural sciences*. A clear and easily accessible insight into this aspect of Carnap's thought can be gotten from the very readable work, *Philosophical Foundations of Physics*. Among the many new ideas which are introduced here, let us mention only one especially important one: the *two-levels theory of scientific languages*. This theory decisively goes beyond the positivistic view according to which all concepts and statements in the natural sciences are ultimately based on what is observable and perceptible. Carnap was able to show that in addition to the observational language, which was the only base language which positivism allowed, there was also a *theoretical language* which, in contrast to the observation language, could be only partially understood. The fundamental concepts and principles of the theoretical language can only be *partially* interpreted, and even then only in an indirect way by means of so-called rules of correspondence. *Most important of all, the meanings of concepts which occur in a theoretical language, are seen to be dependent on the theory formulated in this language*. So, for example, not only does modern physics propose hypotheses concerning elementary particles different from those of classical physics, but the very meaning of the term 'elementary particles' is entirely different in each of the two theories.

Carnap's central aim, his *theory of inductive inference*, has thus far gone unmentioned. One can get a general idea of Carnap's position here by contrasting it with that of his opposite number, Karl Popper. As is well known, natural scientists often allege that their procedure is *inductive*, by contrast with that of mathematicians, which is deductive. If one asks what this so-called 'inductive' method is, no clear answer is forthcoming. The philosopher David Hume had already remarked upon the many difficulties raised by the view that in inductive arguments the conclusion follows from the premises. Popper considered these difficulties to be insurmountable, and developed a purely deductive theory of the corroboration of hypotheses. According to Popper, we corroborate hypotheses in the natural sciences by attempting to refute them. When this attempt at refutation fails, we then say that the hypotheses have been corroborated. The corroboration of hypotheses and theories consists not in positive confirmation, but rather in the bare statement of the failure of an attempt at falsification. All talk of inductive confirmation and of the probability of hypotheses is therefore nothing but empty chatter, according to Popper.

Carnap found Popper's skepticism about induction unacceptable. He was convinced, instead, that it made perfectly good sense to say of a hypothesis that it was positively confirmed, or that it was more highly confirmed than some other hypothesis, and that therefore it was entirely reasonable to order the probabilities of alternative hypotheses in accordance with the relative weight of available evidence. In the course of his investigations of the precise sense of these new ideas, Carnap made a remarkable discovery: It became clear that there were strict rules at the basis of this inductive reasoning quite analogous to the rules of logical proof in deductive logic. Therefore Carnap conceived his program of an *inductive logic*. 2500 years before, Aristotle had begun to bring the rules of valid deductive inference to the light of day. Analogously, Carnap set himself the task of precisely formulating the rules of inductive reasoning. He saw this task as extraordinarily important, because in contrast to the very small role played by deductive inference in our private and public lives, much the greatest part of our everyday thinking involves inferences of an inductive, probabilitic sort. With this conception, Carnap became the sharpest theoretical opponent of Karl Popper. Whereas Popper declared induction to be superfluous, and the belief in it nothing better than a superstition, Carnap believed he could demonstrate that an inductive logic was at least as legitimate as deductive logic, and should be placed alongside it.

The difficulties which stood in the way of carrying out this project were so many and so enormous that anyone but Carnap would have despaired of continuing with the task. In 1950, he published the first part of his thought on the theme of induction in the extensive work, *Logical Foun-*

dations of Probability. In this work, there was an unusual combination of original insights and most painstakingly detailed work. Still, in many respects the work proved to be inadequate. Carnap therefore set to work anew.

His further work on inductive logic was interrupted, during the next six years or so, by his simultaneous preoccupation with three other projects. The first was the revised edition of his German work, *Abriss der Logistik (Outline of Logistic)*, requested by his publisher. Since a simple reworking of the original text was out of the question because of all the new results which had been achieved in the meantime, Carnap wrote an entirely new book, *Introduction to Symbolic Logic*, which thereafter became one of the best known and most widely read introductory works in modern logic. Two things set this book apart from most other introductory texts: precise treatment of the semantic and syntactic construction of formal languages; and many applications of logic to special questions in the various sciences.

The second project was a revision of his semantics, in which he wanted both to take into account the criticisms made of his earlier publications, and to report on the latest results of his own investigations. This revision was completed only in manuscript form, and it was circulated among a small group of students and friends.

The third project turned into a gigantic task. In the Schilpp series (*The Library of Living Philosophers*), there was to appear a Carnap volume. He was required not only to give an account of his personal and intellectual development, as well as a sketch of his views on philosophical problems, but, in addition, to reply to other scholars who had contributed articles concerning his work. The large number of contributions and their extraordinary diversity give a most impressive overview of the range of Carnap's creative work. In some 250 pages, Carnap gave concise and intellectually concentrated replies to all of the 26 essays.

In his last 15 years, Carnap dedicated himself exclusively to the construction of his inductive logic. The comprehensive manuscript of many hundred pages, which he repeatedly exposed to constant criticism, and which he reworked again and again, became known only to relatively few colleagues and friends. It has only the slightest resemblance to the abovementioned work on this topic. In place of the original linguistic version, there is instead a *model-theoretic* approach by means of which the range of application of inductive logic is vastly extended. It was Carnap's tragedy that he did not live to see the publication of this, the *magnum opus* of his life's work. He died a few days after sending the proof-read manuscript to the publisher.

Carnap's theory of inductive probability led more often to polemic than to agreement. Still, even his most radical critics always agreed that Carnap's analyses yielded significant contributions to special sets of problems. In the following eleven points, I will introduce the topics of Carnap's researches, all of which can be recognized as highly significant, independent of whatever point of view one may have concerning Carnap's overall project. This explicit mention of these areas of Carnap's research is important, I feel, because *at least eight of these points have remained unknown until now, and are to be found only in the as yet unpublished manuscripts*. Even in the very recently published book, by L. Krauth, *Die Philosophie Carnaps*, these matters could not yet be mentioned. One may expect that after the details become known, not only will a lively discussion ensue, but the picture we have had of the scientific thinker Carnap will have to be revised in certain essential respects.

1. The so-called *Principle of Indifference* of classical Probability Theory presented itself in contradictory form. All the same, the view that "There is something valid in this principle" kept cropping up again and again. Carnap tried to work out the correct and unassailable kernel of this classical idea, in a series of *invariance principles*.

2. An essential feature of ordinary man, as well as of the scientist, is that he *learns from experience*. By means of the *Principle of the Relevance of the Singular Case* as well as by the formulation of the so-called *Reichenbach-Axioms*, Carnap tried to make clear the concept of rational learning from experience.

3. In introducing the two concepts of *credence-function* and *credibility-function*, Carnap contributed in an essential way to the clarification of the foundations of personalistic probability theory as well as of normative decision-theory. Whereas personalists and decision-theorists go no further than the first concept, Carnap showed that the transition from the first to the second is strongly motivated, and quite analogous to the transition from peripheral manifest properties to their basis in permanent dispositions, in the fields of psychology and physics.

4. Carnap goes along with the personalists in the matter of the *Justifi-*

cation of the Axioms of Probability. Yet, it remained for Carnap to give an exact analysis of the relevant concepts of *bet* and *betting-system*, as well as of (*strict*) coherence, and thereby to give the precision of today's logical standards to the justification procedures for this measure.

5. It is noteworthy to remark on the way in which Carnap built his new model-theoretical conception into probability theory. First of all, *models* yield interpretations of entire language systems – (which are therefore, as a rule, much more comprehensive than those which can be expressed in the usual formal languages) – and such models can be described by a single two-place metrical function. These models, or functions, are chosen as *points in a probability space*. Atomic propositions are construed as infinite classes of models; and the class of propositions is identified with the Sigma-particle derived from the class of atomic propositions. The *probability-measure* is defined for the elements of the latter.

By means of this ingenious trick, of being able to get interpretation systems of entire languages from single functions, and of choosing these as points of a probability-space, *modern model-theory*, which is an integral part of mathematical logic, *is, for the first time, brought into connection with modern measure-theory and probability-theory in an unrestricted way*.

6. Carnap didn't pretend to have solved all the problems which come under the heading, 'Justification of induction'. Nevertheless, his deliberations on these questions contributed much to differentiate the problem, and to keep distinct the different forms of reasoning which are under consideration here. Because there is not a single inductive method, but rather a whole continuum of such methods, the justification-problem has to be divided into weaker and stronger questions, according to Carnap. The *weaker* question concerns the justification for the choice of a *stricter subclass* (of all conceivable inductive procedures) which can be characterized by suitable axioms.

The definite article in the phrase '*The* justification of induction' is thus no longer applicable. For in the context of answering the weaker question, no grounds are given for fixing upon any particular C-function, which represents a particular inductive method. Rather, there are only grounds for the adoption of an axiom system for inductive probability. Only with the answer to the *stronger* question is one led to the choice of a *particular inductive* method. In the attempt to answer the weaker question, all sorts of grounds have to be drawn upon already: *inductive intuition* (coherence or strict coherence), *pragmatic considerations, a priori conjectures* on the structure of the world (e.g. those concerning degree of uniformity), *simplicity-considerations*, as well as various *subjective aspects*.

Carnap was convinced that even if one addressed himself only to the weaker question, one would have to go far beyond what personalistic probability theory was able to afford. This theory satisfied itself with the justification of the axioms of probability theory. On Carnap's view, these provided much too weak a basis for inductive reasoning. His view was based on the proof of the statement that for any two logically independent propositions H and E and for any given real number r, one can always find a C-function which will fulfill the Axioms, such that $C(H, E)=r$. This shows that there is evidently *always* an abundance of irrational functions which are admissible, which can be successively eliminated by further Axioms. The intensive search for such additional axioms and their establishment took up a great part of Carnap's researches in his last years.

7. It is generally recognized that Nelson Goodman uncovered a fundamental difficulty with which all variants of the concept of inductive confirmation are afflicted (the so-called Goodman paradoxes). Carnap tried to overcome at least one important aspect of these paradoxes by his distinction between *absolute* and *relative coordinates*. The concept of absolute coordinates presents an important contribution to the problem: "How can we talk about individual things?"

8. Traditionally, the procedures by which qualitative or classificatory concepts are introduced are completely separated and dealt with differently from those by which quantitative concepts are introduced. It was Carnap's achievement to give so generalized an account, in his *theory of families of attributes*, that it could yield ***a unified method for the treatment of quality and quantity***.

9. The concern with families of attributes led Carnap to the sketch of a *language-independent theory of attribute-space*. This is the first attempt to characterise the topological and metrical structures of attribute-spaces by measure-theoretical means. One may expect that the ideas which Carnap developed here, which are fully independent of his probability-theory, will be fruitfully developed in their applications to other disciplines, e.g. sensory psychology and phenomenology.

10. A strong limiting condition which Carnap had to impose in his treatment of languages, in his original linguistic version, was loosened by his concept of *meaning-postulates* or *analyticity-postulates.* Carnap recognized that this concept was still too weak. He therefore introduced *phenomenological base-principles*, which may be characterized, in more traditional terms, as *synthetic a priori propositions.* In a bold generalization of this notion, he conceived of *comparative and quantitative synthetic a priori propositions.*

11. The considerations noted in 6 above would require a good exemplification of the procedures for establishing norms of rational scientific discourse.

This last point leads to an entirely new assessment of the Carnapian project for a probability logic. Originally, as remarked earlier, Carnap was dominated by the view that deductive logic had to be paralleled by inductive logic. The basic notion here, namely the *notion of partial logical deduction* turned out to be ambiguous. The same holds true of the notion which he always used, *degree of confirmation.* Both of these concepts fade more and more into the background. If we take this into consideration together with the innovations noted above, one may then come to see Carnap's lifework, not as the construction of an inductive logic, *but rather in the foundations of personalistic probability theory, and of normative decision theory, and in the task of making them precise, formalizing them and rationalizing them further.*

If one takes an unprejudiced overview of all of Carnap's works, beginning with his *Logical Construction of the World* and on to the latest, as yet unpublished works, one does not arrive at a simple, conclusive picture. There are many significant thinkers who have done fundamental work in one or two branches of inquiry. However, in comparison, one would have to characterize Carnap as a unique phenomenon in the history of thought, in that he did decisive pioneer work *in six completely different fields.*

Unique also was Carnap's combination of originality, insightful and absolute precision, together with endless patience and persistence, characteristics which are indispensable for the solution of the toughest problems.

Though Carnap was 45 when he first went to the United States, he became one of the most influential philosophers in that country. A great part of the discussion in the U.S. on philosophy of science, on 'the meta-science of science', has gotten its most recent stimulation and problem-settings from Carnap. However, this conclusion needs to be clarified further: When one hears of an influential philosopher in Germany, one thinks almost always of the founder of a so-called school of thought; and beyond that, of someone who uses his established social position as a University professor to take on as students to be advanced through habilitation and teaching appointments, only those who take over, reproduce and further develop his own ideas. Carnap had no students in this sense, nor did he want this sort of student. What concerned him was the education of independent, critical investigators, who would be able to contribute to the development of philosophy as an exact science. The attempt to achieve conceptual clarity and distinctness of thought, and to broaden the spirit of rational criticism bound Carnap closely to many philosopher-friends, across all lines of theoretical difference of opinion (e.g. H. Feigl, E. Nagel, C. G. Hempel, N. Goodman, W. V. Quine, Y. Bar-Hillel. Karl Popper should be named here as well, certainly. If one considers Carnap's dispute with Popper in its basic philosophical terms, it comes to no more than a 'small internal family quarrel', which in retrospect appears much less dramatic than it appeared at the time. In their affirmative attitude to a rational philosophy both of these thinkers were completely at one with each other.)

Immediately after Carnap's death, obituaries appeared in various distinguished German newspapers. In all of these, Carnap's book, *The Logical Construction of the World* was seen as his major work. The later works were ignored. For anyone who is in the least acquainted with Carnap's publications, such accounts make clear how terribly provincial German philosophy has become. Nowhere in these press notices was it indicated that this work was, after all, Carnap's *Habilitationsschrift*, written when he was thirty, and which he himself considered as superseded for more than forty years.

Carnap himself illustrated, in his own case, how constant critical debate and confrontation can lead to progress in philosophy. There were always many criticisms of his publications; and Carnap himself brought forth the sharpest and most decisive critical arguments against his own earlier work. This sometimes led to what would appear to an outsider as paradoxical situations, such as that in which N. Goodman (in the Schilpp-volume) attempted to defend some of Carnap's earlier positions, in the

Logical Construction of the World, against Carnap's later formulations.

The quest for objective criticism was also one of the sources of the many intellectual contacts which Carnap sought and found. When he presented his theses for discussion at a meeting, he didn't do it in order to win adherents to his position, but in order to stimulate objections and proposals for revision and improvement. When I presented him with a written list of questions, at my last meeting with him in 1967, he told me beaming, that it gave him the greatest pleasure when students and colleagues took the occasion of meeting with him to bring along and present prepared lists of problems and objections.

In his last years, Carnap's creative work was adversely affected by two factors: his failing memory, and trouble with his eyes. Once he remarked to me, half in earnest, half in jest: "When it comes to forgetfulness, I break all world-records." This pertained especially to his own visual work. He reported how unspeakably tiring it was for him to proof-read his own manuscript, because he had to turn the pages back again and again since he couldn't remember whether he had already said something earlier, or which sort of formalization he had chosen to use earlier. Until the two successful eye operations in 1969, his eye troubles of later years made any reading all but impossible for him.

On the other hand, his capacity for concentration and the acuteness of his thinking remained unimpaired to the end. I got a lively impression of this also in my 1967 meeting with him. I had been carrying on a discussion with him and a former student for about five hours. He vigorously declined my proposal to take a break in the course of the discussion and to go for a stroll. So we talked non-stop for the whole five hours, on all aspects of the philosophy of science, and he developed his ideas in response to the partly informational, partly polemical questions which I raised through this whole time, without the least sign of tiring.

A conversation with Carnap did not proceed as one who had read any of his great works might have presumed it would. Certainly, Carnap was the great 'formalizer' among philosophers; and one might assume therefore that he was also given only to 'abstract formal thinking'. Nothing could be further from the truth. In fact, he was a thoroughly 'straightforward down-to-earth thinker', if one may describe him thus. An accurate impression of his style of reasoning and exposition can be gotten from his

Philosophical Foundations of Physics which was transcribed from tapes of one of his introductory seminars. Even in discussions of probability theory, in the subtleties of which he was more completely at home than anyone else, his exposition never took on a mathematical character. More often, he would picture the situation in simple model-worlds, as for example, in a world in which there are only things with different colors; or he would describe the considerations that a rational diamond-miner would have to take into account, once he had found a precious stone, and then had to decide where to look further. Carnap's super-ego always forced him, of course, to put his thoughts finally in a precise and formalized way. *But the intuition always came first, and precise refinement and formalization always constituted the last step.* He reported to me that he often advised his students not to try for complete precision in their work, including their dissertations. He added, laconically: "Apparently, my advice never seems very convincing, since I myself violate it constantly".

The large number of friendships which Carnap made during his lifetime did not rest only on his philosophical-scientific contacts. Very often it was his everyday, non-scientific concerns which led to personal friendships. Since the first World War, which he spent mainly at the front, Carnap stood politically very far to the left. His reaction to the Marxist critique of philosophical empiricism is worth noting: whereas he criticized traditional metaphysical philosophers very sharply, he did not attack or criticize the Marxists. More often, he defended the lack of attention which he and his empiricist friends gave to economic and sociological problems, on the principle of the division of labor, on the grounds that he and his empiricist friends happened to come out of the natural sciences. At the same time, he emphasized the hope that these questions would become far more central for philosophy of science in the future. He also agreed with the Marxist critics that the contemporary economic and social order leads to a brutal dehumanization of life, and must be changed therefore. His ideal, which he held with cautious optimism, was a *scientific humanism*. He hoped that one day the social sciences would develop so far that it would become possible for man to fashion his social world correctly and rationally, free of any ideology. He supported everything which he thought would lead to worldwide understanding among men. After the first World War, he studied Esperanto, and enthusiastically joined the world-wide Esperantist movement, in the belief that a common world-language, especially if well formed, could contribute to understanding among men. He later supported many peace initiatives. He was in correspondence with Bertrand Russell. Both philosophers were united in their pacifism. The war in Vietnam was among the most frightful experiences for him, in the years before his death.

Carnap's deep understanding of human needs is little known. Whenever he came upon instances of it, he tried to help. He was also much concerned with the Negro question. He was in long term association with and supported a peaceful Negro organization in Los Angeles, which sought to help achieve better education and better life conditions for its members. The last photograph we have of Carnap shows him in the office of this organization, in conversation with various members. He was the only white in the discussion group.

WOLFGANG STEGMÜLLER

SYMPOSIUM:

THEORETICAL ENTITIES IN STATISTICAL EXPLANATION

JAMES G. GREENO

THEORETICAL ENTITIES IN STATISTICAL EXPLANATION

The ideas in this paper are based on an analysis of statistical explanation that uses the information transmitted by a theory [2]. Consider a theory that specifies a probability distribution on the events of some domain, where for purposes of analysis we divide the variables that describe the domain into two sets: $[M]$, a set of variables whose values are to be explained, and $[S]$, a set of variables whose values are used to explain the values of the variables in $[M]$. The information transmitted by the theory is

$$(1) \qquad I(S, M) = H(S) + H(M) - H(S \times M),$$

where $H(S)$ and $H(M)$ are the uncertainties of the events in $[S]$ and $[M]$, respectively, and $H(S \times M)$ is the uncertainty of the joint events.

To illustrate this idea, consider the problem of explaining medical symptoms such as fever, coughing, skin rash and abdominal pain. The kinds of variables available for explaining symptoms are facts in a patient's medical history and information about the patient's recent contact with other people with similar symptoms. For the moment, I will avoid reference to disease entities, and deal only with the data that are available to the physician. I have to strain your imagination a bit to do this, but a theory about skin rash and fever might run something like this: Let M_1 denote a combination of fever and a blotchy skin rash, and let M_2 denote the absence of this pattern of symptoms. Let $[S]$ be a set of three variables: (1) the patient's age, (2) whether he has been in contact with another person showing the symptoms within the last month, and (3) whether he had the symptoms himself at any previous time. If each of these variables had three values, then $[S]$ would partition the domain of people into $3^3 = 27$ sets: for example, one set could include all infants under the age of two months who had been in contact with someone with the symptoms recently and had not shown the symptoms themselves. The theory then would consist of a set of probabilities of the 27 events distinguished in the explanans, and 27 conditional probabilities – a value of $P(M_1 \mid S_i)$ for each $S_i \varepsilon [S]$. These probabilities are sufficient to specify the probabilities of all the joint events, and therefore the overall probability of M_1 is specified also. Facts like "infants less than two months old seldom have the symptoms" and "contact with a person who has the symptoms increases the probability of having them, unless the person has had the symptoms previously himself" would be incorporated in the conditional probabilities, and facts like the proportion of people who have had the symptoms and the proportion of people who are younger than two months of age would be incorporated in the probabilities of the explanans.

Let

$$a_i = P(S_i), \quad c_k = P(M_k), \quad P_{ik} = P(M_k \mid S_i). \tag{2}$$

In relation to the example about symptoms, the a_i are the proportions of people in the various categories specified by age, medical history, and recent contacts. The c_k are the proportions of people who have or do not have the symptoms now. The p_{ik} are the conditional probabilities of having the symptoms, given the various categories of age, medical history, and recent contacts. The quantities in Equation (1) are defined as

$$\begin{aligned} H(S) &= \sum_i - a_i \log a_i, \quad H(M) = \sum_k - c_k \log c_k, \\ H(S \times M) &= \sum_{ik}\sum - a_i p_{ik} \log a_i p_{ik}. \end{aligned} \tag{3}$$

If we think about a theory in relation to its information transmitted, we are immediately led to considering the overall properties of the theory. Information transmitted is a measure of the reduction in uncertainty brought about by the dependencies between the variables. Thus, it is an index of the explanatory power of the theory in relation to the entire domain of events that the theory deals with. In fact, the point of introducing the analysis based on information transmitted was to provide some concepts that would make it reasonable to consider the evaluation of theories, rather than of single explanations. In my opinion, it is more appropriate and useful to consider general properties of theories than it is to deal with the status of single explanations. And the information transmitted seems to capture some of the properties that are desirable for a measure that is used to evaluate a theory. For example, a higher value of information transmitted generally goes with a greater degree of testability in Popper's sense, and a greater degree of predictive usefulness.

In this paper, I want to extend this line of analysis to the consideration of theories that provide statistical explanations and that postulate theoretical entities. (The earlier analysis was limited to relationships between empirical variables.) To carry out this analysis, I need to introduce a structure that is slightly more complex than the one described above.

Consider a domain Ω, partitioned by three sets of variables: $[S]$, a set of empirical variables used as explanans; $[M]$, a set of empirical variables whose values are to be explained; and $[T]$, a set of postulated theoretical variables. The values of the theoretical variables are assumed to be produced by the values of the explanans, and in turn to produce the values of the explananda, according to statistical laws specified by the theory. The sense in which I use the phrase "theoretical variables produce the values of the explananda" is entirely neutral as regards meta-physics. I simply mean that the conditional probabilities of the empirical outcomes – the explananda – given any theoretical state are the same, regardless of the value of the explanans that applies. In other words, I assume that the probability law connecting any given theoretical state with the explananda is independent of the conditions that produced that theoretical state. Under this assumption, the theory consists of three sets of probabilities: (1) a vector of probabilities a_i, where

(4) $$a_i = P(S_i),$$

(2) a set of conditional probabilities q_{ij}, linking the explanans to the theoretical states, where

(5) $$q_{ij} = P(T_j \mid S_i),$$

and (3) a set of conditional probabilities r_{jk}, linking the theoretical states to the explananda, where

(6) $$r_{jk} = P(M_k \mid T_j).$$

First, it may be remarked that these quantities relate to those described at the beginning in a straightforward way. When theoretical variables are not taken into account, a theory is specified by a vector of probabilities of the explanans, the a_i, and a matrix of conditional probabilities linking the explanans and the explananda, the p_{ik}. This matrix is just the product of the matrices of conditional probabilities q_{ij} and r_{jk}. In other words, the probabilities linking the explanans and the explananda are specified by the probabilities defined in equations (5) and (6).

$$(7) \qquad p_{ik} = \sum_{j=1}^{t} q_{ij} r_{jk} .$$

Of course, a theory is testable because data can be used to check its assumptions. In a theory of the kind described here, the data are the empirical values of the a_i and the p_{ik}, and these can be used to test the theory. If the theory specifies numerical probabilities (rather than merely the existence of specified states) then the process of testing the theory is just that of comparing the theoretical values of the a_i and the values of p_{ik} calculated from Equation (7) with empirical values of a_i and p_{ik} that can be obtained by whatever means are available. When the theory just specifies that certain states exist, testability involves the estimation of parameters, and I will leave that discussion for a later time in this paper.

The main question that I want to deal with is the relationship between theoretical variables and the information-theoretical properties of a theory. The situation involves three matrices: one linking the empirical explanans and the theoretical variables, a second involving the theoretical variables and the empirical explananda, and a third, the product of the first two, linking the empirical explanans and the empirical explananda. We can calculate the information transmitted by each of these. As an hypothetical example, consider a theory that specifies four possible values of S and three possible values of M. This might, for example, involve classification of medical information into four categories of medical history and three categories of symptom patterns. In addition, the theory postulates the existence of a theoretical state, such as a disease entity that may be present or absent, giving two values of T. To have a concrete illustration, suppose the values of the a_i, q_{ij}, and r_{jk} are

$$(8) \qquad [a_i] = [.20, .30, .10, .40],$$

$[q_{ij}] =$

	T_1	T_2
S_1	0.70	0.30
S_2	0.60	0.40
S_3	0.30	0.70
S_4	0.20	0.80

$[r_{jk}] =$

	M_1	M_2	M_3
T_1	0.70	0.20	0.10
T_2	0.20	0.20	0.60

The values of information transmitted by these matrices, using natural logarithms, are

$$I(S, T) = 0.10, \quad I(T, M) = 0.16.$$

The implications of this theory for dependencies between explanans and explananda are described by the values of p_{ik}, which are

$$(9) \qquad [p_{ik}] = \begin{array}{c|ccc} & M_1 & M_2 & M_3 \\ \hline S_1 & 0.55 & 0.20 & 0.25 \\ S_2 & 0.50 & 0.20 & 0.30 \\ S_3 & 0.35 & 0.20 & 0.45 \\ S_4 & 0.30 & 0.20 & 0.50 \end{array}$$

Using the values of a_i given earlier, the information transmitted is

$$I(S, M) = 0.03 .$$

An interesting general fact is illustrated by the example. The value of $I(S, M)$ is never larger than either $I(S, T)$ or $I(S, M)$. This claim can be proved as follows. Consider a state-space $[Q] = [S] \times [T]$. The conditional probabilities $P(M_k \mid Q_h)$ will equal the conditional probabilities r_{jk}, hence, the information transmitted $I(Q, M)$ will be the same as $I(T, M)$. (This is established by an argument given in [2] in connection with Equation (17) of that paper.) The state-space $[S]$ is a partition of $[Q]$, and it can be shown that the information transmitted by collapsing two or more of the states of a matrix cannot be larger than the information transmitted by the original matrix.

What this fact means is that the dependencies between the theoretical variables and the empirical variables are stronger in the sense of approaching probabilities of one and zero than are the dependencies between the two sets of empirical variables. I am inclined to believe that this fact is related to the intuitions that we have regarding the desirability of theoretical explanations when the dependencies between empirical variables are statistical. If the theory correctly specifies a set of states that produce the phenomena to be explained in the sense assumed here, then the explanation in terms of the theoretical states is better, in the sense of information transmitted, than is the explanation in terms of the empirical explanans.

On the other hand, the improved quality of the explanation obtained by introducing theoretical states may be merely a 'paper' profit. Without theoretical development, merely introducing theoretical variables may not change the empirical content of the theory. However, further developments are often guided by the postulated properties of theoretical entities.

One kind of development involves the discovery of new empirical variables that can be added to the explanans. Refer back to Equation (8), and imagine that at some stage of scientific investigation the probabilities specified there represent the best available theory about the explananda $[M]$. This is equivalent to saying that the best judgment that can be made on the available evidence is that there are states T_1 and T_2 that cannot be distinguished directly in observations (at least with present technology) that are related to the explananda according to probabilities given by the values of the r_{jk}. One research problem that would be potentially worthwhile in such a situation would be the search for improved knowledge about the conditions that produce the theoretical states. If additional variables that are related to the theoretical states could be discovered, then the likely outcome would be an increase in $I(S, T)$, and a corresponding increase in $I(S, M)$. The limit of this process of obtaining better knowledge about conditions producing the theoretical states is a theory with information transmitted equal to $I(T, M)$.

A second king of research that could be motivated by a state of knowledge such as that described by Equation (8) involves an effort to develop new theoretical variables in order to increase $I(T, M)$. If the conditional probabilities of the explananda, given the theoretical states are substantially different from one and zero, there is a strong presumption that the theoretical description is incomplete and that additional distinctions among theoretical states can be found to reduce the conditional uncertainty of the explananda given the theoretical states.

A third kind of development from a situation like that of equation (8) could be the discovery of new phenomena that can be explained by the empirical explanans and the theoretical variables of the theory. The extension of a theory to additional explananda usually has the effect of increasing the information transmitted – in this case, involving the relationship between theoretical states and the explananda.

Next, I will discuss theories of the form of Equation (8), but with free parameters rather than numerical values of the conditional probabilities q_{ij} and r_{jk}. Theoretical proposals that specify numerical probabilities are very rare. However, a kind of situation that occurs frequently is one in which dependencies among empirical variables are known, and a theorist proposes to explain the dependencies in relation to a set of theoretical states. In its weakest form, this kind of theoretical proposal simply specifies a number of theoretical states and all the conditional probabilities are free parameters. Let n be the number of values taken by the explanans and let m be the number of values taken by the explananda. Then if t is the number of postulated states in the theory, the form of the hypothesis is

$$\text{(10)}\quad \begin{aligned} &\exists\{q_{ij}: i = 1, \ldots, n; \quad j = 1, \ldots, t\} \\ &\exists\{r_{jk}: j = 1, \ldots, t; \quad k = 1, \ldots, m\} \\ &\qquad (\forall i)\,\forall(k)\left[p_{ik} = \sum_{j=1}^{t} q_{ij} r_{jk}\right]. \end{aligned}$$

That is, the theory asserts that it is possible to find a set of probabilities (the q_{ij} and r_{jk}) such that all the values of p_{ik} can be calculated using Equation (7).

An hypothesis of this kind can be tested if the number of free parameters is smaller than the number of quantities that can be obtained from the empirical dependencies. In the weak form of the theory described above, the number of free parameters is $(t-1)n+(m-1)t$. The number of empirical quantities is $(m-1)n$. Then the theory is testable if the following inequality is satisfied:

$$\text{(11)}\quad t < \frac{mn}{m+n-1}.$$

It should be noted that Equation (11) applies only when a theory is stated without constraints on the theoretical parameters. Most frequently, substantive hypotheses about the postulated states impose constraints that reduce the number of free parameters of the theory. The most general form of the kind of theoretical proposal we are discussing specifies a set of free parameters $(\theta_1, \ldots, \theta_s)$ such that each of the conditional probabilities of the theory is a specified function of the parameters. Then the condition for testability is that s must be less than the number of empirical quantities, and the hypothesis has the form

$$\text{(12)}\quad (\exists\theta_1)\ldots(\exists\theta_s)\,(\forall i)\,(\forall k)\left[p_{ik} = \sum_{j=1}^{t} q_{ij}(\theta_1 \ldots \theta_s)\, r_{jk}(\theta_1 \ldots \theta_s)\right].$$

The procedure for testing the hypothesis involves finding estimates of the theoretical parameters that bring the values of p_{ik} implied by the theory as close as possible to the empirical values. Then the degree of approximation between the theoretical and empirical values of p_{ik} can be evaluated using standard statistical techniques.

There is enough structure in these general concepts now so that a realistic example can be introduced. I will describe a theory about the detection of weak signals proposed by Luce [4]. The theory is used in analysis of experiments where on some trials a relatively weak stimulus signal with some fixed energy level is presented along with a noisy input that makes it difficult to tell whether the signal is there or not. The noise has a fixed mean energy level, and on trials when the signal is not presented the noise is presented alone. There are several conditions designed to produce different response biases. These may be produced by varying the payoffs for different kinds of correct responses (identifying signals when they are present or correctly saying that the signal is absent) or by imposing varying penalties for different kinds of errors (missing signals or saying there is a signal when only noise is presented). Another way of producing response bias is to vary the overall proportion of trials when a signal is presented, thus producing higher or lower expectations of the signal. The empirical explananda are the subjects' responses: on each trial a subject says either 'yes' or 'no', depending on whether he judges a signal to have been present or absent. The explanans are the experimental conditions: on each trial, there is some condition of payoff and a priori expectation of a signal, and either the signal is presented or only noise is presented. At the level of data, an experiment can be described as a set of conditional frequencies

$$p_{ik} = P(\text{yes} \mid \text{condition } i)$$

where the conditions are given in the following equation:

$$(13) \qquad [p_{ik}] = \begin{array}{c|cc} & \text{yes} & \text{no} \\ \hline S_1 & p_{11} & p_{12} \\ SN_1 & p_{21} & p_{22} \\ S_2 & p_{31} & p_{32} \\ SN_2 & p_4 & p_{42} \\ \vdots & \vdots & \vdots \\ S_g & p_{(2g-1)1} & p_{(2g-1)2} \\ SN_g & p_{(2g)1} & p_{(2g)2} \end{array}$$

where g is the number of different motivational conditions.

Luce's theory uses the assumption of a threshold of detection and the probability of exceeding the threshold depends only on whether the signal was presented or not. The other theoretical variable is determined by the motivational conditions. Thus, on each trial, the subject is assumed to be in one of $2g$ theoretical states, as specified in Equation (14).

$$(14)\qquad [q_{ij}] = \begin{array}{c|cccccccc} & \bar{D}_1 & D_1 & \bar{D}_2 & D_2 & \cdots & \bar{D}_g & D_g \\ \hline S_1 & 1-q & q & 0 & 0 & \cdots & 0 & 0 \\ SN_1 & 1-p & p & 0 & 0 & \cdots & 0 & 0 \\ S_2 & 0 & 0 & 1-q & q & \cdots & 0 & 0 \\ SN_2 & 0 & 0 & 1-p & p & \cdots & 0 & 0 \\ \vdots & \vdots & \vdots & \vdots & \vdots & & \vdots & \vdots \\ S_g & 0 & 0 & 0 & 0 & \cdots & 1-q & q \\ SN_g & 0 & 0 & 0 & 0 & \cdots & 1-p & p \end{array}$$

$\bar{D}_h$ denotes a state in which the threshold of detection is not exceeded in motivational condition h, and D_h denotes a state in which the threshold is exceeded in motivational condition h. The probability of exceeding the threshold when the signal is presented is p, and the probability of exceeding the threshold when noise is presented alone is q.

It is assumed that the motivational states affect the relationships between the theoretical states and the responses. The states are assumed to be ordered in their tendency to produce biases favoring 'yes' responses; let State 1 denote the condition producing the greatest reluctance to say 'yes'. It is assumed that there is some motivational state f such that

$$(15)\qquad \begin{aligned} P(\text{yes} \mid \bar{D}_h) &= \begin{cases} 0 & \text{for} \quad h \leqslant f \\ t_h & \text{for} \quad h > f \end{cases} \\ P(\text{yes} \mid D_h) &= \begin{cases} u_h & \text{for} \quad h \leqslant f \\ 1 & \text{for} \quad h > f \end{cases} \end{aligned}$$

with u_h and t_h ordered monotonically with the values of h. In order words, for states in which the subject is reluctant to say 'yes', he always says 'no' if the threshold is not exceeded and divides his responses between 'yes' and 'no' randomly when the threshold is exceeded. And for states in which the subject is reluctant to say 'no' he always says 'yes' if the threshold is exceeded and divides his responses between 'yes' and 'no' when the threshold is not exceeded. The matrix of conditional probabilities connecting the theoretical states and the responses is given below.

$$(16)\qquad [r_{jk}] = \begin{array}{c|cc} & \text{yes} & \text{no} \\ \hline \bar{D}_1 & 0 & 1 \\ D_1 & u_1 & 1-u_1 \\ \vdots & \vdots & \vdots \\ \bar{D}_f & 0 & 1 \\ D_f & u_f & 1-u_f \\ \bar{D}_{f+1} & t_{f+1} & 1-t_{f+1} \\ D_{f+1} & 1 & 0 \\ \vdots & \vdots & \vdots \\ \bar{D}_g & t_g & 1-t_g \\ D_g & 1 & 0 \end{array}$$

Recall that the theoretical values of the p_{ik} that can be compared with data are obtained by multiplying the matrices $[q_{ij}]$ and $[r_{jk}]$. In the case of Luce's theory, this yields

$$(17)\qquad [p_{ik}] = \begin{array}{c|cc} & \text{yes} & \text{no} \\ \hline S_1 & qu_1 & 1-qu_1 \\ SN_1 & pu_1 & 1-pu_1 \\ \vdots & \vdots & \vdots \\ S_f & qu_f & 1-qu_f \\ SN_f & pu_f & 1-pu_f \\ S_{f+1} & (1-q)\,t_{f+1}+q & (1-q)\,(1-t_{f+1}) \\ SN_{f+1} & (1-p)\,t_{f+1}+p & (1-p)\,(1-t_{f+1}) \\ \vdots & \vdots & \vdots \\ S_g & (1-q)\,t_g+q & (1-q)\,(1-t_g) \\ SN_g & (1-p)\,t_g+q & (1-p)\,(1-t_g) \end{array}$$

Luce's theory illustrates the kind of situation described in connection with Equation (12). A number of parameters are specified – there are $g+2$ of them – and the theory asserts that the parameters determine the relationships between explanans (in this case, experimental conditions) and theoretical states and explananda (in this case, judgments about whether a signal was present). The parameters therefore specify a theoretical relationship between the explanans and the explananda. In order to obtain a numerical relationship, the parameters must be estimated from data, and the theory is testable if the number of free parameters is less than the number of empirical quantities in the data. In the present example, the number of empirical quantities is $2g$, so the theory is testable whenever g is greater than two.

Numerical values of the parameters must be estimated both to test the theory and to specify the information-theoretical properties of the theory. Putting this in another way, the information transmitted by the theory is a function of the parameter values. To provide a further illustration of the application of information theory to the analysis of statistical explanations, I have calculated the information transmitted by one special case of the theory. Suppose that $g=5$ (that is, five different motivational conditions are used) and the five conditions are used equally often. Furthermore, suppose that signals are presented on one-half of the trials under each motivational condition. This means that each of the states constituting the explanans has probability 0.10, and from this we can calculate

$$H(S) = 2.30.$$

Now, suppose that an experiment is conducted and the estimated values of p and q are 0.60 and 0.30, respectively. Recall that these are the probabilities of exceeding the threshold on signal trials and noise trials, respectively. This permits us to calculate the probabilities of the theoretical states; the probability of each $\bar{D}_h$ is 0.11 and the probability of each D_h is 0.09. We can then calculate

$$H(T) = 2.30, \quad H(S \times T) = 2.94.$$

The remaining parameters are the conditional probabilities of saying 'yes' in the various theoretical states. Suppose it is estimated that $f=3$, and the values of the u_h and t_h parameters are

$$u_1 = 0.40, \quad u_2 = 0.70, \quad u_3 = 1.00, \quad t_4 = 0.50, \quad t_5 = 0.80.$$

It turns out that this implies that subjects will say 'yes' with probability 0.51, and we can calculate

$$H(M) = 0.69, \quad H(T \times M) = 2.55.$$

Combining these values using Equations (1) and (7), we arrive at values of information transmitted as follows:

$$I(S, T) = 1.66, \quad I(T, M) = 0.44, \quad I(S, M) = 0.17.$$

Note again that the information transmitted by the relationships between explanans and explananda is smaller than either of the quantities involving the theoretical states.

Developments motivated by Luce's theory can be used to illustrate the remarks made earlier about the kinds of research that provide improvements in theories of this kind. One kind of development involves applying the theory to more complex experiments. Luce's theory has been applied to studies in which two different signals have been presented on different trials. The experiments involved auditory detection, and the signals were tones of different frequency. This change increases the number of states in the set of explanans, and thus increases the information transmitted by the theory. In addition, the subjects were asked to identify the signal after they judged whether a signal was presented. Thus, instead of just saying 'yes' or 'no', subjects also classified the stimuli as 'high' or 'low' as well. With a more detailed set of explananda, additional explanatory power was obtained. Of course, this experiment was not just cooked up to provide more detail for its own sake; the threshold theory makes the strong prediction that when the threshold is not exceeded (as is the case for all 'no' responses when $h>f$) the subject should have no information about which stimulus was presented, and the experiment was designed in part to test this prediction.

The other kind of development involves adding to the complexity of the theoretical description, guided by data that are not consistent with the simpler theory. Krantz [3] has proposed such a theory, in which an additional state D_h^* is postulated for each motivational state. D^* is a state of strong detection, in which the subject is sure that a signal was presented. In Krantz' theory, then, there are two postulated thresholds. If the lower threshold is not exceeded the subject is in State $\bar{D}_h$, and if the higher threshold is exceeded the subject is in State D_h^*. It is assumed that the subject always judges that a signal was presented when he is in State D_h^*, regardless of the motivational state. The main reason for complicating a theory, of course, is to correct apparent defects that are revealed by failure of data to agree with predictions derived from the theory. But this also has the effect of increasing the explanatory power of the theory in the sense of information transmitted reflected by values of $I(S, T)$ and $I(T, M)$.

This example from the theory of perception illustrates several aspects of the role of theoretical entities in statistical explanation. Other examples could be used for the same purpose, from other areas of psychological theory as well as from other fields. It also may be remarked that the analysis of discrete states does not affect the main features of the analysis. Systems that involve continuous variables can be analyzed from this point of view and their general properties are analogous to those of discrete-state systems, although the analysis of systems with discrete states is easier to understand.

An important feature of the kind of theory that I have been discussing is that it describes a system whose statistical properties do not change over time. In such a stationary system, the main role of theoretical entities seems to be heuristic, in that they guide the development of new empirical and theoretical research and thus facilitate the extension of knowledge. In the remainder of this paper I will discuss systems that are not stationary, and in these systems the use of theoretical entities can lead to a considerable simplification of the statistical structure of a theory in addition to their heuristic value.

The simplest kind of nonstationary system is illustrated by experiments in learning and problem solving. In their simplest form, these experiments consist of repeated trials where a subject is given opportunities to study the material to be learned or is given information relevant to solving the problem. I will impose a relatively stringent condition of uniformity for the purpose of analysis here. I will ignore differences in procedure that occur on different trials. When such differences cannot be neglected, the situation would be analyzed as the concatenation of different experiments.

At the beginning of an experiment, the subject either gives only incorrect responses or he responds correctly with some probability due to guessing, depending on the procedure of the experiment. As the subject proceeds through the experiment, the probability of correct response increases. The subject may reach a state in which he gives only correct responses, or he may reach some asymptotic state in which the probability of correct response is at some level less than one. Thus, the most salient general feature of the experiment is an increase in the probability of correct response from some initial value c_1 (where c_1 may be zero) to some asymptotic level c_∞ (where c_∞ may be one).

The data from a learning or problem solving experiment are sequences of responses given by the subjects. In general, the probability of response on each trial n depends on the sequence of responses given on the preceding n-1 trials. One natural way to consider the experiment, then, is as a sequence of probabilitistic events in which the response on each trial n is an event to be explained, and the sequence of responses on trials 1, 2,..., n-1 is the event available to be used as the explanans.

To illustrate the situation, consider the first four trials of an experiment. The data are hypothetical, and were generated from a theory that will be presented later. In the notation, a correct response is denoted 0 and an error is denoted 1. On trial 1, the subject either gives a correct response or an error, and this provides the first state-space for the explanans. The conditional probabilities $p_{ik,2}$ are just the probabilities of correct response on trial 2. In the following equation, the numbers in parentheses to the left of the first-trial states are the probabilities of those states, and the probabilities of response on trial 2 are in parentheses at the top.

$$(18) \qquad [p_{ik,2}] = \begin{array}{rr|cc} & & (0.40) & (0.60) \\ & & 0 & 1 \\ \hline (0.25) & 0 & 0.40 & 0.60 \\ (0.75) & 1 & 0.40 & 0.60 \end{array}$$

The information-theoretical quantities are

$$H_2(S) = 0.56, \quad H_2(M) = 0.67, \quad I_2(S, M) = 0.00.$$

The reason that no information is transmitted is that the response on trial 2 is independent of the response on trial 1 in this situation.

The responses on trial 3 are related to the sequences of response on trials 1 and 2.

$$(19) \qquad [p_{ik,3}] = \begin{array}{rr|cc} & & (0.52) & (0.48) \\ & & 0 & 1 \\ \hline (0.10) & 00 & 0.70 & 0.30 \\ (0.15) & 01 & 0.40 & 0.60 \\ (0.30) & 10 & 0.70 & 0.30 \\ (0.45) & 11 & 0.40 & 0.60 \end{array}$$

Carrying out the information-theoretical calculations we obtain

$$H_3(S) = 1.24, \quad H_3(M) = 0.69, \quad I_3(S, M) = 0.04.$$

The situation for trial 4 is as follows:

$$
\text{(20)} \qquad [p_{ik,4}] = \begin{array}{ll|ll}
 & & (0.62) & 0.38) \\
 & & 0 & 1 \\
\hline
(0.07) & 000 & 0.914 & 0.086 \\
(0.03) & 001 & 0.400 & 0.600 \\
(0.06) & 010 & 0.700 & 0.300 \\
(0.09) & 011 & 0.400 & 0.600 \\
(0.21) & 100 & 0.914 & 0.086 \\
(0.09) & 101 & 0.400 & 0.600 \\
(0.18) & 110 & 0.700 & 0.300 \\
(0.27) & 111 & 0.400 & 0.600
\end{array}
$$

$$H_4(S) = 1.88, \quad H_4(M) = 0.66, \quad I_4(S, M) = 0.17.$$

Clearly, this kind of calculation could be carried out indefinitely. It leads quickly to a relatively unmanageable system; on trial n the state-space that constitutes the explanans has 2^{n-1} members. On the other hand, the system eventually becomes uninformative. Assuming that the system eventually reaches a stable asymptotic level of response probability, the responses will eventually become independent of preceding sequences. In other words,

$$\lim_{n \to \infty} I_n(S, M) = 0.$$

This fact makes it reasonable to think about a theory of learning in relation to the sum of the values of I_n across trials. Using this line of thinking, our state of knowledge about a learning system would be evaluated in regard to the extent to which the performance of a learning subject could be predicted on the basis of his earlier performance, and this seems like a reasonable way to proceed. For example, it fits with our intuitions about discoveries that in fact count as additions to our knowledge. When responses are measured in more detail, as by a finer classification of errors or by measuring additional properties such as time to respond, the effect is to increase the values of I_n, for the same reason that additional variables added to the explanans and the explananda always increase information transmitted. In addition, when we analyze properties of the sequence of study trials or the information that is presented, thus going beyond the assumption of uniform trials that I have imposed on this analysis, we add new variables to the explanans and thus also increase the values of I_n.

These remarks about learning systems provide a framework for the methodological analysis of nonstationary systems that is consistent with the information-theoretical analysis worked out earlier for stationary systems. My remaining discussion will consider the role of theoretical entities in such systems.

The general structure that I will use for my remaining remarks involves a state-space of postulated variables. On each trial there is a set of postulated states $[T_n]$ related to the explanans $[S_{n-1}]$ and the explananda $[M_n]$ by specified probability laws given as matrices of conditional probabilities $[q_{ij,n}]$ and $[r_{jk,n}]$. Thus on each trial the same kind of structure that we used earlier for stationary systems can be applied to analyze the information-theoretical properties of a non-stationary system.

In principle, the theoretical states may be as complex as the theorist wishes. However, a number of simplifying restrictions are frequently used. The first is that the theoretical state-space $[T_n]$ is a constant that I will denote $[T]$. Secondly, the conditional probabilities of responses given theoretical states are constant, so that $[r_{jk,n}]$ is a constant $[r_{jk}]$. Finally, and perhaps of greatest significance, the sequence of theoretical states that occurs over trials is assumed to be governed by a probability law that is specified by the theory.

Before considering the nature of the probabilities connecting the theoretical states from trial to trial, it may be noted that the existence of any probability law governing the transitions between theoretical states, along with probabilities relating theoretical states and observable responses, are sufficient to specify empirical probabilities of the kind presented in the illustration given above. The sequence of states is a stochastic process with trial outcomes $T_{j1}, T_{j2}, \ldots, T_{jn}, \ldots$. The probability of any sequence of outcomes can be calculated using the transition probabilities of the system. Given a sequence of theoretical states, the probability of any response sequence can be calculated using the probabilities relating the theoretical states and the responses. The conditional probabilities p_{ik} have the form

$$P(M_{k,n} \mid S_{i,n-1}) = P(M_{k,n} \mid M_{k_1,1}, M_{k_2,2}, \ldots, M_{k_{n-1},n-1})$$

$$\frac{P(M_{k_1,1}, \ldots, M_{k_{n-1},n-1}, M_{k_n,n})}{P(M_{k_1,1}, \ldots, M_{k_{n-1},n-1})}.$$

Since the probabilities of all the sequences can be calculated, so can the conditional probabilities.

The nature of the transition probabilities of the system has a fundamental influence on the properties of the system. A desirable situation is one in which the theoretical states have the Markov property. When a theory is Markovian in its postulated states, the probability of any state T_{jn} on trial n depends only on the state of the system on trial $n-1$ and is independent of the sequence of states that occurred on trials 1, ..., $n-2$. That is,

$$P(T_{jn,n} \mid T_{j1,1}, \ldots, T_{jn-1,n-1}) = P(T_{jn,n} \mid T_{jn-1,n-1}).$$

The Markov property represents a kind of independence of history. In a system lacking the Markov property, the future behavior of the system is dependent both on the present state of the system and on its past behavior. If a system has the Markov property its future behavior depends only on its present state. This has far-reaching implications for the analysis of the system. If its states are Markov, then any method that permits us to specify its present state permits us to predict its future behavior, up to the uncertainty imposed by the probability laws that govern the system. If the states of a theory are not Markov, then predictions about future behavior can be improved by obtaining information about states that occurred in the past. If this is the case, then it follows that the description of a system provided by the theory omits important distinctions. Clearly, the future behavior of the system has to depend on its present state, however it arrived there. And the finding that a theory is not Markov in its postulated states is clear evidence that the states do not give a complete description of the system.

Of course, we can assume that any description involving probabilistic relationships among states is incomplete. If a complete description were available, then the behavior of the system would be deterministic. However, the discovery of a Markovian structure provides a basis for further investigation that simplifies the problem of refining the theoretical description. If the best available theory of a process has states with the Markov property, then further investigation can be focussed on distinguishing between relevant subsets of the class of events that are grouped together in the theoretical description. Whatever one finds regarding the subsets of such events can be treated as a simple reclassification of the events. If the states of a system as described by the theory are not Markov, then variables that are relevant to the system's future behavior must be evaluated in relation to the sequence of states in the history of the system, and this will generally involve a considerable cost in theoretical complexity.

I will illustrate these ideas about Markovian theories for non-stationary systems with a theory of simple memorizing. The kind of experiment to which the theory applies involves presentation of pairs of items that are unrelated. The pairs that a subject is asked to memorize might have short words as stimuli and numbers as responses. On each experimental trial, the experimenter presents a word and asks the subject to give the number that he thinks is correct. After the subject responds, the experimenter presents the correct answer. On the first trial, of course, the subject has to guess. However, after the subject has seen all of the pairs he can remember the correct answer on at least some of the tests, and eventually he is able to give the correct answer to all of the items.

The data from an experiment like this are analyzed in the form of sequences of responses given to the individual items. For example, on one item a subject might give the sequence of responses

$$0 \quad 1 \quad 0 \quad 1 \quad 1 \quad 0 \quad 0 \quad 0 \quad \ldots,$$

meaning that he guessed correctly on the first trial, gave an error on trial 2, a correct response on trial 3, errors on trials 4 and 5, and then correct responses from then on.

The simple model that I will describe was first developed by Bower [1]. According to the model, an individual item is learned in an all-or-none fashion. That is, at the beginning of the experiment each item is unlearned. On each study trial, there is some probability that the item becomes learned. This probability is a constant – that is, the probability of learning an item does not increase over trials during the experiment. Once an item is learned, the subject is assumed to remember it for the remainder of the experimental trials. Prior to learning an item, the subject has to guess the answer on each of the item's tests.

Putting this more formally, the theory postulates two states in which we can find an item – U and L for unlearned and learned. Each item begins in state U, and on each trial there is a constant probability called c that the item goes from state U to state L. State L is absorbing – that is, once an item goes into state L it stays there. This set of assumptions can be expressed in standard notation as

$$P(L_1, U_1) = (0, 1),$$

$$(21) \qquad [t_{j_{n-1}j_n}] = \begin{array}{c|cc} & L_n & U_n \\ \hline L_{n-1} & 1 & 0 \\ U_{n-1} & c & 1-c \end{array}$$

where the first equation states the assumption that all the items start in state U, and the second equation states the assumption of a constant probability of a transition from U to L and the assumption that L is an absorbing state.

The final assumption links these ideas about the postulated states to probabilities of response. There are two possible responses on each trial – the subject can be correct or wrong. A correct response is denoted 0 and an error is denoted 1. While an item is in state U there is a probability, assumed to be constant, of a correct response by guessing. After the item goes into state L the probability of a correct response is assumed to be 1. This is stated in the following equation:

$$(22) \qquad [r_{jk}] = \begin{array}{c|cc} & 0 & 1 \\ \hline L & 1 & 0 \\ U & g & 1-g \end{array}$$

The major simplification resulting from the Markov assumption can be seen readily. As can be seen from Equations (19) and (20), predictions about the response on trial n based on previous responses should use the entire response sequence through trial $n-1$. However, if the theory is correct then a prediction about the postulated state of the system on trial n based on previous theoretical states needs only to use the state of the system on trial $n-1$, and the sequence of states before trial $n-1$ can be ignored.

The information-theoretical structure of the theory can be considered in two general ways. The first is closer to the earlier discussion given for stationary systems. On each trial, the sequence of responses given for an item can be considered an empirical explanans, the response on that trial an explanandum, and the postulated theoretical states mediate between the two in the way considered earlier. For the following calculations, I ave taken the probability of learning, c, at 0.20 and the prelearning guessing probability, g, at 0.25. Then, for trial 2, we would have

$$(23)\qquad [q_{ij,2}] = \begin{array}{ll|ll} & & (0.20) & (0.80) \\ & & L_2 & U_2 \\ \hline (0.25) & 0 & 0.20 & 0.80 \\ (0.75) & 1 & 0.20 & 0.80 \end{array}$$

for the relationship between trial-1 responses and theoretical states on trial 2. The relationship between theoretical states and responses on all trials is given by Equation (22) with $g = 0.25$. Combining results given in Equations (23) and (18), the information-theoretical quantities turn out to be

$$(24)\qquad \begin{array}{l} H_2(S) = 0.56, \quad H_2(T) = 0.50, \quad H_2(M) = 0.67, \\ I_2(S, T) = 0.0, \quad I_2(T, M) = 0.22, \quad I_2(S, M) = 0.0. \end{array}$$

In general, the probabilities of theoretical states on trial n depend on the sequences of responses on all previous trials. On trial 3,

$$(25)\qquad [q_{ij,3}] = \begin{array}{ll|ll} & & (0.36) & (0.64) \\ & & L_3 & U_3 \\ \hline (0.10) & 00 & 0.60 & 0.40 \\ (0.15) & 01 & 0.20 & 0.80 \\ (0.30) & 10 & 0.60 & 0.40 \\ (0.45) & 11 & 0.20 & 0.80 \end{array}$$

$$\begin{array}{l} H_3(S) = 1.24, \quad H_3(T) = 0.65, \quad H_3(M) = 0.69, \\ I_3(S, T) = 0.08, \quad I_3(T, M) = 0.33, \quad I_3(S, M) = 0.04. \end{array}$$

A similar situation exists for trial 4, with the explanans consisting of the eight values given in Equation (20). The information-theoretical quantities are

$$(26)\qquad \begin{array}{l} H_4(S) = 1.88, \quad H_4(T) = 0.69, \quad H_4(M) = 0.67, \\ I_4(S, T) = 0.19, \quad I_4(T, M) = 0.38, \quad I_4(S, M) = 0.11. \end{array}$$

An overall impression of the informational structure of the theory may be obtained by examining the sums of information transmitted across trials. For the parameter values used above, the sums for trials 2–13 are

$$\sum_{j=2}^{13} I_j(S, T) = 2.61, \quad \sum_{j=2}^{13} I_j(T, M) = 3.47,$$

$$\sum_{j=2}^{13} I_j(S, M) = 1.74.$$

Eventually, these sums converge but the values for $c = 0.20$ do not converge within the trials for which I carried out calculations. (The calculations were not carried further because an inordinate amount of time is needed for computation of I_j when j becomes large.) However, to provide some indication of the behavior of the statistics, I carried out calculations for higher values of c until the sums did approach asymptotic values. The results of these calculations are in Table I. The main findings of interest in these

TABLE I

Information-theoretical statistics for all-or-none learning

Parameters	$\sum_{j=1}^{\infty} I_j(S, T)$	$\sum_{j=1}^{\infty} I_j(T, M)$	$\sum_{j=1}^{\infty} I_j(S, M)$
$c = 0.40,\ g = 0.25$	0.93	1.84	0.66
$c = 0.40,\ g = 0.50$	0.64	1.09	0.29
$c = 0.60,\ g = 0.25$	0.29	0.97	0.21
$c = 0.60,\ g = 0.50$	0.19	0.58	0.09

calculations are the strong dependence of the amount of information transmitted on the parameter values, and the further illustrations of the fact that the dependence between empirical variables is always less strong than the dependence between either set of empirical variables and the theoretical variables.

The preceding calculations all have to do with the theoretical states considered as mediators between preceding response sequences and the response on trial n. Another view of the situation can be obtained by examining the sequence of theoretical states, without regard for the observable responses. This latter point of view is concerned with uncertainty and information transmitted at the level of the states that are postulated in the theory, and in some ways gives a more direct evaluation of our state of knowledge than the analysis that deals with responses, assuming the decision that the theory represents the best available understanding of the system in question.

The analyses of theoretical sequences are similar to those given earlier in connection with Equations (18)–(20), except that the Markovian structure of the theory permits us to ignore states of the system occurring in the past. For any trial n, the probability of any state depends only on the state on trial $n-1$; in fact, the probabilities of state-to-state transitions are constant. For trial 2, we have

$$(27)\qquad [t_{j_1 j_2}] = \begin{array}{cc|cc} & & (0.20) & (0.80) \\ & & L_2 & U_2 \\ \hline (0.0) & L_1 & 1 & 0 \\ (1.0) & U_1 & 0.20 & 0.80 \end{array}$$

$$H_1(T) = 0.0, \quad H_2(T) = 0.50, \quad H_{12}(T \times T) = 0.50, \quad I_2(T,T) = 0.0.$$

For trial 3, the probabilities given at the top of Equation (27) are the probabilities of the explanans, and the transition probabilities remain unchanged. The state probabilities on trial 3 are $P(L_3)=0.36$, $P(U_3)=0.64$. The information-theoretical quantities are

$$H_3(T) = 0.65, \quad H_{23}(T \times T) = 1.08, \quad I_3(T, T) = 0.07.$$

On trial 4, the state probabilities are $P(L_4)=0.49$, $P(U_4)=0.51$, leading to

$$H_4(T) = 0.69, \quad H_{34}(T \times T) = 0.97, \quad I_4(T, T) = 0.37.$$

Aside from the marked increase in simplicity, compared with the observable sequences, the theoretical sequences also have more information transmitted. For the trials given above, the calculations given earlier for $I(S, M)$ are 0.0, 0.04, and 0.17. The fact that values of information transmitted are higher in the theoretical sequences than in the observable sequences is not an accident. Any matrix of probabilities in the observable responses like those given in equations (18)–(20) can be derived as the products of three matrices:

$$P(S_{n-1}, M_n) = P(S_{n-1}, T_{n-1})\, P(T_{n-1}, T_n)\, P(T_n, M_n).$$

The matrix designated by the central term is the transition matrix for the theoretical states, and we have already seen that the information transmitted by a product is no greater than the information transmitted by any of the matrices multiplied to form the product.

Both the simplicity and information-theoretical advantages of the theoretical structure are illustrated further by the analysis of information transmitted summed over trials. The analysis of this statistic for sequences of observable responses was discussed earlier; each of the calculations presented there required nearly an hour of calculation on a medium-small computer (an IBM 1800 with 16,000 words of core storage and 4 micro-second memory access). The calculations for the theoretical sequences can be done quite easily by hand. In general,

$$I_n(T,T) = H_{n-1}(T) + H_n(T) - H_{n-1,n}(T,T),$$

where

$$\begin{aligned} H_n(T) &= -P(L_n)\log P(L_n) - P(U_n)\log P(U_n) \\ &= [1-(1-c)^{n-1}]\log[1-(1-c)^{n-1}] \\ &\quad - (1-c)^{n-1}\log(1-c)^{n-1}, \end{aligned}$$

$$\begin{aligned} H_{n-1,n}(T,T) &= -P(L_{n-1})\log P(L_{n-1}) \\ &\quad - P(U_{n-1},U_n)\log P(U_{n-1},U_n) \\ &\quad - P(U_{n-1},L_n)\log P(U_{n-1},L_n) \\ &= -[1-(1-c)^{n-2}]\log[1-(1-c)^{n-2}] \\ &\quad - (1-c)^{n-1}\log(1-c)^{n-1} \\ &\quad - c(1-c)^{n-2}\log c(1-c)^{n-2}. \end{aligned}$$

Combining terms and summing across values of n,

$$\begin{aligned} \sum_{j=2}^{\infty} I_j(T,T) &= \log c - \frac{(1-c)}{c}\log(1-c) \\ &\quad - \sum_{k=1}^{\infty}[1-(1-c)^k]\log[1-(1-c)^k] \end{aligned}$$

For values of c of 0.20, 0.40, and 0.60 the sums $\sum I_n$ for the theoretical sequences are 5.67, 3.99, and 1.16. The last two values can be compared with values given in Table I for the observable sequences.

Since there is a simpler structure and greater information transmitted in the theoretical sequence than there is in the sequence of observations, the theory provides an advantageous basis for developing new knowledge. The development of new measurement techniques for increasing dependencies between observations and the theoretical states, and the refinement of theory to provide increased information transmitted in the trial-to-trial transitions both constitute additions to knowledge that are frequent in scientific investigation, and that are explicable within the framework of the present analysis.

CONCLUSIONS

In presenting the information-theoretical analysis of theoretical entities of this paper, a number of rather routine calculations have been carried out. To some extent, the significance of these analyses is just that they can be carried out. The analyses of this paper serve as an existence proof that it is possible to perform analyses that are relevant to evaluating our state of knowledge when we have statistical knowledge about a system. It is universally recognized that theoretical entities can serve as an heuristic aid in development of new empirical knowledge about a system, and in some cases can provide a substantial simplification in the representation of knowledge. These facts are clarified and specified to some extent by the present analysis. We have seen that the information transmitted by dependencies between a set of empirical variables and a set of theoretical variables is generally greater than the information transmitted between sets of empirical variables, and this fact clarifies the usefulness of postulated entities in guiding empirical investigations. The role of theoretical entities in relation to simplicity is especially striking in nonstationary systems, particularly when the theoretical system has the Markov property.

University of Michigan

BIBLIOGRAPHY

[1] Bower, G. H., 'Application of a Model to Paired-Associate Learning', *Psychometrika*, **26** (1961) 255–80.
[2] Greeno, J. G., 'Evaluation of Statistical Hypotheses Using Information Transmitted', *Philosophy of Science* **37** (1970), 279–93.
[3] Krantz, D. H., 'Threshold Theories of Signal Detection', *Psychological Review* **76** (1969) 308–24.
[4] Luce, R. D., 'A Threshold Theory for Simple Detection Experiments', *Psychological Review* **76** (1969) 308–24.

WESLEY C. SALMON*

EXPLANATION AND RELEVANCE: COMMENTS ON JAMES G. GREENO'S 'THEORETICAL ENTITIES IN STATISTICAL EXPLANATION'**

Since Professor Greeno's paper is, to my mind, both philosophically interesting and somewhat technical, it seems to me that the most useful function I can serve as a commentator is to try to provide some of the background necessary to understand the paper, and to attempt to put his discussion in some philosophical perspective. My main aim will be to relate his discussion to the issues that have been at the forefront of the philosophical discussion of scientific explanation in recent years.

It would be hard to dispute the claim that Professor Carl Hempel's work has dominated the discussion of the nature of scientific explanation for the last twenty-two years – since the publication of the classic article with Paul Oppenheim on the logic of explanation.[1] The views expressed in that article, and considerably elaborated in subsequent writings, have not, to be sure, gained universal assent, but they have been the focus of attention, both critical and lauditory. Moreover, these theories of scientific explanation have been articulated with a degree of clarity and precision that makes them *the* standard of comparison for any alternative theory one wishes to advance.

It is important to recognize, of course, that Hempel has always maintained that there is more than one type of scientific explanation; he *never* claimed that the famous deductive-nomological (D-N) model was the only admissible one. From the very beginning he has acknowledged the existence of inductive-statistical (I-S) explanations. However, it is only in the last few years, since the publication of 'Deductive-Nomological vs. Statistical Explanation,'[2] that this latter kind of explanation has received extensive discussion. One can only wonder why many first-rate philosophers of science saw no need to give more than passing mention to statistical explanation. Perhaps they felt that such a close parallel exists between deductive and inductive explanation that it would be a routine matter to make the extension; perhaps, on the contrary, they found the prospect of elaborating the inductive-statistical model as so formidable a task that they shied away from attempting it. I am inclined to believe that the former assumption is nearer to the truth, and can only say that we should have been more clearly cognizant of the very fundamental and farreaching disanalogies between deductive and inductive logic.[3] By now, however, Professor Hempel has written extensively on statistical explanation, and the current journals publish frequent pieces on the subject.

Quite recently, an alternative to Hempel's account of statistical explanation has been emerging, and it contrasts with his account in several crucial ways. It has only very lately begun to be published, but it has been around in the process of development for several years. The first important publication is Professor Richard Jeffrey's 'Statistical Explanation vs. Statistical Inference', which was published in the Hempel *Festschrift*.[4] Another important publication is Professor James Greeno's 'Evaluation of Statistical Hypotheses Using Information Transmitted.'[5] This paper contains some of the background material for the paper he is presenting in this symposium. Just to make it clear how completely the cards are stacked in this panel, I should mention that the lengthiest treatment to date is my 'Statistical Explanation.'[6] Greeno's present paper is a further important contribution to the development of the new alternative.

I should add, before continuing further, that this attribution of a unified approach to the problem of statistical explanation by the members of this symposium is my own doing. The other symposiasts do not necessarily agree with my assessment of the situation; in particular, I do not mean to suggest that they endorse any of my views on explanation. All that I have any right to claim is that I see a unity in the approach, and I have profited greatly from the work of the other two. I hope that they, too, find some area of agreement.

In order to proceed with the characterization of the new alternative, it will be useful to draw several contrasts with Hempel's theory. According to Hempel, an inductive-statistical (I-S) explanation is an argument to the effect that the event to be explained was to be expected by virtue of certain explanatory facts; i.e., the explanandum-event has high inductive probability relative to the explanans. In this respect, an inductive-statistical explanation is like a deductive-nomological explanation, except that we must be content with high probability in place of deductive certainty. The inductive-statistical explanation is a covering law type of explanation;

the general law is a statistical generalization, and the argument that constitutes the explanation is inductive. Hence the name.

If anyone should ask where he might go to learn more about this inductive probability, and the associated inductive logic, that governs inductive-statistical explanation, the obvious answer would be to consult the works of our late master, Professor Rudolf Carnap, whose system of inductive logic far exceeds in scope and rigor any other available one.[7] But when we examine this system of inductive logic, we learn several shocking facts.[8] First, as is very well known, Carnap's inductive logic does not contain any provision for inductive *inferences* in the ordinary sense of that term – the sense involved in Hempel's claim that I-S explanations *are* inductive inferences. Inductive logic, according to Carnap, embodies analytic degree of confirmation statements, but no rules of inductive inference. Hence, at least in Carnap's theory of inductive probability, there simply are no such inferences as are required in Hempel's theory of I-S explanation.

Second, although perhaps there is an extended sense in which Carnap's degree of confirmation statements function as inferences, these degree of confirmation statements relate particular evidence statements with particular hypotheses. Generalizations are not required. In the sense that there is any inductive inference at all in Carnap's theory, it is inference from particulars to particulars without the mediation of universal or statistical generalizations. Hence, Hempel's insistence upon the indispensability of generalizations in covering law explanations is completely out of harmony with the best-known theory of inductive probability.

Third, when we look at the status of generalizations in Carnap's inductive logic, we find that they can never be highly confirmed on any finite amount of evidence; indeed, on such evidence their degree of confirmation is always zero. In this system of inductive logic, there could never be any basis for placing any confidence in the truth of the very generalizations that are required for Hempel's deductive and inductive patterns of explanation. However, it is not fair to place much emphasis upon this point, for Hintikka has shown how to construct systems of inductive probability that allow for the confirmation of generalizations.[9]

Now, I do not mean to assert that Carnap's inductive logic is the only possible theory of inductive probability, or that the foregoing considerations demonstrate the untenability of Hempel's account of explanation.

Instead, I am calling attention to a heuristic point. In thinking seriously about the role of inductive probability in statistical explanation, might one not be led to question the basic idea that an explanation is an argument at all? This is precisely the revolutionary step that Jeffrey took in the aforementioned paper; he concluded, roughly speaking, that the statistical explanation of an event exhibits that event as the result of a stochastic process from which such events arise with some probability whose degree may be high, middling, or even very low. Exhibition of such a process does not constitute an argument at all, let alone an argument to the effect that the explanadum-event has a high probability. Thus, the first, and perhaps most fundamental, point of disagreement with Hempel consists in the denial that a statistical explanation is an argument at all. On this point, Greeno and I are in enthusiastic agreement with Jeffrey.

Carnap's work on inductive logic provides a point of departure for the second basic disagreement with Hempel's account of statistical explanation. In the preface to the second edition of *Logical Foundations of Probability*, Carnap stressed the importance of the distinction between what he called 'concepts of firmness' and 'concepts of increase of firmness'.[10] This is the distinction between degree of confirmation and degree of relevance. Carnap makes a convincing case that a good deal of mischief in discussions of inductive logic has arisen simply from a confusion of the concepts of the two types. It is plausible to suspect that the same confusion has led to difficulty in the theory of statistical explanation. That is, in fact, a main point I have tried to argue in my aforementioned paper – namely, that statistical relevance rather than high probability is the appropriate desideratum for statistical explanations.

Consider, for example, a person who experiences relief from a neurotic symptom while (or shortly after) undergoing psychotherapy. Does the psychotherapeutic treatment explain the remission of the symptom? The answer to this question depends not only upon the probability of the abatement of symptoms during (or shortly after) therapy; rather, it depends upon the relation between the remission rate for patients undergoing a particular type of treatment and the spontaneous remission rate. Even if the probability of the remission of symptoms for patients in psychotherapy were very high, that would have no explanatory value if the spontaneous remission rate were equally high. At the same time, even if the recovery rate for patients were quite low, but still higher than the spontaneous remission rate, the fact that the individual had submitted to treatment would have some explanatory force in relation to his psychic improvement.

The same point, incidentally, applies to deductive-nomological explanations; they, too, can fail by virtue of relevance considerations. Even if a man faithfully consumes the prescribed number of birth control pills, that would have no value in explaining why he did not become pregnant. I do not mean to suggest that a male's avoidance of pregnancy is the kind of event that needs to explanation. The fact that men, unlike women, do not ovulate seems to me to have considerable explanatory force.

Let me mention still another reason for preferring the alternative to Hempel's view. If we insist, with Hempel, that a statistical explanation must embody a high probability, there may well be events that are intrinsically improbable, and which consequently defy explanation according to that model. For example, in the light of present physical theory, the spontaneous radioactive decay of a uranium atom may be due to an alpha particle tunnelling through the potential barrier of the nucleus. As the alpha particle bombards this potential barrier, there is a probability of the order of 10^{-38} that it will escape, and there are no further relevant factors to determine in which instance it tunnels through. On Hempel's account, such low-probability events are in principle incapable of being explained, but on the alternative account, quantum mechanics provides an explanation because it furnishes the facts relevant to the occurrence of the event.

The foregoing two considerations provide the basis for a sharp distinction between Hempel's view and the alternative. Let us dub the alternative account "the statistical-relevance model" or "S-R model" for short.[11] The term "inductive" is deliberately omitted from the title to emphasize that S-R explanations are not arguments or inferences of any sort. The two models can be briefly characterized as follows:

I-S model (Hempel) – An explanation is an *argument* that renders the explanandum-event *highly probable*.

S-R model (Greeno-Jeffrey-Salmon) – An explanation is an assembly of facts *statistically relevant* to the explanandum-event, *regardless of the degree of probability* that results.

In the latter part of the paper I shall say something more specific about the way in which Greeno's approach captures the statistical-relevance aspect of explanation. Before going on to that point, let me mention one further point of contrast between Greeno's approach and Hempel's.

There is some feeling of queasiness, I must admit, in saying that an event is explained when you have shown that according to all relevant factors, its occurrence is overwhelmingly improbable. I am somewhat inclined to attribute this feeling to intuitions that have been well nurtured on two decades of studying Hempel's very persuasive writings, and to say that we simply have to retrain our intuitions. Greeno has a different, and I suspect better, way of handling this matter. Given that a theory has to explain the occurrence and non-occurrence of many different types of events, and that factors relevant to the non-occurrence of an event seem to have a place in the explanation, Greeno suggests that we evaluate the overall explanatory power or value of a theory to explain all of the kinds of events it purports to explain, rather than attempting to evaluate the goodness of a particular explanation of a particular event. One of the attractive features of Greeno's work is that it provides an appealing method of assessing at least one aspect of the explanatory value of a theory. This result is achieved by application of some concepts of information theory.

When we ask what good there is in having an S-R explanation, it is satisfying to be able to say that the invocation of an explanatory theory increases our information. Indeed, information theory even provides a quantitative measure of the amount of increase. Perhaps there are other desiderata for explanatory theories, but increase of information is a nice one. Let us look at the measure in some detail. We shall see that the addition of information accrues as a result of the fact that the explanatory theory provides appropriate relevance relations. Let us look at a simple example. For the moment, what we are calling an 'explanatory theory' may contain only empirical laws that embody statistical relations among observables. Later we shall consider 'genuine theories' that make reference to 'theoretical entities'.

Suppose that the population of Centerville, USA, is equally divided between Democrats and Republicans. Let us call this partition $\{M\}$ (to be thought of as explanandum), and let

$$M_1 = D; \qquad M_2 = R; \qquad \text{where } \{M\} = \{D, R\}.$$

According to our assumption,

$$(1) \qquad p_1 = P(M_1) = P(D) = \tfrac{1}{2}; \quad p_2 = P(M_2) = P(R) = \tfrac{1}{2}.$$

This partition involves the greatest possible degree of uncertainty for a partition into two subclasses, for knowing that a person is a resident of Centerville tells us nothing about whether he is a Republican or a Democrat. In information theory, this uncertainty is measured by

$$(2) \qquad H(M) = \sum_i - p_i \log_2 p_i = 1,$$

where it is sometimes called, with enormous potentiality for confusion, the 'information.'[12] The bifurcation into two equally probable subsets provides the unit of uncertainty (or information) known as the 'bit.' Notice that the uncertainty achieves its maximum value of 1 when $p_1 = p_2$, and it drops to its minimum of zero when either p_1 or p_2 assumes the value 1. If all residents of Centerville were Republicans, there would be no uncertainty whatever with regard to party affiliation.

Now suppose, moreover, that Centerville is split by a set of railroad tracks that run north-south through the town, so that half of the residents live to the east and half live to the west of the tracks. Here we have another partition; let us call it $\{S\}$ (to be thought of as explanans), and let

$$S_1 = E; \quad S_2 = W; \quad \text{where } \{S\} = \{E, W\}.$$

According to our second assumption

$$(3) \qquad p'_1 = P(S_1) = P(E) = \tfrac{1}{2}; \quad p'_2 = P(S_2) = P(W) = \tfrac{1}{2}.$$

Again, the uncertainty is maximal for such a partition:

$$H(S) = 1.$$

The aggregate uncertainty of the two partitions is their sum,

$$H(M) + H(S) = 2.$$

The important question about these two partitions concerns their mutual independence. Let us call the unconditional probabilities $P(S_i)$ and $P(M_j)$ that have already been given the 'marginal probabilities.' Let us then introduce the conditional probabilities $P(M_j, S_i) = p_{ij}$ from the members of the partition $\{S\}$ to the members of the partition $\{M\}$.[13] By definition, the partition $\{M\}$ is independent of the partition $\{S\}$ if the conditional probabilities equal the respective marginal probabilities:

$$(4) \qquad P(M_j, S_i) = P(M_j) \quad \text{for every } i, j.$$[14]

Intuitively, we want to say that the conditional probabilities contribute no further information if the two partitions are statistically independent of one another, but that they can contribute positive information if there is a statistical dependency between them. The general idea is this. If the probability of being a Democrat varies, depending upon the side of the tracks on which the resident lives, then knowledge of this place of residence reduces the uncertainty about his party affiliation. This reduction in uncertainty can, as I shall point out below, be translated into increased predictive success. If, however, $P(D, E) = P(D, W) = P(D)$, then knowledge of place of residence does not provide any information relevant to party affiliation.

In information theory, the quantitative measure of reduction of uncertainty – that is, the "information transmitted" by the theory T – is given by

$$(5) \qquad I_T = H(M) + H(S) - H(S \times M)$$

where

$$(6) \qquad H(S \times M) = \sum_i \sum_j - p_i p_{ij} \log_2 p_i p_{ij}.$$

As Greeno has shown in the already published paper,

$$H(S \times M) = H(M) + H(S)$$

whenever the conditional probabilities equal the corresponding marginal probabilities (i.e., $p_{ij} = p'_j$). Thus, if the partitions are independent the reduction in uncertainty, or the increase in information, is zero. This, of course, agrees with our intuitions. It also means that a partition $\{S\}$ that is statistically irrelevant to a partition $\{M\}$ cannot have any explanatory value with respect to it.

Suppose, however, that the two partitions are not independent, and that place of residence is relevant to party affiliation. In particular, let

$$(7)\quad \begin{aligned} &P(D, E) = P(M_1, S_1) = p_{11} = \tfrac{3}{4}; \\ &\qquad\qquad P(R, E) = P(M_2, S_1) = p_{12} = \tfrac{1}{4} \\ &P(D, W) = P(M_1, S_2) = p_{21} = \tfrac{1}{4}; \\ &\qquad\qquad P(R, W) = P(M_2, S_2) = p_{22} = \tfrac{3}{4}. \end{aligned}$$

then

$$H(S \times M) = -2\left(\tfrac{3}{8}\log_2 \tfrac{3}{8} + \tfrac{1}{8}\log_2 \tfrac{1}{8}\right) \cong 1.81$$

and

$$I_T \cong 0.19.$$

This quantity represents our increase of information by virtue of the conditional probabilities. In standard philosophical terminology, we may consider the marginal probabilities $p(S_i)$ as the initial conditions and the conditional probabilities $P(M_j, S_i)$ as the explanatory theory (laws).

Greeno has shown, moreover, that the increase in information is maximal when all of the conditional probabilities are either zero or one, which corresponds to the situation in which deductive-nomological explanation is possible. If, however, the marginal probabilities in the original partition are also either zero or one, that maximum represents no gain in information. This situation corresponds to the case in which deductive-nomological explanation becomes vacuous through failure of relevance relations, as in the example of the man who takes birth control pills. These considerations show quite clearly that the measure of explanatory value introduced by Greeno is a statistical relevance measure. An explanatory theory can, on his view, have explanatory value even though it may assign low probabilities to explanadum-events, and it may fail to have explanatory value even if it assigns high probabilities to explanandum-events.

There is another way to look at the increase in information that results from a relevant partition.[15] If we select a particular resident of Centerville, and ask for the probability that he is a Democrat, we would be ill-advised to accept the value $\frac{1}{2}$ – say for use as a betting quotient – for the reference class of residents is not homogeneous with respect to party affiliation. We should look instead at his residence, and if he lives on the east side of the track assign the value $\frac{3}{4}$, while if he lives on the west side the value should be $\frac{1}{4}$. Let us look at the matter quantitatively by assigning the value 1 to each Democrat and 0 to each Republican. This is each individual's 'true value.' If we assign the value $\frac{1}{2}$ to each person, the error in each case is

$\frac{1}{2}$ (in absolute value), and if we square to make everything positive, the squared error is $\frac{1}{4}$ for each individual. Obviously, the mean squared error for all N residents is $\frac{1}{4}$. Suppose instead that we assign the value $\frac{3}{4}$ to each resident east of the tracks, and the value $\frac{1}{4}$ to each resident of the west side. The number of residents on each side is $N/2$; $\frac{3}{4} \times N/2$ of the east siders are Democrats, while $\frac{1}{4} \times N/2$ are Republicans. The converse situation obtains the west side. If we assign the value $\frac{3}{4}$ to someone who is a Democrat the error is $\frac{1}{4}$, and the squared error is $\frac{1}{16}$. If we assign the value $\frac{3}{4}$ to a Republican the error is $\frac{3}{4}$, and the squared error is $\frac{9}{16}$. The cumulative squared error for all residents of the east side is $\frac{3}{4} \times N/2 \times \frac{1}{16} + \frac{1}{4} \times N/2 \times \frac{9}{16} = 12N/128$. The same cumulative squared error occurs if we assign the value $\frac{1}{4}$ to each west sider. The total cumulative squared error is $24N/128$, and the mean squared error is $\frac{3}{16}$ $(<\frac{1}{4})$' We see once more how the increase of information due to a relevant partition of the reference class translates into a numerical measure – in this case, one that is rather obviously related to predictive success.

Greeno has chosen to explicate statistical-relevance explanation in terms of information transmitted, while I have chosen to explicate it in terms of the homogeneity of the reference class. As we have seen, each approach leads to a quantitative measure, as well as a qualitative characterization. Qualitatively, I believe that the explications coincide with one another, for they agree that the essence of the explanation is in the relevance relations expressed by the conditional probabilities that relate the explanans-partition to the explanandum-partition. Both of these treatments provide a straightforward answer to a previously recalcitrant problem concerning the utility of scientific explanations. On the present view, one of the values of an S-R explanation is the increase in information, the decrease in uncertainty, and the increase in predictive success provided by genuine explanations.

The main purpose of my comments is to provide some background that may prove useful for the understanding of Greeno's contribution to the present symposium, and up to this point I have done virtually nothing but recount and relate some ideas that have been advanced in previous writings of these symposiasts. Everything that has been said so far applies to explanatory systems that assert relations among observables. Greeno's present paper attempts to extend these considerations to explanatory systems that involve reference to genuinely theoretical entities. Clearly this is an undertaking of first importance. I should like to conclude my remarks by pointing to what seems to me to be the central question that arises in connection with his treatment of this problem.

Suppose, to continue with the previous example about the residents of Centerville, that we interpolate between the observable explanans-partition $\{S\}$ and the observable explanandum-partition $\{M\}$ an additional partition $\{T\}$ in terms of some theoretical characteristic such as 'inherent liberalism' – which I am taking to be some sort of unobservable constitutional factor. This new partition is related probabilistically to the other two. Now we have three sets of marginal probabilities

$$(8) \qquad P(S_i) = p_i; \quad P(T_j) = p_j^*; \quad P(M_k) = p'_k$$

and three sets of conditional probabilities

$$(9) \qquad P(M_k, S_i) = p_{ik}; \quad P(T_j, S_i) = q_{ij}; \quad P(M_k, T_j) = r_{jk}.$$

The question is whether, and in what manner, the system that incorporates the theoretical terms and the additional conditional probabilities provides an increase in information over the system that contains only the observable partitions and the one set of conditional probabilities relating them.

According to the theorem on total probability

$$(10) \qquad P(M_k, S_i) = \sum_j P(T_j, S_i) \times P(M_k, S_i \cdot T_j)$$

for any arbitrary $\{T\}$ we choose. Introduction of such an arbitrary $\{T\}$, along with suitable conditional probabilities to satisfy the theorem on total probability, surely must not be counted as a gain in information, for such an addition by itself is vacuous. There is, nevertheless, some danger that it will create an illusion of additional information, for now we seem to know something about inherent liberalism as well as place of residence and party affiliation. But we really know nothing more, for the explanation of party affiliation by way of 'inherent liberalism' is no more enlightening than was the explanation of the sleep induced by opium in terms of the dormitive power of that drug.

Greeno does impose one condition upon the theoretical partition $\{T\}$ over and above the vacuous condition that it satisfy the theorem on total probability. This condition stipulates that explanatory force of the explanans-partition be fully absorbed in the theoretical partition. In the example, this would mean that party affiliation is to be explained entirely in terms of inherent liberalism, and that in the presence of the theoretical partition, place of residence no longer has any explanatory power. The partition $\{S\}$, though strongly relevant to party affiliation in the absence of the theoretical partition $\{T\}$, is rendered statistically irrelevant in the presence of $\{T\}$. When this happens, we may say that $\{T\}$ *screens off* $\{S\}$ from $\{M\}$.[16] Formally, this condition can be stated as follows:

$$(11) \qquad P(M_k, S_i) = \sum_j P(T_j, S_i) \times P(M_k, T_j)$$

or

$$(12) \qquad P(M_k, S_i \cdot T_j) = P(M_k, T_j).$$

This says that once you know whether a citizen of Centerville is inherently liberal, it does not help you to know on which side of the tracks he lives in order to predict party affiliation. However, if the only way of inferring inherent liberalism is on the basis of place of residence, it is hard to see how the introduction of the theoretical partition represents any genuine increase in information, even if we assert that the screening-off relation obtains.

I am strongly convinced that the screening-off relation has deep significance in the context of statistical explanation. It would be pleasing to find that it also had profound significance in the context of theoretical explanation. I hope that Professor Greeno can tell us more about how it can function in that context.

Indiana University

NOTES

* Although Professor Greeno's paper is the main paper in this symposium, these comments were presented first in order to provide background for his discussion.

** The author wishes to express gratitude to the National Science Foundation for support of research on statistical explanation.

[1] C. G. Hempel and P. Oppenheim, 'Studies in the Logic of Explanation', *Philosophy of Science* **15** (1948) 135–75.

[2] C. G. Hempel, 'Deductive Nomological vs. Statistical Explanation', in *Minnesota Studies in the Philosophy of Science* (ed. by H. Feigl and G. Maxwell), University of Minnesota Press, Minneapolis, 1962.

[3] I called attention to this danger in 'The Status of Prior Probabilities in Statistical Explanation', *Philosophy of Science* **32** (1965) 137–46. See also, W. C. Salmon, 'Statistical Explanation' in *The Nature and Function of Scientific Theories* (ed. by R. Colodny), University of Pittsburgh Press, Pittsburgh, 1971, footnote 7, for an elaboration of some of the salient differences.
[4] N. Rescher (ed.), *Essays in Honor of Carl G. Hempel*, D. Reidel Publishing Co., Dordrecht-Holland, 1969.
[5] *Philosophy of Science* **37** (1970) 279–93.
[6] *Op. cit.*
[7] The features of his theory to which I shall refer are present in *Logical Foundations of Probability*, University of Chicago Press, Chicago, 1950; 2nd edition, 1962.
[8] Some of these were detailed in my remarks, 'Who Needs Inductive Acceptance Rules', in *The Problem of Inductive Logic* (ed. by I. Lakatos), North-Holland Publishing Co., Amsterdam, 1968.
[9] See, for example, J. Hintikka, 'A Two-dimensional Continuum of Inductive Methods', in *Aspects of Inductive Logic* (ed. by J. Hintikka and P. Suppes), North-Holland Publishing Co., Amsterdam, 1966.
[10] *Op. cit.*, pp. xv–xx.
[11] With apologies to Professor Greeno for the associations that already surround the expression 'S-R' in psychological circles.
[12] Logarithms to the base two are used in information theory to get the value one for the bit. Greeno uses natural logarithms in his computations, but that makes no difference in principle, for there is an elementary conversion rule.
[13] In writing the arguments of the probability function in this order I am conforming to Greeno's usage; I usually follow Reichenbach in writing them in the reverse order.
[14] Independence is a symmetrical relation if none of the probabilities equals zero.
[15] The following measure is essentially that used in my paper 'Statistical Explanation' as a measure of degree of inhomogeneity of the reference class.
[16] This happy term is due to Hans Reichenbach who introduced it in his posthumous work *The Direction of Time* University of California Press, Berkeley and Los Angeles, 1956, p. 189. I have used it extensively in 'Statistical Explanation'.

RICHARD C. JEFFREY

REMARKS ON EXPLANATORY POWER

The idea of using information-theoretical concepts to quantify the notion of explanatory power of theories is an attractive one. Where the theory is adequately represented by a probability measure p on Boolean algebra A of propositions, it is natural to think of measuring explanatory power by some function of $H(A)$, the *entropy* or *uncertainty* of p. But recently, Greeno and others have been exploring measures of explanatory power which are suggested by more detailed representations of theories – representations which are themselves suggested by communication-theoretical analogies.

To avoid inessential complications, suppose that the Boolean algebra A is finite, and let 'a' range over the atoms of A, viz., the strongest propositions in A which are not logically false. (Any two atoms are logically incompatible, and any proposition in A is expressible as a disjunction of atoms.) Then we have

$$H(A) = -\sum_{a} p(a) \log p(a)$$

where we set $p(a) \log p(a) = 0$ in case $p(a) = 0$. If p represents our state of knowledge about A, then $H(A)$ is a plausible measure of the paucity of that knowledge: $H(A)$ is greater, the more ignorant we are, about A. Thus, $H(A) = 0$ when p assigns probability 1 to one of the atoms, and thus assigns probability 0 to all others: Our uncertainty is null when we take ourselves to know which atom is true, and thus take ourselves to know the truth values of all propositions in A. And uncertainty is maximum when p assigns the same probability to all atoms. If there are 2^n atoms, and we take logarithms to the base 2, then $H(A)$ assumes its maximum value, n, when each atom has probability 2^{-n}. Thus, maximum uncertainty is greater, the more atoms there are. Since the number N of atoms need not be a power of 2, the general expression for maximum uncertainty is $-\log N$.

Boston Studies in the Philosophy of Science, VIII.

In view of the foregoing properties of entropy, the number

(1) $-\log N - H(A)$

has a certain plausibility as a measure of how much information or knowledge we take ourselves to have, when we take ourselves to know that p is the statistical law governing the phenomena described by the propositions in A. Then there is a certain *prima facie* plausibility in the use of (1) as a measure of the explanatory power of the statistical law p, or of the statistical theory $\langle A, p \rangle$.

Greeno proceeds differently. His basic idea is that we can view the theory $\langle A, p \rangle$ as a description of a natural communication channel, driven by a source S, and yielding outputs M. S and M are subalgebras of A. The propositions in M (for 'explananduM') describe the phenomena which are to be explained, and the propositions in S (for 'explananS') describe the phenomena which are to be appealed to as explanantions. Or perhaps it would be better to put the matter thus: The values which p assigns to propositions in S are to explain the values which p assigns to propositions in M, the explanation being mediated by the conditional probabilities of propositions in M, given propositions in S. In terms of the communication-theoretical analogy, the atoms of S are the input letters; the atoms of M are the output letters; and the conditional probabilities $p(m/s)$, where 'm' and 's' range over the atom of M and of S respectively, define a communication channel over which one can transmit information from S to M. The deterministic case in which $p(m/s)$, is always either 0 or 1 corresponds, in the analogy, to the case in which the channel is noiseless: Each input letter determines a unique output letter. In other cases, the channel is noisy: There will be at least one input letter, s, which can be garbled in transmission so as to give rise to any one of a set of two or more output letters, $m_1, m_2, \ldots$, with probabilities $p(m_1/s) \neq 0, p(m_2/s) \neq 0, \ldots$. The extreme of noise occurs when for each s, $p(m/s)$ is the same for all output letters m. Channel capacity is then zero: No information can be transmitted through such a channel, since the output gives no clue as to what the input might have been.

Greeno offers the mutual information, or transinformation,

(2) $I(S, M) = H(S) + H(M) - H(A)$,

as a measure of the explanatory power of the theory $\langle A, p \rangle$, relative to the choice of the subalgebras S and M as explanans and explanandum, respectively.[1] In (2), $H(S)$ and $H(M)$ are the entropies of the probability measures obtained by restricting p to S and to M, respectively:

$$H(S) = -\sum_{s} p(s) \log p(s), \quad H(M) = -\sum_{m} p(m) \log p(m)$$

Note that (2) is not a direct competitor with (1), for (1) purports to measure explanatory power *tout court*, while (2) purports to measure the theory's power to explain a certain class M of phenomena in terms of a certain other class S. A theory may do brilliantly for one choice of S, M, and very poorly for another. That is all to the good.

Greeno's measure (2) has certain virtues: The mutual information is never negative, and vanishes if and only if S and M are independent, relative to p, in the sense that we always have $p(m/s)=p(m)$. (See Perez, [1].) Furthermore, $I(S, M)$ tends to increase as the channel becomes less noisy while input and output probabilities are held as nearly constant as may be. (Extreme example: M and S have the same number 2^n of atoms, and both entropies are maximum: $H(M)=H(S)=n$. If the channel is maximally noisy, $H(A)=2n$ and thus $I(S, M)=0$, while if the channel is noiseless, $H(A)=n$ and thus $I(S, M)=n$.) Then theories in which the relationship between S and M is deterministic tend to have higher explanatory power than those in which the relationship is probabilistic.

On the debit side of the ledger for Greeno's proposal, note that the mutual information is symmetric in its arguments: $I(S, M)=I(M, S)$. This may be disturbing, for it seems to obliterate the distinction between explanandum and explanans. In the interest of asymmetry, Rosenkrantz ([2], Section VI) normalizes the mutual information: he proposes

$$(3) \qquad I(S, M)/H(M)$$

as a measure of explanatory power which vanishes when (2) does, i.e., when S and M are independent relative to p, and which assumes its maximum value, unity, in the deterministic case where the conditional probabilities $p(m/s)$ assume only the extreme values, 0 and 1.

Then Rosenkrantz's proposal (3) seems to have some advantages over Greeno's (2). On the other hand, it seems a defect of (3) that it assigns the same explanatory power, 1, to all deterministic theories. This suggests further tinkering, e.g. perhaps one ought to multiply Rosenkrantz's (3) by (1), and use

(4) $$I(S, M)[-\log N - H(A)]/H(M)$$

as a measure of the explanatory power of $\langle A, p\rangle$ relative to S as explanans and M as explanandum. Or perhaps a different repair would be better.

How are we to decide among competing explications of the notion of explanatory power? Presumably, by reference to clear cases, where $\langle A, p\rangle$ is intuitively more explanatory relative to $\langle S, M\rangle$ than $\langle A', p'\rangle$ is relative to $\langle S', M'\rangle$. An explication which assigns the higher numerical measure of explanatory power to the intuitively less explanatory theory can then be ruled out. The possibility of computing such numbers for a particular proposed explication is no more than an entrance requirement for the competition, which is met equally by (2), (3), and (4).

In his present paper, Greeno tests his explication (2) against the problem of explaining why, by and large, phenomenological theories seem less explanatory than theories which make essential use of 'theoretical terms'. He observes that the propositions in S and in M may all be 'observational', so that the theoretical burden is being carried by the conditional probabilities $p(m/s)$, in a way that invites analysis. Greeno's analysis involves a third subalgebra, T, the members of which are 'theoretical' propositions. (The scare quotes signal the fact, which Greeno points out, that nothing in his analysis depends on any special view of the nature or even the existence of the observational/theoretical dichotomy: The two terms serve here only as handy counters.) Concerning T, Greeno posits only what Salmon calls the 'screening' property: Conditionally on T, M is independent of S in the sense that we have $p(m/ts)=p(m/t)$ for all atoms, m, t, s of M, T, and S respectively. Thus, T screens S off from M. It is this property which is definitive of the term 'theoretical' in the present context. The analysis would be the same, if the members of T were as accessible to observational testing as are the members of S or of M; but presumably, propositions which one regards as theoretical relative to S and M do have the screening property. (This last consideration does as much to establish the asymmetrical roles of S and M as it does to characterize theoreticity.)

Natural examples are those in which members of S serve as symptoms of members of T, while members of M are effects of members of T. (Here, one must think not of the propositions themselves, but of corresponding events or standing conditions.) Thus, the fact that Czarevitch Alexis was a bleeder is expressed by some proposition in M, let us say, and this circumstance is an effect of another, which is described by a proposition in T: That the Czarina, Alexis's mother, was a carrier of hemophilia. According to genetic theory, the probability of this member of M, conditionally on the cited member of T, is $\frac{1}{2}$. Suppose, now, that S contains the proposition that the Czarina's brother Frederick was a bleeder. Conditionally on that proposition, the probability is $\frac{1}{2}$ that the Czarina was a carrier, according to genetic theory. Frederick's malaise may have been the symptom which made one fear that his sister was a carrier, but Frederick's problem is 'screened off' from Alexis's by the hypothesis that the Czarina was a carrier, for the probability that her son would be a bleeder remains $\frac{1}{2}$, even when the condition is strengthened from 'His mother is a carrier' to 'His mother is a carrier and her brother was a bleeder.'

In terms of the communication-theoretical analogy, Greeno's basic move is to analyze the channel from S to M into two channels – from S to T and from T to M – connected in cascade. This is a worthwhile step, for it increases the amount of information – theoretical lore which can be brought to bear on the analysis of probabilistic theories, and suggests further such steps, e.g. the analysis of time-series of the process of 'transmitting messages' from S to M via T, where today's S may be yesterday's M: The channel contains a feedback loop, making information about past outputs available at the input.

Greeno's cascade analysis is a genuine advance; but it may be thought to tell against his proposed explication (2) of explanatory power, when such alternative proposals as (4) are in the field. The problem is that in computing the mutual information (2) – as in computing Rosenkrantz's normalization, (3) – one views the cascaded channel from S to T and thence to M as a single channel from S to M, as if it were in a 'black box'. In particular, the probabilities $p(t/s)$ and $p(m/t)$ figure nowhere in the calculation. The only relevant probabilities are of form $p(s)$, $p(m)$, and $p(m/s)$, and therefore the explanatory power of the full theory, relative to S and M, is precisely the same as the explanatory power of the phenomenological subtheory which is obtained by restricting p to the product algebra $S \times M$. (The full theory is given by p on the full algebra, $A = S \times T \times M$.)

As Greeno remarks, the screening property ensures that $I(T, M)$ will be at least as great as $I(S, M)$: Tapping into the channel and transmitting from T cannot leave us worse off, and may leave us better off, than we are when transmitting from S. Indeed it accords well with intuition, that $I(T, M) \geqslant I(S, M)$ – a point in favor of Greeno's explication (2). But this point seems cancelled by the insensitivity of (2) to the difference between the full theory and its phenomenological fragment, in view of which $I(S, M)$ seems inadequate as a measure of the explanatory power of $\langle A, p \rangle$, relative to $\langle S, M \rangle$.

But tools for repairing the explication lie ready to hand. The spirit of Greeno's enterprise seems to require that we speak of the explanatory power of the theory $\langle A, p \rangle$ relative to an ordered triple, $\langle S, T, M \rangle$. In place of the quantity $I(S, M)$ proposed in (2), we need some more complicated function, of which the arguments are not simply the first three of the magnitudes

$$(5) \qquad H(S), H(M), H(S \times M), H(T), \\ H(S \times T), H(T \times M), H(A).$$

Such a function is (4) above – among many others which have some *prima facie* plausibility.

In conclusion, let us return to the question, "How are we to decide among competing explications of the notion of explanatory power?" I suspect that what seems to be an embarrassment of riches, in the way of competing, plausible explications, will prove on closer inspection to be a less pleasant sort of embarrassment. Perhaps the various simple, plausible formulas, with their various virtues, will all be vitiated by various defects. A variety of *ad hoc* concoctions may avoid the pitfalls on the short list, but exhibit an unpalatable complexity which paralyzes judgement. If this is so, it may indicate that there is no clear, simple, useful intuitive notion of explanatory power to be explicated. Indeed, the key term is 'useful'. Until we have some idea of how the concept of explanatory power is to function together with other, sometimes equally fuzzy concepts, in an account of scientific methodology, we have no firm basis on which to decide, e.g., whether the symmetry of (2), or its insensitivity to the existence of the theoretical subalgebra T, really vitiate (2) as an explication, or are tolerable oddities which make no difference to the satisfactory functioning of the new concept, in its context. Lacking the context, we are reduced to a sort of ordinary-language analysis which seems unpromising in view of the fact that the notion to be explicated is itself a technical term, having very tenuous links with common usage. But perhaps my guess is wrong: Perhaps there will prove to be a unique one among the simple, plausible functions which conforms well with intuition' If so, that function will emerge as one of the touchstones against which other concepts in scientific methodology are to be tested.

It seems clear that Greeno's concept, (2), can win no such easy victory – nor can Rosenkrantz's (3) or, I suspect, the messier (4). But as far as I know, there may be some well-motivated function of the quantities (5) which does the job,[2] and in exploring such functions one should bear Greeno's novel cascade analysis in mind. The demonstrated possibility of calculating $I(S, M)$ and $I(T, M)$ – especially, for theories like Luce's, in the building of which communication-theoretical materials were liberally and consciously used – can hardly be judged a reason for adopting (2). But indeed, I take Greeno to be offering (2) only as a function which has some of the desired features – as an example of the sort of thing that is being sought. And it is conceivable that when (2) is put to work, in the context of an account of scientific methodology, such oddities as symmetry will prove harmless. Meanwhile, debate may be pointless.

University of Pennsylvania

NOTES

[1] For simplicity, I assume here that A is the product $S \times M$ of S and M, i.e., I assume that each atom of A is the conjunction of an atom of S with an atom of M. Where this is not the case, '$H(A)$' in (2) must be replaced by '$H(S \times M)$'.
[2] For a suggestion along these lines, see the function J in [3], p. 55.

BIBLIOGRAPHY

[1] Perez, A., 'Notions généralisées d'incertitude, d'entropie et d'information du point de vue de la théorie de martingales', *Transactions of the First Prague Conference on Information Theory, Statistical Decision Functions and Random Processes*, Prague, pp. 183–208.
[2] Rosenkrantz, R., 'Experimentation as Communication with Nature', *Information and Inference* (ed. by J. Hintikka and P. Suppes), Dordrecht 1970, pp. 58–93.
[3] Watanabe, S., *Knowing and Guessing*, New York 1969.

SYMPOSIUM:

CAPACITIES AND NATURES

MILTON FISK

CAPACITIES AND NATURES

1. Capacities and the Fine Structure of Entities

The last two decades have witnessed a basic change in the philosopher's working model for capacities. The change has been from a stimulus-response model to a fine-structure model. The stimulus-response model was an attempt to understand capacity propositions by means of conditional propositions in which the antecedents expressed operations in certain circumstances and the consequents expressed realizations of the capacities in question. No attempt is made on this model to relate such conditional propositions to components of the entities with the capacities. An entity with a capacity is just a black box.

Now the conditional proposition to which a capacity proposition is related on the stimulus-response model is a modal conditional. It expresses the idea that the behavior characteristic of the capacity must follow upon those operations in those circumstances for an entity like one with the capacity. Otherwise, the normal inference from a capacity proposition to a counterfactual conditional would be blocked. But then one is led to ask how the stimulus-response model contributes to the understanding of this necessity. The stimulus, in the specified circumstances, hardly accounts for the fact that it is necessarily followed by a certain response. The fine-structure model attempts to complete the stimulus-response model by providing an answer to the question of the source of necessity. After all, in the very concept of a capacity there is the implication that an entity with a capacity shapes the outcomes of actions on it. So unless one pries inside the black box, one fails to get at the ontological roots of capacities by failing to get at the roots of the necessary connections associated with capacities.

The fine-structure model for capacities consists of two parts. First, to say an entity has a certain capacity is to say that there are components of that entity. These components may be parts in the material sense or they may be properties. But saying that an entity has a capacity does not commit one to ascribing any specific components to the entity. Second, to say an entity has a certain capacity is to say that the unspecified components are such that the entity's having them plays a causal role in the realization of the capacity. I shall speak of the having of a component as a condition of an entity. The component and the related condition are distinct, and it is the having of a component and not a component itself that will play a causal role. Salt's condition of being ionic in structure and its condition of having a certain crystal energy are among the causal factors in salt's exercising its capacity to dissolve in water.

The problem of the modal conditional that was given no resolution on the stimulus-response model comes back in a slightly transformed fashion to plague the fine-structure model. That this is the case can be seen by considering the causal part of the fine-structure model. If a causal role is played by the condition of having a component of the fine structure, then there will be a necessary connection that in some way involves that condition and the exercise of the capacity. Or if we prefer to say that the causal role requires that there be some covering law for the specific situation in question, then since holding by law is just holding by physical necessity, the problem of necessity is still not evaded. It is tempting to come to a negative conclusion at this point as regards the fine-structure model. Since the fine-structure model was to handle the modal conditional as the stimulus-response model had failed to do, there is really little advance made in understanding capacities by turning to the fine-structure model. Perhaps it is best simply to admit that capacities are entities over and above properties and parts.

But the stimulus-response model and the fine-structure model belong to two fundamentally different ontological traditions, the former to the tradition of instrumentalism and the latter to that of scientific realism. Within the tradition of scientific realism, the advocate of fine structure in the interpretation of capacities can find the materials he needs to cope with the problem of necessity that arises in the causal part of that interpretation. The scientific realist is committed to explaining the fact that entities obey laws relating operations to responses in terms of the parts and properties of those entities, in other words, in terms of the fine structure of those entities. Such an explanation is deemed satisfactory only if it is also an explanation of the fact that entities must obey the observed regularities, or of the fact that they obey them by law. So fine structure is an advance over operations and responses since it accounts for the modality involved.

If the fine structure did not also account for the modality of the connection it would be superfluous. What role would be left to it if it did not account for the modality of the observed connection? The fine structure would be introduced simply as something that happened to obey certain regularities from which the observed regularity could be deduced. It would then be a matter of sheer contingency that the fine structure obeyed regularities compatible with the operational regularities. But then the fine structure is not really explaining anything. The core of the explanation is shifted entirely to the regularities from which the observed regularity can be deduced. And once this is done we no longer have a mode of explanation that belongs properly to scientific realism. For the instrumentalist too explains by deductions from propositions rather than by appeal to entities. So, for the scientific realist, elements of the fine structure of an entity are explanatory since it is assumed that those regularities associated with the elements of the fine structure that are relevant to the observed regularity to be explained are not just regularities the elements of the fine structure happen to obey, but are ones they must obey. This being so, an appeal to fine structure to explain an observed regularity is sufficient to explain also why entities must obey that regularity.

This can all be extended to still deeper levels of entities. If fine structure accounts for the modal character of observed connections, there is no reason why it cannot account for the modal character of connections involving fine structure. Necessary connections involving atoms can be accounted for in terms of atomic particles. So scientific realism presents us with a general theory of the source of necessity. For any given level of investigation, it is the elements of the fine structure of entities at that level that are the source of the necessity between entities at that level.

With these resources of scientific realism at his disposal, the advocate of the fine-structure model for capacities can answer the criticism that his model shares the fate of the stimulus-response model in that it fails to resolve the problem of the modal conditional. The necessary connection implied by the causal part of his model has its source in a still deeper level of fine structure.

The scientific realist has at least posed the problem of necessity in an ontological way. The idea that an entity should necessarily be what it is because of some component of it is not foreign to the realist as it was to his positivist forebearers. They could not think beyond a notion of necessity that had its source in intentions, that is, in propositions, concepts, or language. I am here accepting the assumption of the consistent realist that the basis for entities' having to be the way they regularly are is their composition. The question is then whether the components making up fine structure – properties and parts – are the right ones for an account of modality.

My case against the account of capacities and of necessities in terms of fine structure has to do with the fact that the account involves a regress. The elements of the fine structure used to account for a given necessity will themselves stand in necessary connections. Otherwise they will be incapable of accounting for the original necessity. At the very least, the relevant elements of the fine structure will have to belong of necessity to the entity that has them. For if it might not have them, it would not have to stand in the connection whose necessity was to be accounted for. But then there will have to be still further complexity of the entity in question to account for the necessities involving the first level of fine structure. There can then be no end to the complexity of the entity that is necessarily anything. But it will be said that this regress is benign. Though not all necessities are accounted for at once, no one necessity is left unaccounted for.

The regress is, it will turn out, vicious. The only plausible way of saving the scientific realist's account of necessity is to add to the components of fine structure another component of entities that is different from both properties and parts. I shall argue that natures satisfy these requirements. But first, some details should be filled in regarding the fine structure model for capacities.

2. Freedom and Necessity

The proposition that lead is malleable is undoubtedly true. But I find myself agreeing with both the stimulus-response model and the fine-structure model that the truth of this proposition does not require that there be a component called malleability that any hunk of lead have. That is, I agree that there are no dispositional properties that entities have. When it is insisted that lead must have malleability as one of its components if the proposition is to be true, it should be pointed out that the same proposition can equally well be expressed in another way. Instead of saying that lead is malleable, one can just as well say that lead is capable of having its shape changed. And 'capable of' plays a role remarkably like 'possible to' or 'necessarily'. The problem of finding the sorts of entities needed for capacity propositions to be true can be treated as the problem of finding the truth conditions for a modal proposition whose modality is 'capable of'. And the general form of this problem is the problem of finding in what contexts the truth conditions are satisfied for the constituent nonmodal proposition. In other words, the problem of finding the truth conditions for the proposition that this hunk of lead is capable of having its shape changed is the problem of finding in what contexts the truth conditions for the non-modal proposition that this hunk of lead is changing its shape.

The proposal made by the fine-structure model is that the contexts are causal. Lead will be changing shape when the circumstances are such that lead's having a certain component, or complex of components, is a factor in causing the change. The components are not themselves capacities, even though, in this way, their presence is the source of capacities of entities of which they are components.

It seems difficult to reconcile this requirement of causality with the obvious fact that many capacities may be exercised freely. We might then distinguish two senses of capacities, the one for causal capacities and the other for freely exercised capacities. But it is not at all evident that a stone's capacity to fall and a man's capacity to talk are different as capacities. So the device of splitting senses would be objectionably *ad hoc*.

Fortunately, there is another avenue to reconciliation. The notion of cause is not limited to that of a cause that in given circumstances is sufficient for the effect. It is also appropriate to speak of a cause of a certain condition when the cause only explains the obtaining of that condition *when* it obtains. Let it be granted that when I talk I do so freely; still there are causal factors involved in my talking. Conditions of the brain explain the talking when it occurs, without necessitating the talking.

Suppose then we are concerned with the capacity of an entity, a, to ϕ. First of all, a will have an unspecified element of fine structure, θ. Second, there will be the causal requirement, to be stated in such a way as to allow for the free exercise of capacities. This can be done by requiring that, given that a does ϕ, a's having θ makes a have ϕ. Only one thing has been left out; the causation is expected to occur only in certain circumstances.

But in different circumstances the same capacity may be realized through different features of the fine structure. So *a* is capable of ϕ-ing when there are circumstances and there are components, θ, such that *a* has θ and in those circumstances, given that *a* does ϕ, *a*'s having θ makes *a* have ϕ.

We emphasized above that the fine-structure model does not avoid the problem of a basis for necessity. But once it has been agreed that the causal part of the fine-structure model involves only the above kind of conditional causality it seems less clear where necessity comes in. For if the causality is only conditional, the exercise of the capacity is not necessitated by anything. But before it is concluded that the fine-structure model is not after all faced with the question of the source of necessity, we should ask if this is the only way necessity can enter in. The answer is clearly no.

The causal part of the model can still be broken down into a factor involving action and a factor involving necessity. As in any case of causation, there is an action other than the action that might be the effect. The condition of having an element of fine structure does not bring about the exercise of the capacity without some action somewhere in the circumstances. And this will not be just any action in the circumstances, but rather one that has the exercise of the capacity as a result. This does not yet bring in necessity, for the result of doing something need not be necessitated. In any event, the notions of action and result are so taken here as to imply no necessary connection between action and result. However, there will be a necessary conditional that mirrors the conditional causality. It will be necessary that if the capacity is exercised then if the entity with the capacity has the causal component of fine structure and the specified circumstances obtain then indeed the capacity is exercised. In other words, $\Box(\phi a \rightarrow ((\theta a \ \&$ the circumstances obtain$) \rightarrow \phi$a$))$. So, to summarize, in circumstances of type K, given that *a* has ϕ, *a*'s having θ makes *a* have ϕ when, first, K contains an action that has *a*'s having ϕ as a result and, second, *a* is necessarily such that if it has ϕ then its having ϕ is implied by its having θ and by K's obtaining. In the final analysis, this is how necessity enters into the account of capacities, and hence this is how the problem of necessity arises in connection with capacities.

The connectives in the conditional proposition to which necessity applies cannot be material conditional connectives. For, then the necessity that when a capacity is exercised its being exercised is implied by the presence of the causal component and the circumstances would be a mere logical necessity. Our conditional would be an instance of the well known law that anything materially implies a true proposition. But the connective is not the strict conditional connective either. For this is a modal connective, and it is not the case that if the capacity is exercised its being exercised necessarily follows from the presence of the causal component and the circumstances. Otherwise, the exercise of the capacity could not be free. The conditional connective is then to be a non-modal one stronger than the material connective. The connective of relevant implication satisfies these demands. But permutation is permissible in a chain of relevant implications, as it is not in a chain of strict implications. So we are led to ask whether it is undesirable that our conditional proposition permutes so as to say that the presence of the causal component and the circumstances implies that the exercise of the capacity follows from itself. It might seem undesirable if we insist on thinking in terms of material or strict implication, for anything materially or strictly implies a tautology of this sort. But we are not thinking in these terms now. It is then possible to view the permuted form as the significant claim that the self-implication property is something an entity has due to the presence of the causal component and the circumstances. Of course, it might have the self-implication property for other reasons as well, but it is important to have this source of the self-implication property pointed out.

3. The Ontology of Parts and its Failure

One form of the fine-structure model for capacities limits fine structure to parts in the material sense. A corresponding form of the realistic account of necessity would attempt to account for necessities by means of material parts to the exclusion of properties and actions. Both the fine-structure part and the causal part of the account of capacities are made to rely simply on parts. Structure and implication cease to be different, both resolving into component parts. This is in many ways an attractive view. John Locke was one of its modern avatars, and he is not without contemporary disciples. According to Locke, the real essence of a substance is that "constitution of the parts of matter" on which the powers of that substance depends.

If parts are the basis for necessities, there will be no simple material parts. As argued above, any part that plays a role in grounding a necessity will itself be subject to necessities and these will then be grounded by parts within parts. So the general thesis that a fine-structure account of necessity admits of no end to the complexity of fine structure applies here in the form that there are no simple material parts. Similarly, since at any level of the analysis the parts appealed to will themselves have capacities to form the structures they do form, there will have to be parts of these parts to account for these capacities.

The only reason for limiting the account of capacities and necessities to material parts is that the background ontology is one according to which entities have no other components than parts. They might also be said to have powers, but their having powers would depend solely on their having material parts. The non-dispositional properties that we think of entities as having are readily reduced to parts. Locke in particular defined the so-called primary qualities in terms of parts.

Methodological advantages can doubtless be claimed for such an ontology of parts, where we include not just particles but also other entities such as fields under the heading of parts. What for example is the basis for inertia, the power of resisting change of state? Is it, perhaps, the property of being material? The ontology of parts would resist such an answer and hold out for an account in terms of parts. It is given support by the relation of mass to energy in Special Relativity. It follows from this relation that the power of resisting is based on potential energy in bodies. Since potential energy is itself a capacity to do work, we end our search as far as the ontology of parts is concerned by noting that this capacity is based on the configuration of parts of the inertial body.

But methodological advantages must be weighed against ontological disadvantages, which are indeed serious. We have spoken so far about entities and their parts. It may be however that since entities have no components other than parts that they are really not entities at all except for convenience of designation. A collection of parts is not an entity; only its parts can be entities. Remember though that there are no simple parts. So parts will not themselves be entities. If, however, there is nothing other than a part that can be an entity, we cannot avoid the conclusion that there is nothing. Being eddies down into ever finer parts and vanishes altogether.

To avoid this unpalatable conclusion, it must be held that there are genuine entities composed of parts, assuming still that there are no simple parts. But will the ontology of parts support the view that there are entities composed of parts? I think it cannot. If all there is to a putative entity is its parts, then there is no basis for the entity that is not also a basis for the existence of all the parts. But without such a basis, and with only a basis that is just sufficient for the parts, there is not a sufficient basis for there being the entity. Of course, if there is an entity that is composed of parts the distinctive thing about it, as opposed to the parts, may be a component that has come about because of the parts. But once such a component exists, it provides a basis for there being an entity rather than just many parts. If powers are not reduced to parts but are regarded as dependent on parts, it might be thought that powers could provide the needed extra to have an entity. But without reducing them to parts, they can be viewed as powers, not of an entity composed of parts, but of the parts to behave together in specified ways. So if there is a unity composed of parts, the entity that is that unity must have some other components than parts. A property, an action, or a nature would fill the bill.

Once we admit, say, properties, in addition to parts, there is not any longer a good reason for basing capacities solely on parts. Even so, the admission of properties into the ontology does not seem to help to resolve a difficulty that faces the fine-structure model. The difficulty is that by basing a capacity on an element of fine structure, whether it be a part or a property, we seem to be requiring that that element itself have a capacity. By basing a's capacity to ϕ on the component θ we imply that θ has the capacity to make a to ϕ. It is essential that θ have this capacity if it is to play the role of grounding the capacity to ϕ. So we really have not come close to having an account of capacities that allows us to say that capacities are not entities.

The solution to this important problem requires that the ontology of components behind this entire discussion be made more explicit. The basic feature of this ontology is that components of entities are not distinct from those entities. An entity is a unity and its components share that unity. This differs markedly from the view that properties of entities are distinct entities to which they are somehow related. However, though a component is not distinct from an entity with it, the condition of having that component is to be treated both as something distinct from the component and as something distinct from the entity with it. Thus one and the same entity can be associated with a multiplicity of distinct events or conditions, and can thereby exist at distinct times.

We should then distinguish the component of fine structure behind a capacity and the condition of having that component. It is not the component that makes the entity with the capacity do something. For it is not components that are causes but their corresponding conditions. As a consequence, it is the condition, not the component, that has the capacity to make the entity exercise its capacity. This capacity, like the capacity of the entity with the condition, is based on the corresponding component. So if *a*'s capacity to ϕ is based on θ, it is also true that the capacity of *a*'s condition of having θ to make *a* to ϕ is based on θ. There is then no unaccounted for capacity of the element of fine structure. And thus the account of capacities by means of fine structure does not require that the elements of fine structure have capacities.

The application of this view to parts can be made as follows. Component parts, unlike component properties, may become distinct entities. As distinct entities they have capacities. If these capacities are based on parts, then they must have had such parts even as components. For by being isolated they do not change their structure. To speak of parts even as components as having capacities can then be construed as a way of attributing to them the structure needed in order to have capacities when they are isolated. This was our basis earlier for saying that if capacities are based solely on parts then there can be no simple parts since parts themselves will have capacities.

4. Natures and essentially repetitive explanations

Our context is the realistic program of accounting for entities' obeying the laws they do and their having the capacities they have in terms of components of those entities. It was argued that any attempt to account for entities' obeying the laws they do in terms of components was, if it was not to collapse into the instrumentalist's attempt to account for entities' obeying laws by deductions from propositions, also an attempt to account for the lawlike or necessary character of the regularities they obey. So the realistic program is also a program for accounting for necessity by composition. We have just seen that an ontology of material parts will not be a satisfactory basis for accounting for modality, since that ontology suffers from serious internal shortcomings. We arrive at the question whether supplementing the ontology of parts with properties and actions gives a satisfactory basis. The question is important in regard to capacities since the realistic account of capacities by fine structure possesses a decided advantage over the instrumentalist account by stimulus and response only if the realist can solve the problem of the basis of necessity. If fine structure fortified to include properties and actions as well as parts cannot provide us with the basis for necessity, there is no escape route along an intentional account of necessity in terms of conceptual content or rules for the use of terms. For to take such a route would be to abandon altogether the realistic program of accounting for entities' obeying the laws they obey in terms of their components.

The view that it is not properties or parts but natures that base necessity can be expressed as the view that for an entity to have a component necessarily is for it to have that component by nature. The truth conditions for necessity propositions are made to involve not only entities and their parts, properties, and actions but also natures. Thus it is true that an entity a necessarily ϕ's if and only if the non-modal proposition that a has ϕ corresponds to the nature of a. To say the proposition corresponds to the nature of a is only to say that what the proposition says a is is what it is the nature of a to be. Clearly this covers only so-called *de re* necessity, but the basic idea can be expanded to cover *de dicto* necessity as well. The *de dicto* claim that it is necessary that the first human born at sea was risible is the claim that the first human born at sea was by nature risible and that anything you take is by nature such that if it is the first human born at sea then it is human and if it is human then it is risible.

The natures to which necessary propositions correspond are not entities distinct from the entities of which they are the natures. Like properties, parts, and actions, they are components of entities. But though other components may be had by nature, they are not natures. It is the nature of a crystal of salt to be ionic, but the property ionic is not the nature of the crystal. A property, or any component other than a nature, is not such that because of what it is it is necessarily had by entities. Rather, because of the nature of an entity it has certain properties necessarily. The property ionic is not the basis for any entity having it necessarily. If it were all entities would have it, but they do not. Rather it is the nature of some entities to be ionic. Even in the case of logical properties, it is not the property that forces entities to have it. Rather entities are such by nature that they have the property. Of course, to deny that other components base their own necessity is not to deny that they base the necessity of other components. We have yet to prove the latter.

A final point before undertaking the proof. If natures are not among the familiar components of entities, then they are not only undiscoverable, they are undescribable. Would it then not be better, if it is to be insisted that there be natures that they be identified with already familiar members of the realistic ontology of components? If natures are anything they are capacities or parts or properties or some combination of these. In response, I point out, first, that if we identify natures with these more familiar entities then the realistic program of accounting for modality by components collapses. And, second, the threat of introducing undescribable entities is an empty one, for it can only mean that an entity is being introduced that is not a property, a part, or an action. Indeed natures are describable as those components of entities to which propositions that are necessary correspond.

It was noted that the fine-structure account of necessity involved a regress. Necessities at any given level are accounted for by what is fine structure for that level. But to do this that fine structure will itself be subject to necessities. And so on. The regress was allegedly benign. I wish now to show that it is in fact vicious.

Imagine an unending series of entities. Each entity in the series is both *A* and *B*. Suppose our task is to give an account of *A*-ness. The account cannot be carried out by staying with just one member of the series. So we are led through the series in giving our account, in the way we are above in giving an account of necessity on the fine-structure view of it.

It is quite important to distinguish two kinds of explanation of *A*-ness that involve the series.

(i) The account of *A*-ness in respect to any given member of the series can be given by the fact that the next member in the series is *B only if* it is also the case that this next member is also *A*. That is, *B* in the next member explains *A* in the preceding member not of itself but only if the condition is satisfied that *B* is accompanied by *A* in the next member. I call such an explanation "essentially repetitive." It is repetitive since in the account of *A*-ness, *A* makes its appearance. It is essentially so since *A* must make its appearance for there to be an account of *A*-ness in the preceding entity. It seems obvious that an essentially repetitive explanation is a vicious regress. For at each stage the feature to be accounted for is indispensible for accounting for itself. Thus at no stage is there a genuine account.

(ii) The other type I call an 'accidentally repetitive' explanation. Here the account of A-ness in respect to any given entity in the series can be given by pointing to the fact that the next member on is *B*. We have assumed that the next member on is also *A*. But *B*-ness explains the *A*-ness of the preceding member whether or not it is associated with another instance of *A*-ness. It may even be the case that being *B* depends on being *A*. Still so long as the explanation by *B* does not rely on *A*-ness, the dependence of *B* on *A* is incidental. An accidentally repetitive explanation is a benign regress, and can then be perfectly satisfactory as an explanation.

Enough has been said so that it is now easy enough to show that the fine-structure account of necessity is an essentially repetitive account. It is desired to account for the necessity of *a*'s being ϕ by means of some element, θ, of the fine structure of *a*. Now this account is possible only if the following two conditions are satisfied. First, it must be the case that *a* has θ of necessity. Why? Well, if a were contingently θ, it could exist without θ as a component, and hence without ϕ as a component. But we had assumed that *a* must be ϕ. So the element of fine structure must itself belong of necessity to the entity of which it is a component. Second, if θ accounts for *a*'s having to ϕ then it is of necessity the case that if *a* has θ then *a* has ϕ. Otherwise, even though *a* has the component of fine structure, θ, necessarily, it could conceivably lack ϕ, but by hypothesis it cannot lack ϕ.

This is quite different from its happening to be the case that the entity subject to a necessity has the fine structure that accounts for this necessity of necessity. On both of the above counts, the explanation does not begin unless the element of fine structure is itself subject to necessity. So not only is the account of necessity repetitive, it is also essentially so, and hence viciously regressive. So the realist's position does not have an advantage over the instrumentalist's on the crucial issue of modal connections. The fine-structure model for capacities leaves the problem of necessity untouched. The factor of necessity in the causal part of the fine-structure model does not itself admit of reduction in terms of fine structure.

From what has been said about natures it follows that the account of necessity by natures is not essentially repetitive. It will of course be the case that for a nature to be what a necessity corresponds to the nature must belong to the entity having it with necessity. The nature must be identical with what has it. Does this not mean that there is an essentially repetitive account? There is essential reliance on necessity but the account is not repetitive. For it is the nature that grounds both the necessity of an entity having a property and the necessity of its being identical with the entity. The necessary identity of the nature with the entity does not, then, lead back to a further component for its explanation. However, the necessity of any other component does lead back to another component, since components other than natures do not force entities to have them. Rather if entities have to have them it is because of the natures of these entities.

But even though there is no repetition through a series, is there not straightforward circularity? The answer is again no. For though the nature can account for necessity only by being itself necessary, it – the nature – is, since it is the source of all the necessity an entity is subject to, the source of this essential necessary identity. Since this is so there is no basis for the charge of circularity.

One more end to tie off. Not only must the nature be necessarily identical with what has it, but also, as in the case of the fine-structure account, there will be a necessary conditional. It will be necessary that if the entity has the nature then it has the property it has by nature. Since, however, the nature grounds this necessity too, the explanation is not repetitive. And it is not circular since, though having this necessity is a condition for the adequacy of the explanation, the nature grounds this necessity and the original one equally well.

I submit then that blending the fine-structure model with the view that natures account for necessity provides a perfectly adequate model for capacities. And it is a model that has the advantage over the stimulus-response model of coping with the problem of necessity. To have a capacity is then to have an element of fine structure, which need not be by nature, and in addition to have by nature the conditional property corresponding to conditional causality. There may well be no end to the complexity of fine structure, but this is not required now by the nature model for capacities and necessity. But if there should be no end to the complexity of fine structure, the necessary connections at any level of the analysis and between any two levels of the analysis will all rest on the nature of the entity with these levels.

Indiana University

ERNAN MCMULLIN

CAPACITIES AND NATURES: AN EXERCISE IN ONTOLOGY

My task in writing a paper to accompany Milton Fisk's complex and provocative one is unusually difficult. His argument is so condensed and at such a high level of abstraction that I could easily devote my entire discussion to a single paragraph of his, to his very first one for example. Instead of doing this, I thought it would be more helpful to say something about the historical background from which his essay derives, and then go on to raise some difficulties regarding his central claim that nature is something over and above part and property. I will end with some suggestions on my own part as to how the categories of *capacity* and *nature* might best be related.

But first it is worth asking: to whom is this paper addressed? Its language and general approach are not those to which readers of the philosophy of science are accustomed. When the author asks whether, for example, "capacities are entities over and above properties and parts", where is one to turn for evidence for or against whatever answer is proposed? What would amount to a proof that capacities are (or are not) entities over and above properties and parts? Will the position adopted on this affect the way in which science would be carried on? Is it perhaps a matter of exploring what is already implicit in the practice of science? Or is the issue a much more general one, prior to the specific methods and theories of science? I think that Dr. Fisk would lean to this last suggestion, but I shall try to suggest at least one reason for taking the first one seriously too.

It is important to have this metaquestion in mind as we proceed. The author's aim is to propose a complex set of distinctions between parts, properties, capacities, conditions, components and natures. Is this to be regarded as a basically stipulative enterprise, an elaborate proposal for one possible consistent usage of a set of terms that are hopelessly vague in ordinary usage? Or is it appropriate to ask for evidence for the system proposed, and if so, of what kind? Can there be compelling reasons for accepting the ontology accompanying a categorial system of this sort?

What happens to the population of our universe if natures are admitted as components of entities over and above parts, properties, and actions, as Dr. Fisk argues they should be?

1. Aristotle

Let us begin with the Aristotle of the *Physics*, and work gradually towards the problem outlined in the paper before us. Aristotle noticed that the explanations commonly given of the regularities perceived in the world around us fall into four different categories, which he labelled form, matter, end and agency. He assumed that the efficacy of the categories in the analysis of physical explanation showed that they are ontologically grounded in distinct though connected aspects of the physical object, specifically in its 'nature', i.e. in the object considered as acting upon, or being acted upon, in explicable ways. The four aspects ('components' in Fisk's sense) taken together constitute nature; no further component is needed. In fact, the components of nature could be reduced to two, form and matter, since agency and end can be understood as rooted in the form-matter composite. The distinction between form and matter itself could be taken in two different ways: as relative to a particular level of explanation, or as absolute. Bronze will be the 'matter' of an explanation of the coming-to-be of a statue. But obviously bronze itself has a form-aspect and can, in other contexts, be further broken down into (understood as) a certain mixture of the four basic physical elements. From the absolute standpoint, however, the matter-aspect of an entity will ultimately have to be without any of the intelligibility of form; this is how Aristotle arrives at his much-disputed notion of a 'primary' matter. Specific regularities of behavior (the 'causal necessities' of Fisk's analysis) will thus be understood in one or other of two ways. If the change is only a partial (qualified, accidental) one, it will be understood by adverting to form and to a specific ('second') matter, provisionally understood as the grounds of the causal regularities. If on the other hand the change is total (unqualified, substantial), one must explain its specificity in terms of the form alone; the matter-factor functions as the guarantor of spatio-temporal continuity, but in a wholly unspecific way.

Aristotle inherited one special problem about the ontological basis of change. The failure of Parmenides' attempt to account for becoming in terms of the very broad categories of Being and Not-Being had led to troublesome paradoxes. Aristotle did not disagree with Parmenides' fundamental assumption that whatever could not be comprehended in the categories of Reason must be regarded as unreal. But he challenged the adequacy of the categorial system proposed by Parmenides, claiming that the distinctions it permitted were simply not enough to take into account some of the most obvious general features of our knowledge of the world. To describe becoming, one needs (Aristotle suggests) three categories: Form, Matter and Privation, where Form replaces Being, Privation replaces Not-Being, and Matter has taken on some of the characteristics of both the Being and the Not-Being of the Parmenidean account. To explicate this last, Aristotle further distinguishes between potentiality and actuality. The 'matter' aspect of a physical entity is the locus of potentiality, i.e. of the ability to become something other than the entity now is. To say of the acorn that it can become an oak, given a 'natural' environment is to say much more than to say simply that it is not-oak. An acorn is also not-mouse, but this does not mean it can become a mouse. So there are certain types of being that an entity is not which are not simply Not-Being in its regard. With regard to the present form-aspect of the entity, they are negation (privation), but with regard to its matter-aspect, there is something there which gives grounds for a direction of becoming which will terminate in one of these other kinds of being.

Now let us draw together some features of this familiar account to see what Aristotle would have to say about Dr. Fisk's problem. Potentiality for Aristotle and capacity for Fisk do not seem to be identical; capacity is not linked with privation as potentiality is. To say that *X* has the potentiality to become *Y* implies, among other things, that it is at present not-*Y*. Whereas to say that *X* has the capacity to *Y* does not apparently imply that it is at present not *Y*-ing. (A lump of salt has the 'potentiality' to dissolve in water only when it is not-dissolved; it has the 'capacity' to dissolve in water whether or not it is not-dissolved at the time at which one is speaking). A potentiality statement has the function of indicating what sorts of thing a particular entity can become (differences of outcome being due to differences of context). Capacity statements (as Dr. Fisk defines them) serve to define the range of actions and responses open to an entity of a particular kind in different contexts, irrespective of which of them is being evinced at the time of speaking. Potentiality responds to the question: what different entity can this entity become?, capacity to the question: what different behaviors can this entity elicit without ceasing to be 'this' entity, i.e. without changing its 'nature'? The contrast is between *becoming* and *behaving*; it involves two somewhat different sorts of 'difference', and different types of predication.

In one case, the form-matter distinction brings out that the same form-term cannot be predicated at the end of the becoming, although the same matter-term can be. (Where the becoming is an unqualified 'substantial' one there will be no *specific* term whereby the 'matter' of the change can be designated.) In the other case, the same classificatory 'nature'-term is predicable no matter whether the capacity is being exercised, or not. The regularities (causal necessities) that are characteristic of becoming are traceable to nature considered both as 'form' and as 'matter' of that particular kind of becoming. If the stress is on capacity, rather, it will not be necessary to resort to the form-matter distinction; capacity can be understood to be rooted in (characteristic of) nature itself, taken as unitary. In short, then, we need the category of potentiality in order to 'understand' how an entity can become something different (i.e. in order to describe adequately the most general traits such an entity would require). Whereas we need the category of capacity in order to 'understand' how the same entity displays a multiplicity of different (though inter-related) modes of behavior at different times. Yet the two categories are evidently very closely linked; in fact, the potentialities of an entity will serve to define its capacities and vice versa.

Since potentiality is rooted in the matter rather than the form-aspect, Aristotle would not regard it as a property in the strict sense. It does not correspond to a predicate, categorically attributed, but rather to a modal predicate. Likewise, Fisk distinguishes capacities from properties, depending on whether a predicate is attributed modally ('*X* can *Y*') or categorically ('*X* is *Y*-ing'). The 'property' here will be an action of some sort, since capacity refers to possible action; for Aristotle, as we have seen, property is, rather, a possible state of being. (And 'potentiality' refers for him to such a state, not to the action whereby the state can be brought about.) Yet since the capacity can be predicated of the entity here and now, why should *it* not be regarded as a property of a special sort? Fisk introduces yet another term for such a property, a 'disposition', and continues to distinguish between properties and capacities, by assuming that the 'properties' of which he is speaking are non-dispositional, i.e. are predicated on the basis of their being exhibited at the time of predication. Thus, although solubility in water can be regarded as a dispositional property of salt, the capacity to dissolve is not a 'property' in the narrower sense in which Fisk uses this term. (The notion of 'exhibiting' on which this property-capacity distinction is based is a very complex one, but it would take us too far afield to explose its ramifications.) Fisk does not regard the attribution of dispositional properties as helpful, and expresses this rather too strongly by saying that 'entities have no dispositional properties' (§ 2). What he means is that such properties do not necessarily have to be invoked; in his view, at least, one can 'just as well' express them in terms of capacities. (Of course, 'just as well' works in both directions; the reader might ask why if a proposition concerning dispositional properties 'can equally well be expressed' in terms of capacities, a proposition concerning capacities could not equally well be expressed in terms of dispositional properties.)

Aristotle speaks of matter and form as distinct 'principles' of things. Later scholastic philosophers used words like 'constituent' and 'component'; Fisk speaks of 'components' in a similar sense. Parts, properties, actions, even natures, are 'components' of an entity, as Fisk uses the term. The 'composition' referred to is not a physical one, involving possible separability of the 'components'. Rather it is between different aspects, considered as the ontological 'grounds' of conceptually irreducible categories. To say that the properties and the capacities of an entity are distinct components (Fisk prefers the term 'distinct' to the term 'different' in this context) is to say that the notion of a property is irreducible to that of a part. When he arrives at his major conclusion, which is that capacities can be understood only by postulating natures as components of entities over and above their parts and properties, what this amounts to is that the notion of a capacity cannot in his view be derived from those of part and property alone.

One term used by Fisk in a rather special sense is 'condition'. Where Aristotle had made substances (entities) the causes of change, Fisk stipulates that 'components' and 'entities' do not play a causal role; it is the 'having of a component' (which he calls a 'condition') rather than the component itself that is to be regarded as a cause. Salt's ionic structure and crystal energy does not cause it to dissolve; rather, it is salt's *posses-*

sion of that structure and energy that causes it to dissolve in certain circumstances. This is a sophisticated point, but one quite crucial (as we shall see) to his argument. A property or a part as such do not explain why something should happen; only the possession of that part or property by an entity with a nature, i.e. the situating of it with other factors that go to make up a natural entity, explains (according to Fisk) a dependable regularity of behavior. This seems quite reasonable. But suppose the entity is specified in terms of parts and their properties considered *as* parts and properties (as a biochemist might specify a DNA molecule, say), do we have to postulate a *further* distinct 'it' which 'possesses' these parts and properties in order to specify the general categories underlying our effort to understand why the entity behaves as it does? Fisk argues that we do. But this is by no means evident, unless the parts and properties are taken as disconnected items having within their definition no hint of causal regularity. But surely this is to import without justification a Humean limitation into the account of components other than 'natures' in order to limit causal necessity to a distinct component called 'nature'. The dubiousness of this move can be brought out by asking what a 'part' would amount to in an explanation if it did not have a causally dependable way of relating with other parts and their properties, i.e. if it did not itself have a 'nature' to begin with. It must be stressed, then, that the adoption of a parts-and-properties ontology does not necessarily exclude the category of nature; my point rather is that this category must not be added as a sort of 'third' to the others. I shall return to this later.

Aristotle saw in a property such as man's rationality the cause (explanation) of man's sense of humor. He did not seek a further explanation for essential properties or for their interconnection in a particular nature, except to suggest that an account in terms of the good (whether of the species or of the cosmos as a whole) in principle underlies all other modes of explanation. The matter-aspect of any particular instance of becoming is that which is 'given', that beyond which one does not press the search for explanation, that whose potentialities can be accepted as the ultimate from which this sort of becoming 'naturally' proceeds. The questioning terminates at it, as also (in a different way) at the form. Thus once one has listed the activities essential to the particular nature under investigation (and this can be done, Aristotle thinks, in a finite way), there does not seem to be much more that one can (or, for that matter, should) do. He expected that the properties of different natures could be interlocked in a hierarchical scheme of genus and differences, lending itself to syllogistic demonstration of the necessity of particular properties. But all such demonstration assumed the self-evidence and ultimacy of such statements of essence as "every man is rational". But how is the necessity of these premises to be assured? The induction-insight to which Aristotle is concluding did not seem to stretch far beyond the over-worked 'man is rational' instance. Could the properties of a composite, such as wood, be explained by those of the elements making it up, and if so, how? The combining of the contraries proper to the sense of touch (hot-cold, dry-moist) to define the elements themselves (water as the cold-moist, etc.) seemed to provide the only instance where one could get beyond the first level of property, and it was by no means obvious what this qualitative reduction of properties amounted to, or how it could be validated.

2. Locke

The growth of 'corpuscularian' views in the 17th century revived the atomist ideal of structural explanation of wholes in terms of parts and their properties. Both parts and properties were to be conceived in terms amenable to a mathematized mechanics. A whole series of thinkers, Bacon, Galileo, Descartes, Hobbes, Boyle, and many others, were led in consequence to distinguish between 'primary' and 'secondary' properties. The basis for this distinction varied; sometimes it was a question of deciding which properties could in principle be reduced to (explained in terms of) others. Or the primary qualities could be defined as those appropriate to mechanical modes of explanation. Or they were the 'objective' properties as opposed to those generated by the subject in sense-knowledge.

It was Locke, perhaps, who best expressed the ontology implicit in the dominant mechanical and corpuscularian philosophies of the century. His starting-point (like that of Hobbes half a century earlier) was the conviction that the only way in which we can conceive bodies to operate to "produce ideas in us ... is manifestly by impulse". But since these bodies:

> may be perceived at a distance by the sight, it is evident some singly imperceptible bodies must come from them to the eyes, and thereby convey to the brain some motion which produces these ideas which we have of them in us.[1]

Thus his hypothesis that bodies are composed of a multitude of imperceptibly small corpuscles rests not upon direct evidence nor on any predictive successes of this model in the science of the day (despite his best efforts, Robert Boyle had never been able to link the corpuscularian theory of matter with any specific chemical reactions), but upon the assumption that all action must ultimately be explicable in the simple terms of mechanical impact: no other kind of action is 'conceivable'. In less than a century, the successes (very limited as they now seem to us) of the mechanics of Galileo and Descartes had sufficed to limit the imagination thus drastically. It is important to notice that the confident new insistence on the all-sufficiency of structural modes of explanation rested not at all on scientific practice but upon a bold (and, as it shortly turned out, unwarranted) extrapolation of the model of action easiest to handle in mathematical-experimental terms to cover all forms of interaction, no matter how apparently irreducible to the meagre categories of impact mechanics.

If the behavior of the corpuscles is to be explicable in terms of percussion alone, then only a small number of properties need be attributed to them: "solidity, extension, figure, motion, or rest, and number" were Locke's choice. These qualities are 'primary' in three rather different senses: they are those that "the mind finds inseparable from every particle of matter," both because division or other alteration of the body can never take them away, and because they are the necessary conditions for impact action; they "really exist in the bodies themselves," whereas all other qualities are:

> no more really in them than sickness or pain is in manna. Take away the sensation of them... and all colors, tastes, odors, and sounds... vanish and cease and are reduced to their causes, i.e. bulk, figure, and motion of parts.[2]

Finally, primary qualities as they exist in the object resemble our ideas of them, whereas "the ideas produced in us by secondary qualities have no resemblance of them at all".

Secondary qualities are thus "nothing in the objects themselves but *powers* to produce various sensations in us by the primary qualities ... of their insensible parts." Locke notes that in consistency one must also allow a third class of qualities, which have exactly the same ontological status as secondary qualities, namely powers of affecting things *other* than the human sense-organs:

The power in fire to produce a new color or consistency in wax or clay by its primary qualities is as much a quality in fire as the power it has to produce in me a new idea or sensation of warmth or burning... by the same primary qualities.

These 'tertiary' qualities differ, however, from the secondary ones in that the action of the former is completely reducible to that of primary qualities. Whereas, "there is no discoverable connection between any secondary quality and those primary qualities which it depends on". In fact, the association of one idea rather than another with a particular secondary quality (i.e. configuration of primary qualities) is ultimately an arbitrary one. The "ideas of the blue color and sweet scent" of the violet are 'annexed' by God to certain impulses brought about by the figures, bulks, motions, of the minute particles of which the violet is composed. But since these qualities "have no similitude" with the ideas of blue and sweet, there is simply no way in which we can discover why a particular idea is brought about in us by a particular corpuscular configuration. On the other hand, the action of the tertiary qualities can in principle be more fully understood, because the powers involved are directed not to the sense-organ but to a body whose reaction can be fully analyzed in mechanical terms:

Did we know the mechanical affections of the particles of rhubarb, hemlock, opium and a man, as a watchmaker does those of a watch... we should be able to tell beforehand that rhubarb will purge, hemlock kill, and opium make a man sleep.[3]

But Locke is convinced that these affections never can be known, and that thus a "science of bodies" is in practice impossible. Such a science would need to reveal necessary connections (co-existence) between the ideas that go to make up real essences; thus to understand gold one would need to see why "a particular sort of yellowness, weight, fusibility, malleability, and solubility in *aqua regia*" coexist in the "unknown substratum" ('substance') with which he had replaced the primary matter of Aristotle. There are two reasons (he thought) why this sort of knowledge is out of reach: first, the minuteness of the constituent corpuscles of bodies prevents us from discovering the particular specifications of their primary qualities; second, even if these were to be known, there is no intuitively certain way (of the kind that a true science would require, according to Locke) of associating particular secondary qualities with specific corpuscular configurations. Thus there is no way of establishing intuitively necessary connections between the secondary qualities themselves. Even though secondary qualities are completely dependent for their reality upon primary qualities, there is in practice no way consequently of arriving at a scientific explanation in a fully reductive sense of the one in terms of the other. A geometrical mechanics of impact is the only discoverable natural science; where structural explanation would be required (as in chemistry and biology) a true science can never be attained.

Locke proposes, therefore, an ontology of corpuscles with their quota of primary qualities. The myriad properties that bodies possess (other than the few primary ones) are treated as powers of acting or being acted upon; the distinctions between these are entirely dependent upon the characteristic primary qualities of the corpuscles composing the two bodies in interaction. Particular essences are regarded as clusters of qualities (ideas). Two quite different sorts of structural explanation are in fact being proposed. In one, the "structure" is a configuration of particles whose interactions with one another are in principle defineable in mechanical terms; in the other, the "structure" is an intuitively connected cluster of qualities inhering in a mysterious substratum. But Locke's own scepticism about the extent to which powers can be explained in terms of primary qualities raises a doubt about the *bona fides* of the reduction that he proposes. Not only is there for the most part "no visible necessary connection" between the simple ideas "whereof our complex ideas of substances are made up," but:

> the ideas that our complex ones of substances are made up of... are those of secondary qualities; which depending all upon the primary qualities of their minute and insensible parts – or if not upon them upon something yet more remote from our comprehension – it is impossible we should know which have a necessary union or inconsistency one with another.[4]

But if this is the case, how can he be so sure that powers *are* reducible to primary qualities? He has to fall back upon his original Cartesian assumption of the inconceivability and therefore the impossibility of interactions other than simple impact ones. If, in fact, he could have brought forward some instances of successful reduction, his argument would have carried much more weight. But not only were there no confirmatory instances available in his day, he gives reasons for supposing that few ever would be.

Fisk describes Locke's ontology as one of parts only "to the exclusion of properties and actions" (§ 3), and goes on to show, without too much difficulty, the inadequacy of such a strong reduction. Locke, he asserts, "defined the so-called primary qualities in terms of parts". Several comments are in order here. First, the 'parts' to which Locke ordinarily refers (as in the last quotation above) are the "minute and insensible" particles of which he supposed all bodies to be composed. These particles possess the primary qualities of extension, solidity, shape, motion or rest, and number, as their essential properties, the properties which make possible the application to them of mechanical modes of analysis. These properties are ultimately rooted in the substance-substratum. Solidity, for instance, or motion are quite clearly *properties* of the parts of which bodies are composed and Locke always describes them as such. It is in this way he arrives at his characteristic notion of a substance-substratum, an "unknown support of those qualities we find existing" in response to his question: "What is it that solidity and extension inhere in?" When Fisk dismisses "ontologies of parts," one is tempted to ask what such an ontology could possibly amount to, were the parts not to have properties also, i.e. were it not to be possible to predicate something of them.

The misunderstanding here could come from either of two sources. Several of the primary qualities are reducible to geometrical terms; one might want to say that extension is 'part' of a body rather than a property of one. Yet this does not seem to be correct, unless one is using 'part' in a very broad sense to include what would ordinarily be regarded as property, in which case the original thesis becomes vacuous. More importantly, there is a well-known ambiguity in Locke's notion of *idea:*

It will be convenient to distinguish ideas as they are ideas or perceptions in our minds and as they are modifications of matter in the bodies that cause such perceptions in us.[5]

Primary qualities are 'ideas', and substances are understood to be made up of 'combinations' of such ideas. If one presses the phenomenalist implication of Locke's choice of terms, all qualities (both primary and secondary) can be regarded as ideas which form 'parts' of the cluster of 'ideas' whose co-existence in a substratum go to make up an essence. In this sense, they *would* be 'parts'. But this part-whole relationship is one of idea to complex idea; it exists in the mind where the combination occurs. It does not warrant our holding that Locke's ontology is one of 'parts' only. The confusion between a quality and the idea of a quality, between a substance and the idea of a substance, admittedly confuses the ontological implications of what Locke is professing. But insofar as an ontology of the causes of ideas can be discerned, it is assuredly one in which the primary qualities figure as properties, not as parts. Furthermore, a question must be raised about the sense in which the "minute parts" *are* really taken as parts by Locke and by reductionists who follow his general approach. The component particles are for him wholes, fully defined in their own right; though they make up larger wholes, they are really not dependent upon these larger wholes in any significant way for their mode of action.

3. Fisk

Fisk goes on to argue that even the admission of properties as well as parts (his "fine-structure model") will not suffice to provide an ontology sufficiently rich to account for the causal necessities corresponding to capacities. His argument is an intricate one, starting from the observation that "the difficulty is that by basing a capacity on an element of fine structure, whether it be a part or a property, we seem to be requiring that that element itself have a capacity" (§ 3), and thus capacities would not be eliminated from the basic furniture of the world after all. His own proposal is that natures must be included among the basic 'components' of entities if capacities are to be accounted for. Natures are, in fact, for him by *definition* "those components of entities to which propositions that are necessary correspond". The reason is that nature is the only component that belongs "with necessity to the entity having it". Whereas "a property, or any component other than a nature, is not such that because of what it is, it is necessarily had by entities. Rather because of the nature of an entity it has certain properties necessarily." A property such as being ionic is not such that any entity has it necessarily: "if it were, all entities would have it." But it is the *nature* of some entities to be ionic. Thus it is "the nature of a crystal of salt to be ionic, but the property ionic is not the nature of the crystal".

There are several questionable moves here. The property ionic *is* part of the nature of salt. Furthermore, it is such that is *necessarily* had, not by all entities obviously, but by salt (among other things). We are no more led to saying that *all* entities must possess a property if it is had necessarily than that all entities must possess a particular nature if it is had necessarily. It is a particular *sort* of (named) entity that has a particular nature necessarily, and the same entity has a set of properties equally necessarily. The necessity here is primarily *de dicto*: to be salt (rather than sugar) necessarily implies having the nature of salt (i.e. this is what it *means* to be salt). Likewise to be salt necessarily implies having a set of properties, among them a particular ionic constitution. The necessity here is that of: "If x is salt, then necessarily x has the nature of salt (and the property, ionic)." Fisk holds that "for an entity to have a component necessarily is for it to have that component by nature". Particular sorts of entities possess certain properties necessarily; if this means that they have them 'by nature', then one must conclude that properties may be had by nature just as much as natures may be. In fact, one might contend that it is properties that primarily *are* had 'by nature'; to say that 'natures' are had 'by nature' has the necessity of a tautology at best.

To say of this latter sort of necessity that an entity is 'subject to it' (§ 4) does not seem helpful. This is the necessity of identity, the necessity with which A is A. It is not the necessity of the causal connections involved in the account of capacities. This is brought out by Fisk's use of such phrases as "the necessity of necessity", or the claim that "the nature must belong to the entity having it with necessity. The nature must be identical with what has it." Of course it must. But is this 'must' the problematic 'must' of the capacity statement: "salt must dissolve when put in water"? Fisk argues that the merit of adopting natures as separate components to account for necessity is that "the necessary identity of the nature with the identity does not lead back to a further component for explanation," whereas "components other than natures do not force entities to have them" (§ 4). By now we can readily remark to this that parts and properties are such that they *do* 'force' entities *of specific kinds* to have them, just as it is only entities of specific kinds that have specific natures necessarily. The notion of explanation that demands a 'necessary identity' of nature with (specific) entity as the only way of halting regress would appear to be of a logical, not an ontological sort.

It is his notion of explanation, then, that leads Fisk to introduce natures as extra components of entities. And this notion, in turn, derives from what may be the crucial premiss in his entire argument: "it is not components that are causes but their corresponding conditions", because it is the condition, i.e. the *having* of a component, and not the component itself that plays a causal role (§ 3, § 1). 'Causes' are what make something do something. It is the possession of ionic structure that makes salt soluble in water, not the ionic structure itself. "As a consequence, it is the condition, not the component, that has the capacity to make the entity exercise its capacity" (§ 3). But now since the 'condition', which is a 'possessing', can be attributed only to the entity as a whole (not to a component such as part or property), it follows that capacity *cannot* be accounted for by means of parts and properties, but only by the condition of possessing that component, which in turn is attributed back to the entity considered as the ground of capacity, i.e. to 'nature'. What this complicated sequence amounts to is that Fisk's stipulation that a 'condition' and not a 'component' such as part or property can be a 'cause' means that by definition, part and property cannot ultimately account for capacity in an acceptably 'causal' way; only 'nature' can – in Fisk's sense of 'cause' and its correlative 'accounting for', that is.

But must one accept this definition of 'cause'? Why should a component, *qua component*, not equally well be taken as a 'cause', i.e. as that which accounts for whatever is under investigation? What 'accounts for' salt's solubility in this sense of that slippery phrase, would be its ionic structure, not of course taken in isolation, but considered within the complex of properties and parts making up salt. Thus capacity can be accounted for by parts and properties, without introducing a separate item 'nature' defined by the so-called 'conditions' characteristic of the entity. Instead, then, of postulating 'nature' as a distinct ground of all the conditions governing an entity, i.e. the entity considered as possessing all the various components attributable to it and thus as locus of the causal necessity associated with capacity, it will suffice to postulate parts and properties (including dispositional properties), considered as 'causes' and thus as the grounds of the causal necessities associated with capacities. To put this in another way, the parts with their properties are now conceived as having a 'nature' as parts of a given whole, of behaving in the context of this complex necessarily in certain regular ways. 'Nature' does not, therefore, have to be introduced as an additional component; it is implicit in the way in which we talk of 'parts' and 'properties' in the context of structural explanation.

4. Structural explanation

It seems, then, that we are back with something like the Lockean model.

But there are some quite crucial differences. First, two centuries of successful structural explanation in chemistry and biology, and even in physics itself, have confounded Locke's prophecies of "incurable ignorance" regarding the "minute parts" of macroscopic bodies. What Locke overlooked was what has come to be called hypothetico-deductive validation; he demanded intuitive certainties, as Descartes had done, in relating the simple ideas or ascribing particular extensions, shapes, etc., to the constituent corpuscles of a particular kind of entity. The successful practice of several centuries has shown, however, that a tentative hypothetical starting-point, an atomic model like the Bohr model, for example, can be indirectly warranted by progressive successful testing. Thus we *can* come to know a good deal about the properties of the "minute parts" of which the familiar bodies of everyday experience are composed, though in a sense of 'know' that Locke would not have countenanced – mainly because he never anticipated it. Structural explanation, i.e. explanation in terms of parts and their properties, in fact, a long time ago proved to be one of the two principal systematic modes of understanding the physical world (the other being genetic explanation, a historical account of how the structures came to be the way they are).

It is, thus, to the practice of science itself that we turn in order to decide upon the adequacy (in the context in which they were originally proposed, i.e. that of physics) of the Aristotelian categories of form and matter or the Lockean ones of corpuscle and primary quality, considered as specifications of the basic ontology of the world. It would be simply inadmissible to exclude the history of science as a major source of evidence, to suppose that we could restrict the issue of ontology to the commonalities of pre-scientific experience. Locke did not do this, but relied heavily – too heavily – on the mechanics of his day in choosing his categories; even Aristotle was influenced by his discoveries in biology and in formal logic. Thus our warrant for proposing parts and their properties as our basic framework rather than, say, clusters of qualities inhering in substances, is not that the inadequacy of the latter can somehow be demonstrated on the basis of ordinary pre-scientific experience. There is no way of showing that such a world could not exist; there does not even seem to be a way of showing that the features of such a world would conflict with those of the world we experience at the commonsense macroscopic level. It is only the continued success of structural explanation (and the failure of the substance-quality-model) over several centuries of scientific progress that indicates where our ontological bets should lie. Thus this is not simply a question of metaphysics nor of philosophy of nature (in the Aristotelian sense); it pertains also to the philosophy of science, specifically to the philosophic consideration of one of the most general methodological traits of science in its historical dimension.

There is one further important respect in which Locke's proposal has proved unacceptable. His theory of primary quality took for granted that one could exhaustively specify these qualities, for once and for all, in the terms appropriate to an already-known mechanics. His great contemporary, Newton, struggled with this same assumption; he never did succeed in reconciling his own most far-reaching theory, that of gravitation, with it. Today we would want to insist that the 'primary qualities' of current mechanics are provisional both in number and in definition; it is this tentativeness, the character of 'model' rather than intuitively clear 'idea', that permits science to progress, to meet anomaly when it arises. The simplistic reductionism of Democritus, Descartes, Hobbes, Locke, does not work because an 'ultimate' mechanics cannot be proposed – nor could it work. Even more important, the 'primary qualities' of mechanics at any given time are not the intuitively 'given' ones of extension or impact that satisfied the corpuscular philosophers. They are sophisticated constructs like energy or magnetic field strength, with some intuitive continuity perhaps with the properties of ordinary experience, but reshaped over and over again to fit the ever more remote data of experiment.

One feature of these 'primary' properties recalls us to the problem from which we began. They are not attributable simply to (nor discoverable simply by the scrutiny of) 'parts', as extension was attributable to the 'minute parts' in the eyes of Descartes or Locke. It was only by studying the behavior of the hydrogen atom in magnetic fields, or when emitting light, that the notion of energy levels – or even the very sophisticated quantum concept of energy itself – came to be attributed to the postulated parts of which the hydrogen atom is composed. What I am proposing, therefore, is not a reductionist model of explanation, strictly speaking. Even though it explains in terms of part and property, the 'parts' and 'properties' it calls on are not discoverable independently of their occurrence in particular complexes. No amount of study of free electrons, no theory of free electrons based on such study only, has been able to tell us what would happen when electrons combine in shells around a nucleus in stable atomic configurations. What we have, then, is an explanation of the part in terms of the whole at least as much as the reverse. The ontology implicit in natural science as we know it, therefore, is one of part and property considered precisely *as* parts and properties of complex wholes. It is only in the context of these wholes that most of the capacities of the parts are actualized.

What, then, are we to make of capacities in this ontology? Are they something over and above? Are they to be eliminated? One of the most debatable features of Fisk's paper is his assumption that capacities may not be allowed among the basic categories of an ontology, but must somehow be accounted for in the non-modal terms of nature, part and (non-dispositional) property. But why must this be so? Why should we not have parts, properties *and* capacities? Or, more economically, why can we not propose parts and properties, both non-dispositional and dispositional? Fisk assumes that capacity statements must be reducible to statements about structures categorically possessed by an entity of a given nature. But why should one not argue the reverse: that all categorical property-statements can be grounded in a set of modal capacity-statements? Locke made his secondary qualities into 'powers' or capacities. But, as Berkeley was later to point out, he ought to have done the same for his primary qualities. Solidity can surely be regarded as a power, a property whose 'exercise' is elicited only in special contexts. It was the intuitively categorical character of the geometrical qualities of extension and shape that presumably led Locke, as it had everyone else of his time, to suppose that primary qualities were somehow fully exhibited, fully predicable, at every instant; there was no latency about them, no dependence for their evocation upon context, no mysterious conditional that might never be realized.

Apart from the difficulties that could have been raised against this view even in Locke's own time, we must note a series of even more troublesome ones of more recent origin. The concepts of contemporary science are incurably dispositional. To put this in a rather stronger way, it is not at all clear that the distinction between dispositional and non-dispositional, between capacity and property (in Fisk's sense) can be satisfactorily drawn in the context of contemporary science. Fisk notes that inertia is a "power of resisting change of state" [6]; energy, mass, momentum, force, potential, stress, resistance, can all likewise be explicated in 'power' terms. (If one prefers, they can still, of course, be expressed as operationally specified, and continuously 'possessed', properties.) Even the notion of part no longer enjoys the clearly 'categorical' character it did in the seventeenth century. To say that the orbital electron is 'part' of the hydrogen atom is not like saying that a mainspring is part of a watch. There is no way of defining the position or momentum of an intra-atomic particle; in fact, even the notion of 'particle' itself is only analogously predicable, since entities like electrons and nucleons do not obey the classical statistics that well-defined 'parts' should. To say that the orbital electron occupies a particular energy level must be interpreted as a modal statement about "what would happen if ...".

The attempt, therefore, to construct a 'categorical' ontology of parts and non-dispositional properties, and to express all statements about capacities in terms of these, finds very little support in the practice of contemporary physical explanation. [7] Fisk begins his paper with the words:

> The last two decades have witnessed a basic change in the philosopher's working model of capacities. The change has been from a stimulus-response model to a fine-structure model.

And he goes on to show the inadequacy of the 'black-box' S–R model of explanation current among the positivists and logical empiricists of 1920–1950. But the main reason to reject this model is surely that it at no time corresponded to what those actually engaged in the search for physical explanations were *doing*. That it could be said to have been "the philosopher's working model" of two decades ago is sad testimony indeed to the divorce between philosophy and science characteristic of that period, and the consequent *a priorism* of philosophies of science that took their inspiration even more directly from the constructive possibilities of formal logic than did the physics of Aristotle.

To sum up, then, instead of an ontology of parts, properties, and natures, in which capacities are regarded as secondary and reducible precisely as were Locke's "powers", we are proposing an ontology of parts and properties, in which the dimensions of capacity and nature are already contained in the notions of part and property. The causal necessities are rooted in the natures of the parts and properties themselves, regarded *as* parts and properties. At any given time, there is a limit to the extent to which the behavior of physical entities can be explained in terms of parts and properties and these in terms of yet further substructures. There is thus a necessity in a sense unaccounted for at the end of every physical explanation. This is the modern analogue of the matter-aspect of explanation in Aristotle's physics, the bronze whose own properties are not up for question in the explanation of how the statue comes to be. The way is always open, therefore, to the postulation of yet finer structures in the light of new evidence, and thus the locating of the unexplained necessity at a yet lower level.

But when we have shown how the operations of a complex 'necessarily' follow from the properties of its parts, we still may only have the necessity of logical implication. The ontological 'necessities' governing the behavior of complex entities are no more 'explained' by showing that this behavior can be inferred from a knowledge of parts and properties, than is the necessity evinced in the regularities of behavior of the parts themselves 'unexplained' if not further reduced. Fisk is right in appealing to 'nature' here, but it is nature as exhibited in the dependable regularities of causal succession at all levels equally, that of complex as well as that of part. It is these regularities of nature that underlie the logical relations of necessary implication that characterize the postulational structure of physical explanation. If asked whether we need to postulate natures over and above parts and properties in order to account for these necessary connections, we may appeal to the practice of physical explanation. No such components seem to occur either in science or in pre-scientific explanation. Nor are they required, provided that the other explanatory elements called upon (part and property) are regarded from the beginning as 'natural', as interacting 'by nature' in a regular and dependable way within complexes, in the context of which alone can their capacities be fully exercised.

University of Notre Dame

NOTES

[1] *Essay Concerning Human Understanding*, Book I, Chapter 8. The quotations in the next few paragraphs come from the same chapter, except where otherwise noted.
[2] Book IV, Chapter 3.
[3] *Loc. cit.*

[4] *Loc. cit.*

[5] Book I, Chapter 8.

[6] He traces it further back, via the mass-energy relations of Special Relativity, to the notion of potential energy. This is to take 'resistance' a little too literally, just as Newton did in his notion of a *vis insita*. Potential energy is not involved in the explication of the concept of inertia; it *is* involved in the sort of resistance Locke characterized as the quality of solidity.

[7] The attempts of John Wheeler and others to reformulate contemporary mechanics in purely geometric terms (masses for instance appear as "wormholes" in space) might be regarded as an attempt to return to a Cartesian-type mechanics comfortably free of overtly dispositional concepts. But, of course, the "geometry" of their geometrodynamics is by no means that of Euclid and Descartes; it involves dynamical terms that would be impossible to situate within the old dichotomies of power and primary property. See the Epilogue to *The Concept of Matter in Modern Thought* (ed. Ernan McMullin, University of Notre Dame Press, revised edition, 1971), especially § 6: 'Matter and space'.

BRUCE AUNE

FISK ON CAPACITIES AND NATURES

Fisk begins his discussion by asserting, that the FS model attempts to complete the SR model by accounting for the necessity with which, according to the SR model, a capacity will be exercised when certain operations are performed. He argues that the FS model ultimately fails in this task, and he seeks to improve upon the FS approach by grounding the relevant necessity in an entity's nature – a component that defenders of the FS model ignore or misinterpret. For the most part my comments will concern the general strategy of his paper rather than details of the arguments he offers.

Since the major focus of Fisk's paper is on the necessity with which, according to his description of the SR model, a capacity is exercised under certain conditions, it is worth emphasizing that this idea is pretty implausible. To have a capacity to do something is only to be capable of doing it. But although doing what one is capable of doing may be granted to result from some cause, it is far from clear that the cause can always (or even generally) be specified in 'operational' terms. Unlike an owl or cat, I am capable of laughing; yet it is extremely doubtful whether there is a specific condition describable in operational terms that would *necessitate* my laughter.

The significance of this point is at least twofold. First, the SR model as described by Fisk has at best a very limited applicability; an operationalist can't expect to discover as many (physically) necessary conditionals as he might believe. Second, a basic motive for adopting the FS model is not to account for necessary SR conditionals but to express one's belief that there are specific conditions (of fine structure) which account for the fact that capacities are regularly exercised by certain *kinds* of things and materials under fairly specific conditions – these conditions often being most reliably identified by reference to fine structure. Thus, to make sure that the liquid in which you are trying to dissolve a certain material is a good sample of the kind of liquid you want, it will often be wise to have some of it analyzed by a competent chemist.

Boston Studies in the Philosophy of Science, VIII.

I have other reasons for doubting that proponents of the FS model have the aim Fisk credits them with – that is, the aim of accounting for the necessity of SR conditionals. For one thing, some of the best known defenders of the model do not seem to have this aim. Quine, for example, accepts the FS model, but his aim is apparently to avoid familiar problems with subjunctive (particularly, counterfactual) conditionals.[1] If I understand him correctly, he would not grant that the SR model affirms a necessity that deserves to be accounted for. For another thing, the FS model does not specify a property of fine structure that even appears to account for an SR necessity. It merely asserts that there is some property or other of fine structure that, when known, would account for the truth of an SR conditional. And this is no explanation; at most it is an expression of the belief that an appropriate explanation can be found – when we learn more about fine structure.

All this is not to deny an important point that Fisk in effect emphasizes – namely, that the FS approach is patently inadequate as a general treatment of capacities. Few philosophers would deny that the world is built up from fundamental entities, though they might well deny that we are currently able to identify them. But these fundamental entities, which as fundamental lack fine structure, have capacities as well as dispositions. The most the FS approach can do, therefore, is to point toward a FS explanation of the capacities of nonfundamental entities. It can have nothing useful to say about the capacities of fundamental entities and therefore nothing useful to say about capacities generally.

Although (if I am right) proponents of the FS model are not specifically concerned to account for the necessity of SR conditionals, they certainly believe that at least the regularities affirmed by some of these conditionals can in principle be explained by reference to the fine structure of the entities concerned. Since, however, an entity's fine structure is ultimately to be understood in terms of the lawful behavior of fundamental entities, proponents of the model cannot hope for more than an explanation of one regularity by reference to other, more fundamental regularities. Does this not show that they cannot ultimately account even for the regularities affirmed by the SR model? The answer, I suppose, is 'Yes', but then it is far from clear that such ultimate explanations are even possible. To parody Wittgenstein, explanations have to come to an end somewhere. As I see it, the fundamental laws about the world are brute facts; if known, they would be unexplained explainers. It might be irrational ever to stop seeking explanations, however fine-grained our picture of the world may appear. But we cannot expect to have explanations for the basic laws concerning basic realities.

Fisk, of course, believes that reference to a thing's nature may bring explanations to an end. His views on this matter are hard to evaluate because they are hard to understand. He says that 'As necessary ϕ' means 'As ϕ by nature,' but it is not clear that 'ϕ-ing by nature' means anything more than exhibiting a regularity distinctive of the kind of thing in question. If this is so, Fisk's account would not really be an improvement over the standard empiricist alternative. To convince us that he has succeeded in accounting for the necessity of a SR conditional in a way that others have not, Fisk will have to take further steps in elucidating his conception of a nature.

I should add in this connection that Fisk's treatment of causation is as dark as his treatment of a nature. The darkness is introduced by his remarks on free actions. "It is appropriate," he says, "to speak of a cause of a certain condition when the cause only explains the obtaining of that condition *when* it obtains." This claim is not exact, since a cause could hardly explain the obtaining of a condition when it does not obtain. His idea, however, is that if *B* occurs in certain circumstances *C*, *A* may be its cause even though the presence of *A* in *C* would not insure the presence of *B*. That is, if *B* does obtain in *C*, then *A* is its cause; but the presence of the cause *A* in *C* need not bring about *B*. But this is what is so obscure: How could *A* be the cause of *B* if *A* is not sufficient to bring about *B* when appropriate circumstances obtain? What does the word 'cause' mean in his account? This needs clarification.

A related point needing clarification arises from Fisk's description of the necessity involved in a FS treatment of capacities. Fisk formulates this necessity with a connective of relevant implication as follows:

$$(1)\ \Box(\phi a \rightarrow [\theta a \text{ \& the circumstances obtain.} \rightarrow \phi a])$$

But this formula is equivalent to the following (he says) in a system of relevant implication:

$$(2)\ \Box(\theta a \text{ \& the conditions obtain.} \rightarrow [\phi a \rightarrow \phi a]).$$

Although (2) would hold necessarily if it were a material conditional,

Fisk says that the presence of the special connective allows us to view it as "the significant claim that the self-implication property is something an entity has due to the presence of the causal component and the circumstances."

I cannot see that this claim is significant. Could an entity fail to have the self-implication property? The answer is 'No'; the formula '$(\phi a \rightarrow \phi a)$' is a logical truth. Of course, the statement (2) is not a theorem in a system of relevant implication, but this is only because its consequent is not considered a logically relevant consequence of what is affirmed in its antecedent. Since '$(\phi a \rightarrow \phi a)$' necessarily holds of any object *a*, it seems erroneous to say that its truth might *depend* on the presence of some causal component in certain circumstances. It seems equally erroneous to say that '$(\phi a \rightarrow \phi a)$' is a *relevant* consequence of the stated premiss; if the inference of '$(\phi a \rightarrow \phi a)$' from that premiss is not warranted on logical grounds, it doesn't seem warranted at all.

One of Fisk's chief reasons for introducing his peculiar notion of causality into his analysis of capacities was to leave room for human freedom. Since it seemed to him 'objectionably *ad hoc*' to distinguish two kinds of capacities, one for free agents and one for causal contexts, he offered a single analysis that appeared generally applicable. It seems to me, however, that his analysis requires some supplementation if it is to fit cases of freely exercised capacities. The difficulty is that his analysis requires only that some causal component or other be responsible for exercising the capacity in certain circumstances. This does not seem sufficient in the case of freely exercised capacities, for some arcane neurological component might trigger off the appropriate behavior even though the agent had no intention whatever of producing that behavior. If I found myself laughing due to some obscure neurological condition, I might strongly object to the idea that I was freely exercising my capacity to laugh.

At the beginning of my comments I expressed some doubts about the plausibility of the SR model as described by Fisk. I want to end my comments by offering some additional remarks on the necessities this model involves.

Fisk observes that the conditionals he has in mind are expressed in the subjunctive mood. This is obviously true of what philosophers call 'dispositions'; to say that something is water-soluble is to say that it would dissolve if it were put in water. On the first page of his paper Fisk says that the conditional involved in a capacity statement must be necessary because the latter warrants a counterfactual conditional. His reasoning here is mistaken.

Many counterfactual conditionals express mere contingencies. If I have reason to believe that Jones has coins in his pocket, I might say to him: "If your hand were in your pocket, you would be touching coins." Clearly, this counterfactual does not express a necessity; the claim it expresses is as contingent as that expressed by "If you had more money, you would (probably) be happier." The fact is, counterfactual conditionals may express a very wide range of claims, from mere conditional possibilities or weak probabilities to firm contingencies and various kinds of necessities.

I have already said that necessary SR conditionals are pretty hard to come by at the relatively gross level of ordinary observation. But whether they are rare or even plentiful, the task of explaining why things satisfy them does not seem to differ appreciably from that of explaining why things conform to merely statistical regularities. When we affirm a physically necessary conditional, we express our confidence (as Hume would say) that no exceptions can be expected to occur. This confidence may have no basis in a belief about fine structure. We may indeed succeed in explaining the necessity by reference to fine structure, but this merely amounts to explaining (I believe) why *no* exceptions to the regularity can reasonably be anticipated. This task is not different in principle from that of offering a fine-structure explanation of why certain things satisfy a purely statistical hypothesis. If I am right about this, there is no special problem in accounting for physical necessities. The modality of a causal conditional is merely a reflection of our attitude toward the regularity the conditional affirms.[2]

University of Massachusetts, Amherst

NOTES

[1] W. V. O. Quine, *Word and Object*, Cambridge, Mass. 1960, pp. 222–5.

[2] A similar approach to causal conditionals can be found in W. Sellars, 'Counterfactuals, Dispositionals, and the Causal Modalities', in *Minnesota Studies in the Philosophy of Science*, Volume II (ed. by H. Feigl *et al.*) Minneapolis 1958, pp. 225–308.

SYMPOSIUM:

HISTORY OF SCIENCE AND ITS RATIONAL RECONSTRUCTION

IMRE LAKATOS

HISTORY OF SCIENCE AND ITS RATIONAL RECONSTRUCTIONS*

TABLE OF CONTENTS

INTRODUCTION

"Philosophy of science without history of science is empty; history of science without philosophy of science is blind". Taking its cue from this paraphrase of Kant's famous dictum, this paper intends to explain *how* the historiography of science should learn from the philosophy of science and *vice versa*. It will be argued that (a) philosophy of science provides normative methodologies in terms of which the historian reconstructs 'internal history' and thereby provides a rational explanation of the growth of objective knowledge; (b) two competing methodologies can be evaluated with the help of (normatively interpreted) history; (c) any rational reconstruction of history needs to be supplemented by an empirical (socio-psychological) 'external history'.

The vital demarcation between normative-internal and empirical-external is different for each methodology. Jointly, internal and external historiographical theories determine to a very large extent the choice of

* The notes are to be found on pp. 122–34. It is to be regretted that they could not be printed at the foot of each page, because they form an integral part of the paper (Ed.).

Boston Studies in the Philosophy of Science, VIII.

problems for the historian. But some of external history's most crucial problems can be formulated only in terms of one's methodology; thus internal history, so defined, is primary, and external history only secondary. Indeed, in view of the autonomy of internal (but not of external) history, external history is irrelevant for the understanding of science.[1]

1. RIVAL METHODOLOGIES OF SCIENCE; RATIONAL RECONSTRUCTIONS AS GUIDES TO HISTORY

There are several methodologies afloat in contemporary philosophy of science; but they are all very different from what used to be understood by 'methodology' in the seventeenth or even eighteenth century. Then it was hoped that methodology would provide scientists with a mechanical book of rules for solving problems. This hope has now been given up: modern methodologies or 'logics of discovery' consist merely of a set of (possibly not even tightly knit, let alone mechanical) rules for the *appraisal* of ready, articulated theories.[2] Often these rules, or systems of appraisal, also serve as 'theories of scientific rationality', 'demarcation criteria' or 'definitions of science'.[3] Outside the legislative domain of these normative rules there is, of course, an empirical psychology and sociology of discovery.

I shall now sketch four different 'logics of discovery'. Each will be characterised by rules governing the (scientific) *acceptance* and *rejection* of theories or research programmes.[4] These rules have a double function. First, they function as *a code of scientific honesty* whose violation is intolerable; secondly, as hard cores of *(normative) historiographical research programmes*. It is their second function on which I should like to concentrate.

A. *Inductivism*

One of the most influential methodologies of science has been inductivism. According to inductivism only those propositions can be accepted into the body of science which either describe hard facts or are infallible inductive generalisations from them.[5] When the inductivist *accepts* a scientific proposition, he accepts it as provenly true; he *rejects* it if it is not. His scientific rigour is strict: a proposition must be either proven from facts, or – deductively or inductively – derived from other propositions already proven.

Each methodology has its specific epistemological and logical problems. For example, inductivism has to establish with certainty the truth of 'factual' ('basic') propositions and the validity of inductive inferences. Some philosophers get so preoccupied with their epistemological and logical problems that they never get to the point of becoming interested in actual history; if actual history does not fit their standards they may even have the temerity to propose that we start the whole business of science anew. Some others take some crude solution of these logical and epistemological problems for granted and devote themselves to a rational reconstruction of history without being aware of the logico-epistemological weakness (or, even, untenability) of their methodology.[6]

Inductivist criticism is primarily sceptical: it consists in showing that a proposition is unproven, that is, pseudoscientific, rather than in showing that it is false.[7] When the inductivist historian writes the *prehistory* of a scientific discipline, he may draw heavily upon such criticisms. And he often explains the early dark age – when people were engrossed by 'unproven ideas' – with the help of some 'external', explanation, like the socio-psychological theory of the retarding influence of the Catholic Church.

The inductivist historian recognizes only two sorts of *genuine scientific discoveries*: *hard factual propositions* and inductive *generalisations*. These and only these constitute the backbone of his *internal history*. When writing history, he looks out for them – finding them is quite a problem. Only when he finds them, can he start the construction of his beautiful pyramids. Revolutions consist in unmasking [irrational] errors which then are exiled from the history of science into the history of pseudoscience, into the history of mere beliefs: genuine scientific progress starts with the latest scientific revolution in any given field.

Each internal historiography has its characteristic victorious paradigms.[8] The main paradigms of inductivist historiography were Kepler's generalisations from Tycho Brahe's careful observations; Newton's discovery of his law of gravitation by, in turn, inductively generalising Kepler's 'phenomena' of planetary motion; and Ampère's discovery of his law of electrodynamics by inductively generalising his observations of electric currents. Modern chemistry too is taken by some inductivists as having really started with Lavoisier's experiments and his 'true explanations' of them.

But the inductivist historian cannot offer a *rational* 'internal' explanation for *why* certain facts rather than others were selected in the first instance. For him this is a *non-rational, empirical, external* problem. Inductivism as an 'internal' theory of rationality is compatible with many different supplementary empirical or external theories of problem-choice. It is, for instance, compatible with the vulgar-Marxist view that problem-choice is determined by social needs;[9] indeed, some vulgar-Marxists identify major phases in history of science with the major phases of economic development.[10] But choice of facts need not be determined by social factors; it may be determined by extra-scientific intellectual influences. And inductivism is equally compatible with the 'external' theory that the choice of problems is primarily determined by inborn, or by arbitrarily chosen (or traditional) theoretical (or 'metaphysical') frameworks.

There is a radical brand of inductivism which condemns all external influences, whether intellectual, psychological or sociological, as creating impermissible bias: radical inductivists allow only a [random] selection by the empty mind. Radical inductivism is, in turn, a special kind of *radical internalism*. According to the latter once one establishes the existence of some external influence on the acceptance of a scientific theory (or factual proposition) one must withdraw one's acceptance: proof of external influence means invalidation:[11] but since external influences always exist, radical internalism is utopian, and, as a theory of rationality, self-destructive.[12]

When the radical inductivist historian faces the problem of why some great scientists thought highly of metaphysics and, indeed, why they thought that their discoveries were great for reasons which, in the light of inductivism, look very odd, he will refer these problems of 'false consciousness' to psychopathology, that is, to external history.

B. *Conventionalism*

Conventionalism allows for the building of any system of pigeon holes which organises facts into some coherent whole. The conventionalist decides to keep the centre of such a pigeonhole system intact as long as possible: when difficulties arise through an invasion of anomalies, he only changes and complicates the peripheral arrangements. But the conventionalist does not regard any pigeonhole system as provenly true, but only as 'true by convention' (or possibly even as neither true nor false). In *revolutionary* brands of conventionalism one does not have to adhere forever to a given pigeonhole system: one may abandon it if it becomes unbearably clumsy and if a simpler one is offered to replace it.[13] This version of conventionalism is epistemologically, and especially logically, much simpler than inductivism: it is in no need of valid inductive inferences. Genuine *progress* of science is cumulative and takes place on the ground level of 'proven' facts;[14] the *changes* on the theoretical level are merely instrumental. Theoretical 'progress' is only in convenience ('simplicity'), and not in truth-content.[15] One may, of course, introduce revolutionary conventionalism also at the level of 'factual' propositions, in which case one would accept 'factual' propositions by decision rather than by experimental 'proofs'. But then, if the conventionalist is to retain the idea that the growth of 'factual' science has anything to do with objective, factual truth, he must devise some metaphysical principle which he then has to superimpose on his rules for the game of science.[16] If he does not, he cannot escape scepticism or, at least, some radical form of instrumentalism.

(It is important to clarify the *relation between conventionalism and instrumentalism.* Conventionalism rests on the recognition that false assumptions may have true consequences; therefore false theories may have great predictive power. Conventionalists had to face the problem of comparing rival false theories. Most of them conflated truth with its signs and found themselves holding some version of the pragmatic theory of truth. It was Popper's theory of truth-content, verisimilitude and corroboration which finally laid down the basis of a philosophically flawless version of conventionalism. On the other hand some conventionalists did not have sufficient logical education to realise that some propositions may be true whilst being unproven; and others false whilst having true consequences, and also some which are both false and approximately true. These people opted for 'instrumentalism': they came to regard theories as neither true nor false but merely as 'instruments' for prediction. Conventionalism, as here defined, is a philosophically sound position; instrumentalism is a degenerate version of it, based on a mere philosophical muddle caused by lack of elementary logical competence.)

Revolutionary conventionalism was born as the Bergsonians' philosophy of science: free will and creativity were the slogans. The code of scientific honour of the conventionalist is less rigorous than that of the inductivist: it puts no ban on unproven speculation, and allows a pigeonhole system to be built around *any* fancy idea. Moreover, conventionalism does not brand discarded systems as unscientific: the conventionalist sees much more of the actual history of science as rational ('internal') than does the inductivist.

For the conventionalist historian, major discoveries are primarily inventions of new and simpler pigeonhole systems. Therefore he constantly compares for simplicity: the complications of pigeonhole systems and their revolutionary replacement by simpler ones constitute the backbone of his internal history.

The paradigmatic case of a scientific revolution for the conventionalist has been the Copernican revolution.[17] Efforts have been made to show that Lavoisier's and Einstein's revolutions too were replacements of clumsy theories by simple ones.

Conventionalist historiography cannot offer a *rational* explanation of why certain facts were selected in the first instance or of why certain particular pigeonhole systems were tried rather than others at a stage when their relative merits were yet unclear. Thus conventionalism, like inductivism, is compatible with various supplementary empirical-'externalist' programmes.

Finally, the conventionalist historian, like his inductivist colleague, frequently encounters the problem of 'false consciousness'. According to conventionalism, for example, it is a 'matter of fact' that great scientists arrive at their theories by flights of their imaginations. Why then do they often claim that they derived their theories from facts? The conventionalist's rational reconstruction often differs from the great scientists' own reconstruction – the conventionalist historian relegates these problems of false consciousness to the externalist.[18]

C. *Methodological Falsificationism*

Contemporary falsificationism arose as a logico-epistemological criticism of inductivism and of Duhemian conventionalism. Inductivism was criticised on the grounds that its two basic assumptions, namely, that factual propositions can be 'derived' from facts and that there can be valid inductive (content-increasing) inferences, are themselves unproven and even demonstrably false. Duhem was criticised on the grounds that comparison of intuitive simplicity can only be a matter for subjective taste and that it is so ambiguous that no hard-hitting criticism can be based on it. Popper, in his *Logik der Forschung*, proposed a new 'falsificationist' methodology.[19] This methodology is another brand of revolutionary conventionalism: the main difference is that it allows factual, spatio-temporally singular 'basic statements', rather than spatio-temporally universal theories, to be accepted by convention. In the code of honour of the falsificationist a theory is scientific only if it can be *made* to conflict with a basic statement; and a theory must be eliminated if it conflicts with an accepted basic statement. Popper also indicated a further condition that a theory must satisfy in order to qualify as scientific: it must predict facts which are *novel*, that is, unexpected in the light of previous knowledge. Thus it is against Popper's code of scientific honour to propose unfalsifiable theories or 'ad hoc' hypotheses (which imply no *novel* empirical predictions) – just as it is against the [classical] inductivist code of scientific honour to propose unproven ones.

The great attraction of Popperian methodology lies in its clarity and force. Popper's deductive model of scientific criticism contains empirically falsifiable spatio-temporally universal propositions, initial conditions and their consequences. The weapon of criticism is the *modus tollens*: neither inductive logic nor intuitive simplicity complicate the picture.[20]

(Falsificationism, though logically impeccable, has epistemological difficulties of its own. In its 'dogmatic' proto-version it assumes the provability of propositions from facts and thus the disprovability of theories – a false assumption.[21] In its Popperian 'conventionalist' version it needs some (extra-methodological) 'inductive principle' to lend epistemological weight to its decisions to accept 'basic' statements, and in general to connect its rules of the scientific game with verisimilitude.[22])

The Popperian historian looks for great, 'bold', falsifiable theories and for great negative crucial experiments. These form the skeleton of his rational reconstruction. The Popperians' favourite paradigms of great falsifiable theories are Newton's and Maxwell's theories, the radiation formulas of Rayleigh, Jeans and Wien, and the Einsteinian revolution; their favourite paradigms for crucial experiments are the Michelson-Morley experiment, Eddington's eclipse experiment, and the experiments of Lummer and Pringsheim. It was Agassi who tried to turn this naive falsificationism into a systematic historiographical research programme.[23]

In particular he predicted (or 'postdicted', if you wish) that behind each great experimental discovery lies a theory which the discovery contradicted; the importance of a factual discovery is to be measured by the importance of the theory refuted by it. Agassi seems to accept at face value the value judgments of the scientific community concerning the importance of factual discoveries like Galvani's, Oersted's, Priestley's, Roentgen's and Hertz's; but he denies the 'myth' that they were chance discoveries (as the first four were said to be) or confirming instances (as Hertz first thought his discovery was).[24] Thus Agassi arrives at a bold prediction: all these five experiments were successful refutations – in some cases even *planned* refutations – of theories which he proposes to unearth, and, indeed, in most cases, claims to have unearthed.[25]

Popperian internal history, in turn, is readily supplemented by external theories of history. Thus Popper himself explained that [on the positive side] (**1**) the main *external* stimulus of scientific theories comes from unscientific 'metaphysics', and even from myths (this was later beautifully illustrated mainly by Koyré); and that [on the negative side] (**2**) facts do *not* constitute such external stimulus – factual discoveries belong completely to internal history, emerging as refutations of some scientific theory, so that facts are only noticed if they conflict with some previous expectation. Both theses are cornerstones of Popper's *psychology* of discovery.[26] Feyerabend developed another interesting *psychological* thesis of Popper, namely, that proliferation of rival theories may – *externally* – speed up *internal* Popperian falsification.[27]

But the external supplementary theories of falsificationism need not be restricted to purely intellectual influences. It has to be emphasized (*pace* Agassi) that falsificationism is no less compatible with a vulgar-Marxist view of what makes science progress than is inductivism. The only difference is that while for the latter Marxism might be invoked to explain the discovery of *facts*, for the former it might be invoked to explain the invention of *scientific theories*; while the choice of facts (that is, for the falsificationist, the choice of 'potential falsifiers') is primarily determined internally by the theories.

'False awareness' – 'false' from the point of view of *his* rationality theory – creates a problem for the falsificationist historian. For instance, why do some scientists believe that crucial experiments are positive and verifying rather than negative and falsifying? It was the falsificationist

Popper who, in order to solve these problems, elaborated better than anybody else before him the cleavage between objective knowledge (in his 'third world') and its distorted reflections in individual minds.[28] Thus he opened up the way for my demarcation between internal and external history.

D. *Methodology of Scientific Research Programmes*

According to my methodology the greatest scientific achievements are research programmes which can be evaluated in terms of progressive and degenerating problemshifts; and scientific revolutions consist of one research programme superseding (overtaking in progress) another.[29] This methodology offers a new rational reconstruction of science. It is best presented by contrasting it with falsificationism and conventionalism, from both of which it borrows essential elements.

From conventionalism, this methodology borrows the licence rationally to accept by convention not only spatio-temporally singular 'factual statements' but also spatio-temporally universal theories: indeed, this becomes the most important clue to the continuity of scientific growth.[30] The basic unit of appraisal must be not an isolated theory or conjunction of theories but rather a '*research programme*', with a conventionally accepted (and thus by provisional decision 'irrefutable') '*hard core*' and with a '*positive heuristic*' which defines problems, outlines the construction of a belt of auxiliary hypotheses, foresees anomalies and turns them victoriously into examples, all according to a preconceived plan. The scientist lists anomalies, but as long as his research programme sustains its momentum, he may freely put them aside. *It is primarily the positive heuristic of his programme, not the anomalies, which dictate the choice of his problems.*[31] Only when the driving force of the positive heuristic weakens, may more attention be given to anomalies. The methodology of research programmes can explain in this way *the high degree of autonomy of theoretical science*; the naive falsificationist's disconnected chains of conjectures and refutations cannot. What for Popper, Watkins and Agassi is *external*, influential metaphysics, here turns into the *internal* 'hard core' of a programme.[32]

The methodology of research programmes presents a very different picture of the game of science from the picture of the methodological falsificationist. The best opening gambit is not a falsifiable (and therefore consistent) hypothesis, but a research programme. Mere 'falsification' (in Popper's sense) must not imply rejection.[33] Mere 'falsifications' (that is, anomalies) are to be recorded but need not be acted upon. Popper's great negative crucial experiments disappear; 'crucial experiment' is an honorific title, which may, of course, be conferred on certain anomalies, but only *long after the event*, only when one programme has been defeated by another one. According to Popper a crucial experiment is described by an accepted basic statement which is inconsistent with a theory – according to the methodology of scientific research programmes no accepted basic statement *alone* entitles the scientist to reject a theory. Such a clash may present a problem (major or minor), but in no circumstance a 'victory'. Nature may shout *no*, but human ingenuity – contrary to Weyl and Popper[34] – may always be able to shout louder. With sufficient resourcefulness and some luck, any theory can be defended 'progressively' for a long time, even if it is false. The Popperian pattern of 'conjectures and refutations', that is the pattern of trial-by-hypothesis followed by error-shown-by-experiment, is to be abandoned: no experiment is crucial at the time – let alone before – it is performed (except, possibly, psychologically).

It should be pointed out, however, that the methodology of scientific research programmes has more teeth than Duhem's conventionalism: instead of leaving it to Duhem's unarticulated common sense[35] to judge when a 'framework' is to be abandoned, I inject some hard Popperian elements into the appraisal of whether a programme progresses or degenerates or of whether one is overtaking another. That is, I give criteria of progress and stagnation within a programme and also rules for the 'elimination' of whole research programmes. A research programme is said to be *progressing* as long as its theoretical growth anticipates its empirical growth, that is, as long as it keeps predicting novel facts with some success (*'progressive problemshift'*); it is *stagnating* if its theoretical growth lags behind its empirical growth, that is, as long as it gives only *post-hoc* explanations either of chance discoveries or of facts anticipated by, and discovered in, a rival programme (*'degenerating problemshift'*).[36] If a research programme progressively explains more than a rival, it 'supersedes' it, and the rival can be eliminated (or, if you wish, 'shelved').[37]

(*Within* a research programme a theory can only be eliminated by a better theory, that is, by one which has excess empirical content over its predecessors, some of which is subsequently confirmed. And for this replacement of one theory be a better one, the first theory does not even have to be 'falsified' in Popper's sense of the term. Thus progress is marked by instances verifying excess content rather than by falsifying instances;[38] empirical 'falsification' and actual 'rejection' become independent.[39] Before a theory has been modified we can never know in what way it had been 'refuted', and some of the most interesting modifications are motivated by the 'positive heuristic' of the research programme rather than by anomalies. This difference alone has important consequences and leads to a rational reconstruction of scientific change very different from that of Popper's.[40])

It is very difficult to decide, especially since one must not demand progress at each single step, when a research programme has degenerated hopelessly or when one of two rival programmes has achieved a decisive advantage over the other. In this methodology, as in Duhem's conventionalism, there can be no instant – let alone mechanical – rationality. *Neither the logician's proof of inconsistency nor the experimental scientist's verdict of anomaly can defeat a research programme in one blow.* One can be 'wise' only after the event.[41]

In this code of scientific honour modesty plays a greater role than in other codes. One *must* realise that one's opponent, even if lagging badly behind, may still stage a comeback. No advantage for one side can ever be regarded as absolutely conclusive. There is never anything inevitable about the triumph of a programme. Also, there is never anything inevitable about its defeat. Thus pigheadedness, like modesty, has more 'rational' scope. *The scores of the rival sides, however, must be recorded*[42] *and publicly displayed at all times.*

(We should here at least refer to the main epistemological problem of the methodology of scientific research programmes. As it stands, like Popper's methodological falsificationism, it represents a very radical version of conventionalism. One needs to posit some extra-methodological inductive principle to relate – even if tenuously – the scientific gambit of pragmatic acceptances and rejections to verisimilitude.[43] Only such an 'inductive principle' can turn science from a mere game into an epistemologically rational exercise; from a set of lighthearted sceptical gambits pursued for intellectual fun into a – more serious – fallibilist venture of approximating the Truth about the Universe.[44])

The methodology of scientific research programmes constitutes, like any other methodology, a historiographical research programme. The historian who accepts this methodology as a guide will look in history for rival research programmes, for progressive and degenerating problem shifts. Where the Duhemian historian sees a revolution merely in simplicity (like that of Copernicus), he will look for a large scale progressive programme overtaking a degenerating one. Where the falsificationist sees a crucial negative experiment, he will 'predict' that there was none, that behind any alleged crucial experiment, behind any alleged single battle between theory and experiment, there is a hidden war of attrition between two research programmes. The outcome of the war is only later linked in the falsificationist reconstruction with some alleged single 'crucial experiment'.

The methodology of research programmes – like any other theory of scientific rationality – must be supplemented by empirical-external history. No rationality theory will ever solve problems like why Mendelian genetics disappeared in Soviet Russia in the 1950's, or why certain schools of research into genetic racial differences or into the economics of foreign aid came into disrepute in the Anglo-Saxon countries in the 1960's. Moreover, to explain different speeds of development of different research programmes we may need to invoke external history. Rational reconstruction of science (in the sense in which I use the term) cannot be comprehensive since human beings are not *completely* rational animals; and even when they act rationally they may have a false theory of their own rational actions.[45]

But the methodology of research programmes draws a demarcation between internal and external history which is markedly different from that drawn by other rationality theories. For instance, what for the falsificationist looks like the (regrettably frequent) phenomenon of irrational adherence to a 'refuted' or to an inconsistent theory and which he therefore relegates to *external* history, may well be explained in terms of my methodology *internally* as a rational defence of a promising research programme. Or, the successful *pre*dictions of novel facts which constitute serious evidence for a research programme and therefore vital parts of internal history, are irrelevant both for the inductivist and for the falsificationist.[46] For the inductivist and the falsificationist it does not really matter whether the discovery of a fact preceded or followed a theory: only their logical relation is decisive. The 'irrational' impact of the historical coincidence that a theory happened to have *anticipated* a factual discovery, has no internal significance. Such anticipations constitute 'not proof but [mere] propaganda'.[47] Or again, take Planck's discontent with his own 1900 radiation formula, which he regarded as 'arbitrary'. For the falsificationist the formula was a bold, falsifiable hypothesis and Planck's dislike of it a non-rational mood, explicable only in terms of psychology. However, in my view, Planck's discontent can be explained internally: it was a rational condemnation of an '*ad hoc*$_3$' theory.[48] To mention yet another example: for falsificationism irrefutable 'metaphysics' is an external intellectual influence, in my approach it is a vital part of the rational reconstruction of science.

Most historians have hitherto tended to regard the solution of some problems as being the monopoly of externalists. One of these is the problem of the high frequency of *simultaneous discoveries*. For this problem vulgar-Marxists have an easy solution: a discovery is made by many people at the same time, once a social need for it arises.[49] Now what constitutes a 'discovery', and especially a major discovery, depends on one's methodology. For the inductivist, the most important discoveries are factual, and, indeed, such discoveries are frequently made simultaneously. For the falsificationist a *major* discovery consists in the discovery of a theory rather than of a fact. Once a theory is discovered (or rather invented), it becomes public property; and nothing is more obvious than that several people will test it simultaneously and make, simultaneously, (minor) factual discoveries. Also, a published theory is a challenge to devise higher-level, independently testable explanations. For example, given Kepler's ellipses and Galileo's rudimentary dynamics, simultaneous 'discovery' of an inverse square law is not so very surprising: a problem-situation being public, simultaneous solutions can be explained on *purely internal* grounds.[50] The discovery of a new problem however may not be so readily explicable. If one thinks of the history of science as of one of rival research programmes, then most simultaneous discoveries, theoretical or factual, are explained by the fact that research programmes being public property, many people work on them in different corners of the world, possibly not knowing of each other However, really *novel, major, revolutionary* developments are rarely invented simultaneously. Some alleged simultaneous discoveries of novel programmes are seen as having been simultaneous discoveries only with false hindsight: in fact they are *different* discoveries, merged only later into a single one.[51]

A favourite hunting ground of externalists has been the related problem of why so much importance is attached to – and energy spent on – *priority disputes*. This can be explained only *externally* by the inductivist, naive falsificationist, or the conventionalist; but in the light of the methodology of research programmes some priority disputes are vital *internal* problems, since in this methodology *it becomes all-important for rational appraisal which programme was first in anticipating a novel fact and which fitted in the by now old fact only later*. Some priority disputes can be explained by rational interest and not simply by vanity and greed for fame. It then becomes important that Tychonian theory, for instance, succeeded in explaining – only *post hoc* – the observed phases of, and the distance to, Venus which were originally precisely anticipated by Copernicans;[52] or that Cartesians managed to explain everything that the Newtonians *pre*dicted – but only *post hoc*. Newtonian optical theory explained *post hoc* many phenomena which were anticipated and first observed by Huyghensians.[53]

All these examples show how the methodology of scientific research programmes turns many problems which had been *external* problems for other historiographies into internal ones. But occasionally the borderline is moved in the opposite direction. For instance there may have been an experiment which was accepted *instantly* – in the absence of a better theory – as a negative crucial experiment. For the falsificationist such acceptance is part of internal history; for me it is not rational and has to be explained in terms of external history.

Note. The methodology of research programmes was criticised both by Feyerabend and by Kuhn. According to Kuhn: '[Lakatos] must specify criteria which can be used *at the time* to distinguish a degenerative from a progressive research programme; and so on. Otherwise, *he has told us nothing at all*'.[54] Actually, I *do* specify such criteria. But Kuhn probably meant that '[my] standards have practical force only if they are combined with a *time limit* (what looks like a degenerating problemshift may be the beginning of a much longer period of advance)'.[55] Since I specify no such time limit, Feyerabend concludes that my standards are no more than *'verbal ornaments'*.[56] A related point was made by Musgrave in a letter containing some major constructive criticisms of an earlier draft, in which he demanded that I specify, for instance, at what point dogmatic adherence to a programme ought to be explained 'externally' rather than 'internally'.

Let me try to explain why such objections are beside the point. One may rationally stick to a degenerating programme until it is overtaken by a rival *and even after*. What one must *not* do is to deny its poor public record. Both Feyerabend and Kuhn conflate *methodological* appraisal of a programme with firm *heuristic* advice about what to do.[57] It is perfectly rational to play a risky game: what is irrational is to deceive oneself about the risk.

This does not mean as much licence as might appear for those who stick to a degenerating programme. For they can do this mostly only in private. Editors of scientific journals should refuse to publish their papers which will, in general, contain either solemn reassertions of their position or absorption of counterevidence (or even of rival programmes) by *ad hoc*, linguistic adjustments. Research foundations, too, should refuse money.[58]

These observations also answer Musgrave's objection by separating rational and irrational (or honest and dishonest) adherence to a degenerating programme. They also throw further light on the demarcation between internal and external history. They show that internal history is self-sufficient for the presentation of the history of disembodied science, including degenerating problemshifts. External history explains why some people have false beliefs about scientific progress, and how their scientific activity may be influenced by such beliefs.

E. *Internal and External History*

Four theories of the rationality of scientific progress – or logics of scientific discovery – have been briefly discussed. It was shown how each of them provides a theoretical framework for the rational reconstruction of the history of science.

Thus the internal history of *inductivists* consists of alleged discoveries of hard facts and of so-called inductive generalisations. The internal history of *conventionalists* consists of factual discoveries and of the erection of pigeonhole systems and their replacement by allegedly simpler ones.[59] The internal history of *falsificationists* dramatises bold conjectures, improvements which are said to be *always* content-increasing and, above all, triumphant 'negative crucial experiments'. The *methodology of research programmes*, finally, emphasizes long-extended theoretical and empirical rivalry of major research programmes, progressive and degenerating problemshifts, and the slowly emerging victory of one programme over the other.

Each rational reconstruction produces some characteristic pattern of rational growth of scientific knowledge. But all of these *normative* reconstructions may have to be supplemented by *empirical* external theories to explain the residual non-rational factors. The history of science is always richer than its rational reconstruction. *But rational reconstruction or internal history is primary, external history only secondary, since the most important problems of external history are defined by internal history.* External history either provides non-rational explanation of the speed, locality, selectiveness etc. of historic events *as interpreted* in terms of internal history; or, when history differs from its rational reconstruction, it provides an empirical explanation of why it differs. But the *rational* aspect of scientific growth is fully accounted for by one's logic of scientific discovery.

Whatever problem the historian of science wishes to solve, he has first to reconstruct the relevant section of the growth of objective scientific knowledge, that is, the relevant section of 'internal history'. As it has been shown, what constitutes for him internal history, depends on his philosophy, whether he is aware of this fact or not. Most theories of the growth of knowledge are theories of the growth of disembodied knowledge: whether an experiment is crucial or not, whether a hypothesis is highly probable in the light of the available evidence or not, whether a problemshift is progressive or not, is not dependent in the slightest on the scientists' beliefs, personalities or authority. These subjective factors are of no interest for any internal history. For instance, the 'internal historian' records the Proutian programme with its hard core (that atomic weights of pure chemical elements are whole numbers) and its positive heuristic (to overthrow, and replace, the contemporary false observational theories applied in measuring atomic weights). This programme was later carried through.[60] The internal historian will waste little time on Prout's *belief* that if the 'experimental techniques' *of his time* were 'carefully' applied, and the experimental findings properly interpreted, the anomalies would *immediately* be seen as mere illusions. The internal historian will regard this historical fact as a fact in the second world which is only a caricature of its counterpart in the third world.[61] *Why* such caricatures come about is none of his business; he might – in a footnote – pass on the externalist the problem of why certain scientists had 'false beliefs' about what they were doing.[62]

Thus in constructing internal history the historian will be highly selective: he will omit everything that is irrational in the light of his rationality theory. But this normative selection still does not add up to a fully fledged rational reconstruction. For instance, Prout never articulated the 'Proutian programme': the Proutian programme is not Prout's programme. *It is not only the ('internal') success or the ('internal') defeat of a programme which can only be judged with hindsight: it is frequently also its content.* Internal history is not just a *selection* of methodologically interpreted facts: it may be, on occasions, their *radically improved version.* One may illustrate this using the Bohrian programme. Bohr, in 1913, may not have even thought of the possibility of electron spin. He had more than enough on his hands without the spin. Nevertheless, the historian, describing with hindsight the Bohrian programme, should include electron spin in it, since electron spin fits naturally in the original outline of the programme. Bohr might have referred to it in 1913. Why Bohr did not do so, is an interesting problem which deserves to be indicated in a footnote.[63] (Such problems might then be solved either internally by pointing to rational reasons in the growth of objective, impersonal knowledge; or externally by pointing to psychological causes in the development of Bohr's personal beliefs.)

One way to indicate discrepancies between history and its rational reconstruction is to relate the internal history *in the text*, and indicate *in the footnotes* how actual history 'misbehaved' in the light of its rational reconstruction.[64]

Many historians will abhor the idea of *any* rational reconstruction. They will quote Lord Bolingbroke: 'History is philosophy teaching by example'. They will say that before philosophising 'we need a lot more examples'.[65] But such an inductivist theory of historiography is utopian.[66] *History without some theoretical 'bias' is impossible.*[67] Some historians look for the discovery of hard facts, inductive generalisations, others for bold theories and crucial negative experiments, yet others for great simplifications, or for progressive and degenerating problemshifts; all of them have *some* theoretical 'bias'. This bias, of course, may be obscured by an eclectic variation of theories or by theoretical confusion: but neither eclecticism nor confusion amounts to an atheoretical outlook. What a historian regards as an external problem is often an excellent guide to his implicit methodology: some will ask why a 'hard fact' or a 'bold theory' was discovered exactly when and where it actually was discovered; others will ask why a 'degenerating problemshift' could have wide popular acclaim over an incredibly long period or why a 'progressive problemshift' was left 'unreasonably' unacknowledged.[68] Long texts have been devoted to the problem of whether, and if so, why, the emergence of science was a purely European affair; but such an investigation is bound to remain a piece of confused rambling until one clearly defines 'science' according to some normative philosophy of science. One of the most interesting problems of external history is to specify the psychological, and indeed, social conditions which are necessary (but, of course, never sufficient) to make scientific progress possible; but in the very formulation of this 'external' problem *some* methodological theory, *some* definition of science is bound to enter. History of *science* is a history of events which are selected and interpreted in a normative way.[69] This being so, the hitherto neglected problem of appraising rival logics of scientific discovery and, hence, rival reconstructions of history, acquires paramount importance. I shall now turn to this problem.

2. Critical comparison of methodologies: history as a test of its rational reconstructions

Theories of scientific rationality can be classified under two main heads.

(1) *Justificationist methodologies* set very high epistemological standards: for classical justificationists a proposition is 'scientific' only if it is *proven*, for neojustificationists, if it is *probable* (in the sense of the probability calculus) or *corroborated* (in the sense of Popper's third note on corroboration) to a proven degree.[70] Some philosophers of science gave up the idea of proving or of (provably) probabilifying scientific theories but remained dogmatic empiricists: whether inductivists, probabilists, conventionalists or falsificationist, they still stick to the provability of 'factual' propositions. By now, of course, all these different forms of justificationism have crumbled under the weight of *epistemological and logical criticism*.

(2) The only alternatives with which we are left are *pragmatic-conventionalist methodologies*, crowned by some global principle of induction. Conventionalist methodologies first lay down rules about 'acceptance' and 'rejection' of factual and theoretical propositions – without yet laying down rules about proof and disproof, truth and falsehood. We then get *different systems of rules of the scientific game*. The inductivist game would consist of collecting 'acceptable' (not proven) data and drawing from them 'acceptable' (not proven) inductive generalisations. The conventionalist game would consist of collecting 'acceptable' data and ordering them into the simplest possible pigeonhole systems (or devising the simplest possible pigeonhole systems and filling them with acceptable data). Popper specified yet another game as 'scientific'.[71] Even methodologies which have been epistemologically and logically discredited, may go on functioning, in these emasculated versions, as guides for the rational reconstruction of history. But these *scientific games* are without any

genuine epistemological relevance *unless* we superimpose on them some sort of metaphysical (or, if you wish, 'inductive') principle which will say that the game, as specified by the methodology, gives us the best chance of approaching the Truth. Such a principle then turns the pure conventions of the game into fallible conjectures; but without such a principle the scientific game is just like any other game.[72]

It is very difficult to criticise conventionalist methodologies like Duhem's and Popper's. There is no obvious way to criticise either a game or a metaphysical principle of induction. In order to overcome these difficulties I am going to propose a new theory of how to appraise such methodologies of science (the ones, which – at least in the first stage, before the introduction of an inductive principle – are conventionalist). I shall show that methodologies may be criticised without any direct reference to any epistemological (or even logical) theory, and without using directly any logico-epistemological criticism. The basic idea of this criticism is that *all methodologies function as historiographical (or meta-historical) theories (or research programmes) and can be criticised by criticising the rational historical reconstructions to which they lead.*

I shall try to develop this historiographical method of criticism in a dialectical way. I start with a special case: I first 'refute' falsificationism by 'applying' falsificationism (on a normative historiographical meta-level) to itself. Then I shall apply falsificationism also to inductivism and conventionalism, and, indeed, argue that all methodologies are bound to end up 'falsified' with the help of this Pyrrhonian *machine de guerre*. Finally, I shall 'apply' not falsificationism but the methodology of scientific research programmes (again on a normative-historiographical meta-level) to inductivism, conventionalism, falsificationism and to itself, and show that – on this meta-criterion – methodologies can be constructively criticised and compared. This normative-historiographical version of the methodology of scientific research programmes supplies a general theory of how to compare rival logics of discovery in which (in a sense carefully to be specified) *history may be seen as a 'test' of its rational reconstructions.*

A. *Falsificationism as a Meta-criterion: History 'falsifies' Falsificationism (and any other Methodology)*

In their purely 'methodological' versions scientific appraisals, as has already been said, are *conventions* and can always be formulated as a definition of science.[73] How can one criticise such a definition? If one interprets it nominalistically,[74] a definition is a mere abbreviation, a terminological suggestion, a tautology. How can one criticise a tautology? Popper, for one, claims that his definition of science is 'fruitful' because 'a great many points can be clarified and explained with its help'. He quotes Menger: 'Definitions are dogmas; only the conclusions drawn from them can afford us any new insight'.[75] But how can a definition have explanatory power or afford new insights? Popper's answer is this: 'It is only from the consequences of my definition of empirical science, and from the methodological decisions which depend upon this definition, that the scientist will be able to see how far it conforms to his intuitive idea of the goal of his endeavours'.[76]

The answer complies with Popper's general position that conventions can be criticised by discussing their 'suitability' relative to some purpose: 'As to the suitability of any convention opinions may differ; and a reasonable discussion of these questions is only possible between parties having some purpose in common. The choice of that purpose ... goes beyond rational argument'.[77] Indeed, Popper never offered a theory of rational criticism of consistent conventions. He does not raise, let alone answer, the question: *'Under what conditions would you give up your demarcation criterion?'*[78]

But the question can be answered. I give my answer in two stages: I propose first a naive and then a more sophisticated answer. I start by recalling how Popper, according to his own account[78a], arrived at his criterion. He thought, like the best scientists of his time, that Newton's theory, although refuted, was a wonderful scientific achievement; that Einstein's theory was still better; and that astrology, Freudianism and twentieth century Marxism were pseudo-scientific. His problem was to find a definition of science which yielded these *'basic judgments'* concerning particular theories; and he offered a novel solution. Now let us consider the proposal that *a rationality theory – or demarcation criterion – is to be rejected if it is inconsistent with an accepted 'basic value judgment' of the scientific élite*. Indeed, this meta-methodological rule (*meta-falsificationism*) would seem to correspond to Popper's methodological rule (falsificationism) that a scientific theory is to be rejected if it is inconsistent with an ('empirical') basic statement unanimously accepted by the scientific community. Popper's whole methodology rests on the contention that there exist (relatively) singular statements on whose truth-value scientists can reach unanimous agreement; without such agreement there would be a new Babel and 'the soaring edifice of science would soon lie in ruins'.[79] But even if there were an agreement about 'basic' statements, if there were no agreement about how to appraise scientific achievement relative to this 'empirical basis', would not the soaring edifice of science equally soon lie in ruins? No doubt it would. While there has been little agreement concerning a *universal* criterion of the scientific character of theories, there has been considerable agreement over the last two centuries concerning *single* achievements. While there has been no *general* agreement concerning a theory of scientific rationality, there has been considerable agreement concerning whether a particular single step in the game was scientific or crankish, or whether a particular gambit was played correctly or not. A general definition of science thus must reconstruct the acknowledgedly best gambits as 'scientific': if it fails to do so, it has to be rejected.[80]

Then let us propose tentatively that *if a demarcation criterion is inconsistent with the 'basic' appraisals of the scientific élite, it should be rejected.*

Now *if* we apply this quasi-empirical meta-criterion (which I am going to reject later), Popper's demarcation criterion – that is, Popper's rules of the game of science – has to be rejected.[81]

Popper's basic rule is that the scientist must specify in advance under what experimental conditions he will give up even his most basic assumptions. For instance, he writes, when criticising psychoanalysis: '*Criteria of refutation* have to be laid down beforehand: it must be agreed which observable situations, if actually observed, mean that the theory is refuted. But what kind of clinical responses would refute to the satisfaction of the analyst *not merely a particular analytic diagnosis but psychoanalysis itself?* And have such criteria ever been discussed or agreed upon by analysts?'[82] In the case of psychoanalysis Popper was right: no answer has been forthcoming. Freudians have been nonplussed by Popper's basic challenge concerning scientific honesty. Indeed, they have refused to specify experimental conditions under which they would give up their basic assumptions. For Popper this was the hallmark of their intellectual dishonesty. But what if we put Popper's question to the Newtonian scientist: 'What kind of observation would refute to the satisfaction of the Newtonian not merely a particular Newtonian explanation but Newtonian dynamics and gravitational theory itself? And have such criteria ever been discussed or agreed upon by Newtonians?' The Newtonian will, alas, scarcely be able to give a positive answer.[83] But then if analysts are to be condemned as dishonest by Popper's standards, Newtonians must also be condemned. Newtonian science, however, in spite of this sort of 'dogmatism', is highly regarded by the greatest scientists, and, indeed, by Popper himself. Newtonian 'dogmatism' then is a 'falsification' of Popper's definition: it defies Popper's rational reconstruction.

Popper may certainly withdraw his celebrated challenge and demand falsifiability – and rejection on falsification – only for systems of theories, including initial conditions and all sorts of auxiliary and observational theories.[84] This is a considerable withdrawal, for it allows the imaginative scientist to save his pet theory by suitable lucky alterations in some odd, obscure corner on the periphery of his theoretical maze. But even Popper's mitigated rule will show up even the most brilliant scientists as irrational dogmatists. For in large research programmes there are always known anomalies: normally the researcher puts them aside and follows the positive heuristic of the programme.[85] In general he rivets his attention on the positive heuristic rather than on the distracting anomalies, and hopes that the 'recalcitrant instances' will be turned into confirming instances as the programme progresses. On Popper's terms the greatest scientists in these situations used forbidden gambits, *ad hoc* stratagems: instead of regarding Mercury's anomalous perihelion as a falsification of the Newtonian theory of our planetary system and thus as a reason for its rejection, most physicists shelved it as a problematic instance to be solved at some later stage – or offered *ad hoc* solutions. This methodological attitude of treating as (mere) *anomalies* what Popper would regard as (dramatic) counterexamples is commonly accepted by the best scientists. Some of the research programmes now held in highest esteem by the scientific community progressed in an ocean of anomalies.[86] That in their choice of problems the greatest scientists 'uncritically' ignore anomalies (and that they isolate them with the help of *ad hoc* stratagems) offers, at least on our metacriterion, a further falsification of Popper's methodology. He cannot interpret as rational some most important patterns in the growth of science.

Furthermore, for Popper, working on *an inconsistent system* must invariably be regarded as irrational 'a self-contradictory system must be rejected... [because it] is uninformative... No statement is singled out... since all are derivable'.[87] But some of the greatest scientific research programmes progressed on inconsistent foundations.[88] Indeed in such cases the best scientists' rule is frequently: '*Allez en avant et la foi vous viendra*'. This anti-Popperian methodology secured a breathing space both for the infinitesimal calculus and for naive set theory when they were bedevilled by logical paradoxes.

Indeed, if the game of science had been played according to Popper's rule book, Bohr's 1913 paper would never have been published because it was inconsistently grafted on to Maxwell's theory, and Dirac's delta functions would have been suppressed until Schwartz. All these examples of research based on inconsistent foundations constitute further 'falsifications' of falsificationist methodology.[89]

Thus several of the 'basic' appraisals of the scientific *élite* 'falsify' Popper's definition of science and scientific ethics. The problem then arises, to what extent, given these considerations, can falsificationism function as a guide for the historian of science. The simple answer is, to a very small extent. Popper, the leading falsificationist, never wrote any history of science; possibly because he was too sensitive to the judgment of great scientists to pervert history in a falsificationist vein. One should remember that while in his autobiographical recollections he mentions Newtonian science as the paradigm of scientificness, that is, of falsifiability, in his classical *Logik der Forschung* the falsifiability of Newton's theory is nowhere discussed. The *Logik der Forschung*, on the whole, is dryly abstract and highly ahistorical.[90] Where Popper does venture to remark casually on the falsifiability of major scientific theories, he either plunges into some logical blunder,[91] or distorts history to fit his rationality theory. If a historian's methodology provides a poor rational reconstruction, he may either misread history in such a way that it coincides with his rational reconstruction, or he will find that the history of science is highly irrational. Popper's respect for great science made him choose the first option, while the disrespectful Feyerabend chose the second.[92] Thus Popper, in his historical asides, tends to turn anomalies into 'crucial experiments' and to exaggerate their immediate impact on the history of science. Through his spectacles, great scientists accept refutations readily and this is the primary source of their problems. For instance, in one place he claims that the Michelson-Morley experiment decisively overthrew classical ether theory; he also exaggerates the role of this experiment in the emergence of Einstein's relativity theory.[93] It takes a naive falsificationist's simplifying spectacles to see, with Popper, Lavoisier's classical experiments as refuting (or as 'tending to refute') the phlogiston theory; or to see the Bohr-Kramers-Slater theory as being knocked out with a single blow from Compton; or to see the parity principle 'rejected' by 'counterexample'.[94]

Furthermore, if Popper wants to reconstruct the provisional acceptance of theories as rational on *his* terms, he is bound to ignore the historical fact that most important theories are born refuted and that some laws are further explained, rather than rejected, in spite of the known counterexamples. He tends to turn a blind eye on all anomalies known before the one which later was enthroned as 'crucial counter-evidence'. For instance, he mistakenly thinks that 'neither Galileo's nor Kepler's theories were refuted before Newton'.[95] The context is significant. Popper holds that the most important pattern of scientific progress is when a crucial experiment leaves one theory *unrefuted* while it refutes a rival one. But, as a matter of fact, in most, if not in all, cases where there are two rival theories, both are known to be simultaneously infected by anomales. In such situations Popper succumbs to the temptation to simplify the situation into one to which his methodology is applicable.[96]

Falsificationist historiography is then 'falsified'. But if we apply the same meta-falsificationist method to inductivist and conventionalist historiographies, we shall 'falsify' them too.

The best logico-epistemological demolition of inductivism is, of course, Popper's; but even if we assumed that inductivism were philosophically (that is, epistemologically and logically) sound, Duhem's historiographical criticism falsifies it. Duhem took the most celebrated *'successes' of inductivist historiography*: Newton's law of gravitation and Ampère's electromagnetic theory. These were said to be two most victorious applications of inductive method. But Duhem (and, following him, Popper and Agassi) showed that they were not. Their analyses illustrate how the inductivist, if he wants to show that the growth of actual science is rational, must falsify actual history out of all recognition.[97] Therefore, if the rationality of science is inductive, actual science is not rational; if it is rational, it is not inductive.[98]

Conventionalism – which, unlike inductivism, is no easy prey to logical or epistemological criticism[99] – can also be historiographically falsified. One can show that the clue to scientific revolutions is not the replacement of cumbersome frameworks by simpler ones.

The Copernican revolution was generally taken to be the *paradigm of conventionalist historiography*, and it is still so regarded in many quarters. For instance Polanyi tells us that Copernicus's 'simpler picture' had 'striking beauty' and '[justly] carried great powers of conviction.'[100] But modern study of primary sources, particularly by Kuhn,[101] has dispelled this myth and presented a clear-cut historiographical refutation of the conventionalist account. It is now agreed that the Copernican system was 'at least as complex as the Ptolemaic'.[102] But if this is so, then, if the acceptance of Copernican theory was rational, it was not for its superlative objective simplicity.[103]

Thus inductivism, falsificationism and conventionalism can be falsified as rational reconstructions of history with the help of the sort of historiographical criticism I have adduced.[104] Historiographical falsification of inductivism, as we have seen, was initiated already by Duhem and continued by Popper and Agassi. Historiographical criticisms of [naive] falsificationism have been offered by Polanyi, Kuhn, Feyerabend and Holton.[105] The most important historiographical criticism of conventionalism is to be found in Kuhn's – already quoted – masterpiece on the Copernican revolution.[106] The upshot of these criticisms is that all these rational reconstructions of history force history of science into the Procrustean bed of their hypocritical morality, thus creating fancy histories, which hinge on mythical 'inductive bases', 'valid inductive generalisations', 'crucial experiments', 'great revolutionary simplifications' etc. But critics of falsificationism and conventionalism drew very different conclusions from the falsification of these methodologies than Duhem, Popper and Agassi did from their own falsification of inductivism. Polanyi (and, seemingly, Holton) concluded that while proper, rational scientific appraisal can be made in *particular* cases, there can be no *general* theory of scientific rationality.[107] *All* methodologies, *all* rational reconstructions can be historiographically 'falsified': science *is* rational, but its rationality cannot be subsumed under the general laws of any methodology.[108] Feyerabend, on the other hand, concluded that not only can there be no general theory of scientific rationality but also that there is no such thing as scientific rationality.[109] Thus Polanyi swung towards conservative authoritarianism, while Feyerabend swung towards sceptical anarchism. Kuhn came up with a highly original vision of irrationally changing rational authority.[110]

Although, as it transpires from this section, I have high regard for Polanyi's, Feyerabend's and Kuhn's criticisms of extant ('internalist') theories of method, I drew a conclusion completely different from theirs. I decided to look for an improved methodology which offers a better *rational* reconstruction of science.

Feyerabend and Kuhn immediately tried to 'falsify' my improved methodology in turn.[111] I soon had to discover that, at least in the sense described in the present section, my methodology too – and any methodology whatsoever – *can* be 'falsified', for the simple reason that no set of human judgments is completely rational and thus no rational reconstruction can ever coincide with actual history.[112]

This recognition led me to propose a new *constructive* criterion by which methodologies *qua* rational reconstructions of history might be appraised.

B. *The Methodology of Historiographical Research Programmes. History – to Varying Degrees – Corroborates Its Rational Reconstructions*

I should like to present my proposal in two stages. First, I shall amend slightly the falsificationist historiographical meta-criterion just discussed, and then replace it altogether with a better one.

First, the slight amendment. If a universal rule clashes with a particular 'normative basic judgment', one should allow the scientific community time to ponder the clash: they may give up their particular judgment and submit to the general rule. 'Second-order' – historiographical – falsifications must not be rushed any more than 'first order' – scientific – ones.[113]

Secondly, since we have abandoned naive falsificationism in *method*, why should we stick to it in *meta-method?* We can easily replace it with a methodology of scientific research programmes of second order, or if you wish, a methodology of historiographical research programmes.

While maintaining that a theory of rationality has to try to organise basic value judgments in universal, coherent frameworks, we do not have to reject such a framework immediately merely because of some anomalies or other inconsistencies. We should, of course, insist that a good rationality theory must anticipate further basic value judgments unexpected in the light of its predecessors or that it must even lead to the revision of previously held basic value-judgments.[114] We then reject a rationality theory only for a better one, for one which, in this 'quasi-empirical' sense, represents a *progressive shift* in the sequence of research programmes of rational reconstructions. Thus this new – more lenient – meta-criterion enables us to compare rival logics of discovery and discern growth in 'meta-scientific' – methodological – knowledge.

For instance, Popper's theory of scientific rationality need not be rejected simply because it is 'falsified' by some actual 'basic judgments' of leading scientists. Moreover, on our new criterion, Popper's demarcation criterion clearly represents progress over its justificationist predecessors, and in particular, over inductivism. For, contrary to these predecessors, it rehabilitated the scientific status of falsified theories like phlogiston theory, thus reversing a value judgment which had expelled the latter from the history of science proper into the history of irrational beliefs.[115] Also, it successfully rehabilitated the Bohr-Kramers-Slater theory.[116] In the light of most justificationist theories of rationality the history of science is, at its best, a history of *pre*scientific preludes to some *future* history of science.[117] Popper's methodology enabled the historian to interpret more of the *actual* basic value judgments in the history of science as rational: in *this* normative-historiographical sense Popper's theory constituted progress. In the light of better rational reconstructions of science one can always reconstruct more of actual great science as rational.[118]

I hope that my modification of Popper's logic of discovery will be seen, in turn – on the criterion I specified – as yet a further step forward. For it seems to offer a coherent account of *more* old, isolated basic value judgments; moreover, it has led to new and, at least for the justificationist or naive falsificationist, surprising basic value judgments. For instance, according to Popper's theory, it was irrational to retain and further elaborate Newton's gravitational theory after the discovery of Mercury's anomalous perihelion; or again, it was irrational to develop Bohr's old quantum theory based on inconsistent foundations. From my point of view these were perfectly rational developments: some rearguard actions in the defence of defeated programmes – even after the so-called 'crucial experiments' – are perfectly rational. Thus my methodology leads to the reversal of those historiographical judgments which deleted these rear-guard actions both from inductivist and from falsificationist party histories.[119]

Indeed, this methodology confidently predicts that where the falsificationist sees the instant defeat of a theory through a simple battle with some fact, the historian will detect a complicated war of attrition, starting long before, and ending after, the alleged 'crucial experiment'; and where the falsificationist sees consistent and unrefuted theories, it predicts the existence of hordes of known anomalies in research programmes progressing on possibly inconsistent foundations.[120] Where the conventionalist sees the clue to the victory of a theory over its predecessor in the former's intuitive simplicity, this methodology predicts that it will be found that victory was due to empirical degeneration in the old and empirical progress in the new programme.[121] Where Kuhn and Feyerabend see irrational change, I predict that the historian will be able to show that there has been rational change. The methodology of research programmes thus predicts (or, if you wish, 'postdicts') novel historical facts, unexpected in the light of extant (internal and external) historiographies and these predictions will, I hope, be corroborated by historical research. If they are, then the methodology of scientific research programmes will itself constitute a progressive problemshift.

Thus progress in the theory of scientific rationality is marked by discoveries of novel historical facts, by the reconstruction of a growing bulk of value-impregnated history as rational.[122] In other words, the theory of scientific rationality progresses if it constitutes a 'progressive' historiographical research programme. I need not say that no such historiographical research programme can or should explain *all* history of science as rational: even the greatest scientists make false steps and fail in their judgment. Because of this *rational reconstructions remain for ever submerged in an ocean of anomalies. These anomalies will eventually have to be explained either by some better rational reconstruction or by some 'external' empirical theory.*

This approach does not advocate a cavalier attitude to the 'basic normative judgments' of the scientist. 'Anomalies' may be rightly ignored by the internalist *qua* internalist and relegated to external history only as long as the internalist historiographical research programme is *progressing*; or if a supplementary empirical externalist historiographical programme absorbs them *progressively*. But if in the light of a rational reconstruction the history of science is seen as increasingly irrational *without* a progressive externalist explanation (such as an explanation of the degeneration of science in terms of political or religious terror, or of an antiscientific ideological climate, or of the rise of a new parasitic class of pseudoscientists with vested interests in rapid 'university expansion'), then historiographical innovation, proliferation of historiographical theories, is vital. Just as scientific progress is possible even if one never gets rid of scientific anomalies, progress in rational historiography is also possible even if one never gets rid of historiographical anomalies. The rationalist historian need not be disturbed by the fact that actual history is more than, and, on occasions, even different from, internal history, and that he may have to relegate the explanation of such anomalies to external history. But this unfalsifiability of internal history does not render it immune to constructive, but only to negative, criticism – just as the unfalsifiability of a scientific research programme does not render it immune to constructive, but only to negative, criticism.

Of course, one can criticise internal history only by making the historian's (usually latent) methodology explicit, showing how it functions as a historiographical research programme. Historiographical criticism frequently succeeds in destroying much of fashionable externalism. An 'impressive', 'sweeping', 'far-reaching' external explanation is usually the hallmark of a weak methodological substructure; and, in turn, the hallmark of a relatively weak internal history (in terms of which most actual history is either inexplicable or anomalous) is that it leaves too much to be explained by external history. When a better rationality theory is produced, internal history may expand and reclaim ground from external history. The competition, however, is not as open in such cases as when two rival scientific research programmes compete. Externalist historiographical programmes which supplement internal histories based on naive methodologies (whether aware or unaware of the fact) are likely either to degenerate quickly or never even to get off the ground, for the simple reason that they set out to offer psychological or sociological 'explanations' of methodologically induced fantasies rather than of (more rationally interpreted) historical facts. Once an externalist account uses, whether consciously or not, a naive methodology (which can so easily creep into its 'descriptive' language), it turns into a fairy tale which, for all its apparent scholarly sophistication, will collapse under historiographical scrutiny.

Agassi already indicated how the poverty of inductivist history opened the door to the wild speculations of vulgar-Marxists.[123] His falsificationist historiography, in turn, flings the door wide open to those trendy 'sociologists of knowledge' who try to explain the further (possibly unsuccessful) development of a theory 'falsified' by a 'crucial experiment' as the manifestation of the irrational, wicked, reactionary resistance by established authority to enlightened revolutionary innovation.[124] But in the light of the methodology of scientific research programmes such rearguard skirmishes are perfectly explicable *internally*: where some externalists see power struggle, sordid personal controversy, the rationalist historian will frequently find rational discussion.[125]

An interesting example of how a poor theory of rationality may impoverish history is the treatment of degenerating problemshifts by historiographical positivists. [126] Let us imagine for instance that in spite of the objectively progressing astronomical research programmes, the astronomers are suddenly all gripped by a feeling of Kuhnian 'crisis'; and then they all are converted, by an irresistible *Gestalt*-switch, to astrology. I would regard this catastrophe as a horrifying *problem*, to be accounted for by some empirical externalist explanation. But not a Kuhnian. All he sees is a 'crisis' followed by a mass conversion effect in the scientific community: an ordinary revolution. Nothing is left as problematic and unexplained.[127] The Kuhnian psychological epiphenomena of 'crisis' and 'conversion' can accompany either objectively progressive or objectively degenerating changes, either revolutions or counterrevolutions. But this fact falls outside Kuhn's framework. Such historiographical anomalies cannot be formulated, let alone be progressively absorbed, by his historiographical research programme, in which there is no way of distinguishing between, say, a 'crisis' and 'degenerating problemshift'. But such anomalies might even be predicted by an externalist historiographical theory based on the methodology of scientific research programmes that would specify social conditions under which degenerating research programmes may achieve socio-psychological victory.

C. *Against Aprioristic and Antitheoretical Approaches to Methodology*

Finally, let us contrast the theory of rationality here discussed with the strictly aprioristic (or, more precisely, 'Euclidean') and with the anti-theoretical approaches.[128]

'Euclidean' methodologies lay down *a priori general rules* for scientific appraisal. This approach is most powerfully represented today by Popper. In Popper's view there must be the constitutional authority of an *immutable statute law* (laid down in his demarcation criterion) to distinguish between good and bad science.

Some eminent philosophers, however, ridicule the idea of statute law, the possibility of any valid demarcation. According to Oakeshott and Polanyi there must be – and can be – no statute law at all: only case law. They may also argue that even if one mistakenly allowed for statute law, statute law too would need authoritative interpreters. I think that Oakeshott's and Polanyi's position has a great deal of truth in it. After all, one must admit (*pace* Popper) that until now all the 'laws' proposed by the apriorist philosophers of science have turned out to be wrong in the light of the verdicts of the best scientists. Up to the present day it has been the scientific standards, as applied 'instinctively' by the scientific *élite* in *particular* cases, which have constituted the main – although not the exclusive – yardstick of the philosopher's *universal* laws. But if so, methodological progress, at least as far as the most advanced sciences are concerned, still lags behind common scientific wisdom. Is it not then *hubris* to try to impose some *a priori* philosophy of science on the most advanced sciences? Is it not *hubris* to demand that if, say, Newtonian or Einsteinean science turns out to have violated Bacon's, Carnap's or Popper's *a priori* rules of the game, the business of science should be started anew?

I think it is. And, indeed, the methodology of historiographical research programmes implies a pluralistic system of authority, partly because the wisdom of the scientific jury and its case law has not been, and cannot be, fully articulated by the philosopher's statute law, and partly because the philosopher's statute law may occasionally be right when the scientists' judgment fails. I disagree, therefore, both with those philosophers of science who have taken it for granted that general scientific standards are immutable and reason can recognise them *a priori*,[129] and with those who have thought that the light of reason illuminates only particular cases. The methodology of historiographical research programmes specifies ways both for the philosopher of science to learn from the historian of science and *vice versa*.

But this two-way traffic need not always be balanced. The statute law approach should become much more important when a tradition degenerates[130] or a new bad tradition is founded.[131] In such cases statute law may thwart the authority of the corrupted case law, and slow down or even reverse the process of degeneration.[132] When a scientific school degenerates into pseudo-science, it may be worthwhile to force a methodological debate in the hope that working scientists will learn more from it than philosophers (just as when ordinary language degenerates into, say, journalese, it may be worthwhile to invoke the rules of grammar).[133]

D. *Conclusion*

In this paper I have proposed a 'historical' method for the evaluation of rival methodologies. The arguments were primarily addressed to the philosopher of science and aimed at showing how he can – and should – learn from the history of science. But the same arguments also imply that the historian of science must, in turn, pay serious attention to the philosophy of science and decide upon which methodology he will base his internal history. I hope to have offered some strong arguments for the following theses. First, each methodology of science determines a characteristic (and sharp) demarcation between (primary) internal history and (secondary) external history and, secondly, both historians and philosophers of science must make the best of the critical interplay between internal and external factors.

Let me finally remind the reader of my favourite – and by now well-worn – joke that history of science is frequently a caricature of its rational reconstructions; that rational reconstructions are frequently caricatures of actual history; and that some histories of science are caricatures both of actual history and of its rational reconstructions.[134] This paper, I think, enables me to add: *Quod erat demonstandum.*

London School of Economics

NOTES

* Earlier versions of this paper were read and criticized by Colin Howson, Alan Musgrave, John Watkins, Elie Zahar, and especially John Worrall.

The present paper further develops some of the theses proposed in my (1970). I have tried, at the cost of some repetition, to make it self-contained.

[1] 'Internal history' is usually defined as intellectual history; 'external history' as social history (cf. e.g. Kuhn (1968)). My unorthodox, new demarcation between 'internal' and 'external' history consitutes a considerable problemshift and may sound dogmatic. But my definitions form the hard core of a historiographical research programme; their evaluation is part and parcel of the evaluation of the fertility of the whole programme.
[2] This is an all-important shift in the problem of normative philosophy of science. The term 'normative' no longer means rules for arriving at solutions, but merely directions for the appraisal of solutions already there. Thus *methodology* is separated from *heuristics*, rather as value judgments are from ought statements. (I owe this analogy to John Watkins.)
[3] This profusion of synonyms has proved to be rather confusing.
[4] The epistemological significance of scientific 'acceptance' and 'rejection' is, as we shall see, far from being the same in the four methodologies to be discussed.
[5] '*Neo*-inductivism' demands only (provably) highly probable generalisations. In what follows I shall only discuss classical inductivism; but the watered down neo-inductivist variant can be similarly dealt with.
[6] Cf. p. 107.
[7] For a detailed discussion of inductivist (and, in general, justificationist) criticism cf. my (1966).
[8] I am now using the term 'paradigm' in its pre-Kuhnian sense.
[9] This compatibility was pointed out by Agassi on pp. 23–27 of his (1963). But did he not point out the analogous compatibility within his own falsificationist historiography; cf. above, pp. 98–9.
[10] Cf. e.g. Bernal (1965), p. 377.
[11] Some logical positivists belonged to this set: one recalls Hempel's horror at Popper's casual praise of certain external metaphysical influences upon science (Hempel, 1937).
[12] When German obscurantists scoff at 'positivism', they frequently mean radical internalism, and in particular, radical inductivism.
[13] For what I here call *revolutionary conventionalism*, see my (1970), pp. 105–6 and 187–9.
[14] I mainly discuss here only one version of revolutionary conventionalism, the one which Agassi, in his (1966), called 'unsophisticated': the one which assumes that factual propositions – unlike pigeonhole systems – can be 'proven'. (Duhem, for instance, draws no clear distinction between facts and factual propositions.)
[15] It is important to note that most conventionalists are reluctant to give up inductive generalisations. They distinguish between the '*floor of facts*', the '*floor of laws*' (i.e. inductive generalisations from 'facts') and the '*floor of theories*' (or of pigeonhole systems) which classify, conveniently, both facts and inductive laws. (Whewell, the conservative conventionalist and Duhem, the revolutionary conventionalist differ less than most people imagine.)
[16] One may call such metaphysical principles 'inductive principles'. For an 'inductive principle' which – roughly speaking – makes Popper's 'degree of corroboration' (a conventionalist appraisal) the measure of Popper's verisimilitude (truth-content minus falsity-content) see my (1968a), pp. 390–408 and my (1971a), § 2. (Another widely spread 'inductive principle' may be formulated like this: "What the group of trained – or up-to-date, or suitably purged – scientists decide to *accept* as 'true', is true.")
[17] Most historical accounts of the Copernican revolution are written from the conventionalist point of view. Few claimed that Copernicus' theory was an 'inductive

generalisation' from some 'factual discovery'; or that it was proposed as a bold theory to replace the Ptolemaic theory which had been 'refuted' by some celebrated 'crucial' experiment.

For a further discussion of the historiography of the Copernican revolution, cf. my (1971b).

[18] For example, for non-inductivist historians Newton's '*Hypotheses non fingo*' represents a major problem. Duhem, who unlike most historians did not over-indulge in Newton-worship, dismissed Newton's inductivist methodology as logical nonsense; but Koyré, whose many strong points did not include logic, devoted long chapters to the 'hidden depths' of Newton's muddle.

[19] *In this paper I use this term to stand exclusively for one version of falsificationism, namely for 'naive methodological falsificationism', as defined in my* (1970), *pp.* 93–116.

[20] Since in his methodology the *concept* of intuitive simplicity has no place, Popper was able to use the term 'simplicity' for 'degree of falsifiability'. But there is more to simplicity than this: cf. my (1970), pp. 131ff.

[21] For a discussion cf. my (1970), especially pp. 99–100.

[22] For further discussion cf. pp. 108–09.

[23] Agassi (1963).

[24] An experimental discovery is *a chance discovery in the objective sense* if it is neither a confirming nor a refuting instance of some theory in the objective body of knowledge of the time; it is *a chance discovery in the subjective sense* if it is made (or recognised) by the discoverer neither as a confirming nor as a refuting instance of some theory he personally had entertained at the time.

[25] Agassi (1963), pp. 64–74.

[26] Within the Popperian circle, it was Agassi and Watkins who particularly emphasized the importance of unfalsifiable or barely testable '*empirical*' theories in providing *external* stimulus to later properly *scientific* developments. (Cf. Agassi, 1964 and Watkins, 1958.) This idea, of course, is already there in Popper's (1934) and (1960). Cf. my (1970), p. 184; but the new formulation of the difference between their approach and mine which I am going to give in this paper will, I hope, be much clearer.

[27] Popper occasionally – and Feyerabend systematically – stressed the catalytic (*external*) role of alternative theories in devising so-called 'crucial experiments'. But alternatives are not merely catalysts, which can be later removed in the rational reconstruction, they are *necessary* parts of the falsifying process. Cf. Popper (1940) and Feyerabend (1965); but cf. also Lakatos (1970), especially p. 121, footnote 4.

[28] Cf. Popper (1968a) and (1968b).

[29] The terms 'progressive' and 'degenerating problemshifts', 'research programmes', 'superseding' will be crudely defined in what follows – for more elaborate definitions see my (1968b) and especially my (1970).

[30] Popper does not permit this: 'There is a vast difference between my views and conventionalism. I hold that what characterises the empirical method is just this: our conventions determine the acceptance of the *singular*, not of the *universal* statements' (Popper, 1934, Section 30).

[31] The falsificationist hotly denies this: 'Learning from experience is learning from a refuting instance. The refuting instance then becomes a problematic instance'. (Agassi, 1964 p. 201). In his (1969) Agassi attributed to Popper the statement that 'we learn from experience by refutations' (p. 169), and adds that according to Popper one can learn *only* from refutation but not from corroboration (p. 167). Feyerabend, even in his (1969), says that '*negative instances suffice in science*'. But these remarks indicate a very

one-sided theory of learning from experience. (Cf. my (1970), p. 121, footnote 1, and p. 123.)

[32] Duhem, as a staunch positivist within philosophy of science, would, no doubt, exclude most 'metaphysics' as unscientific and would not allow it to have any influence on science proper.

[33] Cf. my (1968a), pp. 383–6, my (1968b), pp. 162–7, and my (1970), pp. 116ff. and pp. 155ff.

[34] Cf. Popper (1934), Section 85.

[35] Cf. Duhem (1906), Part II, Chapter VI, § 10.

[36] In fact, I define a research programme as degenerating even if it anticipates novel facts but does so in a patched-up development rather than by a coherent, pre-planned positive heuristic. I distinguish three types of *ad hoc* auxiliary hypotheses: those which have no excess empirical content over their predecessor ('*ad hoc*$_1$'), those which do have such excess content but none of it is corroborated ('*ad hoc*$_2$') and finally those which are not *ad hoc* in these two senses but do not form an integral part of the positive heuristic ('*ad hoc*$_3$'). Examples for an *ad hoc*$_1$ hypothesis are provided by the linguistic prevarications of pseudosciences, or by the conventionalist stratagems discussed in my (1963–4), like 'monsterbarring', 'exceptionbarring', 'monsteradjustment', etc. A famous example of an *ad hoc*$_2$ hypothesis is provided by the Lorentz-Fitzgerald contraction hypothesis; an example of an *ad hoc*$_3$ hypothesis is Planck's first correction of the Lummer-Pringsheim formula (also cf. p. 103). Some of the cancerous growth in contemporary social 'sciences' consists of a cobweb of such *ad hoc*$_3$ hypotheses, as shown by Meehl and Lykken. (For references, cf. my (1970), p. 175, footnotes 2 and 3.)

[37] The rivalry of two research programmes is, of course, a protracted process during which it is rational to work in either (*or, if one can, in both*). The latter pattern becomes important, for instance, when one of the rival programmes is vague and its opponents wish to develop it in a sharper form in order to show up its weakness. Newton elaborated Cartesian vortex theory in order to show that it is inconsistent with Kepler's laws. (Simultaneous work on rival programmes, of course, undermines Kuhn's thesis of the psychological incommensurability of rival paradigms.)

The progress of one programme is a vital factor in the degeneration of its rival. If programme P_1 constantly produces 'novel facts' these, by definition, will be anomalies for the rival programme P_2. If P_2 accounts for these novel facts only in an *ad hoc* way, it is degenerating by definition. Thus the more P_1 progresses, the more difficult it is for P_2 to progress.

[38] Cf. especially my (1970), pp. 120–1.

[39] Cf. especially my (1968a), p. 385 and (1970), p. 121.

[40] For instance, a rival theory, which acts as an *external* catalyst for the Popperian falsification of a theory, here becomes an *internal* factor. In Popper's (and Feyerabend's) reconstruction such a theory, after the falsification of the theory under test, can be removed from the rational reconstruction; in my reconstruction it has to stay within the internal history lest the falsification be undone. (Cf. note 27.)

Another important consequence is the difference between Popper's discussion of the Duhem-Quine argument and mine; cf. on the one hand Popper (1934), last paragraph of section 18 and Section 19, footnote 1; Popper (1957b), pp. 131–3; Popper (1963a), p. 112, footnote 26, pp. 238–9 and p. 243; and on the other hand, my (1970), pp. 184–9.

[41] For the falsificationist this is a repulsive idea; cf. e.g. Agassi (1963), pp. 48ff.

[42] Feyerabend seems now to deny that even this is a possibility; cf. his (1970a) and especially (1970b) and (1971).

[43] I use 'verisimilitude' here in Popper's technical sense, as the difference between the truth content and falsity content of a theory. Cf. his (1963a), Chapter 10.
[44] For a more general discussion of this problem, cf. pp. 108–09.
[45] Also cf. p. 94, 96, 98, 106, 120.
[46] The reader should remember that in this paper I discuss only naive falsificationism; cf. note 19.
[47] This is Kuhn's comment on Galileo's successful *pre*diction of the phases of Venus (Kuhn, 1957, p. 224). Like Mill and Keynes before him, Kuhn cannot understand why the historic order of theory and evidence should count, and he cannot see the importance of the fact that Copernicans *pre*dicted the phases of Venus, while Tychonians only explained them by *post hoc* adjustments. Indeed, since he does not see the importance of the fact, he does not even care to mention it.
[48] Cf. note 36.
[49] For a statement of this position and an interesting critical discussion cf. Polanyi (1951), pp. 4ff and pp. 78ff.
[50] Cf. Popper (1963b) and Musgrave (1969).
[51] This was illustrated convincingly, by Elkana, for the case of the so-called simultaneous discovery of the conservation of energy; cf. his (1971).
[52] Also cf. note 47.
[53] For the Mertonian brand of functionalism – as Alan Musgrave pointed out to me – priority disputes constitute a *prima facie* disfunction and therefore an anomaly for which Merton has been labouring to give a general socio-psychological explanation. (Cf. e.g. Merton 1957, 1963 and 1969.) According to Merton "scientific *knowledge* is not the richer or the poorer for having credit given where credit is due: it is the social *institution* of science and individual men of science that would suffer from repeated failures to allocate credit justly" (Merton, 1957, p. 648). But Merton overdoes his point: in important cases (like in some of Galileo's priority fights) there was more at stake than institutional interests: the problem was whether the Copernican research programme was progressive or not. (Of course, not all priority disputes have scientific relevance. For instance, the priority dispute between Adams and Leverrier about who was first to discover Neptune had no such relevance: whoever discovered it, the discovery strengthened the same (Newtonian) programme. In such cases Merton's external explanation may well be true.)
[54] Kuhn (1970), p. 239; my italics.
[55] Feyerabend (1970), p. 215.
[56] *Ibid.*
[57] Cf. note 2.
[58] I do, of course, *not* claim that such decisions are necessarily uncontroversial. In such decisions one has to use also one's *common sense*. Common sense (that is, judgment in *particular* cases which is not made according to mechanical rules but only follows general principles which leave some *Spielraum*) plays a role in all brands of non-mechanical methodologies. The Duhemian conventionalist needs common sense to decide when a theoretical framework has become sufficiently cumbersome to be replaced by a 'simpler' one. The Popperian falsificationist needs common sense to decide when a basic statement is to be 'accepted', or to which premise the *modus tollens* is to be directed. (Cf. my (1970), pp. 106ff.) But neither Duhem, nor Popper gives a blank cheque to 'common sense'. They give very definite guidance. The Duhemian judge directs the jury of common sense to agree on comparative simplicity; the Popperian judge directs the jury to look out primarily for, and agree upon, accepted basic statements which clash with

accepted theories. My judge directs the jury to agree on appraisals of progressive and degenerating research programmes. But, for example, there may be conflicting views about whether an accepted basic statement expresses a *novel* fact or not. Cf. my (1970), p. 156.

Although it is important to reach agreement on such verdicts, there must also be the possibility of appeal. In such appeals inarticulated common sense is questioned, articulated and criticised. (The criticism may even turn from a criticism of law interpretation into a criticism of the law itself.)

[59] Most conventionalists have also an intermediate inductive layer of 'laws' between facts and theories; cf. note 15.

[60] The proposition "the Proutian programme was carried through" looks like a 'factual' proposition. But there are no 'factual' propositions: the phrase only came into ordinary language from dogmatic empiricism. *Scientific 'factual' propositions* are theory-laden: the theories involved are 'observational theories'. *Historiographical 'factual' propositions* are also theory-laden: the theories involved are methodological theories. In the decision about the truth-value of the 'factual' proposition, 'the Proutian programme was carried through,' two methodological theories are involved. First, the theory that the units of scientific appraisal are research programmes; secondly, some *specific* theory of how to judge whether a programme was 'in fact' carried through. For all these considerations a Popperian internal historian will not need to take any interest whatsoever in the *persons* involved, or in their beliefs about their own activities.

[61] The 'first world' is that of matter, the 'second' the world of feelings, beliefs, consciousness, the 'third' the world of objective knowledge, articulated in propositions. This is an age-old and vitally important trichotomy; its leading contemporary proponent is Popper. Cf. Popper (1968a), (1968b) and Musgrave (1969) and (1971a).

[62] Of course what, in this context, constitutes 'false belief' (or 'false consciousness'), depends on the rationality theory of the critic: cf. pp. 94, 96 and 98. But no rationality theory can ever succeed in leading to 'true consciousness'.

[63] If the publication of Bohr's programme had been delayed by a few years, further speculation might even have led to the spin problem without the previous observation of the anomalous Zeeman effect. Indeed, Compton raised the problem in the context of the Bohrian programme in his (1919).

[64] I first applied this expositional device in my (1963–4); I used it again in giving a detailed account of the Proutian and the Bohrian programmes; cf. my (1970), pp. 138, 140, 146. This practice was criticised at the 1969 Minneapolis conference by some historians. McMullin, for instance, claimed that this presentation may illuminate a *methodology*, but certainly not real *history*: the text tells the reader what ought to have happened and the footnotes what in fact happened (cf. McMullin, 1970). Kuhn's criticism of my exposition ran essentially on the same lines: he thought that it was a specifically *philosophical* exposition: "a *historian* would not include *in his narrative* a factual report which he knows to be false. If he had done so, he would be so sensitive to the offence that he could not conceivably compose a footnote calling attention to it." (Cf. Kuhn, 1970, p. 256.)

[65] Cf. L. P. Williams (1970).

[66] Perhaps I should emphasize the difference between on the one hand, *inductivist historiography of science*, according to which *science* proceeds through discovery of hard facts (in nature) and (possibly) inductive generalisations, and, on the other hand, the *inductivist theory of historiography of science* according to which *historiography of science* proceeds through discovery of hard facts (in history of science) and (possibly)

inductive generalisations. 'Bold conjectures', 'crucial negative experiments', and even 'progressive and degenerating research programmes' may be regarded as 'hard historical facts' by some inductivist historiographers. One of the weaknesses of Agassi's (1963) is that he omitted to emphasize this distinction between scientific and historiographical inductivism.

[67] Cf. Popper (1957b), Section 31.

[68] This thesis implies that the work of those 'externalists' (mostly trendy 'sociologists of science') who claim to do social history of some scientific discipline without having mastered the discipline itself, and its internal history, is worthless. Also cf. Musgrave (1971a).

[69] Unfortunately there is only one single word in most languages to denote history_1 (the set of historical events) and history_2 (a set of historical propositions). Any history_2 is a theory and value-laden reconstruction of history_1.

[70] That is, a hypothesis h is scientific only if there is a number q such that $p(h, e) = q$ where e is the available evidence and $p(h, e) = q$ can be *proved*. It is irrelevant whether p is a Carnapian confirmation function or a Popperian corroboration function as long as $p(h, e) = q$ is allegedly proved. (Popper's third note on corroboration, of course, is only a curious slip which is out of tune with his philosophy: cf. my (1968a), pp. 411–7.)

Probabilism has never generated a programme of historiographical reconstruction; it has never emerged from grappling – unsuccessfully – with the very problems it created. As an epistemological programme it has been degenerating for a long time; as a historiographical programme it never even started.

[71] Popper (1934), Sections 11 and 85. Also cf. the comment in my (1971a), footnote 13.

The methodology of research programmes too is, in the first instance, defined as a game; cf. especially pp. 99–100.

[72] This whole problem area is the subject of my (1968a), pp. 390ff, but especially of my (1971a).

[73] Cf. Popper (1934), Sections 4 and 11. Popper's definition of science is, of course, his celebrated 'demarcation criterion'.

[74] For an excellent discussion of the distinction between nominalism and realism (or, as Popper prefers to call it, 'essentialism') in the theory of definitions, cf. Popper (1945), vol. II, chapter 11, and (1963a), p. 20.

[75] Popper (1934), Section 11.

[76] *Ibid.*

[77] Popper (1934), Section 4. But Popper, in his *Logik der Forschung* never specifies a *purpose* of the game of science that would go beyond what is contained in its rules. The thesis that the *aim* of science is *truth*, occurs only in his writings since 1957. All that he says in his *Logik der Forschung* is that the quest for truth may be a psychological *motive* of scientists. For a detailed discussion cf. my (1971a).

[78] This flaw is the more serious since Popper himself has expressed qualifications about his criterion. For instance in his [1963a] he describes 'dogmatism', that is, treating anomalies as a kind of 'background noise', as something that is 'to some extent necessary' (p. 49). But on the next page he identifies this 'dogmatism' with 'pseudoscience'. Is then pseudoscience 'to some extent necessary'? Also, cf. my (1970), p. 177, footnote 3.

[78a] Cf. Popper (1963), pp. 33–7.

[79] Popper (1934), Section 29.

[80] This approach, of course, does not imply that we *believe* that the scientists 'basic judgments' are unfailingly rational; it only means that we *accept* them in order to criticise universal definitions of science. (If we were to add that no such *universal*

definition has been found and no such *universal* definition will ever be found, the stage would be set for Polanyi's conception of the lawless closed autocracy of science.)

My meta-criterion may be seen as a 'quasi-empirical' self-application of Popperian falsificationism. I introduced this 'quasi-empiricalness' earlier in the context of mathematical philosophy. We may abstract from *what* flows in the logical channels of a deductive system, whether it is something certain or something fallible, whether it is truth and falsehood or probability and improbability, or even moral or scientific desirability and undesirability: it is the *how* of the flow which decides whether the system is negativist, 'quasi-empirical', dominated by *modus tollens* or whether it is justificationist, 'quasi-Euclidean', dominated by *modus ponens*. (Cf. my (1967).) This 'quasi-empirical' approach may be applied to *any* kind of normative knowledge: Watkins has already applied it to ethics in his (1963) and (1967). But now I prefer another approach: cf. note 122.

[81] It may be noted that this metacriterion does not have to be construed as psychological, or 'naturalistic' in Popper's sense. (Cf. his (1934), Section 10.) The definition of the 'scientific *élite*' is not simply an empirical matter.

[82] Popper (1963a), p. 38, footnote 3; my italics. This, of course, is equivalent to his celebrated 'demarcation criterion' between [internal, rationally reconstructed] science and non-science (or 'metaphysics'). The latter may be [externally] 'influential' and has to be branded as pseudoscience only if it declares itself to be science.

[83] Cf. my (1970), pp. 100–1.

[84] Cf. e.g. his (1934), Section 18.

[85] Cf. my (1970), especially pp. 135ff.

[86] *Ibid.*, pp. 138ff.

[87] Cf. Popper (1934), Section 24.

[88] Cf. my (1970), especially pp. 140ff.

[89] In general Popper stubbornly overestimates the immediate striking force of purely negative criticism. "Once a mistake, or a contradiction, is pinpointed, there can be no verbal evasion: it can be proved, and that is that" (Popper, 1959, p. 394). He adds: "Frege did not try evasive manoeuvres when he received Russell's criticism." But of course he did. (Cf. Frege's *Postscript* to the second edition of his *Grundgesetze*.)

[90] Interestingly, as Kuhn points out, "a consistent interest in historical problems and a willingness to engage in original historical research distinguishes the men [Popper] has trained from the members of any other current school in the philosophy of science" (Kuhn 1970, p. 236). For a hint at a possible explanation of the apparent discrepancy cf. note 129.

[91] For instance, he claims that a perpetual motion machine would 'refute' (on his terms) the first law of thermodynamics (1934, Section 15). But how can one interpret, on Popper's own terms, the statement that '*K* is a perpetual motion machine' as a 'basic', that is, as a spatio-*temporally* singular statement?

[92] I am referring to Feyerabend's (1970) and (1971).

[93] Cf. Popper (1934), Section 30 and Popper (1945), Vol. II, pp. 220–1. He stressed that Einstein's problem was how to account for experiments 'refuting' classical physics and he "did not... set out to criticise our conceptions of space and time." But Einstein certainly did. His Machian criticism of our concepts of space and time, and, in particular his operationalist criticism of the concept of simultaneity played an important role in his thinking.

I discussed the role of the Michelson-Morley experiments at some length in my (1970).

Popper's competence in physics would never, of course, have allowed him to distort the history of relativity theory as much as Beveridge, who wanted to persuade economists to an empirical approach by setting them Einstein as an example. According to Beveridge's falsificationist reconstruction, Einstein 'started [in his work on gravitation] from facts [which refuted Newton's theory, that is,] from the movements of the planet Mercury, the unexplained aberrancies of the moon' (Beveridge, 1937). Of course, Einstein's work on gravitation grew out from a 'creative shift' in the positive heuristic of his special relativity programme, and certainly not from pondering over Mercury's anomalous perihelion or the moon's devious, unexplained aberrancies.

[94] Popper (1963a), pp. 220, 239, 242–3 and (1963b), p. 965. Popper, of course, is left with the problem why 'counterexamples' (that is, anomalies) are not recognised immediately as causes for rejection. For instance, he points out that in the case of the breakdown of parity "there had been many observations – that is, photographs of particle tracks – from which we might have read off the result, but the observations had been either ignored or misinterpreted" (1963b, p. 965). Popper's – external – explanation seems to be that scientists have not yet learned to be sufficiently critical and revolutionary. But is not it a better – and internal – explanation that the anomalies *had* to be ignored until some progressive alternative theory was offered which turned the counterexamples into examples?

[95] *Op. cit.*, p. 246.

[96] As I mentioned, one Popperian, Agassi, did write a book on the historiography of science (Agassi, 1963). The book has some incisive critical sections flogging inductivist historiography, but he ends up by replacing inductivist mythology by falsificationist mythology. For Agassi *only* those facts have scientific (internal) significance which can be expressed in propositions which conflict with some extant theory: only their discovery deserves the honorific title 'factual discovery'; factual propositions which *follow from* rather than *conflict with* known theories are irrelevant; so are factual propositions which are *independent of* them. If some valued factual disovery in the history of science is known as a confirming instance or chance discovery, Agassi boldly predicts that on *close* investigation they will turn out to be refuting instances, and he offers five case-studies to support his claim (pp. 60–74). Alas, on *closer* investigation it turns out that Agassi got wrong all the five examples which he adduced as confirming instances of his historiographical theory. In fact all the five examples (in our normative meta-falsificationist sense) 'falsify' his historiography.

[97] Cf. Duhem (1906), Popper (1948) and (1957), Agassi (1963).

[98] Of course, an inductivist may have the temerity to claim that genuine science has not yet started and may write a history of extant science as a history of bias, superstition and false belief.

[99] Cf. Popper (1934), Section 19.

[100] Cf. Polanyi (1951), p. 70.

[101] Kuhn (1957). Also cf. Price (1959).

[102] Cohen (1960), p. 61. Bernal, in his (1954), says that "[Copernicus's] reasons for [his] revolutionary change were essentially philosophic and aesthetic [that is, in the light of conventionalism, scientific];" but in later editions he changes his mind: "[Copernicus's] reasons were mystical rather than scientific."

[103] For a more detailed sketch cf. my (1971b).

[104] Other types of criticism of methodologies may, of course, be easily devised. We may, for instance, apply the standards of each methodology (not only falsificationism) to itself. The result, for most methodologies, will be equally destructive: inductivism

cannot be proved inductively, simplicity will be seen as hopelessly complex. (For the latter cf. end of note 106.)

[105] Cf. Polanyi (1958), Kuhn (1962), Holton (1969), Feyerabend (1970) and (1971). I should also add Lakatos (1963–4), (1968b), and (1970).

[106] Kuhn (1957). Such historiographical criticism can easily drive some rationalists into an irrational defence of their favourite falsified rationality theory. Kuhn's historiographical criticism of the simplicity theory of the Copernican revolution shocked the conventionalist historian Richard Hall so much that he published a polemic article in which he singled out and re-asserted those aspects of Copernican theory which Kuhn himself had mentioned as possibly having a claim to higher simplicity, and ignored the rest of Kuhn's – valid – argument (Hall, 1970). No doubt, simplicity can always be defined for *any* pair of theories T_1 and T_2 in such a way that the simplicity of T_1 is greater than that of T_2.

For further discussion of conventionalist historiography cf. my (1971b).

[107] Thus Polanyi is a conservative rationalist concerning science, and an 'irrationalist' concerning the philosophy of science. But, of course, this meta-'irrationalism' is a perfectly respectable brand of rationalism: to claim that the concept of 'scientifically acceptable' cannot be further defined, but only transmitted by the channels of 'personal knowledge', does not make one an outright irrationalist, only an outright conservative. Polanyi's position in the philosophy of natural science corresponds closely to Oakeshott's ultra-conservative philosophy of political science. (For references and an excellent criticism of the latter cf. Watkins (1952)). Also cf. pp. 120–122.

[108] Of course, none of the critics were aware of the exact logical character of meta-methodological falsificationism as explained in this section and none of them applied it completely consistently. One of them writes: 'At this stage we have not yet developed a general theory of criticism even for scientific theories, let alone for theories of rationality: therefore if we want to falsify methodological falsificationism, we have to do it before having a theory of how to do it' (Lakatos, 1970, p. 114).

[109] I used the critical machinery developed in this paper against Feyerabend's epistemological anarchism in my (1971b).

[110] Kuhn's vision was criticised from many quarters; cf. Shapere (1964 and 1967) Scheffler (1967) and especially the critical comments by Popper, Watkins, Toulmin, Feyerabend and Lakatos – and Kuhn's reply – in Lakatos and Musgrave (1970). But none of these critics applied a systematic *historiographical* criticism to his work. One should also consult Kuhn's 1970 *Postscript* to the second edition of his (1962) and its review by Musgrave (Musgrave, 1971b).

[111] Cf. Feyerabend (1970a, 1970b and 1971); and Kuhn (1970).

[112] For instance, one may refer to the actual immediate impact of at least *some* 'great' negative crucial experiments, like that of the falsification of the parity principle. Or one may quote the high respect for at least *some* long, pedestrian, trial-and-error procedures which occasionally precede the announcement of a major research programme, which in the light of my methodology is, at best, 'immature science'. (Cf. my (1970), p. 175; also cf. L. P. Williams's reference to the history of spectroscopy between 1870 and 1900 in his (1970)). Thus the judgment of the scientific élite, on occasions, goes also against *my* universal rules too.

[113] There is a certain analogy between this pattern and the occasional appeal procedure of the theoretical scientist against the verdict of the experimental jury; cf. my (1970), pp. 127–31.

[114] This latter criterion is analogous to the exceptional 'depth' of a theory which clashes

with some basic statements available at the time and, at the end, emerges from the clash victoriously. (Cf. Popper's, 1957a) Popper's example was the inconsistency between Kepler's laws and the Newtonian theory which set out to explain them.
[115] Conventionalism, of course, had performed this historic role to a great extent before Popper's version of falsificationism.
[116] Van der Waerden had thought that the Bohr-Kramers-Slater theory was bad: Popper's theory showed it to be good. Cf. Van der Waerden (1967), p. 13 and Popper (1963a), pp. 242ff; for a critical discussion cf. my (1970), p. 168, footnote 4 and p. 169, footnote 1.
[117] The attitude of some modern logicians to the history of mathematics is a typical example; cf. my (1963–4), p. 3.
[118] This formulation was suggested to me by my friend Michael Sukale.
[119] Cf. my (1970), Section 3(c).
[120] Cf. my (1970), pp. 138–73.
[121] Duhem himself gives only one explicit example: the victory of wave optics over Newtonian optics (1906), Chapter VI, § 10 (also see Chapter IV, § 4). But where Duhem relies on intuitive 'common sense', I rely on an analysis of rival problemshifts (cf. my (1972)).
[122] One may introduce the notion of '*degree of correctness*' into the meta-theory of methodologies, which would be analogous to Popper's empirical content. Popper's empirical 'basic statements' would have to be replaced by quasi-empirical 'normative basic statements' (like the statement that 'Planck's radiation formula is arbitrary').

Let me point out here that the methodology of research programmes may be applied not only to norm-impregnated historical knowledge but to any normative knowledge, including even ethics and aesthetics. This would then supersede the naive falsificationist 'quasi-empirical' approach as outlined on Note 80.
[123] Cf. text to note 9. (The term 'wild speculation' is, of course, a term inherited from inductivist methodology. It should now be reinterpreted as 'degenerating programme'.)
[124] The fact that even degenerating externalist theories have been able to achieve some respectability was to a considerable extent due to the weakness of their previous internalist rivals. Utopian Victorian morality either creates false, hypocritical accounts of bourgeois decency, or adds fuel to the view that mankind is totally depraved; utopian scientific standards either create false, hypocritical accounts of scientific perfection, or add fuel to the view that scientific theories are no more than mere beliefs bolstered by some vested interests. This explains the 'revolutionary' aura which surrounds some of the absurd ideas of contemporary sociology of knowledge: some of its practitioners claim to have unmasked the bogus rationality of science, while, at best, they exploit the weakness of outdated theories of scientific rationality.
[125] For examples cf. Cantor (1971) and the Forman-Ewald debate (Forman, 1969 and Ewald, 1969).
[126] I call '*historiographical positivism*' the position that history can be written as a completely *external* history. For historiographical positivists history is a purely empirical discipline. They deny the existence of objective standards as opposed to mere beliefs about standards. (Of course, they too hold beliefs about standards which determine the choice and formulation of their historical problems.) This position is typically Hegelian. It is a special case of *normative positivism*, of the theory that sets up might as the criterion of right. (For a criticism of Hegel's ethical positivism cf. Popper (1945), Vol. I, pp. 71–2, Vol. II, pp. 305–6 and Popper (1961).) Reactionary Hegelian obscurantism

pushed values back completely into the world of facts; thus reversing their separation by Kantian philosophical enlightenment.

[127] Kuhn seems to be in two minds about objective scientific progress. I have no doubt that, being a devoted scholar and scientist, he *personally* detests relativism. But his *theory* can either be interpreted as denying scientific progress and recognising only scientific change; or, as recognising scientific progress but as 'progress' marked solely by the march of actual history. Indeed, on his criterion, he would have to describe the catastrophe mentioned in the text as a proper 'revolution'. I am afraid this might be one clue to the unintended popularity of his theory among the New Left busily preparring the 1984 'revolution'.

[128] The technical term 'Euclidean' (or rather 'quasi-Euclidean') means that one starts with universal, high level propositions ('axioms') rather than singular ones. I suggested in my (1967) and (1962) that the 'quasi-Euclidean' versus 'quasi-empirical' distinction is more useful than the '*a priori*' versus '*a posteriori*' distinction.

Some of the 'apriorists' are, of course, empiricists. But empiricists may well be apriorists (or, rather, 'Euclideans') on the meta-level here discussed.

[129] Some might claim that Popper does *not* fall into this category. After all, Popper defined 'science' in such a way that it should include the refuted Newtonian theory and exclude unrefuted astrology, Marxism and Freudianism.

[130] This seems to be the case in modern particle physics; or according to some philosophers and physicists even in the Copenhagen school of quantum physics.

[131] This is the case with some of the main schools of modern sociology, psychology and social psychology.

[132] This, of course, explains why a good methodology – 'distilled' from the mature sciences – may play an important role for immature and, indeed, dubious disciplines. While Polanyiite academic autonomy should be defended for departments of theoretical physics, it must not be tolerated, say, in institutes for computerised social astrology, science planning or social imagistics. (For an authoritative study of the latter, cf. Priestley (1968).)

[133] Of course, a critical discussion of scientific standards, possibly leading even to their improvement, is impossible without articulating them in general terms; just as if one wants to challenge a language, one has to articulate its grammar. Neither the conservative Polanyi nor the conservative Oakeshott seem to have grasped (or to have been inclined to grasp) the *critical* function of language – Popper has. (Cf. especially Popper (1963a), p. 135).

[134] Cf. e.g. my (1962), p. 157 or my (1968a), p. 387, footnote 1.

REFERENCES

Agassi, J. (1963), *Towards an Historiography of Science*.

Agassi, J. (1964), 'Scientific Problems and their Roots in Metaphysics', in *The Critical Approach to Science and Philosophy* (ed. by M. Bunge), pp. 189–211.

Agassi, J. (1966), 'Sensationalism', *Mind* **75**, 1–24.

Agassi, J. (1969), 'Popper on Learning from Experience', in *Studies in the Philosophy of Science* (ed. by N. Rescher), pp. 162–71.

Bernal, J. D. (1954), *Science in History*, 1st Edition.

Bernal, J. D. (1965), *Science in History*, 3rd Edition.

Beveridge, W. (1937), 'The Place of the Social Sciences in Human Knowledge', *Politica* **2**, 459–79.
Cantor, G. (1971), 'A Further Appraisal of the Young-Brougham Controversy', in *Studies in the History and Philosophy of Science*, forthcoming.
Cohen, I. B. (1960), *The Birth of a New Physics*.
Compton, A. H. (1919), 'The Size and Shape of the Electron', *Physical Review* **14**, 20–43.
Duhem, P. (1905), *La théorie physique, son objet et sa structure* (English transl. of 2nd (1914) edition: *The Aim and Structure of Physical Theory*, 1954).
Elkana, Y. (1971), 'The Conservation of Energy: a Case of Simultaneous Discovery?', *Archives Internationales d'Histoire des Sciences* **24**, 31–60.
Ewald, P. (1969), 'The Myth of Myths', *Archive for the History of Exact Science* **6**,72–81.
Feyerabend, P. K. (1964), 'Realism and Instrumentalism: Comments on the Logic of Factual Support', in *The Critical Approach to Science and Philosophy* (ed. by M. Bunge), pp. 280–308.
Feyerabend, P. K. (1965), 'Reply to Criticism', in *Boston Studies in the Philosophy of Science* **2** (ed. by R. S. Cohen and M. Wartofsky), pp. 223–61.
Feyerabend, P. K. (1969), 'A Note on Two "Problems" of Induction', *British Journal for the Philosophy of Science* **19**, 251–3.
Feyerabend, P. K. (1970a), 'Consolations for the Specialist', in *Criticism and the Growth of Knowledge* (ed. by I. Lakatos and A. Musgrave), pp. 197–230.
Feyerabend, P. K. (1970b), 'Against Method', in *Minnesota Studies for the Philosophy of Science* **4**.
Feyerabend, P. K. (1971), *Against Method* [expanded version of Feyerabend (1970b)].
Forman, P. (1969), 'The Discovery of the Diffraction of X-Rays by Crystals: A Critique of the Critique of the Myths', *Archive for History of Exact Sciences* **6**, 38–71.
Hall, R. J. (1970), 'Kuhn and the Copernican Revolution', *British Journal for the Philosophy of Science* **21**, 196–97.
Hempel, C. G. (1937), Review of Popper (1934), *Deutsche Literaturzeitung*, pp. 309–14.
Holton, G. (1969), 'Einstein, Michelson, and the "Crucial" Experiment', *Isis* **6**, 133–97.
Kuhn, T. S. (1957), *The Copernican Revolution*.
Kuhn, T. S. (1962), *The Structure of Scientific Revolutions*.
Kuhn, T. S. (1968), 'Science: The History of Science', in *International Encyclopedia of the Social Sciences* (ed. by D. L. Sills), Vol. **14**, pp. 74–83.
Kuhn, T. S. (1970), 'Reflections on my Critics', in *Criticism and the Growth of Knowledge* (ed. by I. Lakatos and A. Musgrave), pp. 237–78.
Lakatos, I. (1962), 'Infinite Regress and the Foundations of Mathematics', *Aristotelian Society Supplementary Volume* **36**, 155–84.
Lakatos, I. (1963-4), 'Proofs and Refutations', *The British Journal for the Philosophy of Science* **14**, 1–25, 120–39, 221–43, 296–342.
Lakatos, I. (1966), 'Popkin on Skepticism', in *Logic, Physics and History* (ed. by W. Yourgrau and A. D. Breck), 1970, pp. 220–3.
Lakatos, I. (1967), 'A Renaissance of Empiricism in the Recent Philosophy of Mathematics', in *Problems in the Philosophy of Mathematics* (ed. by I. Lakatos), pp. 199–202.
Lakatos, I. (1968a), 'Changes in the Problem of Inductive Logic', in *The Problem of Inductive Logic* (ed. by I. Lakatos), pp. 315–417.
Lakatos, I. (1968b), 'Criticism and the Methodology of Scientific Research Programmes', *Proceedings of the Aristotelian Society* **69**, 149–86.
Lakatos, I. (1970), 'Falsification and the Methodology of Scientific Research Programmes', in *Criticism and the Growth of Knowledge* (ed. by I. Lakatos and A. Musgrave).

Lakatos, I. (1971a), 'Popper on Demarcation and Induction' in *The Philosophy of Sir Karl Popper* (ed. by P. A. Schilpp), forthcoming. (Available in German in *Neue Aspekte der Wissenschaftstheorie* ed. by H. Lenk.)

Lakatos, I. (1971b), 'A Note on the Historiography of the Copernican Revolution', forthcoming.

Lakatos, I. (1972), *The Changing Logic of Scientific Discovery*, forthcoming.

Lakatos, I. and Musgrave, A. (1970), *Criticism and the Growth of Knowledge*.

McMullin, E. (1970), 'The History and Philosophy of Science: a Taxonomy', *Minnesota Studies in the Philosophy of Science* **5**, 12–67.

Merton, R. (1957), 'Priorities in Scientific Discovery', *American Sociological Review* **22**, 635–59.

Merton, R. (1963), 'Resistance to the Systematic Study of Multiple Discoveries in Science', *European Journal of Sociology* **4**, 237–82.

Merton, R. (1969), 'Behaviour Patterns of Scientists', *American Scholar* **38**, 197–225.

Musgrave, A. (1969), *Impersonal Knowledge: A Criticism of Subjectivism*, Ph. D. thesis, University of London.

Musgrave, A. (1971a), 'The Objectivism of Popper's Epistemology', in *The Philosophy of Sir Karl Popper* (ed. by P. A. Schilpp), forthcoming.

Musgrave, A. (1971b), 'Kuhn's Second Thoughts', *British Journal for the Philosophy of Science* **22**, pp. 287–97.

Polanyi, M. (1951), *The Logic of Liberty*.

Polanyi, M. (1958), *Personal Knowledge, Towards a Post-Critical Philosophy*.

Popper, K. R. (1935), *Logik der Forschung*.

Popper, K. R. (1940), 'What is Dialectic?', *Mind* **49**, 403–26; reprinted in Popper (1963), pp. 312–35.

Popper, K. R. (1945), *The Open Society and Its Enemies*, Vol. I–II.

Popper, K. R. (1948), 'Naturgesetze und theoretische Systeme', in *Gesetz und Wirklichkeit*, (ed. by S. Moser), pp. 65–84.

Popper, K. R. (1963), 'Three Views Concerning Human Knowledge', in *Contemporary British Philosophy* (ed. by H. D. Lewis), 1957, pp. 355–88; reprinted in Popper (1963), pp. 97–119.

Popper, K. R. (1957a), 'The Aim of Science', *Ratio* **1**, 24–35.

Popper, K. R. (1957b), *The Poverty of Historicism*.

Popper, K. R. (1959), *The Logic of Scientific Discovery*.

Popper, K. R. (1960), 'Philosophy and Physics', *Atti del XII Congresso Internazionale di Filosofia* **2**, 363–74.

Popper, K. R. (1961), 'Facts, Standards, and Truth: A Further Criticism of Relativism', *Addendum* to the Fourth Edition of Popper (1945).

Popper, K. R. (1963a), *Conjectures and Refutations*.

Popper, K. R. (1963b), 'Science: Problems, Aims, Responsibilities', *Federation Proceedings* **22**, 961–72.

Popper, K. R. (1968a), 'Epistemology Without a Knowing Subject', in *Proceedings of the Third International Congress for Logic, Methodology and Philosophy of Science* (ed. by B. Rootselaar and J. Staal), Amsterdam, pp. 333–73.

Popper, K. R. (1968b), 'On the Theory of the Objective Mind', in *Proceedings of the XIV International Congress of Philosophy*, Vol. **1**, pp. 25–33.

Price, D. J. (1959), 'Contra Copernicus: A Critical Re-estimation of the Mathematical Planetary Theory of Ptolemy, Copernicus and Kepler', in *Critical Problems in the History of Science* (ed. by M. Clagett), pp. 197–218.

Priestley, J. B. (1968), *The Image Men.*
Scheffler, I. (1967), *Science and Subjectivity.*
Shapere, D. (1964), 'The Structure of Scientific Revolutions', *Philosophical Review,* 383–84.
Shapere, S. (1967), 'Meaning and Scientific Change', in *Mind and Cosmos* (ed. by R. G. Colodny), pp. 41–85.
Van der Waerden, B. (1967), *Sources of Quantum Mechanics.*
Watkins, J. W. N. (1952), 'Political Tradition and Political Theory: an Examination of Professor Oakeshott's Political Philosophy', *Philosophical Quarterly* **2**, 323–37.
Watkins, J. W. N. (1958), 'Influential and Confirmable Metaphysics', *Mind* **67**, 344–65.
Watkins, J. W. N. (1963), 'Negative Utilitarianism', *Aristotelian Society Supplementary* **37**, 95–114.
Watkins, J. W. N. (1967), 'Decision and Belief', in *Decision Making* (ed. by R. Hughes), pp. 9–26.
Watkins, J. W. N. (1970), 'Against Normal Science', in *Criticism and the Growth of Knowledge* (ed. by I. Lakatos and A. Musgrave), pp. 25–38.
Williams, L. P. (1970), 'Normal Science and its Dangers', in *Criticism and the Growth of Knowledge* (ed. by I. Lakatos and A. Musgrave), pp. 49–50.

THOMAS S. KUHN

NOTES ON LAKATOS

I. INTRODUCTION

The invitation which has brought me here to comment on Professor Lakatos' paper has given me much pleasure, for I have long been an admirer of his work, particularly of his early four-part paper, 'Proofs and Refutations'. That does not mean, of course, that we have often agreed, but I have enjoyed the arguments that resulted and looked forward to this one. My pleasure, furthermore, was considerably enhanced when I discovered that Lakatos was going to be able to confound all precedent, his own and others, by getting this paper to me well in advance. It is a privilege few commentators are given, and I am correspondingly grateful.

All that I could have said before opening Lakatos' manuscript – in fact, I did so in letters to both Lakatos and Roger Buck. Reading it has only increased my satisfaction, but in an unanticipated way. As with some earlier Lakatos papers, I have had trouble with translation. Phrases like "the methodology of research programs" are not part of my familiar mode of communication; phrases like 'internal' and 'external history', although familiar, are used by Lakatos in novel and unexpected ways. I believe, however, that I have managed the translation, though perhaps without assimilating the language. As I have done so and simultaneously caught the spirit of his enterprise, I have been surprised and pleased at how congenial I find his present views. I conclude, finally, that I have read no paper on scientific method which expresses opinions so closely paralleling my own, and I am necessarily encouraged by that discovery, for it may mean that in the future I shall not be quite as alone in the methodological arena as I have been in the past. The resemblance between our views ought also, of course, to disqualify me as a commentator. One of my critics rather than I should be standing here, and if I had seen the difficulty in time, one would be. Since I did not, I shall have to do my best to play the critic. It is therefore fortunate that my agreement with Lakatos, however far it extends, is less than total.

II. PARALLELS

Before turning to the points at which we part company, I shall have to enumerate briefly and globally the areas in which our views coincide. There is, I think, no other way to isolate our difference, or, since 'difference' may not be the right word, to discover those portions of his paper in which Lakatos says things that I could never make my own.

Among our areas of agreement is the one Lakatos describes as meta-methodological or meta-historical. No historian, whether of science or some other human activity, can operate without preconceptions about what is essential, what is not. Those preconceptions do, if the historian deals with science, play an important role in determining what he takes to be 'internal', what 'external' in Lakatos' sense. Agassi has previously made the same point very effectively, and I welcome Lakatos' extension of it. I think of myself as having argued the converse even earlier, suggesting that failure to fit historical data provides grounds for criticizing a current methodological position. Lakatos has not, I shall shortly argue, yet altogether seen how to develop a philosophical basis for that converse, but I am not sure I have done better and am correspondingly gratified by his attempt.

That much agreement is probably not remarkable, but its extension from meta-methodology to substantive methodology is – or so it seems to me. I have, for example, repeatedly emphasized that the important scientific decisions – usually described as a choice between theories – are more accurately described as a choice between 'ways of doing science', or 'between traditions', or between 'paradigms.' Lakatos' insistence that the unit of choice is a 'scientific research program' seems to me to make the identical point.

Again, in discussing research conducted within a tradition, under the guidance of what I once called a paradigm, I have repeatedly insisted that it depends, in part, on the acceptance of elements which are not themselves subject to attack from within the tradition and which can be changed only by a transition to another tradition, another paradigm. Lakatos, I think, is making the same point when he speaks of the 'hard core of research programs,' the part which must be accepted in order to do research at all and which can be attacked only after embracing another research program.

Finally, though it does not exhaust our areas of agreement, I would point to Lakatos' emphasis on what he calls the 'degenerating stage' in the evolution of a research program, the stage in which it ceases to lead to new discoveries, in which *ad hoc* hypotheses accrue to it, and so on. I cannot myself tell the difference between what he has to say about this important stage and what I have said about the role of crisis in scientific development. Lakatos clearly does, but I get no help at all from the passages where he refers to them: for example, a reference late in his paper to "the Kuhnian psychological epiphenomenon of 'crisis'." [p. 120]

You will see, I think, why I speak of parallels and why I find them so encouraging. But they leave a puzzle. Why, if these parallels are real, is Lakatos so unable to see them? That he does not do so is illustrated by the phrase just quoted, and there are many others of the same sort in his paper. Undoubtedly part of the difficulty is the obscurity of my original presentation, something I can only regret. But I think that there is a deeper source, and it points to the areas in which we disagree or at least seem to.

Scattered through Lakatos' paper are a number of remarks like the following: Kuhn has, Lakatos suggests, come "up with a highly original vision of irrationally changing rational authority." [p. 116] Elsewhere he says, "When Kuhn and Feyerabend see irrational change, I predict that historians will be able to show there has been rational change." [p. 118] These reiterated contrasts between my irrationality and Lakatos' rationality isolate the difference which Lakatos sees between our views. For him it is apparently so deep that he remains blind to our close parallels. I shall argue that, even in suggesting the contrast, he is missing the point both of his present work and of my own.

I have never, in fact, accepted the description of my views as a defense of irrationality in science, but I have usually understood its source, seen why my critics thought the description apt. In this case, however, I cannot even do that. Considering the extent of the parallels between our views, Lakatos' use of terms like 'irrational' is, I think, only a mouthing of shibboleths. Either we are both defenders of irrationality, which I join him in doubting, or else, as I suppose, we are both trying to change a current notion of what rationality is. Arguments to that effect make up the balance of my remarks, though the issue in that form will not be entirely explicit until my conclusion.

III. 'INTERNAL' AND 'EXTERNAL'

Let me start by commenting on Lakatos' use of the terms 'internal' and 'external history'. In an early footnote he points out that the distinction is quite standard among historians of science but that he is using it in a new way. I am not, quite obviously, the man to be critical of a colleague who adapts an old term to his own purposes. What I think Lakatos does not realize, however, is how little need there is in this case to strain someone else's usage. The main virtue of the transition in terms is, I suspect, that it facilitates an unconscious sleight of hand.

In standard usage among historians, internal history is the sort that focuses primarily or exclusively on the professional activities of the members of a particular scientific community: What theories do they hold? What experiments do they perform? How do the two interact to produce novelty? External history, on the other hand, considers the relations between such scientific communities and the larger culture. The role of changing religious or economic traditions in scientific development thus belongs to external history, as does its converse. Among other standard topics for the externalist are scientific institutions and education, as well as the relations between science and technology. The internal-external distinction is not always hard and fast, but there is wide consensus in its application among historians. That consensus proves, I believe, at once implicitly vital and explicitly irrelevant to Lakatos' argument.

Obviously there is much overlap between normal usage and Lakatos'. In both, such factors as religion, economics, and education are external; Newton's Laws, Schrödinger's equation, and Lavoisier's experiments are internal. If there were no readily available alternatives, Lakatos' preemption of these terms would therefore be appropriate. But they would strain normal usage, for Lakatos' internal history is far narrower than that of the historian. It excludes, for example, all consideration of personal idiosyncrasy, whatever its role may have been in the choice of a theory, the creative act which produced it, or the form of the product which resulted. By the same token, it excludes such historical data as the failure of the man who creates a new theory and of his entire generation to see in that theory consequences which a later generation found there, a point I shall need to discuss further below. And, finally, it excludes consideration of mistakes or of what a later generation will see as having been mistakes and will accordingly feel constrained to correct.

Historical data of these sorts are all central and essential for the internal historian of science. Often they provide his most revealing clues to what occurred. Since Lakatos insists they be excluded from internal history, I wonder why he adopts the term. Could he not easily instead have spoken of rational history, or better, of history constructed from the rational elements in a science's development? I think that is what, most fundamentally, he means: the 'internal' in Lakatos' sense and in this context is closely equivalent to 'rational' in the ordinary sense. Furthermore, Lakatos' 'internal' carries with it from the ordinary use of 'rational' an all-important characteristic: as a criterion of selection it is prior to the pursuit of history and independent of it.

If that is right, then it is, of course, apparent why Lakatos does change terms. If 'internal' were an independent term unequivocally applied, as it is for the historian, then one could hope to learn something about rational methodology from the study of internal history. But if 'internal history' is simply the rational part of history, then the philosopher can learn from it about scientific method only what he puts in. Lakatos' meta-methodological method is in danger of reducing to tautology.

IV. LAKATOS THE HISTORIAN

As developed so far, my argument applies completely only to the first half of Lakatos' paper. That is the part in which he sets up his version of the internal-external distinction and then shows how what one takes to be internal and external changes with the choice of a prior methodological position. The second part of the paper is, of course, different. There he suggests that the choice of a methodology supplies a meta-historical research program. The actual attempt to apply such a program to historical data may show that the program is degenerating. As a result, a new methodology may arise and be accepted. I myself believe that exactly that can and does happen. Yet I wonder why Lakatos should expect it to. Given what he has made of the internal-external distinction, and given also his conception of what a historian does, no such effect is possible. Lakatos, I now want to argue, skirts as close to tautology in the second half of his paper as in the first.

Midway through the paper, for example, he remarks: "History of *science* (meaning here internal history) is a history of events which are selected and interpreted in a normative way." [p. 108] With that point I would thoroughly agree if it meant only that all historians necessarily select and interpret their data. But Lakatos, when he introduces the term 'normative' means something else. He has previously suggested that it is "philosophy of science [which] provides normative methodologies" [p.91] to the historian. His point is not simply that the historian selects and interprets, but that prior philosophy supplies the whole set of criteria by which he does so. If that were the case, however, there would be no way at all in which the selected and interpreted data could react back on a methodological position to change it.

Fortunately for Lakatos' point, other selective principles are available to the historian in addition to prior concepts of methodology. His narrative must, for example, be continuous in the sense that one event must lead into or set up the next; one may not skip about. In addition, his story must be plausible in the sense that men and institutions must behave in recognizable ways. It is legitimate to criticize a historian's narrative by saying: That cannot be what occurred, for only a madman would behave that way, and we have been given no reason to believe that the king was mad. Finally, and for present purposes most important, history must be constructed without doing violence to the data available for selection and interpretation. Only if these and other internal criteria of the historian's craft are used, can the results of historical research react back on and change the philosophical position with which the historian began.

My concern with Lakatos' paper is that it throws all these criteria away, thus depriving history of any philosophical function. For example, just before the last passage quoted, Lakatos writes: "One way to indicate discrepancies between history and its rational reconstruction is to relate the internal history *in the text*, and indicate *in the footnotes* how actual history 'misbehaved' in the light of its rational reconstruction." [p. 107] A recently published paper (his contribution to *Criticism and The Growth of Knowledge*) indicates what he means. In his text he tells a succession of straightforward stories, then in the footnotes he adds: that, of course, is not quite what happened; rather it is what would have happened if people had behaved rationally as they should. A somewhat different and equally informative example is contained in his present paper. Rational reconstruction, Lakatos suggests, can properly attribute the idea of electron spin to Bohr in 1913. Probably, he concedes, Bohr did not think of it then, but it was compatible with the research program implied by the Bohr atom. In fact, however, as Lakatos surely knows, Bohr was quite skeptical of the idea of spin even in 1925. That is not because Bohr was irrational. Instead, Lakatos, by once more discarding evidence which does not fits his prior principle of rationality, has misconstrued Bohr's program. If one constructs it properly from the evidence, one discovers that spin fits it very badly. From which program, Bohr's or Lakatos' misconstruction, ought philosophical analysis begin?

What I am trying to suggest, in short, is that what Lakatos conceives as history is not history at all but philosophy fabricating examples. Done in that way, history could not in principle have the slightest effect on the prior philosophical position which exclusively shaped it. That is not to say that historical reconstruction is not intrinsically a selective and interpretative enterprise, nor that a prior philosophical position has no role as a tool for selection and interpretation. But it is to insist that, in the only sort of history which can hold philosophical interest, a prior philosophical position is not the only selective principle and also that it is not, as a selective principle, inviolate. When one's historical narrative demands footnotes which point out its fabrications, then the time has come to reconsider one's philosophical position.

V. HISTORY AND IRRATIONALITY

Why is it, I now ask in conclusion, that Lakatos feels the need to protect himself from real history? Why does he provide a parody in its place? My best guess is that he fears that history, if taken seriously as an independent discipline, may lead him to the position he attributes to me; the view that science is fundamentally an irrational enterprise. As a hypothesis about causes and motives that can only be a guess, and nothing very important depends on its being correct. But what his paper does make unequivocally clear is his belief that I have been led to defend irrationality by taking seriously aspects of history which he seeks a basis for omitting or rewriting.

As I have said before, both here and elsewhere, I do not for a moment believe that science is an intrinsically irrational enterprise. What I have perhaps not made sufficiently clear, however, is that I take that assertion not as a matter of fact, but rather of principle. Scientific behavior, taken as a whole, is the best example we have of rationality. Our view of what it is to be rational depends in significant ways, though of course not exclusively, on what we take to be the essential aspects of scientific behavior. That is not to say that any scientist behaves rationally at all times, or even that many behave rationally very much of the time. What it does assert is that, if history or any other empirical discipline leads us to believe that the development of science depends essentially on behavior that we have previously thought to be irrational, then we should conclude not that science is irrational but that our notion of rationality needs adjustment here and there.

That position, so long as it remains abstract, is one with which Lakatos seems to agree. Whether or not he has managed it altogether correctly, the entire last half of his paper argues that historical study, properly done, can alter the line between the internal and external. In consequence, he says, it can change our notion of scientific rationality as well. Having taken that position, he may properly, of course, reject my views on substantive grounds; because I may have made historical, logical, or philosophical mistakes, as I doubtless have. What he may not do, but nevertheless does, is reject them *simply* or *merely* because my conclusions from history attribute an essential role to behavior he thinks irrational. Arguments of that sort contradict the core of his present methodological position.

So far I have argued the irrelevance of Lakatos' charge of irrationality on grounds of principle. Let me now try to make a similar point on substantive grounds. I began these comments by suggesting that Lakatos' present position has grown very close to my own. I shall close by suggesting that in key respects the parallel between our views goes even further than I then allowed. There are, I think, three main grounds on which charges of irrationality have been levelled at me. Two of these Lakatos now concedes, one explicitly, the other implicitly. The third he rejects in a footnote aside, ignoring in the process one of the most active and exciting areas in contemporary philosophy.

The first source, I think, of the charge that I make science an irrational enterprise is my insistence that the choice between paradigms (or theories, for present purposes) cannot be compelled by logic and experiment alone;

in these matters there is no such thing as proof, no point at which the opponent of a newer view violates a rule of science, begins to behave unscientifically. Lakatos makes exactly the same point repeatedly. "One may rationally stick to a degenerating program until it is overtaken *and even after*." [p. 104] "One *must* realize that one's opponent, even if lagging badly behind, may still stage a comeback." [p. 101] "No advantage for one side can ever be regarded as absolutely conclusive." [p. 101] If this be irrationality – as Lakatos has occasionally supposed in the past – then we are both guilty.

An even more frequent reason for the charge of irrationality has been my insistence that ultimately the choice between paradigms is a community decision, that what passes for proof, verification, or falsification in the sciences has not occurred until an entire community has been converted or re-formed about a new paradigm. On this point my views were not originally so clearly expressed as I should like, and they have in any case evolved since. What I should like to have said, however, is very close to what Lakatos now does say, though I am far from sure he realizes its consequences.

Throughout his paper Lakatos refers to the importance in scientific decision-making of what he calls a "code of scientific honesty" or a "code of scientific honor". [p. 92] When he distinguishes his position from the one to which he objects, he makes remarks like: "What one must *not* do is to deny [a research program's] poor public record", [p. 104] or "The scores of the rival sides... must be recorded and publicly displayed at all times." [p. 105] Elsewhere he speaks of answering colleagues's objections "by separating rational and irrational (or honest and dishonest) adherence to a degenerating research program." [p. 105]

Lakatos' views cannot, however, be distinguished from mine or anyone else's in this way. On the contrary, he and I come closest at just these points. Who does he suppose believes that science could continue if scientists were dishonest? If I have been defending the irrational, it has not been by defending lies. In fact, Lakatos' references to honesty, to a '*public* record', or to a score that must be '*recorded*' and '*publicly* displayed' suggest that he too is thinking of theory-choice as a community activity which would be impossible unless public records of this sort were kept. When the individual may decide alone, nothing of the sort is needed. Finally, and most important, Lakatos' emphasis on a code of honor carries him even further in the same direction, for a code consists of *values* not of rules, and values are intrinsically a community possession.

However obscurely presented, my own position has from the start been that the choice between theories (and also the identification of anomalies, a process which raises similar problems) has to be made by a very special sort of community; otherwise there would be no science. Much of what is special about such communities is, I have tried also to argue, the shared values of their members – they must prefer the simple to the complex, the natural to the *ad hoc*, the fruitful to the sterile, the precise to the vague, and so on – a very usual list. Without such values the community's decisions would be different, and something other than science would result. I have also argued, however, that these values do not carry with them a set of criteria sufficient to dictate unequivocally their application in concrete cases. To a considerable extent they are acquired from the study of examples of past applications rather than by learning rules about how they are to be applied. Two men who employ the same values when choosing between competing theories may therefore differ vehemently about which theory is to be preferred. Only the man who says for example – theory *A* is simpler than theory *B*; the two are in other respects the same; nevertheless I prefer *B* – only a man who makes decisions of that structure violates what Lakatos calls the scientist's code of honor.

I am left, I think, with only one other source for the charge that I make science irrational – my discussion of incommensurability, which Lakatos brushed aside in a footnote. Since the hour is late, and he has given me no handle, I shall attempt only the following rejoinder here: Anyone who supposes that the points at which Feyerabend and I have aimed in introducing 'incommensurability' into our discussions of theory-choice are either trivial or obviously mistaken must simultaneously brush aside much of the contemporary literature on radical translation. I cannot think that that should be lightly done.

Princeton University

HERBERT FEIGL

RESEARCH PROGRAMMES AND INDUCTION

At the risk of being ostracized (if not annihilated) by the community of Popperians present, I wish to remark that Professor Lakatos is – and, I think – cannot help being, a second-level inductivist. If Professor Kuhn has pointed out (most eruditely) that science quite frequently is in a rut, and occasionally gets out of it (and into a new one), then Professor Lakatos appraises problem and theory shifts, and methodological innovations in the sciences, in the light of his criteria of 'progress' or 'degeneration'. There can be little doubt that he wishes to serve (at least) in a critical and/or advisory capacity to scientists. But he can do that only if he 'places his bets', i.e., conjectures as to the fruitfulness of a method, and along with it of a theory engendered or supported by such a method along the lines of success or failure, whichever may be plausibly indicated. I find Professor Lakatos's refutations of simple inductivism (here he agrees with Popper), as well as of simple falsificationism (here he disagrees with a caricature of the early Popper), completely convincing. But if he is to fulfill the critical and/or advisory functions, what else can he do but watch the course of the 'shifts' and *extrapolate*?!

Sir Karl has put 'the fear of God' into (most) inductivists. I think I am one of the few who resisted his persistently brilliant and persuasive arguments. (Cf. my contribution 'What Hume Might Have Said to Kant' in the Popper *Festschrift* [M. Bunge, ed., *The Critical Approach to Science and Philosophy*, Free Press of Glencoe, Collier-Macmillan, New York and London, 1964].) To be sure, Hume once and for all made clear that a justification of inductive inference is apt to beg the question (or what is tantamount: be viciously circular, or rely on an infinite regress). Nevertheless, there is such a thing as *confirmation* in the sciences, (often even very striking ones – scores of examples could be cited from the history of scientific theorizing). If the Popperians insist, I can put my point in terms of *corroboration*, i.e., of non-refutation despite strenuous and/or ingenious efforts at severe criticism. But this seems rather oblique in such cases as the observational verification (!) of the existence of further planets like

Boston Studies in the Philosophy of Science, VIII.

Neptune and Pluto; or of the lattice structure of crystals as confirmed by the experiments of von Laue and Bragg; or the disclosure of the etiology of general paresis by the discovery of the spirochete treponema pallidum; etc., etc., etc. But I am not adamant about *formulations*.

Perhaps the following vindication of induction (or if Professor Lakatos insists, only of *second* level inductions) will be acceptable – even if not congenial at first blush – to Popper and his disciples: Consider observation, experimentation, as well as statistical investigations as obtaining finite (usually rather small) samples from possibly unlimited universe. Of course, we have learned from Popper, Feyerabend *et al.* that such sampling procedures do not even begin to make sense except in the light of background theories (or as the philosophers usually put it: 'presuppositions'). Such background theories or assumptions, along with relevant evidence, provide the prior probabilities if we are to apply something like Bayes' theorem in the estimation of the probability of hypotheses under consideration. Now, it seems 'rational' (in one of the four or five philosophically important senses of this slippery term!) to assume – until further notice! – that the samples thus obtained are approximately *representative* of the 'population' from which they are 'drawn'. And this for the simple reason that assuming them to be *un*representative would be utterly arbitrary, irrational (or an expression of complete cognitive skepticism). Samples from an unlimited universe can be unrepresentative in an unlimited number of ways. Hence I still think despite the very ingenious, but defective criticisms of J. J. Katz (*The Problem of Induction and its Solution*, University of Chicago Press, 1962) that H. Reichenbach, though he was a first (as well as second to *n*th) level inductivist, was essentially right; I admit: not with his straight rule, but with his justification of induction. My own vindication was developed independently, in an early form in my doctoral dissertation (Vienna 1925–27). If any procedure works, i.e., yields correct extrapolations (etc.), the inductive and/or the hypothetico-deductive methods will do the same. And this can be shown to be true *deductively*.

It should hardly be necessary for an empiricist (like myself) to stress adherence to the policy of the open mind. Even the best established laws of nature (i.e., our formulations of the regularities in nature) may be faulty because of some as yet unknown but consequential parameters which may vary from one space-time region to another. Such a situation would still be compatible even with a basic methodological as well as ontological determinism. But once a fundamental indeterminism is accepted (as in the prevalent interpretations of quantum physics), it becomes even easier to conceive of much greater degrees of *dis*order in nature – such that the only successful induction might be reduced to the rather modest conclusion: 'The mess will continue to prevail'. But even that would be the conclusion of an inductive inference!

Finally, a few words about 'background theories'. There are theories and theories. (I prefer in some cases rather to speak of 'sets of interconnected empirical laws', but since this is partly a terminological question, I shall not insist – for the present purpose.) Some background assumptions (laws, hypotheses, theories) are so well 'secured' (corroborated, confirmed – put it as you please) that *in the context* of the given problem and its investigation they are (until further notice!) not called into question. Thus, for example, the optics of microscopes, telescopes, and spectroscopes is currently assumed without hesitation by (respectively) microbiologists, astrophysicists or cosmologists. Of course (being an empiricist – *horribile dictu*!) I not only admit, but insist, that even those background assumptions must be kept open – in principle – for revision. In many cases the background theory (or set of assumptions) consists merely in the views of common sense that we usually take unquestioningly for granted in the affairs of ordinary life. To provide just one example (out of hundreds that could be cited): If a psychologist undertakes to examine (test) the correctness of some parts of psychoanalytic theory, he will assume that it is (genidentically!) the *same* person whom he keeps putting on the couch three times a week for a year or two, (and makes tape recordings of the interviews). In some extremely exceptional cases (think of certain mystery stories!) it may, of course, come to pass that just this assumption should be questioned. Similarly, the general frame of ordinary realism is certainly the unquestioned presupposition of almost all scientific observation and experimentation. We assume that we return to the (genidentically) same laboratory, measuring instrument, or apparatus; that our sense organs keep functioning in roughly the same way as on previous occasions; that, quite generally the knowing subjects (or, if you prefer, the perceiving and responding organisms – or machines!) are embedded in a world that they (largely) never made, and of which they are special parts.

In sum: the revised inductivism here submitted provides a justification (vindication) for inductive extrapolations (and interpolations). These inferences not only do occur, but are indispensable for the growth of knowledge. Probabilities for such inferences require background assumptions whose vindication must ultimately be secured in the same manner. There is no other way than the one expressed in the well-known proverbs: 'Look (at least a little) before you leap!' and 'If at first you don't succeed, try, try again'. The sort of vindication sketched above is the one step we can take beyond Hume and skepticism. (Of course, a good deal of further logico-analytic scrutiny will be needed – and some of it is already promisingly under way – in regard to the probability or degree of substantiation, as well as the simplicity of hypotheses. In this connection, relevant aspects of decision theory are vitally important, but not philosophically as fundamental as is the general issue of vindication.)

Minnesota Center for the Philosophy of Science

RICHARD J. HALL

CAN WE USE THE HISTORY OF SCIENCE TO DECIDE BETWEEN COMPETING METHODOLOGIES?

How are we to decide between the different methodologies that have been proposed for science? In particular, how are we to decide between inductivism, conventionalism, falsificationism, and research programism? Lakatos says that we should use history of science to help us decide. He proposes, or at any rate seems to propose, the following criterion: Given competing methodologies we should prefer that methodology according to which more of the actual history of science is internal and rational, and more of scientists' own judgments about science are correct. This criterion presupposes the following proposition, which Lakatos also asserts: The different methodologies lead to determinable and different dividing lines between internal (rational) history of science and external (empirical) history of science. I have doubts about both of these, the criterion and the proposition presupposed by it, and therefore I have doubts about whether history of science can be used as Lakatos proposes to judge between different methodologies.[1] I shall indicate what my doubts are, starting first with the proposition that the different methodologies lead to determinable and different dividing lines between internal and external history of science.

I

On the first page, Lakatos says, "The vital demarcation between normative-internal and empirical-external is different for each methodology." And on the last page he says, "... each methodology of science determines a characteristic (and sharp) demarcation between (primary) internal history and (secondary) external history ...." In the intervening pages he attempts to show how the demarcation is drawn by the different methodologies, at least for a few specific cases. This is no inessential part of Lakatos' argument, for it is obvious that if we are to use the history of science to judge between the different methodologies, in accordance with his criterion, the different methodologies must lead to at least fairly well defined, and different, demarcations between internal and external history of science. However, I believe that at present these different methodologies do not determine demarcations between internal and external history of science, whether they will in the future being a matter of some doubt. Let us consider two methodologies as examples, Lakatos' own methodology of research programs and inductivism.

In Lakatos' methodology we have 'research programs' with 'progressive' and 'degenerating problemshifts'. Let us suppose these notions were precisely defined so that we could tell exactly what the different research programs were in the history of science and when they were progressing and when degenerating. (Obviously being able to do this would also require fairly complete historical knowledge, which we may not have in all cases.) The question arises exactly how long it is rational for a scientist to stick with a degenerating research program. Lakatos quite definitely states that scientists should not necessarily drop a research program at the first sign of trouble. But on the other hand they certainly should drop it if things go badly enough for it. So exactly how long should a scientist stick with a degenerating research program? Lakatos gives no rules for this and in fact admits that:

> It is very difficult to decide, especially since one must not demand progress at each single step, when a research programme has degenerated hopelessly or when one of two rival programmes has acieved a decisive advantage over the other. In this methodology, as in Duhem's conventionalism, there can be no instant – let alone mechanical – rationality. Neither the logician's proof of inconsistency nor the experimental scientist's verdict of anomaly can defeat a research programme in one blow. Once can be 'wise' only after the event. (p. 101)[2]

How, then, would Lakatos determine whether a scientist's adherence to a degenerating research program was rational? By seeing if the scientist was honest, to himself and others, about the fact that the research program was degenerating. 'It is perfectly rational to play a risky game: What is irrational is to deceive oneself about the risk." (p. 104) But surely honesty and rationality are quite different things. Everyone knows people who are dishonest but quite rational, and others who are honest but somewhat irrational. Similarly, we can quite well imagine a scientist irrationally clinging to a degenerating research program, without kidding himself about the fact that it was degenerating (perhaps he has some emotional tie to the research program stemming from the old days). So I cannot accept Lakatos' equating of rationality with honesty, and I am forced to conclude that on Lakatos' methodology we have no way of telling which actions of scientists were rational and which not. Therefore, we have no way of evaluating Lakatos' claim, for example, that adherence of scientists to the Newtonian research program after the discovery of the anomaly in Mercury's orbit was rational according to the research program methodology but not according to Popper's methodology. Frankly, I think the methodology of research programs will have to be spelled out in *much* greater detail before we will be able to start to make sense out of such claims.

As a second example I should like to consider inductivism. At the outset it must be said that it is not clear who Lakatos has in mind when he attacks inductivism since he mentions hardly any names. However, one can hardly go wrong in taking Carnap to be an inductivist and in any case, if Lakatos' theses don't hold for Carnap's methodology, that surely casts doubt on them; so let us look at Carnap's inductivist methodology. Which actions of scientists does Carnap reconstruct as rational and which not? The question is difficult to answer. In fact I should say it is impossible to answer. It is instructive to see why. On Carnap's methodology for rational action we have at least three components: an inductive logic, a utility function for the person involved, and a set of acceptance rules.[3] With at least the first two components we run into problems in applying the methodology to actual situations in the history of science. The first component, the inductive logic, has only been worked out for fairly simple languages, nothing as complex as the actual language of physics. Therefore, we can't determine what degree of confirmation Einstein's predictions had, say, in 1917, and whether they were more highly confirmed or less highly confirmed by the available evidence than Newton's. And with regard to the second component, how are we to determine the utility function of some scientist who lived 200 hundred years ago and of whose actions we know only a few? Indeed, the third component is not completely unproblematic, for although Carnap adopted the 'maximize the utility' rule, one can imagine situations in which one might want to adopt a more complicated rule, such as a 'maximin' rule or a 'maximax' rule – this, while still using inductive logic and therefore still being an inductivist, presumably.[4] In short, on all three counts it seems quite impossible to say which actions of scientists Carnap's methodology would reconstruct as rational and which not.

Because it will help clarify some issues, I would like to discuss a little further the second component – the utility function – of Carnap's methodology. I said that we would have to reconstruct a given scientist's utility function as a necessary first step in determining whether his actions were rational. But perhaps we do not need to reconstruct the scientist's total utility function. Perhaps we are not interested in whether his actions were rational *all his goals considered*, but rather in whether they were rational from the restricted point of view of his scientific goals. It could make a difference and so we need an answer from Lakatos. Lysenko's actions, for example which might not have been rational from the restricted point of view of his scientific goals, might well have been quite rational given the goals of life, liberty, and the pursuit of happiness. I suspect that Lakatos is interested in the more restricted kind of rationality – scientific rationality, let us call it. Limiting the problem to one of whether a scientist's actions were scientifically rational might simplify the problem some. Instead of having to reconstruct the scientist's total utility function we might only need a part of it – that part which was relevant to his scientific goals (his scientific utility function, let us call it). But now a further question poses itself for us: Are we really interested in the *given scientist's beliefs* about the goals of science. Are we interested in *his* scientific utility function? Or are we interested in whether his actions are rational given *our* views on the goals of science and therefore *our* scientific utility function? It could make a difference' In fact it probably would in a case like the dispute between Bohr and Einstein over quantum mechanics. On Einstein's views of the goals of science, on Einstein's scientific utility function, Einstein's actions would probably be rational; but on ours, at least if ours is like Bohr's, they might not be. Again we need an answer from Lakatos. But let us take the simplest course and use *our* views as to the goals of science, our scientific utility function, so that we bypass all problems in reconstructing some past scientist's utility function. Our troubles are not over. For what are our views as to the goals of science; what is our scientific utility function? Hempel points out the difficulties in answering this question:

> What will have to be taken into account in constructing or justifying inductive acceptance rules for pure scientific research are the obectives of such research or the importance attached in pure science to achieving certain kinds of results. What objectives does pure scientific research seek to achieve? Truth of the accepted statements might be held to be one of them. But surely not truth at all costs. For then, the only rational decision policy would be never to accept any hypothesis on inductive grounds since, however well supported, it might be false.

> Scientific research is not even aimed at achieving very high probability of truth, or very strong inductive support, at all costs. Science is willing to take considerable chances on this score. It is willing to accept a theory that vastly outreaches its evidential basis if that theory promises to exhibit an underlying order, a system of deep and simple systematic connections among what had previously been a mass of disparate and multifarious facts.
>
> It is an intriguing but as yet open question whether the objectives, or the values, that inform pure scientific inquiry can all be adequately characterized in terms of such theoretical desiderata as confirmation, explanatory power, and simplicity and, if so, whether these features admit of a satsifactory combination into a concept of purely theoretical or scientific utility that could be involved in the construction of acceptance rules for hypotheses and theories in pure science.[5]

However these problems may turn out for the inductivist, it is clear that at present we can say virtually nothing about how the inductivist methodology would draw the line between internal and external history of science.

(While on inductivism, I feel compelled to add a somewhat extended parenthetical note defending inductivism against some of Lakatos' attacks. Lakatos states very plainly at the beginning of his paper that the methodologies he is talking about are not methodologies in the older sense of rules for discovering or thinking up theories in the first place, but rather methodologies for evaluating and accepting theories already present (p. 92). But then he reverts to talking about and criticising inductivism as if it were some sort of method for starting with facts and arriving at theories. In particular, he claims (1) that inductivism must relegate all scientific theories that preceded the presently accepted ones to the era of pre-science or pseudo-science (p. 93), (2) that inductivism cannot give an internal explanation for the selection of facts that actually occurs in science (p. 93), (3) that it is against the inductivist code of honor to propose unproven theories (p. 95, p. 96) and (4) that inductivism has been historically falsified, e.g. by Duhem (p. 115). All of this is plain false of inductivism as a method of assessing theories and hypotheses already proposed. Then Lakatos goes on to appeal to the logico-epistemological troubles with inductivism, in particular with justifying the principle of induction, and he suggests that these troubles have demolished the position. But later on he himself forcefully claims that any methodology – conventionalism, falsificationism, or his own methodology of research programs – has got to assume some sort of principle of induction if it is not to be a mere game with no relevance to epistemology (p. 109). It is true that he says he himself will use induction as an 'extra-methodological' principle (p. 101), whatever that means, but it is not clear how this legitimizes it for him and not for the inductivist. In any case, his claim (p. 108) that inductivism, along with other forms of justificationism, has crumbled under epistemological and logical criticism is pretty clearly empirically false – to use his own terminology, even if inductivism has been *refuted*, which isn't obvious, it certainly hasn't been *rejected*.)

Returning to the main line of the argument, it seems to me that none of the methodologies that Lakatos discusses have been worked out with sufficient precision to determine exactly where they would draw the line between internal and external history of science. One last point with regard to this problem. Even if we had a very precisely worked out methodology, it seems clear that we would need to conjoin it with an extensive historical knowledge of the historical situation in question. For certainly any reasonable methodology will be conditional rather than categorical, that is, will make what's rational to do dependent on the conditions present. In general, the conditions that we need to know will probably be fairly complex, including at least the total set of relevant beliefs of the scientist in question. I doubt that we have this complete anhistorical knowledge about very many events in the history of science, so that again determining where a given methodology draws the line between internal and external history of science would seem to be impossible.

II

Let us suppose, however, that we could determine where this line would be drawn by several different methodologies. Let us suppose, further, that the line would be drawn at different places by the different methodologies. Let us turn to Lakatos' criterion, which tells us how we should use history of science to decide between the different methodologies. If I understand Lakatos correctly, he is saying that we should prefer the methodology that reconstructs more of the actual history of science as internal and rational, and more of scientists' own judgments (basic value judgments) about science as correct. Lakatos says, for example, "In the light of better rational reconstructions of science one can always reconstruct more of actual great science as rational." (p. 117) And again, "When a better rationality theory is produced, internal history may expand and reclaim ground from external history." (p. 119) And Lakatos says that Popper's methodology is preferable to inductivism, and Lakatos' own methodology preferable to Popper's, because in each case the first methodology includes more of actual science as internal and agrees with more of scientists' basic value judgments.

Such a criterion has a great deal of plausibility about it. Surely, preanalytically, we'd be inclined to say that most of what scientists do within science is fairly rational. Or, to put it the other way around, we would be inclined, preanalytically, to have doubts about a proposed methodology which made out most of scientists' actions within science to be irrational. And further, we would generally be inclined to accept scientists' own basic value judgments about what is good science and what isn't, and we would be inclined to dismiss a methodology which disagreed violently with these judgments.

But just in putting it this way we can see a problem. Nobody would want to say that *all* of science is rational and that *all* of scientists' judgments about science are correct. We all recognize that there are Lysenkoesk episodes in the history of science – episodes that nobody would require a methodology of science to show to be internal and rational (scientifically rational, that is). Lakatos, of course, realizes this. He says,

> The methodology of research programmes – like any other theory of scientific rationality – must be supplemented by empirical-external history. No rationality theory will ever solve problems like why Mendelian genetic disappeared in Soviet Russia in the 1950's, or why certain schools of research into genetic racial differences or into the economics of foreign aid came into disrepute in the Anglo-Saxon countries in the 1960's. Moreover, to explain different speeds of development of different research programmes we may need to invoke external history. Rational reconstruction of science (in the sense in which I use the term) cannot be comprehensive since human beings are not *completely* rational animals; and even when they act rationally they may have a false theory of their own rational actions. (p. 102)

But this seems to imply the possibility of a methodology that includes too much of science as internal, a rational reconstruction that makes too much of science rational. In particular, it would be easy to construct a methodology that reconstructs Lysenko's actions as scientifically rational and internal. The following methodology should do the trick: It is rational to accept a progressing research program in preference to a degenerating one unless the government strongly encourages you to accept the degenerating one. (Obviously this methodology could be fixed up if necessary.) Should we accept this methodology because it includes as internal some of science that Popper and Lakatos relegate to external history? Of course not. Should we accept this methodology because it construes the basic value judgments of Lysenko and cohorts as correct? Of course not. But then I don't see how we are going to be able to accept without qualification Lakatos' criterion which directs us to accept the methodology which reconstructs more of science as rational and more of scientist's basic value judgments as correct.[6]

So how are we to decide between competing methodologies and does history of science have a role in such a decision? I think this much at least can be said: If a methodological rule in which we don't have too much prior confidence conflicts with the practice of a great many acclaimed scientists, then we should call the rule into question; and if the practice of a few minor scientists conflicts with a methodological rule which has a great deal of initial plausibility, then we should call the practice of those scientists into question. There must be a sort of double feedback which allows us both to temper flights of methodological fancy by appeal to actual practice, and also to correct scientific mal-practice by appeal to good methodology. There is nothing new or profound about this kind of view. Goodman said much the same thing about the rules of deductive logic,[7] and no doubt others said it before him. Lakatos himself realizes that there must be such an interplay rather than a one way appeal to the actual practice of scientists when he says, near the end,

> And, indeed, the methodology of historicographical research programmes implies a pluralistic system of authority, partly because the wisdom of the scientific jury and its case laws has not been, and cannot be, fully articulated by the philosopher's statute law, and partly because the philosopher's statute law may occasionally be right when the scientists' judgment fails. I disagree, therefore, both with those philosophers of science who have taken it for granted that general scientific standards are immutable and reason can recognize them a priori, and with those who have thought that the light of reason illuminates only particular cases. The methodology of historiographical research programmes specifies ways both for the philosopher of science to learn from the historian of science and vice versa. (p. 121)

Here, then, we have the beginnings of a plausible way to use history of science to help us decide between competing methodologies. As it stands (in the paragraph immediately preceding) it is too vague and general to be of much use in concrete cases. But perhaps it can be worked out in more detail. And then if the various methodologies can be worked out in such a way that we can tell when they conflict with the practice of scientists and when they don't, we might be in a position to use the history of science to help choose between methodologies. At present, I don't think we are even close to being in such a position.

Michigan State University

NOTES

[1] I also have doubts about the tenability of the distinction between internal and external history of science. But for the purposes of this paper, I shall simply accept that purported distinction.
[2] Page numbers in parentheses refer to Lakatos' article in this volume.
[3] See, for example, R. Carnap, 'The Aim of Inductive Logic', in *Logic Methodology and Philosophy of Science* (ed. by E. Nagel, P. Suppes and A. Tarski), Stanford University Press, Palo Alto, 1962.
[4] C. G. Hempel, *Aspects of Scientific Explanation*, The Free Press, New York, 1965, pp. 466–7.
[5] C. G. Hempel, 'Recent Problems of Induction', in *Mind and Cosmos* (ed. by Robert G. Colodny), University of Pittsburgh Press, Pittsburgh, 1966, p. 131.
[6] It may be that on no methodology can we construe all the basic value judgments of scientists as correct because these judgments may be inconsistent among themselves. But in such a case why should we automatically prefer the methodology that agrees with the most of these judgments? It is probably true that in the case of Lysenko, the majority of the scientists stayed (scientifically) rational but it could have been the other way around, as Lakatos admits in his astronomy-to-astrology example (p. 120).
[7] N. Goodman, *Fact, Fiction, and Forecast*, The Bobbs-Merrill Co., Inc., Indianapolis 1965, pp. 63–64.

NORETTA KOERTGE

INTER-THEORETIC CRITICISM AND THE GROWTH OF SCIENCE

This paper is intended to be a small contribution to a future comprehensive Theory of Scientific Growth. I take it that such a theory would give an idealized description of the repeating patterns of growth found within the history of science and show how these developmental patterns are different from those found in the case of theories such as witchcraft, on the one hand, and the patterns found in the growth of the 'practical arts', such as pottery-making, on the other. The theory would go on to explain why such patterns might be expected to produce scientific knowledge by pointing out the critical forces which are operating and the rationality of the responses to them. In short, an adequate philosophical theory in this area should not only give the kinetics of scientific growth, but also the dynamics of that process.

Here I will argue that mono-theoretic accounts of the growth of science – i.e., those accounts which claim that the most important critical processes take place within the context of a *single* theory or a *single* research programme – have serious shortcomings when viewed as a comprehensive theory of scientific development. In this connection, I will discuss Lakatos' position.

I will then stress the importance of clashes between theories, both for a correct description of the actual growth of science and for an understanding of the nature of strong criticism within science. And I will describe in detail one important pattern of growth which involves a clash between two theories which are largely complementary – *the pattern of dialectical ascent*.

1. AN EXPOSITION OF LAKATOS' METHODOLOGY OF SCIENTIFIC RESEARCH PROGRAMMES

By far the most sophisticated mono-theoretic account we have to date is that offered by Lakatos' 'Methodology of Scientific Research Programmes'.[1] Although many theories are generated within a Lakatosian research programme, his model is essentially a mono-theoretic one because it does not stress the importance of dialectic between radically different theories or research programmes.

I view Lakatos' position as an attempt to cope with the following problem: Kuhnian 'normal science' exists – or at least something similar to it does – and it seems to figure importantly within the history of science. Yet 'normal science' appears to violate the rules of good scientific practice which have been suggested by Popper and others. Therefore, we must ask, "What is the rationality of normal science?" Or, as I would prefer to phrase it, "What are the critical forces operating within normal science?"

Lakatos answers this by giving a characterization of what he calls a 'scientific research programme' and arguing for the rationality of research programmes which are in a progressive phase.

A Lakatosian research programme consists of three parts:[2]

(i) A 'hard core' of theory. (Two examples are Newton's Laws and Bohr's Quantum Postulates.)

(ii) A 'negative heuristic'. These are methodological principles which have two functions: First, they protect the hard core from experimental refutation (i.e., they instruct us to modify auxiliary hypotheses, not to change the hard core). Secondly, they rule out radically different sorts of explanatory attempts (e.g., trying to use classical oscillators to explain atomic spectra within Bohr's research programme).

iii) A 'positive heuristic'. This is a plan of how to improve the sophistication of the theory's explanatory models; (e.g., in the Newtonian programme one first calculated orbits of the planets using point masses, then mass balls, then considering the inter-planetary effects, and so forth.) Lakatos says one function of the positive heuristic is to

> [save] the scientist from becoming confused by the ocean of anomalies... the scientist's attention is riveted on building his models following instructions which are laid down in the positive part of his programme. He ignores the *actual* counterexamples, the available 'data'.[3]

In time, the research programme generates a series of off-spring theories, each one sharing the same hard core, but each one contradicting its predecessors because of the variation in their auxiliary hypotheses. In the typical case there will be known counterexamples to each of the offspring theories.

All research programmes have the above characteristics. However, only some will be in what Lakatos calls a 'progressive phase' and it is *progressive* research programmes which contribute significantly to the advancement of science.

We are to judge the research programme to be in a progressive phase if each successive member of the series of theories adds novel corroborated content to the system. For example, if by using Sommerfeld's extension of Bohr's theory we not only account for the major spectral energies, but are also able to calculate the fine structure, then the programme is in a progressive phase.

Note that for a programme to be in a progressive phase it is not strictly necessary that a single old anomaly be explained. Although it may happen that old problems are cleared up, all that is required is that some new content be added to the theory. I quote Lakatos:

> Our considerations show that the positive heuristic forges ahead with almost complete disregard of 'refutation': it may seem that it is the 'verifications' rather than the refutations which provide the contact points with reality. [In a footnote he explains that by verification he means "a corroboration of excess content in the expanding programme".] ... it is the 'verifications' which keep the program going, recalcitrant instances notwithstanding.[4]

In short, it is rational to protect the hard core of a theory as long as these protective stratagems continue to yield interesting new predictions.[5]

Thus the proponent of a research programme can give the following very plausible sounding argument: "Granted that the existence of anomalies shows that there are imperfections in our present theory. But this does not require us to assume that there is anything wrong with the heart of our theory, the hard core. Moreover, by working with it we have succeeded in extending our scientific system into new areas. Let us continue."

Thus rests the case *for* a progressive research programme. What arguments can someone who believes the hard core to be false bring forward? According to Lakatos, there is no way for someone outside a progressive research programme to argue *effectively* against it. It is useless to generate new counterexamples or point to a build-up of anomalies because there are no Kuhnian crises in his account. We cannot even point to a more successful rival programme. He grants that if one is opportunistic (e.g., a member of a government funding committee), one *may* abandon a slowly progressing research programme for another more successful programme, but one need not.[6] All a would-be critic of a research programme can do is sit around and wait for it to hit a 'degenerating phase'. It is only if the last several theoretical modifications have been unsuccessful and if the positive heuristic is exhausted (i.e., if there remain no obvious extensions of the theory), that we may *cogently* argue that it is time to abandon the programme.

II. CRITICISM OF LAKATOS' POSITION

I will argue that, contrary to Lakatos, *there are ways of directly criticizing a research programme* (namely, with certain kinds of alternative theories) and that we cannot understand the growth of science without focusing on these stronger forces. I also want to maintain that his requirements for progressive research programmes are too weak even in situations where there is no direct criticism from outside theories. I will deal with the latter point first.

Lakatos says that "with sufficient resourcefulness, and some luck, any theory can be defended 'progressively' for a long time, even if it is false".[7] To see how easily a research programme can be kept in a 'progressive phase' let us consider the programme of fundamentalist Biblical criticism:

The hard core of this theological enterprise contains the statement: "Everything the Bible says is literally true." The negative heuristic protects the theory against fossil data indicating that men may have been on earth for more than 6000 years, and other counterexamples. Following the positive heuristic, one initiates a program of exegesis, linguistic analysis, and historical study which is designed to remove any apparent contradictions within the Bible itself. The program is very successful, not only in detecting mistranslations within the Bible, but also in making new predictions which are confirmed by independent historical investigations; (to cite one example, the broad wall mentioned in Nehemiah 3:8 and 12:38 has been uncovered in Israel recently).

The above example satisfies all of Lakatos' requirements for a progressive research programme. (As we shall see it is similar to one of his own examples – the progressive programme based on Prout's hypothesis.) I personally find both the Biblical exegesis example and the Prout case intuitively unacceptable as instances of sustained scientific progress.

We can rule out such cases by strengthening the requirements on a progressive research programme in two ways:

(1) We can demand that each theoretical modification within the

research programme not only make successful new predictions but also that the programme become more coherent and unified as time progresses. As William Whewell observed:

... we have to notice a distinction which is found to prevail in the progress of true and false theories. In the former class all the additional suppositions *tend to simplicity* and harmony; the new suppositions resolve themselves into old ones, or at least require only some easy modification of the hypothesis first assumed: the system becomes more coherent as it is further extended... [But with false theories] the new suppositions are something altogether additional; – not suggested by the original scheme; perhaps difficult to reconcile with it. Every such addition adds to the complexity of the hypothetical system... [8]

There are passages which indicate that Lakatos has such a strong requirement in mind.[9] However, one of his examples of a progressive research programme which was prematurely abandoned, Prout's Hypothesis,[10] does not satisfy my strong requirement. In that case, improved chemical analysis resulted in the correction of some atomic weights which made them closer to integers. But other atomic weights remained stubbornly non-integral. Certainly there was no 'convergence' in the programme, no feeling that one now had a deeper understanding of the relationships between phenomena.

(2) The second way we can rule out non-intuitive counterexamples is by requiring that the series of models or modifications laid down by the positive heuristic be strongly suggested by the conjunction of the hard core and the known boundary conditions Lakatos points out, for example, that Newton's Laws *themselves* dictated that a sophisticated theory of planetary motion would include the perturbing effects of other planets. It often seems that he wants to require this of a positive heuristic.[11] However, one of his three major examples violates such a requirement. Holders of Prout's hypothesis had no systematic programme for analytic chemists. All they could say was: "Keep on purifying until you get integral atomic weights!" In the case of Bohr's theory the positive heuristic was much stronger than Prout's but still weaker than Newton's. An illustration of this point: Newton would have been *very* surprised if calculations including perturbations had not been closer to observational data than the simple calculations had been. However, there was *no* reason for Bohr to expect from the very beginning that elliptical orbits would be better than circular.

Since there are passages where Lakatos endorses the requirements of heuristic power and coherent growth, it might be thought that my re-remarks above should be taken only as a small historical criticism of one of his examples. Perhaps so. In any case, this illustrates a fundamental tension within his theory. In order to argue that it is rational to hold a theory in the face of many, persisting counterexamples, Lakatos must be able to point to some very strong positive features of the research programme of which the theory is a part (e.g., its heuristic power, its coherence, and its predictive power). This is motivation to put strong requirements on progressive research programmes. However, he also wishes his theory to form the conceptual framework for a full account of the history of great science. And since few programmes in the history of science had such a fully developed positive heuristic as Newton's (even Bohr's programme looks much more like piece-meal tinkering), there is a tendency to weaken the requirements on progressive research programmes so as to cover more of the actual history of science. But let us now turn to what I consider to be a more important process in science – the process of dialectical ascent.

III. THE PATTERN OF DIALECTICAL ASCENT

I maintain that often the most important challenge facing a scientific research programme is not that posed by the existence of unexplained experimental anomalies or the lack of novel predictions, but that provided by another successful research programme which has a hard core inconsistent with the hard core of the given programme. I will also argue that it is very often the case that major advances in science come when these conflicts are resolved. Hence we must reject any methodology which inhibits such criticism. Lakatos permits the "dialectic of speculative conjectures and empirical refutation" within one research programme. We must also allow *dialectic between research programmes.*

Feyerabend has already ably discussed the general advantages that accrue from working with a plurality of theories. He points out that different theories tend to encourage the building of different instruments and the performing of different experiments. There will be less chance of neglecting data or of accounting for recalcitrant data with *ad hoc* hypotheses. Moreover, since 'comprehensive' theories cannot be eliminated by a direct confrontation with 'the facts' (as an example he cites witchcraft), he concludes that whereas unanimity of opinion may be fitting for a church, or for the willing followers of a tyrant, or some other kind of 'great man', variety of opinion is a methodological necessity for the sciences ...[12]

Most of Feyerabend's examples involve theories which deal with roughly similar domains of phenomena. However, I wish to emphasize the importance of another type of pluralistic criticism – the case where the conflicting theories originate in separate fields of scientific inquiry but are nevertheless mutually relevant; in fact, they are inconsistent. The conflict is eventually resolved by a deeper theory which assimilates them both and unifies them in a content-increasing way. Typically, the deeper theory corrects both of its predecessors and stands in a correspondence relation to them. My primary concern here is not to explicate the relation between the unifying theory and its predecessors, but to emphasize the important role of the clash between theories in the growth of science. Obviously I think the formal difficulties suggested by the Incommensurability Thesis can be overcome.[13]

Before explaining why the conflict between what we might call 'quasi-complementary' theories ('complementary' because most of the phenomena covered by one theory are qualitatively different from those covered by the other theory, 'quasi-complementary' because some of their claims are inconsistent) is important, let me give an historical illustration of what I shall call the 'pattern of dialectical ascent'.[14] This will also serve to demonstrate the inadequacy of the Lakatos position as a framework for understanding the history of science.

Let us re-examine the case study which Lakatos presents in most detail – Bohr's research programme. Lakatos discusses the 'marvelously fast progress' of the programme and stresses that all of this 'breathtaking success' was possible even though the theory had 'inconsistent foundations'. (The theory denied that the accelerating electron radiated energy but then used classical electromagnetic theory to calculate the magnetic moment provided by the orbital electrons.) Eventually, he says, "... even this great programme ... petered out. *Ad hoc* hypotheses multiplied and could not be replaced by content-increasing explanations."[15] And soon Bohr's old quantum theory was replaced by the new research programme of wave mechanics. This constitutes the entire historical framework provided by the methodology of scientific research programmes.

However, upon investigating the broader historical context we find that while physicists were assuming that the electron moved in planetary orbits around the nucleus in order to calculate atomic spectra, the chemists had proposed a static, geometric model of the atom which gave a remarkably simple, though somewhat rough, account of the bonding and structure of both inorganic and organic compounds. A mature version of the so-called 'electron-pair theory' was given by G. N. Lewis in 1916 and Langmuir further developed it while Lewis was involved in World War I.[16]

Besides the assumption that electrons were stationary, the theory had two additional fundamental postulates, the Rule of Eight and the Rule of Two. According to the Rule of Eight, the most stable configuration for an atom within a molecule is attained when it is surrounded by eight electrons. Such a configuration can be gained either through the transfer or the sharing of electrons. According to the Rule of Two, bonding electrons are more stable if they occur in pairs. For most atoms, this suggested a tetrahedral configuration with an electronpair at each vertex.[17]

There is only time here to hint at what an important theoretical achievement this was in chemistry.[18] It resolved the long-standing conflict between a Berzelius-type electrovalent theory and the various theories of covalency. It made sense out of weak electrolytes (compounds which seemed to be neither ionic nor covalent). It explained the 'inductive effects' of electronegative substituents in organic compounds. It gave the structural formulae for molecules such as fluoboric acid which had previously been considered to be 'incongruities'. It predicted that water should be a bent molecule (confirmed by Vorländer in 1922), that there should be ionic hydrides, and that transition metals formed compounds by losing non-valence electrons. In addition, there was fairly direct physical evidence for the Rule of Two arising out of investigations of the paramagnetic and diamagnetic behavior of elements.

The interesting features of this case which exhibit the pattern of dialectical ascent are these:

(a) The period 1920–25 was what Pauling has called "the decline of the old quantum theory".[19]

(b) In the middle of the same period (1923), G. N. Lewis published his *Valence and the Structure of Atoms and Molecules*, marking a high point for the electron-pair theory.

(c) In this work Lewis explicitly discussed the conflict in the two theories:

These two views [seem] to be quite incompatible, although it is the same atom that is being investigated by chemist and by physicist. If the electrons are to be regarded as taking an essential part in the process of binding atom to atom in the molecule, it [seems] impossible that they could be actuated by the simple laws of force, and travelling in the orbits, required by the planetary theory. The permanence of atomic arrangements even in very complex molecules, is one of the most striking of chemical phenomena. Isomers maintain their identity for years, often without the slightest appreciable transformation.[20]

Lewis also tried to reconcile the two views by suggesting that each electron-pair moved in orbits such that its *average* position corresponded to the *fixed* position assigned by the static theory of the atom[21] but he did not work out the details.

(d) A deeper synthesis was given by wave mechanics which, among other things, corrected and extended the old quantum theory of atomic spectra (for example, allowing the calculation of intensities) and which also corrected and explained (through the work begun by Heitler and London) the old theory of the chemical bond.

I submit that for an adequate understanding of the development of wave mechanics we must take into account such interactions between theories.

IV. THE SIGNIFICANCE OF INTER-THEORIC CRITICISM

I will conclude with three general theses about dialectical ascent:

(1) An historical thesis: Many of the greatest advances in the development of science have come when a new theory resolves the conflict between quasi-complementary old theories by providing a content-increasing unification.

Here I can only mention some historical cases which appear to be examples of dialectical ascent – it would be necessary to investigate the exact nature of the conflict in each case.[22] In some cases it may be that two complementary theories only become strictly inconsistent if we make a plausible extension of one or both of them. Perhaps the most familiar recent examples are the conflict between Lorentz-invariant electromagnetic theory and Galileo-invariant classical mechanics[23] or time-reversible classical mechanics and time-reversible classical thermodynamics. In nineteenth century chemistry, there were conflicts between the laws of inorganic and organic compounds. To go back earlier still, Stahl thought that one of his major contributions was to show that the chemistry of metals was not fundamentally different from the chemistry of non-metals. Sometimes scientists have attempted to paper over the conflicts between theories by giving one of them an instrumentalist interpretation – most notably, perhaps, in the case of Copernican astronomy and Aristotelian mechanics.

(2) A philosophical thesis: Such patterns of growth are found because the problem posed by conflicting quasi-complementary theories is a particularly deep one.

One measure of the seriousness of the problem presented by the discovery of a contradiction within our scientific system is the empirical content of the set of statements we would have to discard in order to resolve the inconsistency. (Of course, I am not recommending we actually resolve the problem in this manner.) Thus the conflict between a theory of high explanatory content and an isolated experimental datum does not pose as deep a problem as the conflict between a theory and a conjunction of lots of different anomalies. (This partially explains how Kuhnian crises operate.) And the problem posed by finding a contradiction in the union of two quasi-complementary theories of high explanatory content is more serious still. (Note that the conflict between alternative theories which cover largely overlapping domains does not pose as deep a problem on this account because discarding one of them would not result in as great a loss of explained content.)

Another important factor in gauging the seriousness of the problem posed by an inconsistency is the availability and plausibility of a hypothesis which would resolve it. (This is why Lakatos is right in saying that the problem of anomalies is not serious as long as there are untried articulations of the theory which may cover them.) In the case of conflicting alternative theories covering roughly the same domains, there is one fairly attractive solution – drop one of them. However, in the case of conflicting quasi-complementary theories, the problem is severe. We cannot drop either of them without losing a lot of explanatory power. And there is no research programme around to tell us how to generate auxiliary hypotheses to resolve the conflict. Just as the discovery of a paradox is a growing point in philosophy, so the discovery of an inconsistency between two largely independent, well-established theories is a growing point in science.[24]

(3) A methodological thesis: Since science grows by solving problems, the major methodological error a community of scientists can make is to ignore deep problems.[25]

Sometimes inter-theoretic conflicts are deliberately neutralized by giving one theory an instrumentalist interpretation. (Thus I would argue in favor of methodological realism because it increases the possibilities for criticism and hence for growth.)

Another criticism-reducing strategy is to deny the contradiction by a device which I call 'protective partitioning'.[26] This sometimes consists in arguing that the questions which the two theories are designed to answer are different. (Thus, in the Bohr-Lewis case, followers of this strategy would point out that Lewis was asking 'chemical' questions, Bohr 'physical' ones.)

Other times it is pointed out that the concepts in the two theories are different. (Since we speak of 'training' a rat but 'educating' a child, some educational theorists would argue that on these *a priori* grounds we can prove that rat psychology can never be relevant to theories of human learning.)

One task of the methodologist is to criticize such stratagems. Admittedly, it is not always easy to see the relationships between theories. One productive way to find out if theories *are* inconsistent is by the use of thought experiments. Thought experiments are more useful in locating a contradiction than 'crucial experiments' are in resolving it. If a 'crucial experiment' tells against a theory we may succeed in adjusting auxiliary hypotheses to save the refuted theory. However, direct inconsistencies between the cores of the theories cannot be explained away in this fashion.

A final summarizing remark: Lakatos has said that theories tend to develop 'protective belts' which the small arrows of scattered, isolated counterexamples *cannot* dent. I have shown that such a theory can be attacked if one uses a projectile of comparable size and weight. Direct criticism can be provided by a conflicting theory of comparable empirical content and explanatory power.

Indiana University

NOTES

[1] I. Lakatos, 'Falsification and the Methodology of Scientific Research Programmes', in *Criticism and the Growth of Knowledge* (ed. by I. Lakatos and A. Musgrave), Cambridge University Press, 1970, pp. 91–195.
[2] *Ibid.*, pp. 133–7.
[3] *Ibid.*, p. 135.
[4] *Ibid.*, p. 137.
[5] Sometimes Lakatos interprets the criterion of novelty so as to make it very weak. He says, for instance, that the Bohr theory's prediction of the Balmer series was in some sense a 'novel fact' because while Balmer observed that "hydrogen lines obey the Balmer formula," Bohr predicted that "the differences in energy levels in different orbits of the hydrogen electron obey the Balmer formula". (*Ibid.*, p. 156). (The contrast is between 'line' and 'difference' in orbital energies.) He puts the general point this way: "... *we should certainly regard a newly interpreted fact as a new fact, ignoring the insolent priority claims of amateur fact collectors.*" (*Ibid.*, p. 157. Italics in the original.) In my criticism of Lakatos' position I will try to ignore this lapse into the we-live-in-a-different-world-after-a-revolution syndrome. If he wants to count a research programme as progressive when it is found to predict a known result, he should just say so.
[6] I. Lakatos, 'History and its Rational Reconstructions', this volume, p. 104.
[7] *Ibid.*, p. 100.
[8] William Whewell, *The Philosophy of the Inductive Sciences*, Facsimile of the 2nd. ed., 1847, Johnson Reprint Co., London, 1967, Vol. 2, pp. 68–9.
[9] See Lakatos, 'Methodology of Scientific Research Programmes', p. 175.
[10] *Ibid.*, p. 140.
[11] *Ibid.*, p. 175.
[12] P. K. Feyerabend, 'Explanation, Reduction and Empiricism', in *Minnesota Studies in the Philosophy of Science*, Vol. III (ed. H. Feigl and G. Maxwell), University of Minnesota Press, Minneapolis, 1962, p. 71.
[13] Holders of the Incommensurability Thesis would question whether quasi-complementary theories can be compared and in what sense we can say they are 'subsumed' or 'corrected- and-explained' by the later synthesis. Aren't they merely discarded in favor of an incommensurable world view?

I hesitate to delve into this can of worms, but I will make two remarks, First, I would deny that communication difficulties because of incommensurability of *concepts* have ever been of any practical significance in the history of science. There have been debates over rules of evidence, over what problems are important, and over the best methods of resolving them. But these have been honest disagreements over *matters of fact* and it seems to me that one of the most important properties of the scientific community is its ability to cut through verbal disagreement and get to the significant issues. I might add that this is a long-standing tradition in science. In one stock example of purported incommensurability, the case of phlogiston vs oxygen, we can hardly say there was a breakdown in communication when in both England and in France there were popular textbooks and monographs which congently and critically compared the two views. For example, in Fourcroy's textbook (the title of the English translation was: "Elementary lectures on chemistry and natural history. Containing a methodical abridgement of all the chemical knowledge acquired to the present time; with a comparative view of the doctrine of Stahl, and that of several modern chemists..."), we find a good discussion of the pros and cons and the relationship between the two theories.

I am not denying that one often finds *claims* in other spheres that debate is impossible because of conceptual incommensurability. (In the London School of Economics library there is a report by the U.S. Senate Committee on the Judiciary Subcommittee to investigate the Administration of the Internal Security Act and other Internal Security Laws which is entitled: 'Wordmanship: Semantics as a Communist Weapon'.) I do deny that this is a tactic used in science and I fervently hope the tradition will continue.

On the other hand, I agree that earlier philosophical analyses of the relationships between theories (such as what Feyerabend calls the 'layer model') have been incorrect Often in order to compare theories one needs to construct a comparison theory which contains bridging postulates of various kinds and of varying complexity.

Some of the postulates will match variables in the two theories whose values coincide within a restricted domain. Some will match expressions within the theories which have the same limited systematic import; i.e., they play similar explanatory roles. (For example, 'phlogiston lost' played the same role in many explanatory arguments within the phlogiston theory as 'oxygen gained' did in Lavoisier's chemistry). Other statements in the comparison theory will make explicit the difference between ontological presuppositions of the two theories, the models or metaphors they employ, their rules of evidence and their heuristic potential. Feyerabend is right – normally, one cannot reduce one theory to the other. However, within a supra-comparison theory, *which must be created for the purpose*, once can compare the predictive accuracy of the theories and their explanatory power. And one can also debate their heuristic promise, or their differing rules of evidence.

[14] I am interpreting the term 'dialectic' in the rational fashion suggested by Popper in 'What is Dialectic?', *Conjectures and Refutations*, Routledge and Kegan Paul, London, 1963, p. 315. It seems an appropriate term because I am concerned with historical cases involving a triad of theories, two of which are on the same 'level' and are in conflict. The third theory which replaces them gives a unified explanation of the phenomena explained by the previous two theories and preserves many of their features, but at the same time modifies them in important respects. A detailed characterization of both the conservative and revolutionary aspects of successive theories in the history of science is given in N. Koertge, 'A Study of the Relations Between Scientific Theories: A Test of the General Correspondence Principle' (unpublished Ph.D. dissertation, University of London, 1969).

[15] I. Lakatos, 'Methodology of Scientific Research Programmes', p. 153.

[16] This account of the new developments in chemistry is based on Kenneth Pitzer's introduction to G. N. Lewis, *Valence and the Structure of Atoms and Molecules*, Facsimile of the 1923 edition Dover, New York, 1966.

[17] *Ibid.*, p. 79.

[18] See *Ibid.*, Chapter 6.

[19] I. Pauling and E. B. Wilson, Jr., *Introduction to Quantum Mechanics*, McGraw Hill, New York, 1935, p. 17.

[20] Lewis, *Valence*, p. 55.

[21] *Ibid.*, p. 56.

[22] Of course, the most interesting examples are those in which the inter-theoretical conflicts have not been resolved. See L. Tisza, 'The Conceptual Structure of Physics', *Reviews of Modern Physics* **35** (1963), 151–85.

[23] However, see P. Havas, 'Four-Dimensional Formulations of Newtonian Mechanics and Their Relation to the Special and the General Theory of Relativity', *Reviews of Modern Physics* **36** (1964), 938–65.

[24] There are also interesting cases in which formal 'flaws' or 'incongruities' within our total theoretical structure (they are not actual inconsistencies) can serve as 'growing points' for a new theory. See H. R. Post, 'Correspondence, Invariance and Heuristics', paper presented at the British Society for the Philosophy of Science, University College, London, April 26, 1965 and 'Rules of Discovery: In Praise of Conservative Induction' (forthcoming).

[25] A possible objection to my position might run like this: "Let it be granted that there have existed contradictions between important theories before a revolution and that the theory which comes afterwards unifies and connects them. This is still an inadequate basis for a dynamic theory of scientific growth. Was the scientific community aware of the inter-theoretic contradition? Did it *cause* a crisis? Did they actively try to resolve it? Unless the answer is yes, you cannot claim to have discovered a driving force within science."

To answer this point, one would, of course, like to rely on more detailed historical investigations. However, I think it likely that there are cases where no one realized that an inter-theoretic contradiction existed until after the theory which resolved it was already in hand. Popper describes such a state of affairs by contrasting the objective problem-situation with the perceived problem-situation. It seems to me that one should conclude that one job of methodology is to make us (scientists) more sensitive to these kind of strains within our total body of knowledge.

In general, a methodologist has two jobs – one is to invent or discover within science new varieties of criticism, new ways of generating problems – Popper's characterization of a severe test is one example of such an activity. The other job for the methodologist is to argue against criticism-reducing strategems – an example of this is Feyerabend's argument against the position that the observation language for physics is fixed and must be that of classical physics.

[26] N. Koertge, 'Theoretical Pluralism, Criticism, and Education', forthcoming in the Proceedings of a Conference on New Directions in Philosophy of Eduction, held at the Ontario Institute for Studies in Education, Toronto, 1970.

IMRE LAKATOS

REPLIES TO CRITICS*

INTRODUCTION

The arguments my critics produce have made me realise that I fail to stress sufficiently forcefully one crucial message of my paper. This message is that my 'methodology', older connotations of this term notwithstanding, only *appraises* fully articulated theories (or research programmes) but it presumes to give advice to the scientist neither about how to *arrive* at good theories nor even about which of two rival programmes he should work on.[1] My 'methodological rules' explain the rationale of the acceptance of Einstein's theory over Newton's, but they neither command nor advise the scientist to work in the Einsteinian and not in the Newtonian research programme.

If at all, I give advice only on two points. First that a public record should be kept of the appraisals which includes a list of known anomalies and inconsistencies. Secondly, that the rigid commandments of superseded methodologies should be ignored. So, for instance, when I claim that a research programme may progress in an ocean of anomalies, I simply claim that my methodological criteria (as opposed to, say, Popper's) can appraise two rival programmes as 'good' and 'better' even though both are infested with anomalies. Thus I do *not* claim that when a research programme is progressing scientists *must not* pay attention to anomalies. Anomalies, like an odd-shaped tree at the seaside, or a serpent seen in a dream, may provide psychological stimulus for progress. Also, when it turns out that, on my criteria, one research programme is 'progressing' and its rival is 'degenerating', this tells us only that the two programmes possess certain objective features but does not tell us that scientists must work only in the progressive one. (Indeed, as I constantly stress, degenerating research programmes can always stage a comeback i.e. lead to further progress in knowledge. But this would, of course, be impossible if no scientist 'worked' on the programme.) Thus, the demarcation between the logic and the psychology of discovery is upheld.

My critics also seem to have missed my through-going *methodological instrumentalism*. In my view all hard cores of scientific programmes are likely to be false [2] and therefore serve only as powerful imaginative devices to increase our knowledge of the universe. This brand of instrumentalism is, however, consistent with realism and I hold that the succession of progressive research programmes constantly superseding each other is likely to produce theories with ever-increasing verisimilitude.[3]

But if all hard cores may be false, whether one *believes* them or not is a psychological irrelevancy. *My methodology is not at all concerned with beliefs*. Some scientists may feel the psychological urge to believe that the 'hard cores' of their programmes are true; they may develop a Kuhnian 'commitment' to them. In the objective appraisal of theories or research programmes such weaknesses of the human psyche are ignored. While upholding the view that the supreme aim of science is the pursuit of truth, one must be aware that the path towards Truth leads through ever-improving false theories. It is therefore naive to believe either that one particular step is already part of the Truth or even that one is on the right path.[4]

I. THE MYTH OF INTERTHEORETIC CRITICISM: A REPLY TO PROFESSOR KOERTGE

Professor Koertge's main line of attack is directed against the 'softness' of the standards I proposed for judging scientific research programmes. Feyerabend and Kuhn, of course, have already claimed that they have so little bite, that they are so 'liberal' that they are not even worth stating. Kuhn sees their weakness primarily in the impossibility of instant application. As he put it: "Lakatos must specify criteria which can be used *at the time* to distinguish a degenerative from a progressive research programme ... Otherwise, he has told us nothing at all." [5] And, as Feyerabend put it with glee: "Scientific method, as softened up by Lakatos, is but an ornament which makes us forget that a position of 'anything goes' has in fact been adopted." [6]

Professor Koertge's reaction is very different. She agrees (or nearly agrees) with Feyerabend and Kuhn that my standards are too soft. But instead of rejoicing she wants to *harden* the objective, 'superparadigmatic' standards, in order to restore the power of hard-hitting criticism, in order to show that the life of science is not quite as liberal as I picture it. My sympathies are with her: but I do not think she has succeeded in achieving her aim.

Professor Koertge proposes *two major improvements* to the methodology of scientific research programmes.

(1) She proposes to raise the standards by which a programme can be said to be progressive. To this end she proposes two additional requirements: (a) that the positive heuristic must organically cohere with the hard core, and (b) that the construction of auxiliary belts must lead to an *increase* in coherence; that 'the programme becomes *more* coherent and unified as time progresses'.

As a historiographical confirmation of these new requirements she adduces the example of Prout's research programme. On my terms this programme was progressive throughout the nineteenth century. On hers it was not: there is a lack of organic coherence between the hard core of Prout's programme on the one hand and its positive heuristic on the other. Since Prout's programme was in ill repute among most chemists of the period, Professor Koertge claims that this fact supports her stronger requirements.

I am not so sure. First, protective belts *are bound to* contain some auxiliary hypotheses which are completely alien to the hard core: Newton's theory of atmospheric refraction is a typical example.[7] Secondly, increase in coherence is a concept even vaguer than coherence. I have replaced Duhemian 'coherence' with 'positive heuristic', where the stress is less on the aesthetic quality of simplicity than on the requirement that not only the experiments but also the theories should be, as it were, planned beforehand. The chain of problems to be solved is anticipated in the programme. Coherence is, anyway, an untenable requirement as soon as inconsistent foundations and 'creative shifts' are permitted: progress on inconsistent foundations only dramatises incoherence and creative shifts (seen as such only with hindsight) represent, at least momentarily, a decrease in coherence. Whewell, whom Professor Koertge quotes, was rather naive in his hope that increase in coherence was a sign of getting nearer to, and possibly even attaining, Truth.[8]

However, Professor Koertge's stricture led me to realise two points which I had not previously noticed. First, I now see that any 'creative shift' is *ad hoc* in my sense; but I give good marks to the former and bad marks to the latter.[9] I think this problem can easily be solved in a non- *ad hoc* way within the methodology of research programmes. Furthermore, I agree that Professor Koertge is right that in appraising research programmes one has to take into account the different power (I prefer this term to degree of 'coherence') of their positive heuristics. Newton's programme, I agree, is far superior, on this *a priori* ground, to Prout's. I suspect that this point too may well have interesting consequences.

(2) Professor Koertge's *main* criticism is of my alleged '*monotheoretical approach*' for whose replacement, by her '*dialectical ascent*', she argues. Thus, in her view, my rational reconstruction of science is, in an important sense, still 'monotheoretical'. But is it?

In Kuhn's 1962 view major fields of science are, and must be, always dominated by *one* single supreme paradigm. My (Popperian) view allows for simultaneously growing rival research programmes. In *this* sense – and I am sure Professor Koertge will agree – no Popperian approach is 'monotheoretical'. But then what does Professor Koertge disagree with? Since she does not distinguish sharply the logic of discovery from the psychology of discovery or possibly some vague rational heuristic,[10] her text remains ambiguous and I have to try different interpretations of her charge of 'monotheoreticality'.

(a) She might have meant that my *appraisal* of a scientific research programme R_1 is completely independent of the development of a rival R_2; that is, my appraisal of whether R_1 is progressive or degenerating depends only and exclusively on R_1. But this is not so. If R_2 progresses, it is bound to slow down the progress of R_1 *since* R_2 *will anticipate some novel facts faster than* R_1. Indeed, R_1 without R_2 may be seen as progressive but against the background of R_2 it may be degenerating, the signposts of progress are anticipated novel facts: a rival programme may eat them away .Without Einstein's programme Newton's could still be seen as progressing. This follows from my definitions. My appraisal is then not monotheoretical.

(b) Then what does she mean? I would not like to think that she means to defend naive falsificationism and that she holds that at least *some* corroborated falsifying hypotheses *do* falsify, namely those with relatively high empirical content: 'major projectiles' may well knock out a programme even if minor ones do no harm. But such weakened naive falsificationism is just as wrong as the original, since my historiographical criticism applies to the one with exactly the same force as it does to the other. (Incidentally, Professor Koertge could not have chosen a worse example for her case than Lewis's theory of valence. This 'projectile' was never of serious 'size and weight'. Lewis's theory of valence of 1916 – involving stable, non-orbiting electrons – was pressed by Bohr's vastly successful programme from above and by many more anomalies than confirmations from below. It was a small, lame, barely progressive programme. Lewis, correctly, was very modest about it and was very careful, as Professor Koertge admits, not to present it as a criticism of Bohr's approach. Bohrians, also correctly, did not pay any attention to it whatsoever. But the new quantum theory opened up new avenues to solve the problem of why some neutral atoms can form a stable pair, some cannot. This challenge was met primarily by Heitler's and London's classical paper of 1927, which, incidentally, does not mention Lewis's name at all. The progress of the new theory of valency with its rich positive heuristic immediately overshadowed the old semi-empirical formulas. A huge theoretical literature on the subject evolved with papers by Wigner, Hund, Slater, Pauling and others. Lewis's dotted figures may have speeded up this spectacular development by a few weeks; just as Balmer's formula speeded up Bohr's old programme by a few weeks.)

(c) Perhaps Professor Koertge wants to introduce a sort of heuristic rule that scientists' psychological confidence *must* be shaken by major rivals. They *must* worry about it.

I, of course, do not prescribe to the individual scientist what to try to do in a situation characterised by two rival progressive research programmes: whether to try to elaborate one or the other or whether to withdraw from both and try to supersede them with a Great Dialectical Leap Forward. Whatever they *have* done, I can judge: I can say whether they have made progress or not. But I cannot advise them – and do not wish to advise them – about exactly what to worry and about in which direction they should seek progress. However, if Professor Koertge's 'dialectical' approach consists of the advice that in the face of two rival programmes one *must* drop both and speculate about both, my sympathies are completely with Kuhn: "The scientist who pauses to examine *every* anomaly (major or minor) he notes will seldom get significant work done".[11] Or: "It is *often* better to do one's best with the tools at hand than to pause for contemplation of divergent approaches."[12]

As long as Professor Koertge's thesis is merely that some proliferation of different approaches – and of problem appreciation – is on the whole likely to be better for the growth of knowledge than none, of course, I agree. On this issue we are both Popperians and anti-Kuhnians. But if her moral sense finds it impossible not to give full attention to the voice of criticism and makes it *mandatory* to divert one's attention from the positive heuristic of the programme to [minor or major] anomalies, I disagree.

(d) But Professor Koertge may well want more than to give tentative heuristic advice. And, I am afraid, the likeliest interpretation of her stricture is the one explicit in her introduction. She is dissatisfied by my methodology because it has "serious shortcomings when viewed as a comprehensive theory of scientific growth", as a "dynamic of the critical forces operating".[13] To my mind these 'shortcomings' of my methodology have to be seen as its merits. I never presumed to offer a comprehensive theory of the growth of science, and, indeed, the very idea makes me shudder. It reminds me of Hegel, Spengler, science planning and of historicism in general. Here I agree with Popper's argument which led him to conclude that "we must reject the possibility of a [scientific] *theoretical history*"[14], and above all the possibility of a scientific theoretical history *of science*. Freedom and creativity are inconsistent with a 'dialectical dynamics of critical forces'.

II. MISUNDERSTANDINGS: A REPLY TO PROFESSOR RICHARD HALL

I am afraid that the criticisms by Professor Hall hinge on misreadings of my text. He makes two major points.

(1) "Lakatos says that we should use history of science to help us to decide [between rival methodologies]. He proposes, or at any rate seems to propose, the following criterion: Given competing methodologies we should prefer that methodology according to which more of the actual history of science is internal and rational, and more of scientists' own judgments about science are correct".[15] But this is incorrect: indeed, I abhor this view. Since it leaves 'scientist' undefined, this rule would give *carte blanche* to any pseudoscientific group to set itself up as supreme authority. On the contrary, I devoted the whole last section of my paper (Section 2C) to a plea for *dual authority* in methodology, where I argue

that some of the scientists' 'basic value judgments' can and should be overthrown, especially "when a tradition degenerates or a new bad tradition is founded".[16] Moreover, right from the beginning I emphasize the far-reaching analogy between scientific and methodological research programmes[17] and stress that just as empirical 'basic statements' can be overruled by theory, normative 'basic judgments' can be overruled by methodology.[18] Professor Hall is fully aware of these passages which he regards as inconsistent with my alleged main position which he distils from two sentences by wanton misinterpretation.[19]

(2) Professor Hall then takes me to task for another Lakatosian thesis which he formulates as follows: "The different methodologies lead to determinable and different dividing lines between internal (rational) history of science and external (empirical) history of science".[20] This I happily acknowledge as a correct rendering of my position. But Hall tries to refute it by two counterexamples. He sets out to show that neither my nor Carnap's methodology provides such a dividing line: one can carry out concrete appraisals neither (a) with the help of my logic of research programmes nor (b) with the help of Carnap's inductive logic.

(a) According to Professor Hall "we have no way of evaluating Lakatos's claim that adherence of scientists to the Newtonian research programme after the discovery of the anomaly in Mercury's orbit was rational according to the research programme methodology". So what? I also claim that trying to devise an alternative to Newton is rational too, after the discovery of the Mercury anomaly *or before*. I do not demarcate between different groups of scientists but between different research programmes. My real *methodological* claim is that Mercury's anomalous orbit does not undermine Newton's programme, nor does it vitiate its progressive character. And this appraisal is sharp and unambiguous. But, even after having realised that Hall missed my distinction between appraisal and advice[21], I was still amazed to read that according to me "scientists should drop [a research programme] if things go badly enough for it."[22] Surely, one of my crucial – and repeatedly made – points is exactly the negation of such a *universal* rule.

(b) Professor Hall, however, is completely right in claiming that Carnap's methodology (that is, his inductive logic) cannot draw a dividing line between internal and external history. I happily agree with him. Carnapian inductive logic never progressed far enough out of the mire of the logical and epistemological problems it created for itself to be able to offer a *historiographically helpful* system of appraisals. But classical inductivism, conventionalism, falsificationism, the methodology of research programmes *did*. They *can* be compared historiographically. Thus I was fully aware of Professor Hall's exception which does not vitiate my thesis in the least: it only shows the poverty of inductive logic.

(What puzzles me most is how Professor Hall ever came to believe that I think that Carnap's inductive logic can be historiographically helpful. I explicitly state the contrary.[23] I could not help being amused by his misreading of my passages about inductivism. For Professor Hall takes several pages to explain to me Carnap's true position. He introduces his lengthy exposition of Carnap's 'inductivist methodology' by the statement that "it is not clear who Lakatos has in mind when he attacks inductivism since he mentions hardly any names. However, one can hardly go wrong in taking Carnap to be an inductivist."[24] Alas, I define the inductivist position as asserting that "only those propositions can be accepted into the body of science which either describe hard facts or which are infallible inductive generalisations from them".[25] This, of course, is Newton's position and not Carnap's. I am not so illiterate as to attribute to Carnap such a totally un-Carnapian position.[26] But I do say that while Newton's classical inductivist methodology was historiographically very influential, Carnap's neo-classical probabilistic methodology has been totally uninfluential[27]: no historian of science ever claimed that a major scientific hypothesis h was abandoned in favour of a rival h' because it was found, applying Carnap's measure functions, that $p(h', e) \gg p(h, e)$, since Carnap's available measure functions cannot be defined for any real scientific theory.)

NOTES

* These replies, unfortunately, could not discuss my commentators' essays in proper detail: both space and time were severely limited. I have tried to concentrate on the essentials. Moreover, because of the vagaries of the British Postal Service, I received Professor Kuhn's and Professor Feigl's most interesting comments so late that I was unable to prepare my replies in time for printing in this volume. I hope to have them published shortly.

In my *Replies* I refer time and again to my paper 'Falsification and the Methodology of Scientific Research Programmes' in Lakatos and Musgrave (eds.), *Criticism and the Growth of Knowledge*. For brevity I shall refer to it as 'Lakatos (1970)'.

[1] Cp. especially above p. 92 and note 2 on p. 123. For instance, Tarski's 'methodology' appraises mathematical proofs for validity but offers no heuristic for how to

arrive at valid proofs. On this point also cp. my 'Popper on Demarcation and Induction', footnote 5.

[2] Cp. *Nature* **188**, (1960) 458.

[3] For my 'instrumentalism' cp. especially Lakatos (1970), pp. 184–9. I am afraid these pages were missed by many readers. Perhaps I should not have put them, as I did, into an *Appendix*.

[4] In these matters I follow Bolzano, Frege and Popper. I gave the relevant references to Popper's important recent papers in footnote 1 of p. 180 of Lakatos (1970). My readers will be bound to misunderstand my paper (and, for instance, confuse my research programmes with Kuhn's paradigms) without having studied them.

[5] Kuhn, 'Reflections on my Critics', in Lakatos and Musgrave (eds.), *Criticism and the Growth of Knowledge*, p. 239.

[6] Feyerabend, 'Consolations for the Specialist', *op. cit.*, p. 229.

[7] Cp. Lakatos (1970), p. 130, footnote 5.

[8] Whewell, of course, would have shrunk back from the idea that an inconsistent programme might be in any sense 'coherent', let alone lead us towards Truth.

[9] A 'creative shift' is an empirically progressive development of a programme which was not anticipated at the outset. But I define '*ad hoc*$_3$' in exactly the same way. (Cp. above, p. 103 and note 36, and Lakatos (1970), p. 137 and p. 175, footnotes 2 and 3.)

[10] Cp. the references above, note 1.

[11] Kuhn, *Structure of Scientific Revolutions*, 1962, p. 82, my italics.

[12] Kuhn, 'The Essential Tension: Tradition and Innovation in Scientific Research', in Taylor and Barron (eds.), *Scientific Creativity, its Recognition and Development*, 1963, p. 341.

[13] Cp. above, p. 160.

[14] Popper, *The Poverty of Historicism*, 1957, p. X.

[15] Cp. above, p. 151. He repeats this statement on p. 158.

[16] Cp. above, p. 122.

[17] Cp. above, pp. 116–20.

[18] Cp. above, pp. 116–22.

[19] He quotes these two sentences out of context, above, p. 156. The qualifying normative term 'great' in the first sentence and the cautious word 'may' in the second sentence escaped his attention. Of course, the relevance of these qualifying terms only transpires from the context and the long adjoining exposition, which, however, he curiously regards as aberration from, instead of interpretation of, these sentences.

[20] Cp. above, p. 151.

[21] Cp. above, pp. 92 and note 2 on p. 123.

[22] Cp. above, p. 152.

[23] Cp. p. 128, note 70.

[24] Cp. above, p. 153.

[25] Cp. above, p. 92.

[26] Cp. my monograph on Carnap's inductive logic in Lakatos (ed.), *The Problem of Inductive Logic*, pp. 315–417. Professor Hall, incidentally, does not notice – in spite of my explicit definitions of my technical terms – that my 'inductive principles' are not part of 'inductivism'.

[27] Cp. above, p. 128, note 70.

CONTRIBUTED PAPERS

I. Observation

PETER K. MACHAMER

OBSERVATION

Of the many problems involved in discussions of acceptability and justification I wish to consider in this paper what types of sentences can be used in order to justify a given claim or theory in science, what types of sentence can be appealed to in testing (showing to be acceptable or *rejecting*) a theory or claim. Rather than rehashing old critcisms of previous work, or constructing new counters to works already known to be inadequate, I propose to get on with the problem at hand. Necessarily this 'getting on' is in large part a taxonomic and programmatic venture. I wish to suggest, briefly and by examples, that *in fact* there have been, and are, at least four different types of bases to which appeal is made in testing or justifying a theory or claim. These four types are:

(a) Observation reports.
(b) Accepted Philosophical principles.
(c) Sentences taken from theories other than the theory being tested.
(d) Pragmatic arguments of the form: If such and such a procedure is followed (or point of view adopted), then such and such good or bad consequences will result.

Sentences of type (a) – observation reports – are admitted by all to function as bases for tests and justifications. Indeed, they are usually all that are admitted. I shall discuss these at length below.

Philosophic principles of type (b) are an important source of justification and test. Principles of this kind are not treated as being testable themselves. But though such principles are taken as necessary, immune from test, this status can be revoked upon occasion. Such occasions may be provided by any of the other bases of justification. Philosophic principles are (*à la* Popper) demarcated by the attitudes and tenacity of their users, though there is often available the additional criterion of an *apriori* argument in their favor.

Some of these principles have been accepted for ages, and probably will be acceptable for all time, e.g. consistendy and simplicity. Others have been held for a time, and then rejected, e.g. Plato's postulate of the necessity of circular planetary orbits. Some principles now seem to be firmly entrenched, e.g. the principle of efficiency in its modern form of conservation of energy.

The way in which (b)-principles function as bases for justification or test can be briefly illustrated. Copernicus rejected Ptolemy's models of the Moon because they violated Plato's postulate by introducing motion about equant point.[1] Alhazen justified his theory of refraction by an appeal to a form of the principle of efficiency. He claimed that a ray of light entering a denser medium will bend toward a line drawn perpendicular to the interface. This claim he justified on *causal* grounds by arguing that a denser medium offered more resistance to the entering ray, and *because* the ray always *must follow the most efficient path* (the path of least resistance), the ray would bend toward the perpendicular. The perpendicular was held to be the easiest path through the medium.[2] A final example would be provided by any theory about the energy of sub-atomic particles which are released under particle bombardment, for any such theory must assume the principle of the conservation of energy and make use of it in supporting its claim.[3] Uses of appeals to simplicity and consistency are even more well known.

Justifications and tests which involve sentences from other theories, kind (c), are also common in current and historical cases in science. In order to test or justify a theory or claim one might take a sentence from another accepted theory and use it to justify or refute the claim being considered. For example, Galileo appealed to the principles of statics to justify some of his conclusions about dynamical properties.[4] Dalton rejected Berthollet's theory of the molecular structure of elements (i.e. that there were molecules of some elements which had more than one atom) on the basis of his reading of Newton's 'Laws' of elastic fiuids.[5] Finally, thermodynamics was refuted, the second law was shown to be false, on the basis of statistical mechanics.[6]

In pragmatic justifications and tests of kind (d), appeal is made to how a given claim or theory will work out, to what can be done or accomplished by using it. The justification takes the form that such a principle, law, theory, or definition should be adopted, because *if* it is it will be able to do certain things, accomplish certain goals, or solve certain problems (which have not been done, accomplished, or solved before, or which a competing theory cannot do, accomplish, or solve). *After* the acceptance of the theory the justification proceeds by showing how the theory (or part) brings about the desired consequences. A negative version for testing theories in this way is also possible; one might argue against a theory on the grounds that it did not accomplish certain goals which it was expected to accomplish. Galileo in a first draft of *A Discourse on Floating Bodies* borrowed, almost directly, from Archimedes two axioms. By the time of the publication of *Floating Bodies* Galileo had revised his axioms. The justification for the new axioms was the fact that they could be used to account for more cases; they were able to give solutions to a wider variety of problems.[7] Likewise, Newton justified the adoption of his definition of absolute space (which is part of his theory) on the grounds that if such a definition is not accepted, then no account of absolute motion is possible.[8]

The various bases, (a) through (d), which are appealed to in justifying or testing, are not mutually exclusive (and I am not certain they are exhaustive). They all have in common the fact that at any given time, in order to be used they must at that time be accepted by the relevant community.[9] Generally they will not be argued for since they are accepted. But most of them are subject to change, or to dismissal from the basic class. In the future they may not be acceptable as providing a basis for tests or justifications.[10]

OBSERVATION AND THE DATA OF SCIENCE

I shall now consider type (a) justifications and tests: observation reports. First, they must be reports. To make this clear involves drawing a distinction between seeing and observing. Whatever one's analysis of seeing it is clear that one can see things and not be able to say anyting about what he is seeing on the basis of his seeing. That is, it is possible for us to truly say of a person *G* that he sees *X* even though *G* is incapable of saying he sees *X* or is in some sense unaware of *X* at the time he sees it. The classic case of seeing but not noticing the billboards on the pre-Lady Bird highways provide examples, as doing the seeings of infants and animals.

Simple seeing is not a sufficiently strong condition to guarantee the possibility of communicating (information) about what is seen. I shall use the word 'observation' in such a way that this possibility becomes a necessary condition for observing something. In the case where *G* merely sees *X* he, in no sense, needs to be able to go on and say things about *X*, but in observing *X* I shall require that *G* be able to produce some sort of observational report upon suitable occasion. If *G* is incapable of producing such a report, then there would be no way in which his 'observation' could be of use to anyone but himself; there would be no description of *X* which *G* could give. If *G* could not give any description of *X* then he could not even formulate any claim about *X*. To formulate such a claim one must be able to bring *X* under some description, be able to apply some word to what is being observed.[11]

The reason behind making the possibility of reporting a condition of observing is that if the information gained by a process of observation is to be useful as a means of justifying or testing theories or claims, or even used to convey information, the observer must be able to present his information, the results of his work, in communicable form. (condition of inter-subjectivity).[12] The possibility of issuing a report seems to guarantee the necessity of paying sufficient attention to what is being observed, for such attention is a necessary condition for deciding what words to use in the description.

For a report to be used for communication it seems necessary that (i) the report be put in such a form as is conventionally agreed upon by members of the relevant community (i.e. it must be put in a common 'language'), (ii) the report must be, in fact, successfully 'read' by some member(s) of the community, and (iii) the report must be accepted, as at least partly correct or incorrect, by some members of the community. For observation reports to be used as bases for justification and tests, a further condition is necessary, namely (iv) the report must be acceptable to all (or almost all) members of the community. (Otherwise, the report would not belong to the class of data, but rather to the class of problems, reports which are doubted).

The class of observation reports is not demarcated in terms of meanings. Observation sentences are reports of observations (seeing is a necessary condition), which can be used to convey information; such information can, in some cases, be used in testing or justifying various theories or claims. What is peculiar to observation reports is the generic fact regarding the way the information expressed in the sentence was obtained, i.e. trivially, at least part of the information expressed in the report was obtained by observation or observational procedures. The spelling out of this condition would involve making clear how information is gathered by the senses, and how the senses are 'extended' by various instruments. Not all sentences which might be reasonably called observation reports can be used to test a given theory or claim. There are simple cases where what is observed is just irrelevant to the theory or claim under consideration. There are cases in which the observer does not know what he is observing, and cases in which the description of the what is observed fails to exhibit its connection with what is being tested or justified. In such cases as the latter two *an* observation report can be given, but it will be formed from terms which do not bring out the connection. When Galileo first observed the moons of Jupiter he thought that they were merely new stars. His report of what he observed was not capable of being used to support a claim about the existence of moons revolving about Jupiter, nor *via* analogy to the Earth, of supporting the Copernican theory. For Galileo's observation to be able to function in these ways what he observed had to be re-described in such a way as to exhibit its connection with the claim and theory being examined. Galileo later succeeded in accomplishing this. By continued observation he was led to notice that the 'stars' moved, that they sometimes disappeared from sight as if hidden by another body, and that they seemed to move in orbital paths. It was not until four days (so *he* says) after he began observing the moons that he realized they were moons. This realization led to a re-description of the observations which allowed them to be used for testing and justifying. The re-description asserted that Jupiter had moons. And the information conveyed in the later observation reports was used to establish the claim that the Earth was not unique in having satellites, and thus that the Earth's moon did not present an untoward case for the Copernican theory (it was no longer an oddity by virtue of its being the sole planet with another body revolving about it). The ordinary Latin of the 17th century clearly exhibits the required connection. Galileo links 'conversiones planetarum... circa Iovem" to "conversion Planetarum circa Solem" and to "Luna(e) circa Terram". It is the 'movement about' which exhibits the connection.[13]

The mere fact that an observer has issued a report is not enough to allow that report to function as a basis for justifying or testing a particular theory or claim. What is required is that the observer issue a report which is formulated in terms which exhibit their connection to the claim or theory which is being justified or tested. A necessary condition of an observer's issuing such a report is that he knows which terms are applicable to the case at hand. In some cases the required knowledge, either of the terms or of which description will best exhibit the relation, will only be available to the observer *after he has learned the theory which is being examined.*

Dalton, seeking support for his atomic theory, and for his version of the law of multiple proportions, re-described earlier observations in such a way as to bring out the connection which supported his law, Lavoisier had earlier reported that the analysis of carbon dioxide yielded, by weight, 28% carbon and 72% oxygen. Later Clement and Desormes reported that carbon monoxide was, by weight, 44% carbon and 56% oxygen. These reports were not capable of being used, and were not used, by anyone, for the connection between them and the Law was hidden. This connection became apparent when the weight of carbon was held fixed. The fixation of the weight of carbon, and the corresponding redescription, was suggested to Dalton by his rule of greatest simplicity which held that elements would combine with one another, according to 1–1, or 1–2, or 2–1 ratios. Holding the weight of carbon fixed, as 1, the earlier data could be re-described as: carbon dioxide, gaseous oxide of carbon, 1 part carbon to 2.57 parts oxygen, and carbon monoxide, or suboxide, as 1 part carbon to 1.27 parts oxygen. This was a clear exhibition of the 1–2, and 1–1 patterns, and thus supported the view that carbon and oxygen could enter into compounds in proportion to the elements present.[14]

These examples show that the descriptive terms used in presenting an observation report are important, and it is this fact which seems to lie at the heart of much of the talk about theory-ladenness. The terms used in giving an observation report will be terms which the observers knows (has learned) and which he thinks are applicable and appropriate. (Strictly speaking, it is only an observation report which uses terms drawn from a scientific theory which should be called theory-laden).[15]

But how does an observer satisfy himself as to what is applicable and appropriate? Usually the observer, reacting in a way which is the result of his past learning experiences, has no difficulty in deciding which descriptive terms are applicable and which are not. He has learned his words in a variety of situations, and the occurrence of what he takes to be a situation similar to that in which he learned to apply the term is usually sufficient for him to use the word learned in such situations (though of course he may be mistaken). In cases where he thinks that more than one description is applicable, the question of appropriateness arises. This question is answered in terms of the purposes and reasons for the observation. The most appropriate description is that which best fulfills the purpose for which the observation was made; in most cases in science, the word which best exhibits the connection between what is being observed and the claim or theory being considered. If the observation was made with no such purpose then the observer will probably chose the description which he thinks will best convey the information he has gathered. What "best conveys the information" will be decided by taking into account the audience for which the report is intended. If the audience is not considered, he must just pick the description which fits his fancy.

Further elucidation of the nature and function of observation reports can be seen from considering cases in which there is disagreement or doubt about what is being observed. Two observers, both seeing the same thing, may disagree about what that thing is. In such a case each observer will offer a description of what he is observing which he takes to be applicable and appropriate, yet the two descriptions may be incompatible. In a standard kind of case, as noted before, the description offered by each observer will be a function of his purpose in trying to justify or test a claim or theory. At this point, both observers may re-describe what they observe in such a way that both could assent to the new description. This would be a test of whether or not they were both observing the same thing.

In such cases the adequacy of the reports cannot be tested by how well the descriptions exhibit the connections between what is being observed and the theory(ies) involved. There might be two theories involved, one for each observer, and each report might well exhibit a connection to one particular theory.[16] Such exhibition would not, in this kind of case, allow for a decision as to which observation report is correct, or the most adequate. In such cases appeal is usually made to how well further observational reports correspond with descriptions which are entailed by the original observation report, plus additional accepted information about the kind of object or process which the original report claims is being observed. The original observation report may claim, for example, that what is being observed is an *X*. On the basis of accepted facts which are known about *X*'s (such facts being known as part of the common store of knowledge or on the basis of accepted theories in which '*X*' occurs), we can generate conditionals of the form: If it's an *X*, then it will have property ϕ. Or, if it's an *X*, then it will react in such and such a way under conditions *C*. Such conditionals provide the second test of adequacy for observation reports.

In cases where an original report is challenged, as it would be in the case of two observers' disagreeing, appeal is made to these further entailments, and it is on the basis of these entailments that the adequacy of the original report is judged. The consequences of the conditionals generated from the original reports (plus additional knowledge) are tested by matching them against further observation reports (or sometimes by the other methods of justification, i.e. methods (b) through (d).) These further observation reports issue from observations made to check the entailments. Thus the consequents of the conditionals will be checked by the same procedure that is used in any straightforward case. The consequences of the original observation report are now the claims which have to be justified (for they in turn will justify the observation report, if successful); and the new observation reports produced to test or justify this claim must be accepted by the community, or they too would require further justification.

An observer involved in such a situation of disagreement might check both the entailments from his own report – in order to attempt to justify his own claim – and the entailments from his antagonist's report – in order to test and hopefully to refute the adequacy of the other's report. In order for an observer to test the adequacy of *both* reports (in the way just described) there must be agreement that both are meant to be reports about the same thing (and such agreement, as indicated, can be assured by re-description). And, for the results of the test to be acceptable to both parties the two observers must agree, to some appropriate extent, on the entailments generated by *both* reports. Otherwise the antagonist would reject the validity of the first observer's test of his case; and such cases do occur. But in the case where the two observer's can agree, it means that the two must share (to the same extent at least) some knowledge or theory(ies).

If the original observation report is accepted unanimously, then, though such further tests could be performed, they generally will not be. Such further justification is only required where the original report is questioned or doubted. This doubt may arise in situations like those which occur when alternative reports are given, when it is not clear how the report connects with the theory being tested, or when the observer has some reason to think that his procedure of observation was faulty or somehw inadequate. Situations of these types (and there are probably others) I shall call doubt situations. In such doubt situations a process of 'interpretation' is called for. The situations give rise to doubt about what is being observed, and so the observer attempts to allay the doubt by applying further tests. The process of reflecting upon the offered description(s), and trying to decide on the most appropriate description is the process of interpretation. In cases where doubt-situations are not present, no interpretation occurs. The observer will merely offer his description (in the way described above).[17]

An example might be appropriate at this point. We can say that Galileo and Christopher Scheiner both saw the sunspots. But Galileo and Scheiner disagreed about what they were observing. Scheiner reported that he was observing planets or moons around the sun, while Galileo reported, at first, that what he observed were like clouds, and later, that they were spots on the surface of the sun. Galileo, Copernican that he was, could perfectly well understand Scheiner's report even though Scheiner was a defender of the Tychonic system. Galileo understood Scheiner to be saying that he observed a body which was a round object of relatively unchanging shape, and which traveled about a center; a body which was more distant from that center than was the surface of the body which occupies the center. To call something a planet while holding any contemporary planetary theory committed one to this description. Both Galileo and Scheiner could agree on this description, which was an entailment from Scheiner's planet-description.

Though both men further agreed on the basis of drawings and more 'phenomenal' descriptions that they were observing the same objects, they disagreed about the location and nature of these objects. In the face of such disagreements both men tried further tests to justify their respective claims (and to refute their antagonist's claims). Scheiner looked for constancy of shape, and sought after parallactic evidence to show that the what he was observing was above the surface of the sun. These were the tests of adequacy which he performed. Scheiner either ignored or put down as attributable to haziness Galileo's incompatible descriptions. Galileo, on the other hand, claimed that the things he observed were constantly changing shape, and that they were located on the surface of the sun, and therefore these things could not be planets.

Galileo when he first observed the spots was puzzled. He tried to find out what their nature was and the best description he could come up with was an analogy to clouds. Galileo then applied further tests based upon both his and Scheiner's descriptions (original reports). Galileo tried to interpret what he was observing. He reasoned they could not be planets or stars since the exhibited properties were quite unlike those which would have been exhibited by a planet or star. In his *Letters on the Sunspots* he pointed out to Scheiner all the ways in which the spots were unlike stars and planets, especially how they were unlike the Medicean planets. He pointed out that further tests based upon Scheiner's original report did not fit with the observations which were undertaken to test that report: the description of what was seen did not match with the description entailed by the original report.[18]

Accepted observation reports are agreed upon by the community as providing a basis for justifying or testing claims or theories. Such accepted reports form (at least part of) the data of Science, and might appropriately be called *data sentences*. There are other basic data sentences deriving from the other methods of justification. An observation report given in a doubt situation is not a data sentence. It is not until the doubt has been resolved that the description (or its negation) can take its place in the class of data sentences.

Further, schematic points about Observation.

One can understand why the empiricist tradition took the appeal to observation sentences as being the way in which a theory would be tested or justified. The class of data sentences is, in fact, fairly stable. The use of a data sentence provokes no controversy over its truth, and even observation reports are not very often challenged. But the empiricist tradition went amuck by assuming that no other kind of justification or test was ever given or was acceptable; and by assuming that such a class of data sentences was unquestionable, incorrigible, and unchangeable.

Discussion of the first of these assumptions would take us back to questions about the validity of the other forms of justification I mentioned earlier in this paper. Though I won't do so here I should like to argue that not only are such other bases of justification in fact used, they must be used. The second assumption is just false. I have already shown how some observation reports can be questioned (in doubt-situations), and how further tests of adequacy can be generated (by the conditional method). This too would take care of incorrigibility; but there are other arguments available which likewise do the job, (i.e. show that sense-data are private and useless, and then show how it is always possible to make a mistake in describing something.)[19]

The last part of the second assumption, concerning the unchangeability of the class of data sentences, is somewhat unfair to later positivists. But rather than discuss their views I shall attempt to catalogue the various ways in which the class of data sentences can change. First I shall consider types of *extensional change*. One may be means of *new or better instruments*, or by more acute observation, observe *new things* (either new entities, or processes, or new properties of already known entities or processes). Such instrumental change is exemplified by Galileo's use of the telescope to discover the moons of Jupiter, or the phases of Venus. In such cases the original class (pre-change class) of data sentences would remain as a subclass of the new set, while the addition would come through the sentences reporting the newly discovered objects or properties.

The class of data sentences could be increased by a theory which tells us to look for new things. (This may often overlap with the first kind of case.) Such theory based change is exemplified by the well known case of the discovery of Uranus. The role of discoveries, in the two kinds of cases mentioned above, brings out one relation which obtains between discovery and justification. Finally the class of data sentences can be increased by new descriptions of things or processes which had already been observed. Dalton's redescription of the carbon oxides is an example of such a case.

Decrease in the extension of the class of data sentences is also possible. Sometimes an old description of a previously observed entity or process will be shown to be inconsistent with a newly-offered description, and the old description will be rejected. The new description could result from any of the ways mentioned above, use of a new instrument, new use of an old instrument, a 'closer look', the introduction of a new theory, or the new application of an old theory. Also the class of data sentences can be reduced when a theory once accepted by the community is no longer accepted. Thus it was not that impetus descriptions were incompatible with inertial descriptions, it is on other grounds that the inertial theory was held to be justified and the impetus theory discredited.

Finally, it has been claimed by Feyerabend and Kuhn (and others) that the class of data-sentences can also be changed *intensionally*. A previously acceptable data-sentence, such as "the mass is 5 grams," or "The center of the universe is the focus of the planetary orbits" will still count as a data-sentence even after the introduction of a new theory (or the replacement of an old theory by a new one). But, sometimes at least, the meanings of the terms in the data-sentence are changed from one occurrence to another. Thus, in this sense, even though typographically the sentence has not changed, the meaning of the sentence has changed. While there is often a point in holding such a position, it can, I think, easily be over stated, i.e. by holding complete incommensurability. It is certainly true, as Feyerabend has shown, that in many cases of theory change the set of 'pragmatics' associated with a word also changes. His analysis of 'color' and the result of the discovery of the Doppler effect brings this out nicely. But some of the parameters in the set of pragmatics usually do not change. Thus it becomes a matter of degree of change which is to decide questions about whether *the* meaning of the term in question has changed. My present feeling is that the decision to say that there has been a meaning change in any particular case will have to depend upon the detailing of the similarities and differences between the uses of the word in the different theories.[20] This question, as well as many others, e.g. the possibility of the interaction between the different types of bases, cannot be discussed further.[21]

Ohio State University

NOTES

[1] N. Copernicus, *De Revolutionibus*, Book IV, Chapter 2.

[2] Ibn al-Haitham (Alhazen) *Al Manazir* (Perspectiva). Book VIII, 8. For a discussion see A. I. Sabra, 'Explanation of Optical Reflection and Refraction: Ibn-al-Haytham, Descartes, and Newton', *Actes du Dixième Congrès International d'Histoire des Sciences* (Hermann, Paris 1964) p. 532ff, and D. Linddberg 'The Cause of Refraction in Medieval Optics'. *BJHS* 4 (1968) pp. 26ff. The use of this principle was quite common in much of medieval science. It survived as was down through the 17th century; Descartes makes use of a similar assumption in his treatment of reflection and refraction.

[3] An excellent example of this principle, being explicitly used, is given in J. Chadwick, 'The Existence of a Neutron'. *Proc. Royal Soc.* (London), Series A, **136** (1932) pp. 692, reprinted in H. Boorse and L. Motz (eds.), *The World of the Atom*, Vol. II, Basic Books, N.Y., 1966.

Chadwick writes:

"It is evident that we must either relinquish the application of the conservation of energy and momentum in these collisions or adopt another hypothesis about the nature of radiation. If we suppose that the radiation is not a quantum radiation, but consists of particles of mass very nearly equal to that of the proton, all difficulties connected with the collisions disappear. ..." (reprint, p. 1299).

[4] Galileo Galilei, *De Motu*, Drabkin Translation, University of Wisconsin in Press, Madison, 1960, p. 23.

[5] L. Nash, *The Atomic-Molecular Theory*, Harvard Case Studies in History of Science, Harvard UP, 1950, pp. 88 and 17.

[6] The version of the relation between thermodynamics and statistical mechanics differs from what is usually given, e.g. by Nagel who holds that one reduces to the other; (cf. *Structure of Science*, Chapter 11, pp. 338ff.) But Feyerabend has cogently shown that this relation of reduction is not so, and that the second law is falsified, see P. Feyerabend, 'How to Be a Good Empiricist', reprinted in Midditch, *Philosophy of Science*, Oxford U.P., 1969, p. 28.

[7] The first draft of Galileo 'Floating Bodies' can be found in GG IV, pp. 36ff. A translation of the published work is available, in an edition edited by Stillman Drake, translation by Thos. Salusbury (Univ. of Illinois, 1970). The axioms mentioned can be found stated on pp. 6–7 of that translation. For discussion of the case cf. W. Shea. 'Discorso introne alle cose che stanno in su L'acque e che in qualla si muovone', *Saggi su Galileo Galilei* (ed. by Carlo Maccagni), G. Barbera, Firenze, 1967.

[8] I. Newton, *Principia*. Definitions, Lot Scholium. Motte-Cajori Translation, University of California Press. 1934, pp. 6–7.

[9] For purposes relevant to this paper the idea of a community could be construed in terms of those participating in a research program. Where the research program is being carried on by a single individual this condition is trivialized. Indeed the whole condition is not very problematic, it holds merely (to borrow a sort of Lakatos-type terminology) that it is intellectually dishonest to justify particular claims or theories on bases which the justifier does not accept. Possible engineering type counter-examples, like the use of classical equations in bridge building, are not really counter-examples since the justification of these equations will be pragmatic, type (d). Finally, it is worth noting another aspect of 'acceptability'; while what is required of a person in a tradition or program is that he accept the basis, more general anti-privacy requirement is that it be acceptable, given proper training, etc, to individuals at large. This latter acceptability would be guaranteed by using a natural language but may raise problems in more outré types of possible cases.

[10] More needs to be said about these various kinds of bases. Also the processes by which such grounds of justification and test undergo change needs much analysis. What is here is mere hint.

[11] The idea of an observation report where the *G*'s involved are non-human could be conjured up, by taking the 'report' to be appropriate reactions of specifiable kinds. Thus to find out if the cat was observing to birds we would dream up various tests to ensure that the cat had paid sufficient attention to what he saw, etc. The possibility of non-language tests for humans also exists, but, at least in science, they seem uninteresting.

What is interesting and where the start needs extension is with cases of observation which do not involve sight. I don't have anything to say about such cases at this point.

[12] It seems, contrary to Ryle, *Concept of Mind*, Barnes and Noble, New York, 1960, p. 222 that one could observe *X* with no purpose in mind; all that observation requires is that the observer pay sufficient attention, and possess sufficient knowledge, to be able to gain and report some information about *X* on the basis of his observing *X*. Thus I *might* observe something quite by accident, and be quite disinterested in it, and still meet all the conditions for observing. This makes sense of locuteians like "Walking down the Quai I observed the maidens at their sun worship". It is possible (however unlikely) that in such a case I had no purpose in observing them. For more on 'acceptability' see Note 9.

[13] The argument Galileo gives brings this connection out quite clearly, and it is clear that he takes this connection to provide the support for his position. Galileo writes:

"Eximum praeterea praeclarumque habemus argumentum pro scrupolo ab illis demendo, qui Systemate Copernico conversionem Planetarum cirea Solem aequo animo ferentes, adeo perturbantur ab unius Lunae circa Terram latione, interea dum ambo annum orbem circa Solem absolvunt... nunc enim nedum Planetarum unum circa alium convertibilem habemus, dum ambo magnum circa Solem perlustrant orbem, verum quartuor circa Lovem, instar Lunae circa Tellorum sensus nobis vagantes offert Stellas, cum omnes simul cun Iove, 12 annorum spatio, magnum circa Solem permeant orbem. (*Le Opere di Galileo Galilei*, Vol. IIIi, p. 95) for translation see S. Drake, *Discoveries and Opinions of Galileo*, Doubleday Anchor, Garden City, 1957, p. 57.

[14] L. Nash, *The Atomic-Molecular Theory*, *op. cit*. pp. 39–40. Prof. June Z. Fullmer made helpful suggestions about this case, and suggested that the nitrous oxides might provide another, even better example.

[15] Cf. P. Alexander, *Sensationalism and Scientific Explanation*, Alexander distinguishes between 'theory-laden' and 'concept-laden', Routledge, London, 1963.

[16] Compare Hanson's famous case from *Patterns of Discovery*, Cambridge U.P., 1958, p. 5–8, where Tycho and Kepler watch the sun rise.

[17] The theory of further testing offered here owes something to the pragmatists, esp. to G. H. Mead who in his description of the 'act of perception' included such further tests as part of the 'act'. (cf. Mead's *Philosophy of the Act*): The idea of interpretation is discussed by Hanson, *op. cit.*, Chapter 1. He thinks there is never any interpretation.

[18] This process is readily seen in Galileo's third letter; where he considers Scheiner's consequents, and compares the sunspots to the Medicean planets. Drake, *op. cit.*, pp. 134–42.

[19] This argument comes from Austin's 'Other Minds' and 'Sense and Sensibilia'. It is sketched more fully in my 'Recent Work in Perception', *Amer. Philos. Q.* 7 (1970) 4.

[20] Feyerabend's analysis of the color case occurs in his early paper 'An Attempt at a Realistic Interpretation of Experience', *Proc. Aristotelian Society* (1958), 150ff. I am here assuming that the set of 'pragmatics' determine the use of the word in question and thus are responsible for the meaning. I make this assumption knowing it to be simplistic, and as it stands probably inadequate. The set of relevant, meaning-determining parameters will probably not be wholly physical and will have to account for intentions of both speakers and hearers. I don't think this added problem about meaning affects the point concerning continuity which I am here suggesting. This idea of continuity can be found previously in Dudley Shapere's 'Meaning and Scientific Change' in *Mind and Cosmos* (ed. by R. Colodny), University of Pittsburgh, Pittsburgh, 1966,

p. 70. In a more general vein the method of similarities and differences as the method of understanding was first suggested to me by John Wisdom.

[21] Many points in this paper were clarified with the help of discussions with Richard Garner, Bernard Rosen, Larry Laudan, and Ronald Laymon.

BURKE TOWNSEND

FEYERABEND'S PRAGMATIC THEORY OF OBSERVATION AND THE COMPARABILITY OF ALTERNATIVE THEORIES

The line which separates theory from observation, once drawn with some assurance by students of the structure of science, has more recently come under a great deal of criticism. Some critics allow that there is a distinction to be made between theoretical claim and observational record, but maintain that the distinction merely indicates the character of the directions toward the extremes on a continuous spectrum of scientific statements and cannot usefully be taken as a fiduciary mark for the bifurcation of the language of science. Others, among whom is numbered the object of my attention today, Paul Feyerabend, will not admit the legitimacy of the distinction at all in the form in which it has customarily been cast. Under the distinction in this customary form, Feyerabend, it would seem, would want to judge all statements to be theoretical. Observation statements, under his view, are not to be set apart from theory, for they are in fact but a particular kind of theoretical statement. Not all the statements of a theory will be observation statements, but all statements, including those based upon observation, are *statements within the theory*, and thus are theory-relative statements, statements dependent for their meaning on the theoretical context in which they serve.

This theoretical loading of observation terms, Feyerabend considers to be all-pervasive. His claim is not only that one's expectations, values, and beliefs inevitably color what one sees, but, more fundamentally, that the meaning or significance of an observation statement, indeed of any statement, of necessity is dependent upon the conceptual system in which it functions. To hold otherwise, he feels, is to ignore the fundamental distinction between fact and convention.[1] No sentence is intrinsically meaningful. Our pre-theoretical freedom of interpretation is complete and can be limited only through the language we have adopted or, at the next level, by the theories or general points of view which underlie and support this language.[2] Only our conceptual structure and the language it employs can confer meaning upon an observation sentence, and those who have sought a theory-independent base of observation reports have thus pursued a chimera. An observation report without the support of the theory in which it functions is meaningless, for the only possibility of conferring meaning upon it is through its adoption into a role within a conceptual structure. To hold to a belief in a theory-free, core language of observation, then, is, according to Feyerabend, to maintain a belief in a system whose significance is somehow given, phenomenologically or otherwise, and such a system simply does not exist.

Feyerabend's view of the character of the theoretical loading of observation terms thus goes beyond the increasingly acknowledged psychological and perceptual bias which theoretical belief and expectation inevitably introduce into the reports of any observer, and this view, particularly when coupled with Feyerabend's belief in the importance of theoretical alternatives which depart radically from any accepted scientific system, raises several very puzzling questions about such matters as the comparability of competing theories and the explanation or reduction of one theory in terms of another. Standard accounts of these matters rely upon the assumption of a realm of shared meanings among theories, including, although not limited to, a realm of shared language of observation. The deductive account of the nature of inter-theoretical explanation and reduction, for example, clearly has presupposed common terms among theories, the statements of one theory being claimed to follow as special cases from another and more comprehensive theory. If, as Feyerabend claims, the meanings of the terms are relative to the theory in which they serve, then how could the statements of one theory be deduced from those of another? The fact that the same *words* might happen to be used by the two theories could not, of course, serve to allow a legitimate derivation, for the words would have different meanings in the different theories, and the apparent derivation would in fact be spurious.

Now it is possible, and Feyerabend admits this to be so, that the two theories in question might differ only in trivial respects, or that both might share a common link through roles in a third and still more comprehensive theory, and, in cases such as these, legitimate derivations through shared meanings might be possible. But cases such as these are not representative of the general situation, and they fail in particular to represent the relations between major theoretical alternatives or between a major physical theory and its revolutionary successor. These are the cases in which we are most likely to be interested, and in these cases meanings, even observational meanings, will differ substantially under Feyerabend's view. Indeed, for general methodological reasons, Feyerabend maintains that science should strive continually to provide just such substantially differing theoretical structures, and in this situation we must ask: How, if not by shared meanings, are alternative theories to be related and compared?

The answer which Feyerabend provides is given in his pragmatic theory of observation. This theory is, by Feyerabend's own claim, an essential presupposition of his account of the nature of empirical science[3], and one of the critical roles which it plays in that account, and the one on which I wish to focus here, is that of providing a means, consonant with a complete relativity of observational meaning, for the direct, experimental comparison of competing theoretical alternatives. The pragmatic theory of observation, in other words, is offered to restore that point of contact and comparison which seemed to have been abandoned when the existence of any shared core of concepts was denied. I would like now to examine the adequacy of the theory for this role within Feyerabend's general perspective and to do so by a critical look at the way in which the pragmatic theory functions in Feyerabend's explication of the possibility of crucial experiment between radically departing or, as Feyerabend calls them, incommensurable theories. It will be my contention that the pragmatic theory of observation fails to accomplish all that is asked of it in this role and that as a consequence further work on the comparability of theories is necessary.

The pragmatic theory of observation is, as Feyerabend presents it, a causal theory. "An observation sentence," he says, "is distinguished from other sentences of a theory, not, as was the case in earlier positivism, by its *content*; but by the *cause* of its production, or by the fact that its production conforms to certain behavioral patterns."[4] Thus the difference between an observation statement and any other is not one of semantic content, but one of the psychological, physiological, or physical circumstances of its production.[5] The function of the human observer, Feyerabend feels, is in fact nothing more than that of a measuring instrument, and which of his reactions will count as valid measurement response is determined, just as in the case of the voltmeter, not by some extra-theoretical property or intrinsic meaning of the reaction itself, but by the theoretically established causal structure of the relation between the observer-instrument and its surrounding conditions. Successful observing or measuring is determined by theoretical knowledge of a reliable correlation between the indication of the observer or instrument and a particular situation in the world. The indications of the observer, just like those of the instrument, are without significance as empirical test prior to a theoretically given account of their correlations with external situations.

What may seem initially rather puzzling about this view of the human observer should become clearer when it is realized that Feyerabend's use of the term 'theoretical' is an encompassing one which includes the background or common-sense conceptual structures which we employ in our daily affairs. In this extended usage even our knowledge of the correlation between seeing the book on the table and the fact that the book is on the table would count as theoretical. The point which Feyerabend is trying to make in this regard, I believe, is that even such garden-variety knowledge as this is not somehow given, but must be derived from the structure of a conceptual system. In this respect it is not different in principle from any of the more abstruse teachings of the higher-order systems which are customarily termed 'theoretical', and the point stands: it is only through the establishment of an ordering, conceptual system that the reactions of even the human instrument are provided with significance as indicators of a state of the world.

Observability, then, is determined by the pragmatic structure of the observation situation as it is characterized by our theories and not by any other aspect of the meaning or interpretation of the observation term. The situation is summarized by Feyerabend as follows:

> Whether or not a situation *s* is observable for an organism O can be ascertained by investigating the behavior of O, mental (sensations) or otherwise; more especially, it can be ascertained by investigating O's ability to distinguish between *s* and other situations. And we shall say that O is able to distinguish between *s* and situations different from *s* if it can be conditioned such that it (conditionally or unconditionally) produces a specific reaction *r* whenever *s* is present, and does not produce *r* when *s* is absent.[6]

Using this theory of observation, Feyerabend describes the process of empirical test as one of comparing the reactions of the theory with those of the conditioned observer. The 'reactions' of a theory are most conveniently thought of as the sentences which it predicts, but might also be considered as represented by the behavior of a non-sentient robot which is programmed in accordance with the theory.[7] In either case, the theory is adjudged to have met a test successfully when its reactions mimic those of the observer. A successful theory, in other words, is one whose verbal 'behavior' parallels that of the human observer. Feyerabend explicitly denies that it is the interpreted *statements* about the world which are checked against the observed situation, for the pragmatic theory holds that observation is determined by the structure of the physically caused reaction of the observer and is to be considered separately from any interpretation which that caused reaction might be given. Observation is a process of physical interaction of observer and environment and that interaction as an actual physical process is independent of any interpretation which theory may place upon it. The meaning of a predicted observation sentence, then, is superfluous to its role in an empirical test; all that is necessary is that this predicted sentence, this verbal 'behavior' of the theory, should reflect the behavior of the observer.

Feyerabend's solution to the problem of a crucial experiment between radically departing or incommensurable theories is now clear. Although the *meanings* of the observation statements predicted by the alternative theories are not comparable, it is not the meaning of the predictions which constitutes the test, but rather their reflection of the causally produced behavior of the observer. It is not that one theory contradicts what the other says about the state of the world, but that one theory reflects assenting behavior while the other reflects dissenting. Without any appeal whatsoever to shared meanings, competing theories may be subjected to comparative evaluation through a crucial experimental test by determining, under situations in which their behavior is different, which theory parallels the behavior of the observer.

The pragmatic theory of observation as employed by Feyerabend would, if it should prove sound, surely provide some basis for the comparison of alternative theories, no matter how radically the meanings they impart to observation terms might depart. Not only could crucial experiments be accounted for in the manner just discussed, but a kind of inter-theoretical explanation or reduction could be explicated as a demonstration by one theory that its 'behavior' parallels that of another in some particular region of application. But is the pragmatic theory satisfactory as it is employed by Feyerabend? Not fully, I believe. Much of what Feyerabend has to say about the pragmatic or causal structure of observation and its dependence upon theoretical knowledge for meaning or significance is straight to the point, but in applying this account of observation to the process of empirical test, Feyerabend does not, it seems to me, remain faithful to his own insights, and the result is a failure to produce that unproblematic base for inter-theoretical comparison which it was the task of the pragmatic theory to provide.

Let us look, for example, once again at the process of a crucial experimental test between incommensurable theoretical alternatives. Implicit in Feyerabend's account of this process is the assumption that the proponents of both theories would recognize the validity of the reactions or reports of the same observer, or the same class of mutually consistent observers. But the certification of any human behavioral pattern as an *observation* is by Feyerabend's own account a theory-relative task. In a passage quoted previously, Feyerabend claims that the observer is to be conditioned such that he will produce a specific reaction *r* to a particular situation in the world *s*, and will not produce *r* when *s* does not obtain.[8] But he has also claimed, and correctly I believe, that we require a *theory* to teach us what situations we are to expect in the world and what reliable correlations exist between these states and the indications of any instrument.[9] The process of conditioning an observer, then, the process of determining the states of the world and verifying the reliability of the connections between them and the reactions of the observer must itself be a theoretically based procedure. If no theory-neutral basis exists for assessing the training and the verification of validity of an observer, then how can we assume that the radical alternatives will recognize the same behavioral patterns as constituting observations? If it may happen that alternative theories share not a single observational statement, may it not also happen that they share no common characterization whatsoever of the pragmatics of the observation situation itself? And if this be the case, what will be the basis of comparison between them? We may well imagine two incommensurable alternatives, each of which enjoys remarkable success in predicting the reactions of humans in what it certifies as observation situations, but neither of which is able to do much toward predicting the reactions of humans in what the other certifies as observation situations. What could provide a mutually acceptable empirical test under these conditions?

The idea may sound far-fetched, but I suggest that it is no more far-fetched than the idea of the total incommensurability of observation concepts; its possibility can be justified in much the same manner within the structure of the pragmatic theory of observation as Feyerabend has outlined it. As long as observability and the conditioned response of the observer are to be based upon the causal structure of the physical situation, and as long as the characterization of that situation and the determination of its structure are theory-relative activities, so long would the complete incomparability of theoretical alternatives seem to be a possibility. An observation situation as conceived in the pragmatic theory is an empirical situation to be investigated and described just like any other, and the description of an empirical situation is a theoretical task. The nature of the stimuli impinging upon the organism, the nature of the organism itself, the nature of the responses of the organism, the causal structure of the interaction situation – all these are directly relevant to the certification of the situation as an *observation*, and all these are determined only within the context of a *theory*. Two theories which share no common observation statements would share no common characterization of the pragmatics of the observation situation either, and the possibility of a genuine crucial test between them would thus seem to be obviated.

It might be objected at this point that no matter who trains the observer or what characterization of the observation situation is given, the reactions of that observer in any situation are still facts in the world which must be accounted for by any successful theory. This is true of any successful and *complete* theory, for a complete theory would have to be capable of accounting for everything, including the reactions of observers no matter how trained nor in what circumstances. But completeness, of course, even if a possibility, is not to be required of every alternative, and short of completeness there would seem to be room for alternative theories, theories about the same world, to miss entirely a common base. One wouldn't even really know whether to call such theories alternatives, for without any point of comparison one wouldn't know if they were in some sense in conflict or whether they simply represented entirely distinct aspects of the world.

It might further be objected that an undue stress has been placed upon the *conditioned* responses of the observer. Although Feyerabend does describe observability in terms of conditioned reactions, he also at one point appeals to "human experience as an actually existing process" in employing the pragmatic theory as a basis of inter-theoretical comparison.[10] While experience as an 'actually existing process' is certainly shaped to some extent by the prior conditioning of the subject, surely this shaping is not wholesale and some commonality of the effects of the environment on any human subject may be claimed. Perhaps it is these common effects which are the reactions to be reflected by the theory, rather than the various conditioned responses which may be based upon them. To retreat to this level, however, would look suspiciously like an attempt to rest upon the given. If the theoretically based, conditioned response is to be bypassed, how would the predicted reactions of the theory be communicated? Perhaps by direct link to the human nervous system, but even in this case some learned pattern of associations would be necessary to communicate which aspects of the theory-induced experience are to be compared against the experience induced by the environment. Once one has started down the slide of the theoretical relativity of observation which Feyerabend has described, there seems to be no extra-theoretical stopping point to rescue inter-theoretical comparability.

I am sure, however, that Feyerabend had no intention of seeking guarantees of certainty of infallible, objective checks in appealing to the pragmatic theory of observation, and my purpose in probing that theory has not been to show that it fails to provide a guaranteed reference base for empirical science, but to show that it does not save us from the necessity of making *theoretically* based, fallible judgments of commonality of structure in evaluating alternative theories. If there is a way to make empirical comparisons even of radically departing alternative theories, and I am sure that there is, then it must lie in a theoretically based account of some kind of commonality of observational structure between the two theories. Feyerabend has in effect simply assumed this commonality without giving an account of its nature by assuming that the same situation would count as an observational check in both theories. He avoids saying that an observational test is made simply by checking what a theory says to be the case in the world against what we see *is* the case in the world, and he does so, I believe, because he finds this view of observation to assume that the observer is capable of determining what is the case independently of the theory. The observer is caused to react in a certain way by the pragmatic structure of the observation situation, but his reactions do not show him what is the case in the world until he interprets those reactions within a theoretical context. In describing the way in which the pragmatic structure of the observation situation itself is determined, however, Feyerabend seems to allow himself the view he rejects in all other instances. The circumstances which lead to the production of an observation sentence, he says, are themselves open to observation, and "we can therefore determine in a straightforward manner whether a certain movement of the human organism is correlated with an external event and can therefore be regarded as an indicator of this event."[11] But to determine the correlation between a human reaction and an external event is no less a theoretically based task than is to determine the correlation between the statements of a theory and external states of the world. The problem of determining through observation the pragmatic structure of the observation situation is a problem no different in nature from any other observation problem, which is to say that it is a problem which can be resolved only within the context of a theory.

By characterizing the problem of determining the pragmatic structure of the observation situation as 'straightforward' and by tacitly assuming that even incommensurable theories will recognized the same external situation as constituting a valid observational check, Feyerabend has in effect relied upon a theory-neutral or theory-common characterization of important features of the observation situation without providing any justification for this reliance. I believe that it will prove possible to justify such a reliance upon theory-common features of observation in a manner consistent with the main features of Feyerabend's causal theory of observation, but to do so will require further inquiry into the nature of the theories which we use to describe the human instrument and their relation to more general physical theories. In discussing how the pragmatic theory of observation functions in comparing alternative theories, Feyerabend has bypassed by tacit assumption rather than accounted for directly the features of our theoretical structures which allow for comparability, thus leaving a serious gap in his account of the comparative evaluation of incommensurable theoretical alternatives.

University of Hawaii

NOTES

[1] 'Explanation, Reduction, and Empiricism', in *Scientific Explanation, Space and Time, Minnesota Studies in the Philosophy of Science*, Vol. III (ed. by H. Feigl and G. Maxwell), University of Minnesota Press, Minneapolis, 1962, p. 39.
[2] *ibid.*, p. 39.
[3] *ibid.*, p. 40.
[4] *ibid.*, p. 36.
[5] 'Problems of Empiricism', p. 152, in *Beyond the Edge of Certainty: Essays in Contemporary Philosophy of Science, University of Pittsburgh Series in the Philosophy of Science*, Vol. 2 (ed. by R. G. Colodny), Prentice-Hall, Inc., Englewood Cliffs, N.J., 1965.
[6] 'An Attempt at a Realistic Interpretation of Experience', in *Proceedings of the Aristotelian Society*, N. S. **58** (1958) 146.
[7] 'Explanation, Reduction, and Empiricism', p. 94, and 'Problems of Empiricism', p. 214.
[8] Above, p. 5.
[9] 'Explanation, Reduction, and Empiricism', pp. 36–7.
[10] 'Problems of Empiricism', p. 214.
[11] *ibid.*, p. 212.

J. O. WISDOM

OBSERVATIONS AS THE BUILDING BLOCKS OF SCIENCE IN 20TH-CENTURY SCIENTIFIC THOUGHT

Although there continue to be many interpretations of the nature of science, not simply thought up from afar by philosophers, but in the highest ranks of scientific circles, there are only a few that are significantly different from one another and influential (Wisdom, 1971). I shall confine myself to instrumentalism, conventionalism, and induction. (I shall omit even operationalism as being easily subsumable under conventionalism.) The main thesis will be that they constitute different ways of regarding knowledge as rooted in pure observation, and that the philosophy of observationalism they presuppose is false. The falsifiability interpretation does not view knowledge in this way and is not discussed here.

Whatever the shortcomings of these three positions may be, however, firmly held or rejected they may be according as one assesses existing criticisms, these criticisms mainly lead to a judicial appraisal without pointing to anything new. Here I adduce two new criticisms, sketching one concerning the role of truth-value, which is directed against instrumentalism and conventionalism (developed in the paper referred to above), and working up another concerning observations as interpretation-charged, which is directed against all three positions. These criticisms necessitate a radically different view of the relation between theory and observation.

1. The role of truth-value

Instrumentalism and conventionalism relate differently to truth and reality.

According to instrumentalism, a scientific theory is an instrument, and only an instrument, for computation. Thus scientific theories are neither true nor false; they are only more efficient or less efficient instruments. Moreover, a theory has no power to describe the world of fact; suspicion about the question of reality or anything that smacks of metaphysics is the *raison d'être* of instrumentalism.

Conventionalism on the other hand allows of truth-value though only in a Pickwickian sense, such that the conventions assigning the meanings of the terms of a theory are more or less appropriate. Conventionalism, however, though it holds out no hope of describing the world, adopts this attitude not because of suspicion of metaphysics but because of the apparent impossibility in principle of establishing either the truth or the falsity of a theory – and this very real difficulty is its *raison d'être*.

We thus have two types of view. One repudiates the notion of truth-value. The other interprets truth by means of a device. Neither accepts the ordinary notion of truth according to which a true theory is one that says something about the world. (The situation would be radically altered if it ever turned out that the ordinary notion of truth presupposed the conventionalist one at some level.) We may note, however, that instrumentalism amounts to a particular case of conventionalism in which the Pickwickian concept of truth-value is reduced to zero.

Now why should one fuss about the question of whether truth comes into scientific theories, when we can get along apparently quite well without it?

However, there is a good reason why one has to take the question of truth-value seriously. In the more abstract levels of science, if you have a hypothesis or theory, initial or boundary conditions, and a deduction of an empirical generalization, or deduction of a particular occurrence which may be checked by an observation, then if this turns out to be false and you cannot make amendments you scrap the theory. Now you cannot do that logically unless the predictions are false. For whatever doctrine you follow, you have to say at some point that the deduction of the particular statement does not square with observation, which in common parlance means that it is in an ordinary unsophisticated sense false. You may be able to be sophisticated over the concept of truth at all sorts of levels but not at this point. If you allow that in many cases in science a prediction turns out to be false, you have to take the consequences of it. And one consequence of it is very simple. The logical point is that if you have a false conclusion and the deduction is valid, one of the premises must be false. After deciding that the falsity does not lie with one of the minor premisses, provided you can decide this, you have to pinpoint the falsity on the theory. And you have to use the concept of falsity in the same sense in which you talk about the falsity of the prediction. It is necessary, therefore, to bring in truth-value for theories and hypotheses. And this would be a basic reason for ruling out instrumentalism.

Turning to conventionalism, the position is that a deduced prediction would be false in an everyday sense, whereas the theory would be false in the sense of embodying inadequate conventions. So the same argument would apply to a view that interpreted a theory as true by the use of a device as would apply to an interpretation that required no concept of truth at all.

The notion of refutation involves the idea that a conclusion deduced from a theory is false; if a theory is then rejected, this presupposes that the concepts of truth and falsity applicable to a conclusion have to be applicable to the theory. Those who would support, say, instrumentalism would have to deny to deduced conclusions truth-value in this sort of sense; for if conclusions were denied truth-value then a theory that was given up would have to be interpreted as a theory-instrument on the grounds of being a poor instrument but not on the grounds of being false.

Conventionalism, however, could be remedied simply by altering the conventions attached to the premisses. But again if the conclusion is false in the ordinary sense, then the rejected conventions would have to be interpreted as false in the ordinary sense.

The overall point then is that views of the sort considered either lack the concept of truth-value or use it in the sense which cannot be consistently pursued so long as test observations are given a literal truth-value.

2. Induction

The great traditional interpretation of science was of course induction. It was an invention of philosophers, but, remarkably, scientists mostly acquiesced in this philosophical judgement, not noticing that their practice in no way accorded with it. Induction has usually referred to generalizations derived from experience. Theories built up from generalizations have received but scant attention. The salient feature of an induction was that if carefully processed it was *true* (or probably true). Thus, induction is an interpretation that ascribes truth-value and tells you something true (or probable) about the world. Induction is nothing if not a method for arriving at truth and justifying the truth arrived at.

The hard core of all three positions is the ultimacy of observation, in the sense that observations are data and are certain, and that theories are of subordinate import.

3. Functional and Structural Interpretations

We now have two contrasting philosophers of science, and I wish to try to characterize them more closely.

In the first class of interpretations, we have a theory that somehow enables us to predict observations, connect them with one another, pass one to another. This amounts to a way of handling observations. The methods work out differently in detail, but they share this general characteristic. So I would summarize this approach as *observation-manipulation*; alternatively I would call it, to use a term from social anthropology, a *functional* approach to science (insofar as you are not concerned with a societal custom in itself but only with its role in a society). On this approach to scientific theory, you are concerned only with its role in science – not what it is but what it does.

Popper's excellent term 'instrumentalism' covers one of these, but we need a wider term to cover both it and the other approaches considered above.

Now turn again to the other class. Inductivists do not try merely to manipulate observation; they want observations to grow into something, to grow into a truth wider than themselves, in other words they want to make the observations produce theories. The constrast is between *observation-manipulation* and *observational construction*. Induction is concerned solely with the *structure* of scientific theory. I would call this type of interpretation of science *structuralism*.

4. The Common Problem: The Data-Theory Gap

This leads to a further point. The inductive observationalist is deeply concerned with the gap between data and theory, with the question, how do you pass from data to theories? The classical theory of induction came up with an *algorith* as an answer. If you follow Bacon's or Mill's rules, you take a note of the appropriate observations, you turn the handle of the induction machine, and out comes a generalization which is a scientific truth. The idea of an induction from generalizations to a wider theory such as from Kepler's laws to Newton's theory of gravitation has attracted very little attention or discussion. Moreover, induction is naturally accompanied by certain concomitant views such as the 'stomach theory' of the mind (originating in Popper, see Wisdom, 1971) for which the mind is a passive receptable of observational impressions to be mechanically processed. Again induction leaves no room for scientific enterprise because it can be carried on by the Bacon-Mill processing procedures. Induction implies that science is an *algorithm*.[1]

Instrumentalism and conventionalism treat a theory-tool or a theory-convention as a jig-saw puzzle pattern into which observations are to befitted.

Thus an algorithm is the inductionist answer to the problem of the gap. Instrumentalism and conventionalism abolish the gap by questionable devices.

The situation here is conceived much as Agassi (1966a) saw it, though there may be some difference of conclusion and aim. I see the fundamental problem for both parties as that of the gap between data and theory. Traditional inductivists have tried to fill this gap first by some principle of uniformity of nature and later by more sophisticated devices. Not finding such procedures hopeful, conventionalists reacted to the problem of the gap in a different way. In other words, faced with a failure of the structuralist approach, one functionalist attempt would be to interpret a theory as a set of observations, thus closing the gap by denying that it exists. For although conventionalism does ostensibly appear to make a wider claim than can be obtained simply from the data a theory covers, on closer examination it would seem that a theory in the sense of a convention is as wide only as the data that satisfy it.

Thus, these two groups, the functionalist that is mainly the conventionalist, and the structuralist that is mainly the inductivist, are reacting to the same problem of observation as the foundation stone of knowledge.

The position is that there are two groups of interpretations of science. One is the functionalist interpretation, according to which the function of a scientific theory is to manipulate observations. The other is the structuralist interpretation, according to which the structure of a scientific theory is to be constructed out of observations. I wish to consider the metameta-conjecture that both sets are responses to a gap between data (or observations) and theory.

Where there is a gap, which there is no clear way of bridging, there are two obvious devices for handling it: by treating it as non-existent, and by jumping it. The functionalist, I suggest, has taken the former way out; the structuralist the latter.

Both take for granted that the process is 'data → theory', whether it is a heuristic gap to be jumped to make a scientific discovery or a logical gap requiring logical justification (cf. Lakatos 1968). Instead of trying to prove the axiom of parallels, you try a new start by denying the axiom and introducing a different approach to parallels; instead of jumping or justifying the gap from data to theory, you might try a new start by introducing a new view of the gap (though not necessarily by simply reversing the direction to 'theory → data'.)

5. The nature of observation

The relation of data to theory may assume a different hue if we reconsider the nature of observation itself. For, common to observation-manipulation, which is a functionalist approach to observation, and to inductive observationalism, which is a structuralist use of observation, is the ultimacy of observation. In both, observation takes precedence over theory, in that if there is any tension between them it is theory that always and necessarily has to give way. Theory has run no risk for 350 years of being a sacred cow.

In this sense observationalism has been for centuries and still is at the root of scientists' (I do not say scientific) outlook.[2]

This impregnable position bestowed upon observation may be questioned on two counts. (i) There is the simple fact that at times, even though not often, an observation has to give way to a theory: i.e. at times an observation has to be questioned and even rejected in favour of a theory. This must be known to most working scientists, but its metascientific significance is usually overlooked. (ii) There is a much more subtle situation that arises if we are correct in supposing, as Popper (1959) asserted, that every so-called observation is theory-laden or permeated with interpretation, or, otherwise put, that there is no such thing as a 'pure' observation, an objective datum independent of all experience, as it were waiting to be photographed by a human mind. Let us consider this.

With many examples it is obvious that unsuspected interpretation is present. Is this general? It seems to me that an observation is a selection formed in terms of certain ideas the observer brings to bear on the world. There would possibly be little tendency to dispute this, were it not for one point that constantly obtrudes; how can there be interpretation unless there is something uninterpreted to interpret? This difficulty may yield to the idea of layered interpretations: interpretation$_1$ is of an observation that is uninterpreted in the sense of being something upon which interpretation$_1$ is imposed, although in itself it is an interpretation$_2$ of ... etc. This smacks of artificiality, for there seems to be no reason for the dual role played by any particular layer; also the sequence seems to be an actual infinity present before us. However, it can be altered to the following: an ordinary observation$_1$ is an interpretation$_1$ from which we can *abstract* what is, in relation to *it*, a pure observation$_1$; this abstraction is in its turn seen to be an ordinary observation$_2$ in a different context, i.e. to be an interpretation$_2$ from which we can again abstract what is, in relation to *it* a pure observation$_2$; and so on. In this case the layering goes on only so long as we make it go on, i.e. by making abstractions.

Assuming that some such way out of the difficulty exists, my main concern is with a different point – the unexpected consequences of there being no ultimate uninterpreted observation. What effect does it have on the dominance of observation over theory?

With instrumentalism, we should be pushed into the contention that a theory is an instrument for computing entities that are precluded from being pure observations and that are therefore a fusion of instrument and observation, or rather an observation interpreted as an instrument. About this it is to be noted that an instrument for computation is not an entity but an operator; hence if it calculates only another instrument, then it is calculating an operator and not an observation in the sense of something that is the case. Hence instrumentalism cannot be articulated on the basis that an observation is itself a blend with an instrumental component.

Again with conventionalism, we should be making a theory or convention fit not an observation but a fusion of which one component is a convention. Now a convention to make observations tractable is one thing; a convention for handling conventions is another. For a convention describes not an entity but an operation. Hence, as with instrumentalism, conventionalism cannot be articulated.[3]

And, turning to induction, an observational premiss has to be regarded not as pure but as a compound embodying a theoretical component not yet justified by induction. Thus an induction cannot be justified from its premisses without justification of its premisses which require inductive justification. Hence inductive justification presupposes an infinite regress of inductive justification,

Thus a common element of all the functional as well as the structural approaches is the ultimacy of observation.

But we may note that since this is untenable the dominance of observation over theory must go.

It is widely noticed that observation is theory-impregnated, but its full significance seems to be hardly recognised. So it is a matter of some interest and importance to ask what difference it makes to science itself what interpretation is adopted. Consequences of observationalism are:

(i) The shared defeat, the acceptance that reality is unknowable, obliterates the distinction between theoretical science and technology on the instrumentalist and conventionalist interpretations – theory becomes pencil-and-paper technology.

(ii) In a clash between theory and observation, observation *must* always win.

(iii) If a theory cannot in principle represent reality, the (?insoluble) problem arises of how a theory (instrument-convention) can function – can make predictions/patterns of observations (which would not be a problem if a theory represents reality, i.e. is connected with observations in accordance with it).

(iv) For induction theories become algorithms. While some algorithms are ingeniously contrived empirically, it is common for them to be derived from theories (presumed not to be algorithms); hence no non-algorithmic theory would be available for providing algorithms. This would seem to be so even for instrumentalism and conventionalism.

(v) These are meta-scientific considerations. Are there any that bear on the working scientist? More specifically, are there any questions he cannot ask or solutions he cannot use? It would have been improper to postulate the atom and (after agreeing that chemistry could proceed satisfactorily without it simply working 'as if' it existed) it would have been improper to look for it; it would have been improper to postulate 'attractive force', or the electron – our knowledge not only of the world but of science would have been wrongly impoverished. And it is wholly obscure, on these positions, why scientists get so excited about science (Synge, 1970).

Thus the problem of the gap between data and theory becomes transformed by the assumption that observation itself is a blend of observation and theory (or better, is factual interpretation). Once the ultimacy of observation is given up – without insinuating the ultimacy of theory – we may see the structure of science not as a gap to be bridged from data to theory, but as a theoretical structure with observational layers with the boundary shifting according to the problem confronted.

6. Summary and conclusion

As regards instrumentalism, assuming that we cannot do science without the concept of truth (falsity, refutation, etc.), we have:

(a) Refutation of a theory (regarded as an instrument), as a result of a conclusion that is false, is impossible, because an instrument cannot be false, only faulty.

(b) If observations are theory-laden, and if theories are solely instruments, then observations are instrument-laden. Hence the task of science cannot be to find a theory (or instrument) to pass from a pure observation (or non-instrument) to another one.

With conventionalism, we have similarly:

(i) Refutation of a theory (regarded as conventions) as a result of a conclusion that is false is impossible, because a convention cannot be false, only inappropriate.

(ii) If observations are theory-laden, and if theories are solely conventions, then observations are convention-laden. Hence the task of science cannot be to find a theory (or convention) to pass from a pure observation (free of convention) to another one.

Conventionalism could be rescued from both these difficulties if it could be shown that observation-statements are conventional.

Part of this is applicable to induction. Induction requires premisses to be purely observational, uninterpreted; for an interpreted observation would be theory-laden. But if all observations are interpreted, we should lose the characteristic point of induction, of getting theory from pure observation.

This argument, if effective, would refute induction both as inference and as method, but it cannot as yet be regarded as a refutation. It could, however, be put in the form of a dilemma of choice: you cannot accept both induction (inference and method) and observation as theory-laden.

If observations are theory-laden, the approach collapses. While the actual logical procedure might be satisfactorily transformed into a version involving a benign infinite regress of inductive justifications, the *raison d'être* of the method, i.e. the classical empiricist demand for a firm root in experience, would be gone.

Induction is the vault of the scientific (computerised) bank: the vault stocks huge reserves of observations/facts/information; the computer processes this raw material into various forms of currency, according to whether the customer is a pure scientist, an applied scientist, or a technologist. These customers not only make withdrawals of processed currency, they also make huge deposits (of observations/facts/information).

Instrumentalism is a central clearing house, which governs transfers of observations/facts/information from one bank to another.

Conventionalism is the World-Treasury which decides whether to inflate or deflate, whether to revalue the currency, whether to modify its purchasing power – by which the reserves have whatever value it ascribes to them.

Might it not be that these vaults are as useless as Berkeley claimed were the vaults of Amsterdam, full of unused gold? Might it not be that there are other vaults, not known to many meta-scientists, small in size, stocked with a few (maybe a few dozen) theories? That a pure scientist carries some of them in his wallet in the form of film which he looks through when he studies nature? And that sometimes a technologist takes one of these to a shop? And that occasionally a scientist makes a film from something in his mind's eye, and deposits it in the central bank?

York University, Toronto

NOTES

[1] I owe the notion of an algorithm in a very wide sense to Agassi (1966b) though he introduced it in a different setting.

[2] For some very trenchant criticisms of observationalism interpreted in terms of sense-data see Skolimowski (1969). The significance of observationalism was appreciated by Feyerabend (1962) and Agassi (1966a).

[3] Likewise with operationalism, operations would not be pure but would be fusions consisting of operations permeated by theories that inevitably contain concepts not yet operationally defined. Thus operationalism cannot be formulated without infinite regress.

BIBLIOGRAPHY

Agassi, J., 'Sensationalism', *Mind* **75** (1966a), 1–24.

Agassi, J., 'The Need for Corroboration', *Technology and Culture* **7** (1966b).

Feyerabend, P. K., 'Explanation, Reduction and Empiricism', *Scientific Explanation, Space and Time*, in *Minnesota Studies in the Philosophy of Science* Vol. III (ed. by H. Feigl and G. Maxwell), Minneapolis 1962, pp. 28–97.

Lakatos, I., 'Changes in the Problem of Inductive Logic', *London Colloquium on the Philosophy of Science*, Vol. 3, North-Holland, Amsterdam, 1968, pp. 315f.

Popper, K. R., *The Logic of Scientific Discovery*, London and New York 1959.

Skolimowski, H., 'Knowledge, Language, and Rationality', in *Boston Studies in the Philosophy of Science*, Vol. 4, Humanities Press, N.Y., 1969, pp. 184–5.

Synge, J. L., personal communication, 1970.

Wisdom, J. O., 'Four Contemporary Interpretations of the Nature of Science', *Foundations of Physics* **1** (1971).

II. Philosophical Problems of Biology

EDWARD MANIER

FUNCTIONALISM AND THE NEGATIVE FEEDBACK MODEL IN BIOLOGY*

1. Functional analysis in distinct biological contexts

Any study of the philosophical literature dealing with the cluster of topics generally identified as 'functional description', 'functional analysis', and 'teleological explanation' naturally raises the problem of confirming, disconfirming, or at least relating the alternative logical models proposed by philosophers to the actual usage of biologists.[1] A close examination of current biological literature reveals that acceptance or rejection of what philosophers or sociologists might call a 'functionalist' perspective or approach is not significant for the division of biologists into schools or factions. It is certainly not the case that a functionalist perspective distinguishes the wholistic or organismic faction from the molecular biologists. In fact, molecular biologists shift back and forth from 'functional' to simple causal perspectives with considerable indifference. This is particularly true of the literature based upon the so-called operon hypothesis developed by the Nobel laureates Jacob and Monod. According to this hypothesis, some but not all of an organism's genes are code symbols for specific amino acids; distinct 'operator' genes either initiate or inhibit the activity of a specific linear sequence of genes. To establish the distinction of the two types of gene activity requires very careful biochemical and genetic analysis of the straightforward, causal sort associated with those two disciplines. Once established, however, the distinction introduces a functional perspective into the core of molecular biology. Since the operator genes are sentitive to changing conditions of cytoplasm or growth medium, reference to them is a natural element in any discussion of the *regulation* or 'target' orientation, of cellular activity. Consequently, statements such as the following occur with increasing frequency in the literature of molecular biology:

1.1. Feedback inhibition thus acts as a shut-off valve; it allows the cell to tap precursors and energy from the metabolic pool at a rate essential for efficient biosynthesis and no faster (Hartman and Suskind, 1965, p. 84).

In fact, although the formulation of questions and statements concerning organic functions is not distinctive of any one school or faction in biology, such formulations occur in an irreducible variety of contexts and stages in the process of theory construction in biology. As a result, no single logical or semantic schematism can presume to capture the meaning of functional statements or the distinctive form of teleological explanations.

It is possible to distinguish at least three different types of questions about the functional components of organic systems.

1.2. Is a functional system of type *a* to be found in organisms of type *x*? (E.g., Do lower organisms resist infection by anti-body formation?) Or, at the same level, What is the function of item *i* in organisms of type *x*? (What is the function of the thymus in mammals?)

1.3. What is the mechanism of a functional system of type *a*? What is its basic organization or design? Under what conditions does it perform at optimal efficiency? Under what conditions does it break down? (e.g., What is the design of the system which regulates the internal temperature of mammals? What causes this system to break down in instances where a fever is produced?)

1.4. What is the contribution of this functional system to the overall fitness of organisms of type *x*? How did such a system evolve? (e.g., How could a system of warning cries evolve in flocking birds if the sentinels which make the cries are subjected to heavier predation than silent birds?) or What functional systems must be postulated to account for the evolutionary record or organisms of type *x*? (e.g., How can the gene pool of a population be organized so that population will be more likely to survive a capricious sequence of environmental changes?)

Thus the variety of contexts in which statements about organic functions occur include: preliminary identification and description; the analysis of the causal design of the function; and the determination of the role of the function in an over all evolutionary strategy or mechanism. Biologists simply have no reason to attempt to reduce this variety.

Similarly, I can find no persuasive consideration, whether *a priori* or empirical, establishing a singular interpretative model for the causal analysis and explanation of 'target oriented' functional systems. There is no doubt, however, that the 'negative feedback' model of control and regulation is increasingly influential in the biological literature. Space does not allow for the explicit description of the role of this model in the various contexts in which it occurs, but these span a complete range from molecular biology, neural physiology, and the social behavior of animals, to the evolution of distinctively structured population gene pools.

The 'negative feedback' model of goal-directed activity has attracted a significant amount of adverse philosophical criticism since the time of its first appearance. In what follows, I would like to bring the model up to date and re-examine the philosophical commentaries which have been offered of it.

2. The Negative Feedback Model

In its original formulation, the negative feedback schema first outlined by Rosenbluth *et al.* (1943), required that goal-directed systems respond to some 'signal from the goal', on the grounds that if they did not, goal-achievement could hardly be identified as more than a coincidence. This formulation unnecessarily exposed the schema to the objection of the 'missing goal-object': no goal exists or is present to signal in a number of cases where the characterization of an organism as goal-directed system is intuitively acceptable (e.g., a hunting animal for which no prey is available).

Recent versions of the schema replace the requirement of a signal from

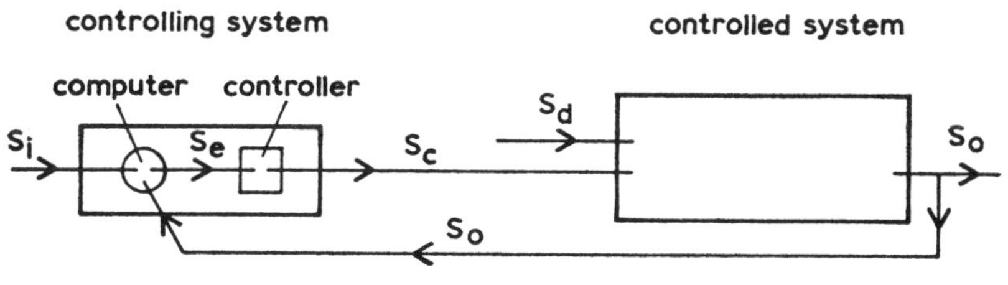

S_i – input signal
S_o – output signal
S_e – error signal
S_c – control signal
S_d – environmental disturbance

(Based on a figure by T. H. Waterman, 1966, 15–45)

the goal with a provision that information concerning actual performance be fed back into a neural center capable of comparing it with information concerning some target, e.g., the path of fleeing prey, or a normal value for some metabolic level. Presumably the neural center utilizes information supplied by previous learning trials or by the organism's genetic heritage. The distinction of 'computer' and 'controller' roughly parallels the common neurological distinction of afferent, efferent, and co-ordinating centers. The controlled system might be a set of co-ordinated muscles or a glandular system. Functions could be ascribed to the computer, the error signal, the controller, the control signal, or the controlled system; i.e., any of the above might serve as values for 'x' in

2.1. A function of x in S is P.

which is elliptical for

2.1.1. A function of x in a system S with organization C is to enable S in environment E to engage in process P or attain state P.

or some still more elaborate variant. As indicated, 'P' may designate either a process or a state; broadly considered, 'P' designates a particular range of values within the domain of possible output signals producible by S. For example, 'P' might designate a particular value for the oxygen/carbon dioxide ratio in arterial blood, a particular level of reproductive activity in flocking birds, or a distinctive limb configuration in a developing embryo.

The model also implies that the components of a functional system can be distinguished as information carrying and calibrating components (I components, including the channels carrying input, error, control, and output signals, as well as the computer and the controller), and operative (O) components generally referred to here as the controlled system.

I would argue that the applicability of the feedback, model to a given organic system implies that that system meets a number of boundary conditions, the examination of which is the principal task of this paper. These boundary conditions are simply verbal characterizations of certain aspects of the feedback model and have the same sort of warrant as the model itself, if they are faithfully and accurately drawn. In order to determine the dimensions of the warrant of the model, it would be necessary to compare the feedback model with other models – perhaps some compatible and employable in tandem with the feedback model to cover phases of organic activity not covered by the feedback model – and others either incompatible or at least competitive with the feedback model and applicable to the same sorts of organic activity as the feedback model itself. This seems to me to be a task where the initiative must rest with the theoretical biologist, but at the moment biologists seem to be devoting more attention to the exploitation of the feedback model than to the development of corollaries or alternatives to it.

The reader should be warned, therefore, that general questions concerning the biological warrant for the feedback model – and, *a fortiori*, any philosophical explication of it – cannot be satisfactorily answered in this paper. The paper assumes, and offers illustrations of, the widespread use of the model in contemporary biological literature; beyond that, it simply seeks to clarify the model by attempting to enumerate the semantic characteristics of the statement 'A function of x in S is P'. – stipulated here to mean that S is a negative feedback system, with x as a particular type of component and P as a specific sort of output.

Boundary Condition I – A Physically Distinct Feedback Loop

One of the most distinctive features of the negative feedback model is its definite division between the components of the system which translate a given stimulus or input into a specific response or output, and the feedback loop which carries the stimulus for compensatory activity if the initial output differs from an established 'target'. For example, if we consider the system organized to homeostatically maintain the internal body temperature of an organism, we know that the output (body temperature) is a function of a number of factors: the ambient temperature, the level of internal metabolic activity, the amount of sweating or panting, superficial vasoconstriction or dilatation and so on. A portion of the output is fed back into the system through thermal sensors probably associated with the thalamus or mid-brain. These sensors in turn are correlated with effector centers which control vasoconstriction, perspiration, respiratory rate, and so on. The interesting mark of the physical distinction of the feedback loop and other system components and inputs is that the output continues to be a lawlike function of the factors comprising the latter even if the former undergoes some temporary or permanent pathology (a fever may intensify because the appropriate sweating response is not elicited or because the heightened metabolic activity required to overcome an infection is not suppressed. I believe that this point was first noted by Beckner, 1968, pp. 139ff). This all implies the following rather complex boundary condition:

2.1. That the function of x in S is P implies that 'P' designates a particular target value from a range of physically possible system outputs; and that some components of S carry or translate information concerning the difference between the output of S at a given time and the value P; that these information carrying (I) components are physically distinct from the operative (O) components which directly determine the output; and that this distinction is reflected in the fact that pathologies of the system making it incapable of achieving P do not imply a disruption of the lawlike relationships between the O components and the output of the system.

No such reference to a feedback loop is at all apparent in the purported paradigm, "The function of the heart beat is to circulate the blood." In fact, the quoted statement is a relatively poor example of a function statement, reflecting usage typical of an elementary phase of identifying functions. In contexts where the explanation of functions or the analysis of the specific design of a functional system are stressed, such statements as "A function of variations in heart rate or blood pressure is the maintenance of an optimal oxygen/carbon dioxide ratio in body tissues" are much more typical.

This feature of the feedback model illuminates the ease with which biologists combine biophysical and biochemical analyses with evolutionary explanations of organic processes. Our discussion so far has taken for granted the fact that functional organic systems are typically directed to attainment of some specific output value, e.g., a normal value for blood pressure, or a normal growth function for an embryo. The feedback model simply assumes that this target value is programmed into the 'computer' (central nervous system, or other discriminative elements) of the functional system.

It would be very interesting to test the thesis that natural selection operates as a comprehensive feedback loop, tending to set the target values of various organic functions. Each generation of parent organisms has as its output a filial generation differing from the parent generation in a variety of ways. This output is recycled into the system when the filial generation reproduces itself. Natural selection 'computes' target values for the organic functions of succeeding generations by assuring that those members of the new parent generation whose functions are best adapted to the available environment produce a disproportionately large percentage of the new filial generation.

Accordingly we might prescribe the following boundary condition (first suggested by Scheffler) as necessary for the biologically complete characterization of an organic function:

2.2. That the function of x in S is P implies that the value of P has been derived through the historical processes of natural selection or individual learning trials.

Boundary Condition II – Autonomy and Individuality

As Beckner has noted, biological functions are not simply passive responses to environmental forces. An internal source of energy is involved. Another way of putting this is to note that an input signal or stimulus typically carries information concerning the environment in a form possessing a small or even insignificant amount of energy, yet it may trigger an output or behavioral routine involving the expenditure of a great deal of energy. Referring to this characteristic by using the term 'autonomy' will probably annoy some philosophers sensitive to the meta-physical freight the term is often obliged to carry. However, I mean to refer to nothing more than the fact that the metabolic organization of biological systems enables them to store energy, and that subsequent stimuli – not necessarily related in time or character to the storage process itself – can trigger the expenditure of any part of these stores.

Naturally, the amount of energy stored in a usable form will vary from organism to organism as a function of a number of factors including the individual's genotype and his recent metabolic history. It is clearly an interesting feature of the functional activity of organisms that it is neither stereotyped nor automatic and that it exhibits what might be called a distinctive biological individuality. Variation in the quality and quantity of stored energy is a significant component of such individuality, but there are others. The feedback model postulates an element in the functional system which responds to a local (and individual) gradient in the internal or external environment by producing changes in the system which heighten the maintenance of the target condition, once it occurs. More accurately, such a system responds to the difference between its output at a given time and the target or normal output. As a result, a well adapted behavioral routine will reflect the specific circumstances of a given individual: genotype, metabolic and behavioral history, and contemporary environmental factors.

2.3. That the function of x in S is P implies that the system laws of S characterize some component of S as responding to the difference between the output of S at a given time and the state or process P by producing changes (in S or in S's relation to its environment) which heighten the probability that P will occur.

This boundary condition obviously is based upon the negative feedback model and the distinctive loop which insures that information concerning system output is recycled into the system where it is compared (in the 'computer') with the standard or target output; if a given output differs from the target output, an error signal stimulates the controlled system to specific compensatory activity.

Boundary Condition III – Plasticity

As Braithwaite observed, one of the most general characteristics of goal-directed organic processes is their plasticity. Reference to such plasticity is a plausible restriction on the use of function statements.

2.4. That the function of x in S is P implies that S is characterized by system laws which result in the attainment of P under a variety of environmental conditions and through the inter-relation of sets of various values of the state variables characterizing the components of S.

Given the circumstances of life in a changing environment, this semantic restriction reflects a general biological interest in the performance of a given system in a range of environments.

The negative feedback model indicates that a particular target output can be attained under a variety of conditions: the exact form of the input signal may vary within a given range (chicks will give the gaping, feeding reaction whenever the nest is tipped slightly and an appropriately colored stimulus object is presented at a given height above eye level); the exact form of the output may be distorted by a number of influences ranging from environmental disturbances ('static') to the below par functioning of the organism; but since information concerning the output is fed back into the system, appropriate compensatory activities will be stimulated until the target output is achieved, or the organism shifts to some other behavioral or physiological routine.

Finally, each of the preceding boundary conditions has as a corollary a general prohibition against the attribution of functions to 'biological atoms', simple or isolated aspects of organisms.

2.5. That the function of x in S is P implies that S is a complex system following system laws which hold for a given range of environments.

This restriction indicates that statements such as "The function of the heart beat is to circulate the blood!" are seriously elliptical. They are incomplete because they omit references to the function-system in which the heart acts, to the complexity of the structure of the heart itself, and to the range of organisms and environments in which the occurrence of such function systems can be explained.

3. Philosophical objections to the negative feedback model

Probably the most common philosophical objection to the negative feedback model has been that it suggests a causal mechanism for a given process but does not explain why the mechanism in question is characterized in *functional* terms. The objectors argue that the model can be applied to a variety of inorganic phenomena which we do not regard as having functions or as being elements in a functional system. The path of a river flowing toward the sea reflects a variable response to environmental obstacles and terminates with the attainment of a definite target. The same thing may be said of an electrical circuit which, having a definite resistance, can be characterized as maintaining the value of a 'target' variable at a standard value; yet "it is hardly true that one would characterize the current in a wire as having the function of maintaining the ratio of electromotive force to current at a constant value."

Condition 2.1 above is responsive to this objection in requiring a physical basis for the feedback loop itself – a physical basis distinct from the straight line causal relationships (expressed by general physical laws) which make the output of the system a function of the relevant state variables (O variables) of the system. This situation can be diagnosed when pathological or experimental modifications of the feedback loop alter the value of the target variable or produce unsystematic fluctuations in the output of the system, even though the system continues to obey the appropriate physical and chemical laws. Such modification is clearly not possible for an electrical circuit regarded as directively organized under Ohm's law. It is physically impossible for such a circuit to be modified so that it no langer maintains the ratio of voltage to current as a constant. The same holds true of the physical forces determining the path of a river flowing toward the sea.

3.2. I have already noted how the current form of feedback model handles the missing goal-object objection.[2]

The model is now flexible enough to incorporate information from an actual goal (a fleeing prey organism) when such information is available; or to respond to information concerning the actual performance of the system by comparing it with 'normal' values of the performance. These normal values may be determined by the individual learning experiences of the organism, or by the evolutionary history of its ancestors. In such case, no signal from the goal itself is necessary to account for the 'goal-directed' character of the functional activity.

3.3. The difficulty of goal-failure is raised not so much against the feedback model itself as against the claim that the feedback model shows that organic phenomenon can be subsumed under the standard version (parity of explanation and prediction) of the nomological model of scientific explanation. So long as one does not claim that any scientifically adequate explanation of an event also enables the prediction of that event's occurrence, the fact of goal-failure is simply not relevant to the adequacy of the feedback model. In fact, the model invites attention to environmental and internal limits within which a functional system can successfully achieve its programmed target output.

3.4. The multiple goals objection does not count against the version of the feedback model outlined above. That objection was originally formulated against a form of the model in which the goal of the controlled activity was described very generally as "a final condition in which the behaving object reaches a definite correlation in time or in space with respect to another object or event." Since functional systems may actually attain an indefinite variety of such final conditions (depending on the attendant circumstances or even the perspective of the observer), the model seems to break down by requiring its user to label any chance performance of the system a goal-directed performance.

Imagine for instance, a clock which runs properly for many years, and then breaks down at twelve o'clock, New Year's Eve – such behavior admirably fits every requirement of purposiveness (Taylor, 1966).

But such behavior does not fit an appropriately amended version of the feedback model. The model requires that information concerning the performance of the system be fed back into the system in some fashion distinct from that in which the momentum and position of a particle at time t are fed into its momentum and position at time $t+\Delta$; i.e., the system must contain a physically distinct feedback loop. The recycled information must be compared with some standard or normal target value, a value determined by the evolutionary history of the system's ancestors or by the past 'learning' performances of the system itself. These requirements place sufficient control on the number of goals which may be attributed to a functional system.

3.5. Finally, one of the most interesting and relevant objections to the feedback model of organic functions is that the model applies to a number of organic processes which it is not appropriate to label 'functional' because they are maladaptive. For example:

3.5.1. A function of a broken blood vessel is the promotion of the growth of new blood vessels.

and

3.5.2. A function of a fish's swim bladder (when it is suddenly raised far above its normal depth) is to produce bursting.

are unacceptable as function statements to some authors, who have taken them as counter-examples to schemata generally comparable to the negative feedback model. No very elaborate rationale has been advanced for this rejection, which apparently is based on the view the statements in question are not idiomatic, or that they incorporate an *odd use* of the term 'function'.

In any case the restriction has been incorporated in several putative conditions for the use of function statements

3.5.3. The function of x is P only if x is not a sufficient condition for any malfunction of the organisms of which it is a part or process (Lehman, 1965).

or

3.5.4. The function of x is P only if in the absence of P, the probability that the organism of which x and P are parts or processes will survive or leave descendants would be smaller than it would, *ceteris paribus*, in the presence of P (Canfield, 1964).

But these restrictions hardly seem justified. Ordinary usage certainly cannot be invoked against the view that a general account of organic functions must include references to functional responses to stress, injury, or disease. Nor can it be a matter of definition (or settled idiomatic practice) that functions must be generally or faultlessly adaptive. On the contrary, the biologist expects such functional units as the swim bladder in fish to have a limited utility, and would not ordinarily feel that he understood their function *unless* he could make predictions about likely failures as well as successes.

Let us imagine a geneticist or biochemist working with a mutant form of Drosophila or Neurospora which is known to have a fitness or selective advantage less than that of the wild type of the same organism. The mutant gene is identified, along with a range of its metabolic, somatic, and behavioral effects. Conclusive evidence is available to indicate that the organism's probabilities of survival and of leaving descendants would in fact be far greater in the absence of that gene and one or all of its effects. Would the geneticist, because of his knowledge of their negative effect on the fitness of the organism, cease referring to the *functions* of that gene and of its effects? This does not seem likely (Hadorn, 1955, p. 277). No alternative terminology is used by biologists to describe or explain such cases.

Moreover, both 3.5.3 and 3.5.4 presume the distinction between 'function' and 'malfunction', but do not leave room for its clarification. It seems likely that any effort to justify the distinction just mentioned will depend upon an elaboration of the notion of function which is open to the possibility that functions may be either adaptive or maladaptive. The unsuitability of any requirement that functions be adaptive can be seen by noting that both 3.5.3 and 3.5.4 are analogous to restrictions which would limit the use of the term 'strategy' in game theory to those sequences which result in a positive pay-off. A game theorist may very well characterize a strategy as a series of moves leading up to a pay-off or a loss, but he will also want to compare the effectiveness of alternative strategies, all of them worthy of the name, with different pay-offs and losses. The evolutionary biologist seems to be in an essentially similar position: he must be prepared to weigh the relative selective advantage or disadvantage of alternative functional solutions to the same problem in adaptation.

4. Summary

First of all, in this paper I understand the term 'functionalism' to refer to a specific scientific bias, usually associated with the investigation of systems which are in some respect self-regulatory, which regards the objects of its study as having a target-oriented or goal-directed character. I note an irreducible variety of contexts in which statements about organic functions occur: preliminary identification and description; interpretation of causal organization or design; assessment of the role of the function in an overall evolutionary strategy.

Secondly, I define or limit the task of semantic explication of statements of the form 'A function of x in S is P.' by the stipulation that in all such statements 'S' shall refer to systems which can be modeled as negative feedback systems; 'P' shall refer to some range of the possible outputs of such systems; and 'x' shall refer to some component of such a system.

The model is then used to guide the formulation of a number of boundary conditions limiting the use of statements concerning organic functions. Some of these boundary conditions specify contexts within which function statements must be open to analysis and explanation. Others identify characteristics which functional systems must possess. No claim is made that the list of boundary conditions is exhaustive, nor that those identified are jointly sufficient to distinguish systems to which the negative feedback model may be applied. The list of boundary conditions is intended to partially explicate the web of conceptual implications characteristic of the feedback model, and to show – in a preliminary way – the kinds of systems to which the model is inapplicable.

I do not claim that the negative feedback model is the only model which can be used to interpret goal-directed biological phenomena, although the widespread and effective use of the model stands as a serious challenge to alternative models. In fact, the whole question of the warrant of this analysis would be more decidable if distinct, equally comprehensive and suggestive alternative models were available. At this point, the most likely alternatives seem to me to be game theory and information theory. Perhaps further investigation will show that there is some orderly relationship among the models characteristic of these theories.

Within this limited context, the boundary conditions here proposed as necessary for the characterization of a system as a functional system are:

(1) target orientation by means of physically distinct system components which insure that information concerning system output will become part of the input for subsequent stages of system activity.

(2) historicity – the target values of the system are set by the historical processes of natural selection or individual learning trials.

(3) individuality and autonomy – the system's output is determined by its individual genotype, metabolic and behavioral history, and by specific and changing features of its current environment; the output may exploit any fraction of the energy available in the system so that the energy actually expended may significantly exceed that of the corresponding input signal or stimulus.

(4) plasticity – the capacity to attain a target condition under a variety of environmental circumstances by means of a variety of internal system states.

(5) complexity – functions are not attributed to biological 'atoms' nor to single entities or processes considered apart from their role in a structured system.

Finally, I argue that previous philosophical criticisms of the negative feedback model of organic functions have been misplaced. The objections considered and disallowed include: (1) the model does not distinguish functional systems from simple causal systems; (2) the model breaks down for examples where the goal or target of the functional system is missing or never exists; (3) the model offers no means of discriminating a 'goal' or 'target' from among the indefinitely large number of terminal states achievable by a functional system; (4) the model must be amended so that functions are positively adaptive by definition.

University of Notre Dame

BIBLIOGRAPHY

Beckner, M., *The Biological Way of Thought*, Berkeley 1968.

Beckner, M., 'Function and Teleology', *Journal of the History of Biology* **2** (1969) 151–164.

Canfield, J., 'Teleological Explanation in Biology', *British Journal for the Philosophy of Science* **14** (1964) 285–95, 327–31.

Canfield, J. (ed.), *Purpose in Nature*, Englewood Cliffs, N. J. 1966.

Darlington, C. D., *Evolution of Genetic Systems*, London 1958.

Fisk, M., 'Analyticity and Conceptual Revision', *Journal of Philosophy* **53** (1966) 627–37.

Gregory, R., 'The Brain as an Engineering Problem', in *Current Problems in Animal Behavior* (ed. by W. Thorpe and O. Zangwill), Cambridge 1961.

Hadorn, E., *Developmental Genetics and Lethal Factors*, London 1961.

Hartman, P. and Suskind, S., *Gene Action*, Englewood Cliffs, N.J. 1966.

Hempel, C. G., 'The Logic of Functional Analysis', in *Aspects of Scientific Explanation*, New York 1965.

Jacob, F. and Monod, J., 'Genetic Regulatory Mechanisms in the Synthesis of Proteins', *Journal of Molecular Biology* **3** (1961) 318–56.

Lehman, H., 'Functional Explanation in Biology', *Philosophy of Science* **32** (1965) 1–20.

Lehman, H., 'On the Form of Explanation in Evolutionary Theory', *Theoria* **1** (1966) 14–24.

Lewontin, R., 'Evolution and the Theory of Games', *Journal of Theoretical Biology* **1** (1961) 328–403.

Manier, E., 'The Experimental Method in Biology: T. H. Morgan and the Theory of the Gene', *Synthese* **20** (1969) 185–205.

Manier, E., '"Fitness" and Some Explanatory Patterns in Biology', *Synthese* **20** (1969) 206–18.

Manier, E., 'Comments on Historical and Functional Aspects of Explanation in Biology', *Journal of the History of Biology* **2** (1969) 206–13.

Nagel, E., 'A Formalization of Functionalism', in *Logic Without Metaphysics*, Glencoe, Ill. 1956.

Nagel, E., *The Structure of Science*, New York 1961.

Quastler, H., 'General Principles of Systems Analysis', in *Theoretical and Mathematical Biology* (ed. by T. H. Waterman and H. J. Morowitz), New York 1965.

Rosenblueth, A., Wiener, N. and Bigelow, J., 'Behavior, Purpose, and Teleology' *Philosophy of Science* **10** (1943) 18–24.

Scheffler, I., 'Thoughts on Teleology', *British Journal for the Philosophy of Science* **9** (1958) 265–84.
Scheffler, I., *The Anatomy of Inquiry*, New York 1963.
Scriven, M., 'Explanation and Prediction in Evolutionary Theory', *Science* **130** (1959) 477–82.
Simpson, G. G., *This View of Life*, New York 1963.
Smart, J. J. C., *Philosophy and Scientific Realism*, New York 1963.
Taylor, R., *Action and Purpose*, Englewood Cliffs, N.J. 1966.
Waddington, C. H., *The Strategy of the Genes*, London 1957.
Waterman, T. H., 'Systems Analysis and the Visual Orientation ol Animals', *American Scientist* **54** (1966) 15–45.
Williams, G., *Adaptation and Natural Selection*, Princeton 1966.
Wynne-Edwards, V., *Animal Dispersion in Relation to Social Behaviour*, London 1962.
Young, J. Z., *A Model of the Brain*, London 1964.

NOTES

* Research supported in part by NSF Grant GS 1236 and the O'Brien Fund, University of Notre Dame.

[1] Broadly speaking, a scientist working within the functionalist perspective imputes a target oriented or goal directed character to the processes, states, and systems which he describes, analyzes, or explains.

[2] The objections considered in 3.2, 3.3, and 3.4 were raised by Scheffler (1958) and Taylor (1966).

WILLIAM C. WIMSATT

SOME PROBLEMS WITH THE CONCEPT OF 'FEEDBACK'

The concept of 'feedback', originally a technical term of cybernetics, has been so widely applied that it has become part of our everyday vocabulary. Though both scientists and philosophers have considered in detail a number of conceptual issues suggested by the perspectives of cybernetics, this basic concept has not yet received the critical attention it deserves.

Thus, while Taylor [1–3] Scheffler [4–5], and others have criticized Rosenblueth, Wiener, and Bigelow [6–8] for trying to define 'teleological' behavior in terms of negative feedback, their objections are directed primarily against the use of 'naive' or 'black box' behavioristic assumptions and the choice of the term 'teleological' to describe the phenomena in question. The concept of feedback is largely accepted by both factions on more or less intuitive grounds. I will use results derived by Kleene [9] in another context to argue that it is not possible to define the concept of feedback solely in terms of the *externally* observable behavior of a purported feedback system.

This result necessitates a look in other directions for possible criteria. Nagel [10], Beckner [11–13] and others [4–5] have proposed analyses for 'self-regulating' and 'goal-directive' systems in terms of their *internal* organization. It seems to be widely supposed that these analyses capture the notion of negative feedback. But I will argue that these criteria fail to distinguish between negative feedback systems and any open system tending to steady-state equilibrium.

I. FEEDBACK AND BEHAVIOR

In attempting to make a place for purpose and teleological language in the scientific description of physical systems, Rosenblueth, Wiener, and Bigelow overtly espouse an almost paradigmatic description of naive behavioristic aims and approaches: ([6], pp. 19–20).

Given any object, relatively abstracted from its surroundings for study, the *behavioristic* approach consists in the examination of the outputs of the object and of the relation

Boston Studies in the Philosophy of Science, VIII.

of this output to the input. By *output* is meant any change produced in the surroundings by the object. By *input*, conversely, is meant any event external to the object that modifies the object in any manner.

The above statement of what is meant by the behavioristic method of study omits the specific structure and intrinsic organization of the object. This omission is fundamental, because on it is based the distinction between the *behavioristic* and the *alternate functional* method of study. In a functional analysis, as opposed to a behavioristic approach, the main goal is the intrinsic organization of the entity studied, its structure and its properties; the relations between the object and its surroundings are relatively incidental.

...By *behavior* is meant any change of an entity with respect to its surroundings... any modification of an object *detectable externally* may be denoted as behavior.

This position is reaffirmed in their reply to Taylor (1) some seven years later: ([7], p. 323).

...if the term 'purpose' is to have any significance in science, it must be recognizable from the nature of the act, not from the study of or from any speculation on the structure and nature of the acting object.

In spite of increasingly widespread dissatisfaction with naive behaviorism in general, and among philosophers, with the proposed analysis of teleological behavior, neither philosopher[1] nor scientist has seen fit to examine the authors' definition and use of the concept of feedback within this framework of behavioristic analysis. The concept of feedback is accepted in the same spirit as it is used – uncritically and intuitively.

For the authors, teleological behavior is behavior controlled by negative feedback, and they characterize feedback as follows: ([6], p. 18).

in a broad sense, it may denote that some of the output energy of an apparatus or machine is returned as input; an example is an electric amplifier with feedback. The feedback is in these cases positive[2] – the fraction of the output which re-enters the object has the same sign as the original input signal. *Positive* feedback adds to the input signals, it does not correct them. The term feedback is also employed in a more restricted sense to signify that the behavior of an object is controlled by the margin of error at which the object stands at a given time with reference to a relatively specific goal. The feedback is then *negative*, that is, the signals from the goal are used to restrict outputs which would otherwise go beyond the goal. It is the second meaning of the term feedback that is used here.

By non-feedback behavior is meant that in which there are no signals from the goal which modify the activity of the object *in the course of the behavior*. ([6], pp. 19–20).

If the authors are going to define teleological behavior in terms of negative feedback, and simultaneously adhere to their behavioristic assumptions, then they must also be prepared to give behavioristic criteria for when a system uses or contains feedback or negative feedback loops. There are two ways in which this might be done:[3] (1) Feedback might be directly observable in the behavior of a system. (2) The presence of some behavioral pattern or sequence might *entail* that the system in question is a feedback system. Since (1) implies (2), but not conversely, and I wish to deny (2), I will concentrate on the second possibility.

Rosenblueth, Wiener, and Bigelow have made it eminently clear that they are not interested in the internal structure of a system, but only in its externally observable behavior.[4] But even the standard diagrams of feedback systems (see Figure 1) characteristically represent feedback loops as wholly or partially internal features of the system.[5]

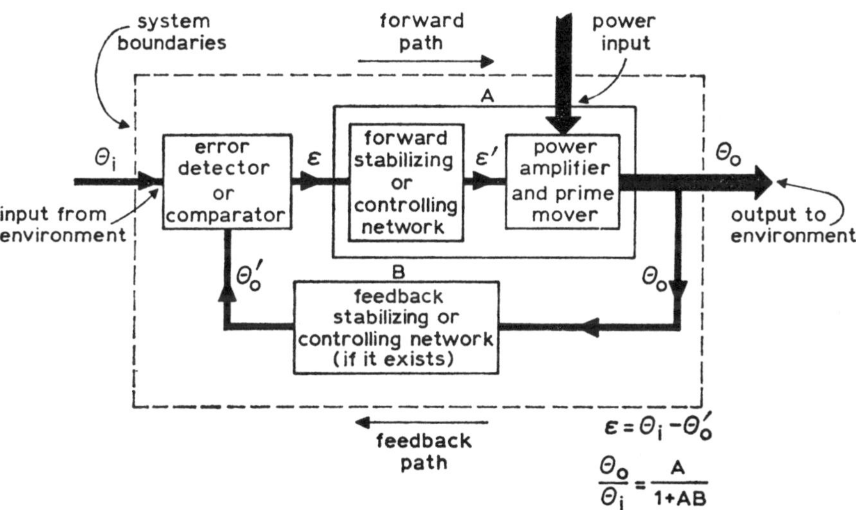

Fig. 1. Diagram of a generalized single-loop negative-feedback mechanism. (Note: This diagram is taken, in slightly modified form from L. K. Frank, G. E. Hutchinson, W. K. Livingston, W. S. McCulloch, and N. Wiener (eds.), *Teleological Mechanisms*; *Annals of the N.Y. Academy of Sciences*, v. 50, art. 4, pp. 187–278 (October 13, 1948). See p. 192. The energy flow of the power input was added to the figure found there.)

Diagrams can be misleading however. Fortunately, there are strong grounds which do not depend upon diagrams for saying that no externally observable behavior patterns could entail that the system in question contains feedback loops.[6]

In 1943, McCulloch and Pitts proposed a simplified logical model of the neuron, since widely known as the 'McCulloch-Pitts' neuron, (16). This 'neuron' is a generalized logical element of two-valued logic, and the aim of McCulloch and Pitts was to develop a model which bore some resemblance to the biological neuron and which, through its immediate connections with symbolic logic, could be easily manipulated to determine the behavior of interconnected networks of these neurons. Their assumptions and some examples of simple networks of these neurons are given in Figure 2.

Their formal results were corrected and extended in 1956 by S. C. Kleene [9]. Among other things, his work implies that any input-output matrix for finite sequences of discrete events with a definite temporal

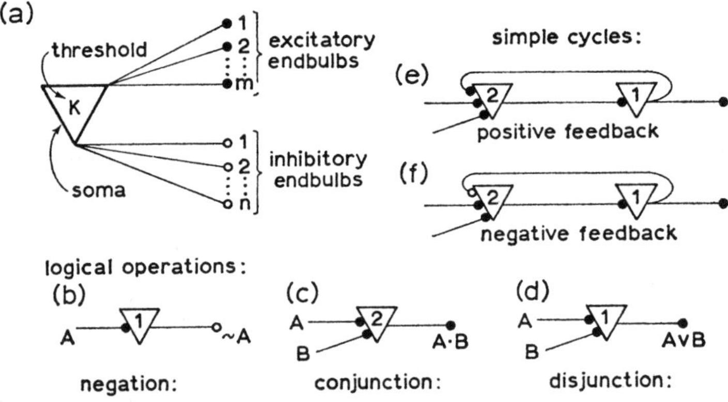

Fig. 2. The McCulloch-Pitts Logical Model of the Neuron. 2a: The generalized McCulloch-Pitts Neuron. It has a 'threshold', k, $(1 \leqslant k < \infty)$; m 'excitatory endbulbs', $(0 \leqslant m < \infty)$; and n 'inhibitory endbulbs', $(0 \leqslant n < \infty)$.

The laws of behavior of these neurons are as follows: (1) If it fires at all, the neuron fires only at integral values of time, i.e., for $t = 1, 2, \ldots$ (2) The conditions for firing of the neuron at t is as follows: (a) No inhibitory endbulb of a neuron which fired at $t-1$ ends on its soma, (b) k or more excitatory endbulbs belonging to neurons which fired at $t-1$ end on its soma.

The basic logical operations can be accomplished by interconnected nets of these neurons. Thus, the operation of negation is performed by the net of figure 2b, the operation of conjunction by that of 2c, and the operation of disjunction by that of 2d. From these facts, it follows that any well-formed formula in the propositional calculus can be represented by operations in a net of McCulloch-Pitts neurons.

Two particularly simple 'circles' or 'cycles' in nets of these neurons are represented in Figures 2e and 2f. Figure 2e is a case of positive feedback and Figure 2f is a case of negative feedback.

starting point such that the events can each be characterized by a finite number of variables, each of which can take only a finite number of values, can be produced by an appropriate net of McCulloch-Pitts neurons.[7] This means, essentially, that any 'black box' input-output behavior matrix (with some qualifications to be discussed later) can be produced by some net of McColloch-Pitt neurons.

Kleene also proved ([9], p. 10, Theorem 1) that any of these input-output matrices can be produced by a net which contains no 'circles'. But 'circles', in McCulloch and Pitts' terminology are what would generally now be called 'feedback loops'. But if any behavior matrix could be realized by a net which contains no such loops, then no input-output matrix *entails* that the system in question contain feedback loops, either positive or negative. It would thus seem to be impossible to define feedback in behavioristic terms.

There are some limitations on this conclusion, none of which seems to be serious: (a) the restriction to a finite number of variables; (b) the restriction of each of these variables to a finite number of values; (c) the assumption that processes are deterministic, rather than probabilistic; (d) the assumption that the event to be described has a well-defined starting time; and, (e) the limitation to behavior sequences of finite length.

Of these, only (e) (for reasons to be discussed shortly) might be thought to have any connection with the concept of feedback. As a result, there are no reasons to believe that dropping these restrictions would make a behavioristic definition of feedback possible.[8] Thus there are intuitive reasons for believing that a behavioristic definition of feedback is impossible for any class of system – even though it may be difficult or impossible to prove with any rigor.

The limitation to behavior sequences of finite length might be thought to be improper for giving a behavioristic definition of feedback for the following reason: The closed nature of a feedback loop makes it possible for the state of the system at a point in the loop to be a causal factor in the state of the system at that place at time Δt_L later, where Δt_L is the time it takes for a signal to propagate all of the way around the closed loop. Thus, the presence of a feedback loop establishes a periodic process, where an event at t_0 has effects on the system at times $(t_0 + n\, \Delta t_L)$ for $n = 1, 2, 3 \ldots$ – a behavior sequence of potentially infinite length. But on the other hand, there are many systems (of which the simple frictionless pendulum is one) whose behavioral description involves a potentially infinite periodic process and which few if any people would say were feedback systems. This alone could thus hardly be a distinguishing feature of feedback systems.

Even with the limitations of this result to deterministic discrete-state[9] systems and to behavior sequences of finite length, a definite counter-example to the claims that it is possible to define feedback generally in behavioristic terms has been provided.

I will assume, on the basis of Kleene's results and the plausibility arguments which I have given that it is not possible to define feedback, either positive or negative, in terms of behavioristic criteria alone. What other alternatives are there?

II. FEEDBACK AND INTERNAL STRUCTURE

In summarizing analyses of goal-directiveness put forth by several writers, Beckner claims ([13], p. 152) that:

> ...there are differences in detail in the accounts of the organization of goal-directive systems, but the acceptable ones have in common (1) some reference to the power of the system to compensate for environmental changes that might impede the system's progress towards the goal, and (2) some reference to the independence of the variables that define the system and its environment... *If conditions* (1) *and* (2) *are satisfied, the system is self-regulating by means of feedback.* [italics added]

He specifically mentions the analyses of Sommerhoff [14] and Nagel [10] in this context, though the characterization given also approximates his own analysis [11] and that of Ashby [15]. Most of these authors appear to believe that they have captured the notion of feedback – or more specifically, that of negative feedback. These analyses differ from that of Rosenblueth, Wiener and Bigelow in that they explicitly make use of general abstract features of the internal organization or structure of systems.

Of the two requirements mentioned, the first presumably captures the 'compensatory' behavior of negative feedback systems. The second is intended to rule out compensatory behavior in systems which are too simple to be regarded as feedback systems, such as simple pendulums. This latter condition basically requires that there be two logically independent but causally related variables in the system in addition to environmental variables.[10] As all of the systems to be considered below meet this requirement of independence, I will not mention it further.

There are two phenomena most frequently invoked in analyses of 'goal-directive' systems. While both are characteristic of feedback control systems, I will show shortly that they are also features of simple open systems in their approach to steady-state equilibrium.

The first is that a system may be capable of reaching the same set of equilibrium values for its state variables from an arbitrarily large range of different initial values of the state variables. (This is the primary feature of Braithwaite's analysis [18], which will not be discussed further here.)[11] Von Bertalanffy has christened this phenomenon 'equifinality' [19], and both he and Kacser [20] have given formal analyses of this phenomenon for open systems of the general type which I will claim are not feedback systems.

The 'compensatory' phenomenon frequently invoked (see Nagel [10], Beckner, [11], Somerhoff [14], and Ashby [15]) is that changes in one (or more) state variable(s) will be partially or wholly compensated for by changes in one or more other state variables in such a way that the original perturbed variable(s) will be wholly or partially restored to their 'normal' equilibrium values.[12] This phenomenon has been described by Kacser [20] as 'buffering' – a word which is a better clue to its real significance than those employed in the analyses mentioned above.

These two features are clearly related. The first describes the effect of the second – that equilibrium can be approached from a variety of initial states, or restored if a deviation occurs. The second gives the mechanism by which this 'equifinality' of non-equilibrium states is brought about: some variables 'take the load' from the 'stressed' variables and allow them to partially or completely approach or regain their 'normal' values.

These features, as well as others of interest can be demonstrated in a simple hydrodynamic model involving the flow of water into and out of a pair of interconnected tanks. (See Figure 3) Two cases will be discussed: one which is a feedback system and one which is not. Both examples possess the two features mentioned above.

The non-feedback system (Figure 3a) consists of two geometrically congruent tanks of constant cross-section, interconnected by an orifice of area a_1, with a water input at the top of one of the tanks at rate $r_I(t)$ and an output of magnitude $r_0(t)$ through an orifice of area a_2 at the bottom of the other tank. Water passes through the connecting orifice at rate r_s (assumed to be positive if flow is from the first tank to the second). r_s is a

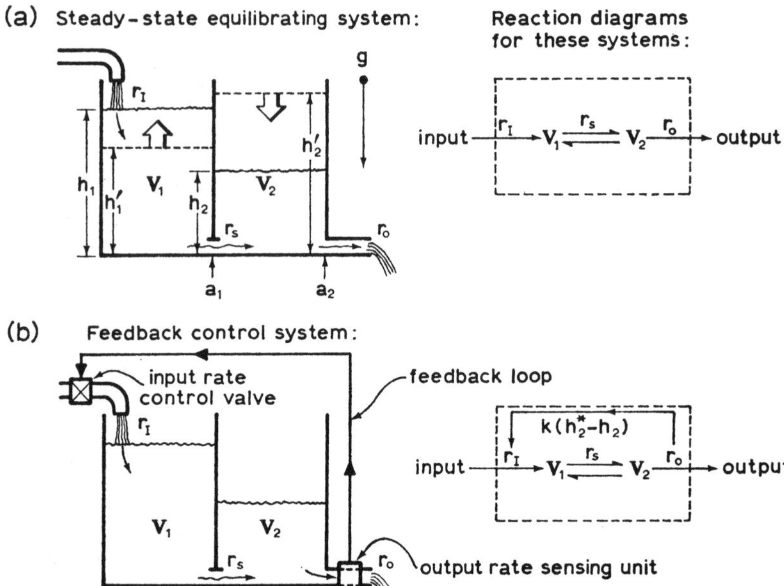

Fig. 3. Hydrodynamic models of feedback and non-feedback systems. 3a: A simple steady-state (non-feedback) system of two components with irreversible input and output. 3b: A Feedback system of two components with irreversible input and output. (Note: The reaction diagrams for these systems indicate the identity and direction of the causal relations between state variables of the system and environment. They are directly analogous to reaction diagrams used in chemical kinetics, which may also be taken as 'causal relation' diagrams.)

function of the water levels in the two tanks, h_1 and h_2, the area of the interconnecting orifice, a_1, the gravitational force, g, and a constant of the orifice c_1.[13] Because of the geometry of the tanks, the volumes of water in the two tanks, V_1 and V_2, is proportional to the water heights, h_1 and h_2, and this constant of proportionality has been included in c_1 and c_2.

Using conservation of mass, and standard physical assumptions, the rate equations for changes in V_1 and V_2 are given by:

$$(1)\qquad dV_1/dt = r_I(t) + ga_1c_1V_2 - ga_1c_1V_1,$$

$$(2)\qquad dV_2/dt = ga_1c_1V_1 - ga_1c_1V_2 - r_O(t).$$

The rate of water transport between the two tanks, r_s, is clearly

$$(3)\qquad r_s = ga_1c_1V_1 - ga_1c_1V_2 = ga_1c_1(V_1 - V_2).$$

The condition of equilibrium of this system is that all of the first-order time derivatives of the state-variables are zero, which is to say that all of the state variables have constant values. The state variables V_1 and V_2 are constant if and only if their inputs equal their outputs, i.e., *iff*:

(4) $r_I = r_s = r_O$.

But $r_O = ga_2c_2V_2$, and substituting for r_O and r_s (from (3) and (4)) yields:

(5) $r_I = ga_1c_1(V_1 - V_2) = ga_2c_2(V_2)$,

as the condition of equilibrium. The factors a_1, a_2, c_1, c_2, and g may be regarded as system parameters which do not vary under the circumstances considered, so that the only variables to consider are r_I, r_O, r_s, V_1 and V_2. From (5), V_2 is proportional to r_1, and the ratio $(V_1 - V_2)/V_2$ is constant at equilibrium. For this to be so, V_1 must be proportional to V_2, and thus also to r_I, and (from (4)), to r_s and r_O.[14]

Suppose that this system starts in a non-equilibrium state, with water levels h_1^1 and h_2^1 in Figure 3a. In this case, V_2 is greater than V_1, although at equilibrium it is less. Thus r_s in equation (3) is negative and there will be a net flow from the second tank to the first until $V_1 = V_2$, regardless of what values r_I and r_O have in the meantime. The direction of flow will then reverse and gradually approach its equilibrium value as V_1 and V_2 approah theirs. This simple open system thus exhibits both of the features supposedly characteristic of goal directive or feedback systems[15] (1) The system tends to approach equilibrium values from arbitrary non-equilibrium states, and (2) the system 'buffers', as an unusually high value of V_2 has been partially compensated for by an increase in V_1. Similar things happen if the system is perturbed the other way from equilibrium or if the value of r_I or r_O is suddenly changed.

Although this system meets the criteria of the analysis mentioned above for feedback systems,[16] I suggest that it is *not* a feedback system. First of all, almost any open system will exhibit this behavior and has this mode of organization. To call this feedback would emasculate that concept. Secondly, there is a closely related system which *is* a feedback system, and the contrast between these two systems illustrates differences in the way in which 'compensation' occurs.

I picked this hydrodynamic example in part because it is a specific example of a general class of systems in which there are reversible reactions between system variables and transport through the system.[17] Such systems are so pervasive in nature that this system cannot be considered to be an isolated case. For example, for each hydrodynamic system of interconnected water tanks, there are isomorphic systems in chemical kinetics and electrical networks which exhibit parallel behavior and abstract organization. They too must intuitively be called feedback systems if we attempt to call such a system a feedback system.

The system of Figure 3a can be easily modified so that it is a feedback system. In Figure 3b, a pressure sensor which measures h_2 (or V_2) is connected to a variable input valve which controls r_I in such a way that $r_I = r_I^* + k(h_2^* - h_2)$, where h_2^* is some 'reference value', r_I^* is the value of r_I for which h_2^* is an equilibrium value in the system without the feedback loop and k is a positive constant[8]. In this manner, deviations of h_2 and h_2^* are counteracted by changes in r_I, and the system is a negative feedback system. If k is a negative constant, then deviations from h_2^* are amplified by changes in r_I, and the system is a positive feedback system.

Corresponding feedback systems are found also in chemical kinetics and electrical networks. Open chemical systems which are autocatalytic and auto-inhibitory are positive and negative feedback systems, respectively, as are electrical amplifiers and regulators.

What is the difference between the types of 'compensation' exemplified in these two classes of cases? I think that the primary difference is that in the former type of case, 'compensation' is always accomplished *via* causal relations which are simple inverses. In hydrodynamic, chemical, and electrical cases of the first type, there is a 'net flow' or 'net reaction' rate between one tank, chemical component, or store of electrons and another. Transport can occur in either direction between them, and the net flow or reaction rate can be considered, algebraically *and physically*, to be the sum of the directed transport rates. The purported 'feedback path' is related to the forward path of the system as the reverse reaction is to the forward reaction in a reversible chemical reaction. These reactions are distinct (for otherwise it would be impossible to have irreversible reactions!) but are specially related, which I will signify by saying that they are *causal inverses* and employ the *same causal mechanism*.

The systems which I have called 'feedback systems' are different however. The 'main' path may involve reversible or irreversible reactions, but in either case, the feedback path is distinct, and can operate even if the system is irreversible in the 'main' path.

Unfortunately, while the distinctness of forward and feedback paths is a necessary condition for being able to speak of feedback, it does not seem to be a sufficient condition. Consider the system of Figure 4a: It is an open system, with two branches of transport from I ('input') to O ('output') – namely, paths $\overrightarrow{IABCO}$ and $\overrightarrow{IADCO}$. All of these reactions are assumed to be reversible, as in a chemical system, and the normal flow direction is assumed to be from I to O along both paths. There are in this system no closed causal (or material transport) loops.

But now suppose that there are energy inputs along branch $\overrightarrow{IADCO}$ at point *C* and *D*, as diagrammed in Figure 4b. This will in general force conversion of some *C* into *D*, and some *D* into *A*, such that there will be material conversion and tranport in a closed causal loop, $\overrightarrow{ABCDAB}$... That is, the direction of flow has been reversed in one of the reaction paths so that a loop is formed out of what had been two parallel paths. I submit (more on intuition than anything else) that this also is not a feedback system, in spite of the fact that there is a closed causal loop and that the forward and reverse paths are not simple causal inverses. This is because it is simply a closed loop of material transport. If the system of Figure 4a is analogized to the hydrodynamic system of 4c, then the system of Figure 4b is directly analogous to that of Figure 4d, where pumps are added to reverse the direction of flow in one of the lines.

A number of other strategies of definition could and have been tried. I have not discussed two of the more standard approaches here – to define feedback either in terms of control, or in terms of a flow of information. This is in part because of a belief that these notions are themselves parasitic upon that of feedback, and will be of no help in trying to define it.[19]

There are several other attempts which could be tried, and I am a (fairly) firm believer that some such attempt will work. Nonetheless, I hope to have shown that offering a definition for feedback is not as trivial or straightforward a task as many writers seem to have assumed. Given the difficulty of defining this concept it cannot automatically be assumed that feedback is a single procise theoretical concept and an objective property of a certain class of systems.

Thus, it might be that different definitions will have to be offered for

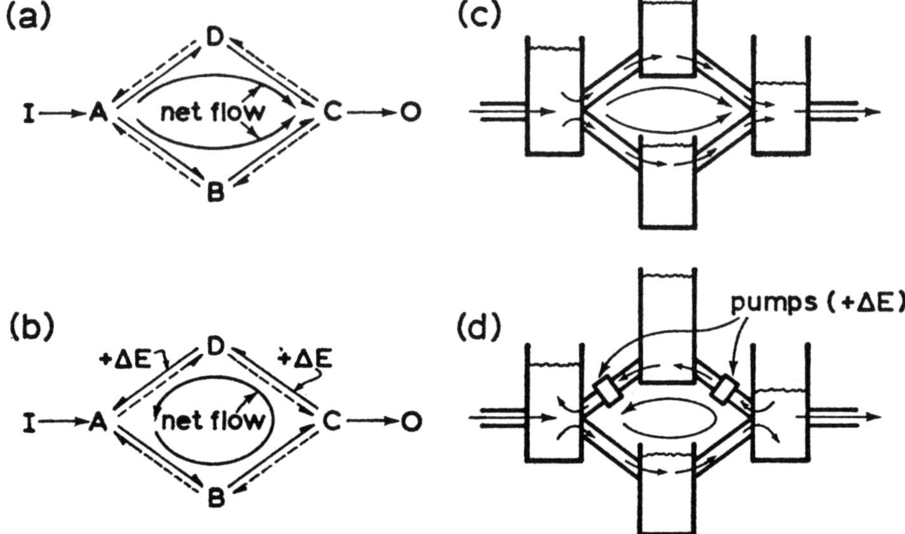

Fig. 4. Some non-feedback mass-transport systems, with and without closed causal loops of mass transport. 4a: Chemical reaction system with net flow from input (I) to output (O) and parallel flow in both branches. 4b: Chemical reaction system with addition of energy in one branch to reverse direction of flow in that branch and generate a closed loop of material transport. 4c: Hydrodynamic system analogous to 4a. 4d: Hydrodynamic system analogous to 4b. (Note: The relative heights of the tanks in Figures 4c and 4d are not intended to be significant – i.e., assume that the bottoms of the four tanks are all at the same level. The heights of the water in the tanks is significant however.

The reactions between chemical components in Figure 4a and 4b are assumed to be reversible. The dominant or net direction of the reactions is given by the solid arrow between components. It is the reversibility of these reactions which makes it possible to reverse the direction of flow between components by adding energy in Figures 4b and 4d.)

different classes of systems. Alternatively, Dick Levins has suggested and argued (in personal conversations) that feedback is, in effect, no more than an artifact of our mode of representation of systems – that it is not an objective property of the systems themselves. If this were so, the concept would appear to have no more than a rough 'heuristic' or 'Gestalt' status – as a useful guide in describing the behavior and organization of some systems.

Whatever the ultimate status of this concept, it is clear that there are disputes in the sciences which presuppose that a distinction between feedback control and simpler forms of equilibration is tenable. Thus, in ecology there are disputes over whether animal numbers are controlled by some internal regulatory feedback mechanism or whether constancies in population sizes are due to external factors or to some simple equilibrium process operating within the population. The common presupposition of both sides is that if there is true internal regulation of population size, this trait has been selected for, but not otherwise. While I think in fact that this is a dubious assumption, this matter need not be discussed here. It is clear that it apparently makes a difference to scientists whether a system can be classified as a feedback system or not. This in itself should offer sufficient justification and motivation for continued closer analyses of this concept.

University of Chicago

ACKNOWLEDGEMENT

Some of the work presented in this paper was done during the tenure of a Woodrow Wilson Dissertation Fellowship at the University of Pittsburgh and a post-doctoral research fellowship with the Committee on Evolutionary Biology at the University of Chicago, supported by the Hinds Fund for Studies in Evolution. I gratefully acknowledge their support.

BIBLIOGRAPHY

[1] Taylor, R., 'Comments on a Mechanistic Definition of Purposefulness', *Phil. Sci.* **17** (1950) 310–318.
[2] Taylor, R., 'Purposeful and Non-Purposeful Behavior: A Rejoinder', *Ibid.*, 327–332.
[3] Taylor, R., *Action and Purpose*, Prentice-Hall, Englewood Cliffs, N.J., 1966.
[4] Scheffler, I., 'Thoughts on Teleology', *Brit. J. Phil. Sci.*, **9** (1959) 265–84.
[5] Scheffler, I., *The Anatomy of Inquiry*, Knopf, New York, 1963, pp. 110–26.
[6] Rosenblueth, A., Wiener, N., and Bigelow, J., 'Behavior, Purpose, and Teleology', *Phil. Sci.* **10** (1943) 18–24.
[7] Rosenblueth, A. and Wiener, N., 'Purposeful and Non-Purposeful Behavior', *Phil. Sci.* **17** (1950) 318–26.
[8] Wiener, N., *Cybernetics* (2nd revised ed.), M.I.T. Press, Cambridge, Mass., 1961.
[9] Kleene, S. C., 'Representation of Events in Nerve Nets and Finite Automata' in *Automata Studies* (ed. by C. E. Shannon and J. McCarthy), Princeton Univ. Press, Princeton, N.J., 1956, pp. 3–42.
[10] Nagel, E., *The Structure of Science*, Harcourt, New York, 1961, Chapter 12.
[11] Beckner, M., *The Biological Way of Thought*, Columbia Univ. Press, New York, 1959.
[12] Beckner, M., 'Teleology', in *The Encyclopedia of Philosophy* (ed. by R. P. Edwards), Macmillan, New York, 1968, pp. 88–91.
[13] Beckner, M., 'Function and Teleology', *J. Hist. Biol.* **2**, 151–64.

[14] Sommerhoff, G., *Analytical Biology*, Oxford Univ. Press, London, 1950.
[15] Ashby, W. R., *Design for a Brain*, Wiley, New York, 1952 (2nd revised ed.), 1960.
[16] McCulloch, W. S. and Pitts, W. H., 'A Logical Calculus of the Ideas Immanent in Nervous Activity', *Bull. Math. Biophysics* **5** (1943) 115–33.
[17] Simon, H. A. and Rescher, N., 'Cause and Counterfactual', *Phil. Sci.* **33** (1966) 323–40.
[18] Braithwaite, R. B., *Scientific Explanation*, Cambridge University Press, London, 1953, Chapter 10.
[19] Bertalanffy, L. von, 'The Theory of Open Systems in Physics and Biology', *Science* **111**, 23–29, and 'An Outline of General Systems Theory', *Brit. J. Phil. Sci.* **1**, 139–64.
[20] Kacser, H., 'Some Physico-Chemical Aspects of Biological Organization', printed as an appendix to C. H. Waddington's *The Strategy of the Genes*, Macmillan, New York, 1957, pp. 191–249.
[21] Grodins, F. S., *Control Theory and Biological Systems*, Columbia University Press, New York, 1963.
[22] Weaver, W., 'Recent Contributions to the Mathematical Theory of Communication', in *The Mathematical Theory of Communication* (ed. by C. E. Shannon and W. Weaver), Univ. of Illinois Press, Urbana, 1949, pp. 95–117.

NOTES

[1] Richard Taylor ((1), p. 317) is an apparent exception. He raises doubts concerning the distinction between feedback and non-feedback behavior, but does not give any hard arguments or try to resolve them and later ((3), p. 239) appears to abandon them.

[2] The term 'positive feedback' clearly applies only to the example of the electric amplifier, not to the 'broad sense' of feedback. Positive and negative feedback are generally understood as species of the general type, 'feedback'.

[3] One can of course fairly simply give good *inductive* grounds for when a system contains what most people would call feedback loops, but these would not do for a *definition* of feedback.

[4] Their definition of behavior quoted above is ambiguous, since it is not clear that "any change of an entity with respect to its surroundings" will always be a "modification of an object which is detectable externally." Since most of their examples involve spatial movements, they are probably misled into thinking that the two formulations are equivalent. The first is *not* a behavioristic definition of behavior, while the second is. I will assume here that they are using the second definition.

[5] Strictly speaking, feedback loops are not always wholly internal features of the system in question, though this depends in part on how the system is defined. If the system is a thermoregulating mammal, then the relevant feedback loops are entirely inside the system. If the system is a heat-seeking missile in the act of homing on a heat-source, then part of the feedback loop 'passes through' the environment. In this case, if the system is taken to include the missile and the heat source and their causal relationships then the feedback loop is again entirely internal to the system.

The choice of just the missile as the system offers little comfort to the behaviorist, however, for part of the feedback loop, (and thus verification that feedback is involved in the heat-seeking behavior) *is* internal to the missile, and is thus inaccessible to the behaviorist. This point is generalizable: at least part of the feedback loop is always internal to the feedback system.

[6] Paradoxically, Wiener himself has in effect acknowledged this for internal structural properties in general. His language is quite misleading however, and he may have led himself down the wrong path. Thus, in his book, *Cybernetics*, ((8), pp. x–xi) he talks about the synthesis of a 'white box' which is an 'equivalent representation' of an unknown 'black box' merely by performing certain operations on the inputs and outputs of the two boxes, and speaks of this procedure as generating an 'analysis' of the black box. Since the white box is a box of known components and structure, this might seem to promise that one could determine the internal structure of a black box without looking inside it.

It is clear from context, however, that 'equivalent representation' merely means, 'has the same input-output function' and that the so-called 'analysis' of the black box is merely a mathematical decomposition of the box's input-output function into a number of mathematical terms or components. As Wiener himself says, later on p. xi, "in this manner we are able to construct a multiple white box which, when it is properly connected to a black box and is subjected to the same random input, will automatically form itself into an *operational* equivalent of the black box, *even though its internal structure may be vastly different*." [italics added]

[7] In general, there are an indefinitely large number of nets of McCulloch–Pitts neurons capable of realizing a given input-output matrix. Kleene merely exhibits a procedure for constructing a net which will do the job when given the input-output matrix.

[8] More can be said in support of this statement:

(1) Dropping (a), (d), or (e) as a restriction would result in definitions which are not operational. Any behaviorist who was also an operationalist would not consider such a move.

(2) Presumably, *anyone would be willing to accept (a) and (d) as limitations.*

(3) Continuous systems and variables can be approximated to any degree of precision for practical purposes by variables having only a finite number of values. Thus, maintaining that (b) was a dangerous limitation would also not be an operationally tenable move. In any case, if classical quantum mechanics is accepted, all natural process are ultimately finitistic. If a behavioristic definition *were* forthcoming only in terms of continuous variables, feedback would be tied to an approximate view of the world and would not be a part of its ultimate furniture.

(4) Condition (c), I think, is clearly irrelevant.

[9] To my knowledge no generalization of this rigor exists for continuous systems, though Grodins (21) discusses a number of cases, and claims (p. 99) that "...the form of the differential equation describing a given system does not reveal whether it is a feedback system or not." A consideration of his cases and construction of others along similar lines allows interpretation of the 'form' of the differential equation to include all of its relevant features: (i) whether it is linear or not; (ii) the order of the equation; (iii) the numerical value or algebraic structure of the coefficients, and (iv) the form of the input function.

[10] All of these authors appear to eschew talk of causal relation for talk about 'epistemic independence.' This concept was defined by Sommerhoff, ([14], p. 86). Since it can be trivially shown that any set of causally related variables are also mutually epistemically independent, I see no reason to introduce that complication here, and I will talk solely about causal relation. I believe that the latter concept has been appropriately explicated for complex systems by Simon and Rescher [17].

[11] This feature is met not only for simple open systems with a constant input, but also for any closed system. Braithwaite's only additional constraint is that the system

does not oscillate through the equilibrium point. Braithwaite's analysis is more simplistic than those discussed here, and he makes no claims to have analyzed the concept of feedback. It is not a plausible analysis for goal-directiveness either however.
[12] These analyses talk about 'goals', 'goal states' or 'sets of goal states', as if they were other than simple or steady-state equilibrium states, but nothing in the analyses prevents the simpler interpretations, and this is one of the primary defects of such analyses.
[13] c_1 will be assumed to be the same for flow in either direction, and is a hydrodynamic constant which reflects properties of the fluid (viscosity only, if the fluid is assumed to be incompressible) and geometrical features of the tank and orifice. For c_1 to be independent of the flow rate, r_s, it must also be assumed that flow is always laminar. Similarly, c_2 is assumed to be independent of r_0, which is a function of h_2, a_2, g, and c_2,
[14] The proportionalities here indicate a third feature which Von Bertalanffy ([19], p. 24) suggests is an important kind of independence of the system from its environment: "... the steady-state ratio of the components depends only upon the system constants, not upon the environmental conditions."
[15] Actually, these conditions are met only if the equilibria of the system are stable, as they are in fact for the system discussed.
[16] In fact, it does *not* meet Beckner's criteria ([11], Chapter 7). Beckner postulates in addition (a) the presence of an internal energy source, and (b) the presence of a 'special hookup', which if removed leaves the state variables causally related but destroys or changes the 'compensation' mechanism. Beckner may have intended the first as a requirement of 'teleological' systems, and not of feedback systems, and I would deny that it is a requirement of the latter. His second requirement is met if a one-way valve preventing flow from the first tank to the second is added to the above system, representing "removal of the special hookup." (Although a piece of apparatus is *added*, the causal relation from V_1 to V_2 is thereby *removed*.) Since V_2 can still affect V_1, Beckner's criterion is met, but the system is still not a feedback system.
[17] It is also a linear system. An interesting criterion for the special case of feedback in linear systems can be offered, but it appears to break down for non-linear systems, and thus fails to be general.
[18] This is not the only possible form of feedback control. The type mentioned above is known as 'proportional control' from the fact that the magnitude of the "restoring force" (given by $k(h_2^* - h_2)$) is proportional to the deviation from equilibrium (given by $(h_2^* - h_2)$). If the system is to be stable, it must also be true that $k < 1$. If $k = 1$, the system will maintain a stable oscillation, and with $k > 1$, the system will "overcompensate" and undergo oscillations of increasing magnitude.
[19] My claim is that it is appropriate to speak of self-controlling processes or of a closed flow of information when and only when it is appropriate to speak of feedback, and that these concepts must either be explicated in terms of feedback or in more anthropomorphic terms which would be of no use here.

The concept of information of course has been given a formal and independent characterization in terms of information theory. But information theory is primarily concerned with the carrying capacity of channels and codes for correcting errors of transmission. It tells nothing about what the information in the feedback loop does to influence the course of the reaction in the main path, or how it does it and this is the feature we need to define feedback. As Warren Weaver [22] says, information theory has nothing to say on how the information is interpreted and what is done with it.

STUART A. KAUFFMAN

ARTICULATION OF PARTS EXPLANATION IN BIOLOGY AND THE RATIONAL SEARCH FOR THEM

With the realization that the grounds upon which an hypothesis comes to be formulated can be considered separately from the grounds upon which it is accepted, it has become popular among some philosophers and scientists to claim that although it may be of psychological interest to understand the genesis of hypotheses, there can be no logic of search or discovery. While it is unclear exactly what is meant by the claim that there can be no logic of search, it *is* clear that this opinion, coupled with anecdotes of Poincaré's sudden solution of a mathematical problem while stepping on a Madrid streetcar, and Kekulé's vision of a snake biting its tail, have left the aura that the generation of an hypothesis is as mysterious as a Gestalt shift in perception of a figure. Perhaps because Gestalt shifts seem to occur without a processes of reasoning, but in some sense, spontaneously, the use of such perceptual shifts as models of hypothesis formation have lent support to the claim that there can be no logic of search. To a practicing scientist, the image of the startling 'shift' and insight might seem overly flattering of the scientist's genius, and the actual generation of hypotheses seem more reasonable and less mysterious. Some of the ways in which the generation of an hypothesis is a rather reasonable affair will be discussed below in conjunction with an effort to examine some of the features of what I am calling articulation of parts explanations, as they occur in biology.

Typical explanations in biology exhibit the manner in which parts and processes articulate together to cause the system to do some particular thing. Examples include accounts of the cardiovascular system, protein synthesizing system, endocrine system, etc. I do not wish to say that articulation of parts explanations occur only in biology, or are the only form of explanation utilized in biology, merely that such explanations are prevalent in biological sciences and exhibit interesting properties which I wish to discuss.

To refer to some types of explanations as articulation of parts explanations suggests that some explanations are not. I suggest that when a

Boston Studies in the Philosophy of Science, VIII.

thing is seen as consisting of a single part, or as a continuum, then explanations of the appropriate aspect of its behavior will not exhibit how parts work together. For example, Newton's first law of motion does not exhibit how parts of a system work together, for there is but a single part – a particle in rectilinear motion. Maxwell's field equations treat an electromagnetic field as a continuum, not as a second order infinity of points. Further, some behaviors of a given system may require an explanation by reference to the interworkings of its parts, while others of its behaviors may not. For example, the manner in which a gasoline engine falls is predictable from an account of the engine as a mass without discriminated parts.

I wish to pursue the following theses:

1. An organism may be seen as doing indefinitely many things, and may be decomposed into parts and processes in indefinitely many ways.

2. Given an adequate description of an organism as doing any particular thing, we will use that description to help us decompose the organism into particular parts and processes which articulate together to cause it to behave as described.

3. For different descriptions of what the organism is doing, we may decompose it into parts in different ways.

4. The use of an adequate description of an organism seen as doing a particular thing to guide our decomposition of it into interrelated parts and processes, and indeed part of the logic of search, is intimately connected with the sufficient conditions for the adequate description. In particular, we can use the sufficient conditions to generate a cybernetic model showing how symbolic parts might articulate together to cause a symbolic version of the described behavior.

5. We can use such a cybernetic model to help find an isomorphic causal model showing how presumptive parts and processes of the real system might articulate to cause the described behavior.

6. Since there may be more than one set of sufficient conditions for the adequate description of the behavior, more than one cybernetic model to account for the behavior may be constructed. Such different cybernetic models will not be isomorphic, and each leads us to decompose the system in a different way. Hence, not only are organisms decomposed into parts to yield articulation of parts explanations in different ways for *different* adequate descriptions of the organism seen as doing diverse

things; but different tentative decompositions can be made of the system seen as doing any one thing, through the use of the diverse sets of sufficient conditions of that adequate description.

7. A successful decomposition leads to an articulation of parts explanation of how the system does what it is seen as doing.

8. We not only use views of what a system is doing to help decompose it into parts, we use information about parts to synthesize new views of what a system is doing.

9. The descriptions of parts and processes of one decomposition need only be compatible with and not deducible from the descriptions of parts and processes of a different composition.

10. There *need* be no 'ultimate' decomposition such that all other decompositions are deducible from it, although there may be such an 'ultimate' decomposition.

1. Perhaps the first point to be made is that there is no uniquely correct view about what an organism is doing. But, in order to achieve an articulation of parts (henceforth, *A* of *P*) explanation about an organism, we must, perforce, be explaining how the parts and processes articulate together so that the system does something. To begin studying such an object, we must come to some initial view about what the system is doing; literally, a view of what is happening. Consider that an organism may be viewed as: a self-replicating system, a developing individual, a parent with similar appearing offspring, a member of an ecosystem, a system exhibiting circadian rhythms, a member of an evolving population, an open thermodynamic system maintaining a locale of low entropy, etc.

Now, not only are multiple views about what a system is doing possible, but also any system may be decomposed into parts in indefinitely many ways, and for any such part, it too can be seen as doing indefinitely many things. Our questions concern the character of the diverse possible decompositions into parts and the relations between parts within one decomposition, and between diverse decompositions.

2. I suggest that we use an adequate description of a view of what a system is doing to help guide our decomposition of the system into a particular set of parts and processes which causally articulate together to cause the system to behave as described. A view of what a system is doing sets the explanandum and also supplies criteria by which to decide whether

or not a proposed portion of the system with some of its causal consequences is to count as a part and process of the system. Specifically, a proposed part will count as a part of the system if it, together with some of its causal consequences, will fit together with the other proposed parts and processes to cause the system to behave as described. In general, it is not possible to decide that a single proposed process is to count as part of the system in isolation from decisions about the adequacy of the suppositions about the remaining parts and processes. Parts and processes are accepted more or less jointly, and with them, the adequacy of a particular articulation of parts explanation in which just those parts and processes are seen as fitting together to yield the behavior in question. Other causal consequences of these parts are then considered irrelevant. I will call such a decomposition of a system a conjugate, coherent decomposition, for it is conjugate to a particular view of what the system is doing, and coherent in that it provides an articulation of parts explanation of how the system does whatever is specified in that particular view of it.

3. Clearly, distinct views of what an organism is doing may lead us to decompose it in distinct ways. Our account of an organism as a device exhibiting circadian rhythms picks out different parts and processes from an account of the effects of chromosomal crossing over on population genetics.

4. I wish to argue that part of what might be called a logic of search involves our use of the sufficient conditions of an adequate description of a view of what an organism does to help find a cybernetic model of the phenomenon, and thence, a causal model. I will consider first a hypothetical case of tissue reaggregation, then an actual description of gastrulation in the chick, and hypotheses generated to account for these phenomena.

Hypothetical Example

Suppose we note that the cells of a sponge can be disaggregated and then allowed to reaggregate, and that, upon reaggregation they always form a particular three dimensional structure in which different specific cell types are at specific loci relative to one another. Sponges are deformed by currents in the water, and we will consider that an adequate description of this hypothetical sponge need only describe cell movement and specify which cell types have which cell types as neighbors, and in which directions, and which cells border internal and external lumens in the final structure. Note that the specification of cell types is made on the basis of some theory, here the cell theory. The observation that the cells reaggregate constitutes a view about what the system is doing, and sets us a question.

There may be more than one set of sufficient conditions for the truth of any description. By sufficient condition I here mean a description of a state of affairs such that from this description and with no further empirical information, the initial description may be deduced. I shall refer to such a description as a descriptive sufficient condition.

Let us assume that in our hypothetical aggregate, each cell type has either no cell as a neighbor in some directions, or only specific other cell types. Thus, we may assume that a description of these restricted adjacency relations, plus an account of the numbers of each type of cell, constitute descriptive sufficient conditions from which, with no further empirical information, we might deduce the initial adequate description.

We can, however, find a different set of descriptive sufficient conditions for the initial description. If the reaggregate is located in a three dimensional coordinate system with some particular cell taken as the origin, then the specification of the coordinates of each cell, and its cell type, is surely a descriptive sufficient condition of the original description, from which that original description can be deduced. I want to argue that each such set of descriptive sufficient conditions can be utilized to help find a model of how the cells manage to reaggregate, and that models derived from different sets of descriptive sufficient conditions pick out different putative parts and processes interacting in different putative articulation of parts explanations.

The power of the strategy of search I shall discuss rests on three features, two logical, one contingent. (1) If a process can be found which is causally sufficient to bring about the state of affairs described in the descriptive sufficient conditions then that process necessarily is causally sufficient to bring about the state of affairs described in the initial description. (2) Any initial description has multiple sets of descriptive sufficient conditions. (3) It is a contingent fact that very often, the ease of finding a causally sufficient process for one set of descriptive sufficient conditions is greater than for other sets of descriptive sufficient conditions. Indeed, sometimes a process to bring about a descriptive sufficient condition is suggested in a transparently obvious way by the descriptive sufficient conditions themselves. The sense of transparently obvious will be brought out below.

The first step, then, in utilizing a descriptive set of sufficient conditions is to suppose that there is a process which is causally sufficient to bring about the state of affairs referred to in the descriptive sufficient conditions. In the hypothetical aggregate we are considering, suppose we are currently using the descriptive sufficient conditions in terms of restricted adjacency relations among the different cell types. We then suppose a process which brings about just these intercell boundries. A particularly obvious choice of process is to assume that only these specific intercell boundaries form bonds between the cells. Since cells move during reaggregation, then remain properly juxtaposed, we also suppose a process which causes movement relative to one another among the cells until an allowed boundary is formed, when relative movement of the two cells stop. These hypothetical processes may now be linked, or articulated together to show how the state of affairs described initially might come about; specifically, cells move about until they come into contact with the appropriate neighbor cell type, when stable bonds are formed, relative movement stops, and the final architecture of the aggregate is generated.

At this stage the model is what I will call purely symbolic, or cybernetic. As yet, no actual causal mechanism is suggested. The cybernetic model provides a set of rules such that a set of symbols behaving as prescribed by the rules, will generate a form isomorphic to the cell assembly. The cybernetic model asserts that if causal mechanisms can be found such that only restricted bounderies form stably and relative motion of cells then ceases, then the observed behavior will occur in the aggregate. The central criterion of adequacy of such a cybernetic model is sufficiency. That is, it must be true that a set of symbolic components behaving as described by the rules in the model would, in fact, result in an aggregate isomorphic to the cell aggregate. Note also that the cybernetic model, while not yet suggesting actual causal mechanisms, exhibits the manner in which the processes of the parts of the system must articulate – namely by mediating the formation of restricted kinds of intercell boundaries. Thus, the cybernetic model, and indeed the adjacency descriptive sufficient conditions themselves, already begins to indicate the relations that are to exist among the parts and processes of the system; furthermore, the cybernetic model coupled with our current knowledge about cells begins to suggest what sorts of causal mechanisms are required. Specifically, mechanisms are

(For an explanation of these figures see text on page 265.)

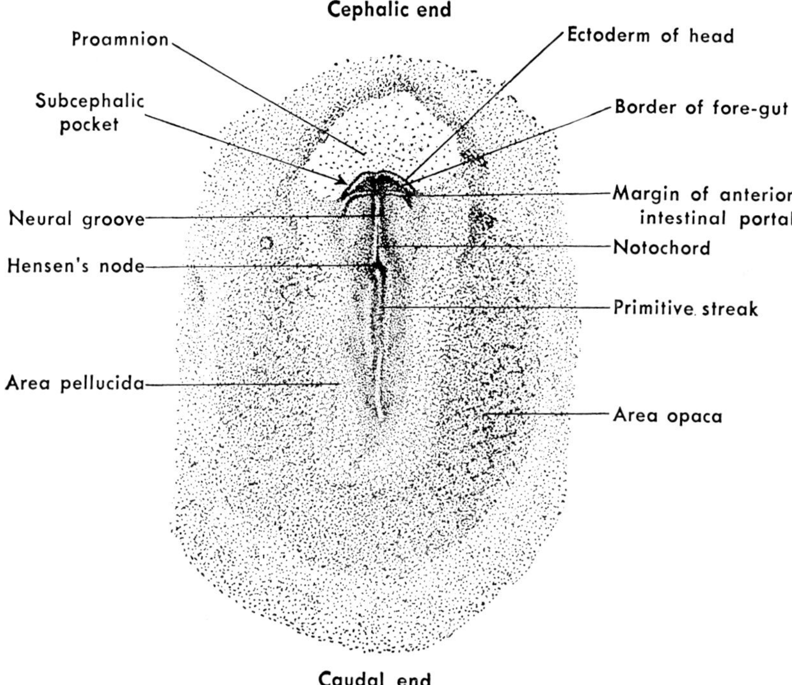

Fig. 1a. 20 hr chick embryo, dorsal view. Redrawn from Patten.

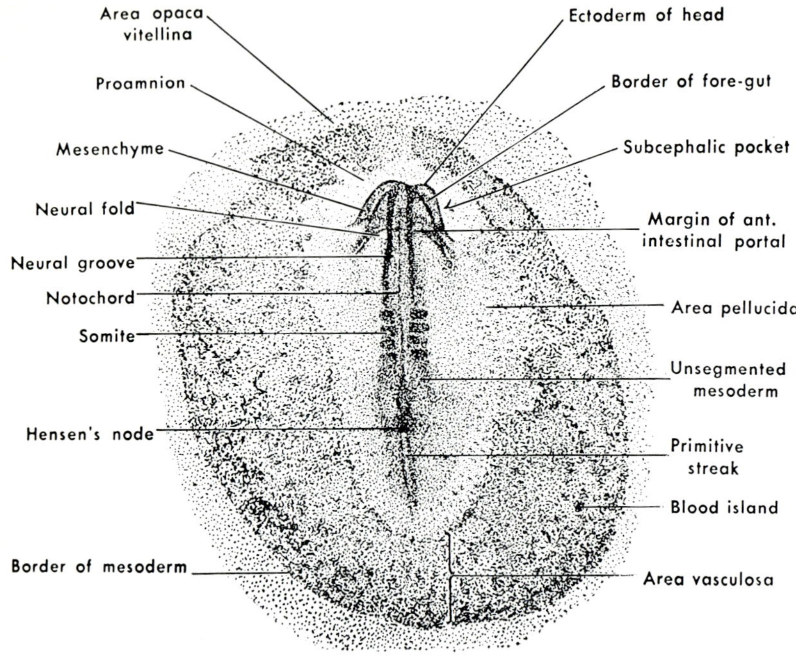

Fig. 1b. 24 hr chick embryo, dorsal view. Redrawn from Patten.

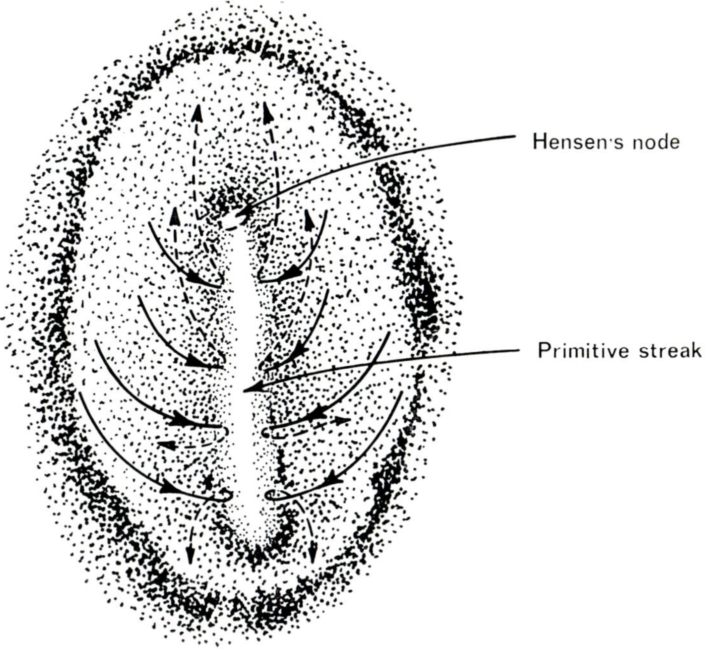

Fig. 1c. Schematic drawing of epiblast cell movement, solid arrows, and movement of mesodermal tissue, dotted arrows.

required which will create specific intercell boundaries, and not allow other intercell boundaries to form. We do not know yet what particular causal mechanism achieves this, but we do know that causal consequences of putative parts of this system which do not yield specific intercell boundaries may be treated as irrelevant causal consequences of the parts of the system. Those causal consequences, that is, processes, will not be regarded as processes of the system, but as irrelevant behaviors.

The descriptive sufficient conditions of the initial adequate description speak, if you will, in the imperative mood. They are an injunction to the scientist to direct his attention to those conditions, for around them it should be possible to build a cybernetic, and later a causal, model to explain the phenomenon. It is at least partially because the sufficient conditions can be used to generate a cybernetic model that the generation of hypotheses is a rather reasonable process.

5. With the cybernetic model in hand, and background knowledge about cells, we can now search for the kind of casual processes which are likely to cause restricted boundary relations – say specific molecules on the membranes of the different cells, each of which only interlocks with the appropriate others. With the suggestion of a specific set of causal mechanisms, the model has become an hypothesis requiring verification. If verified, it will specify which processes – that is, which of the many causal consequences of each of the portions of the system – are to count as processes of the system, and which are irrelevant to this particular account. With the verification of the hypothesis we will have achieved an articulation of parts explanation of how cells manage to reaggregate. Until we are confident of the entire *A* of *P* explanation, we may remain unsure about the claim that any particular process is to be regarded as a process *of* the system.

A cybernetic model is used to help find a causal model isomorphic to it. More than one causal model might be suggested, each isomorphic to the cybernetic model and to one another. The sense of isomorphism intended may be exemplified by supposing that one has designed, on paper, an adding machine. The design is a flow chart of operations performed by as yet symbolic devices. Now an actual machine realizing this design might, for example store numbers by filling and emptying water tumblers, or filling and emptying capacitors. The different physical realizations would be isomorphic to one another and the cybernetic model by virtue of the fact that the parts and processes of one machine can be put into one correspondence with the appropriate part and processes of the other real machines or cybernetic design model. While isomorphic in this sense, the causal mechanisms by which the different physical machines realized the cybernetic design model would differ, and would be described by different causal laws. In the example of the cell aggregate, many different causal means of forming only restricted adjacency relations might be considered.

6. Different sets of descriptive sufficient conditions of the same initial adequate description can be used to generate different, non-isomorphic, cybernetic models, and in turn, non-isomorphic causal models which decompose the organism into different putative parts and processes. For example, specification of the coordinates of each cell, and its cell type, in a three dimensional coordinate system with one cell taken as origin, was a set of descriptive sufficient conditions for the initial adequate description. Utilizing this, we build a cybernetic model by supposing a process which is causally efficacious in generating the state of affairs referred to in the descriptive sufficient condition. In that descriptive sufficient condition, each cell is specified in a coordinate system. A particularly obvious choice of process suggested by this set of sufficient conditions is that there *is* a coordinate system in the reaggregate with some cell as origin, and that each cell 'knows' where it is in the coordinate system and goes to the correct place. We then suppose that an origin cell somehow elaborates a coordinate system and that cells have some means of 'knowing' their location. From this cybernetic model, we make use of knowledge about cells to suggest causal mechanisms and achieve an hypothesis.

Note that the parts and processes specified in this description, and this cybernetic model, cannot be put into one to one correspondence with the parts and processes of the adjacency cybernetic model; here, distance measuring from an origin cell is a process of the system, and formation of restricted types of boundaries between adjacent cells is an irrelevant behavior of the cells. The two cybernetic models are not isomorphic, neither are those causal models which are isomorphic to each, isomorphic to one another across cybernetic models.

7. A successful articulation of parts explanation distinguishes between irrelevant causal consequences and important causal consequences of a

part, thus the explanation not only accounts for the behavior of the whole, it supplies a view of what it is that the parts themselves shall be seen as doing from among the indefinitely many possible things each part might be taken to be doing.

We now consider a more realistic example of hypothesis formation concerning gastrulation in chick embryogenesis (see figures). In its early stages, the chick embryo consists of a double layer, roughly oval, patch of cells on the yolk. The superficial layer is called the epiblast, the layer next to the yolk is the hypoblast. During gastrulation, cells of the epiblast move in an arc backward and medially, on both sides of the patch of cells, and condense in the midline to form the primitive streak. There, the cells sink below the epiblast to the region between the epiblast and hypoblast, and the cells spread laterally and forward between the two initial layers, forming a third, mesodermal layer. Medial flow of the epiblast, invagination, and lateral flow of the mesoderm occur with bilateral symmetry along both sides of the primitive streak. At the head, or rostral end of the primitive streak, a small mass of cells known as Hensen's node forms. With time, the primitive streak becomes shorter, and Hensen's node is carried bodily backward toward the tail, or caudal end of the primitive streak. In the region anterior to Hensen's node, epiblast cells are not moving medially, nor are cells sinking below the epiblast and migrating laterally. That is, Hensen's node is the anterior-most locus in which cells are invaginating to form the mesodermal layer, and as Hensen's node regresses backward, the primitive streak which is the region where invagination is occurring, becomes shorter. Some time after the node regresses, condensations of the mesodermal cells into aggregates to form somites occur bilaterally along both sides of the line that Hensen's node has followed during its regression; that is, on both sides of the line where the primitive streak was prior to its regression. Our problem is to explain the occurrence of bilaterally symmetrical somites in just the loci in which they occur.

Let the foregoing description be considered an adequate description. A different set of descriptive sufficient conditions might be: (1) a description of cell movements as given above. (2) The addition that somites condense only after invagination (sinking below the epiblast) and the other medial and lateral movements of gastrulation have ceased; and further, that somites condense only in those cells which were the last to invaginate to the mesoderm. This description is a sufficient condition of the initial description since it includes the claim that lateral cell movement of the mesoderm stops when invagination stops, and we can therefore deduce from this new description without further empirical information that somites form in bilaterally symmetric lines adjacent to the line of the now regressed primitive streak, as described in the initial description.

Processes capable of bringing about the state of affairs described in these new sufficient conditions will be processes capable of causing the state of affairs initially described. In addition to description of cell movement, this new set of descriptive sufficient conditions refers to: (1) formation of somites only after cessation of movement, (2) aggregation of initially disperse mesodermal cells into local aggregates; (3) the occurrence of these aggregates only among the last cells to invaginate. We need to find processes to bring about these effects. That is, we suppose processes by which (a) cessation of movement allows or causes mesodermal cells to aggregate; (b) mesodermal cells aggregate; and (c) only the last to invaginate aggregate. The descriptive sufficient conditions themselves suggest rather obvious processes. If we suppose there is some process by which mesodermal cells tend to aggregate, then movements of gastrulation might overcome the tendency to aggregate, thus cessation of movement would allow aggregation to occur. We now need a process by which there is a tendency of mesodermal cells to aggregate. A new descriptive sufficient condition for the initial description of aggregation is that originally, mesodermal cells are fairly dispersed, but move about, and when they come together, they stay together. We suppose a process causing the state of affairs described by this new descriptive sufficient condition to occur, say that mesodermal cells are sticky and attach to one another. We now need a process to cause only the last cells to invaginate to aggregate. Since we are to account for aggregation by mesodermal cells sticking together, this process must either be strongest, or most effective among the last cells to invaginate. If we suppose that mesodermal cells elaborate a diffusible 'sticky stuff', then epiblast cells will absorb sticky stuff as they migrate medially over the laterally migrating mesodermal cells. Epiblast cells which have traversed a greater number of mesodermal cells will be stickier than those which have traversed fewer. Epiblast cells arriving at the primitive streak later will be stickier than those arriving earlier, thus the last to arrive will be the stickiest, and the aggregation will occur in the proper place.

We now have a cybernetic model utilizing the three processes; that movement disrupts the formation of aggregates, that mesodermal cells tend to stick together, and that countercurrent movement of the cell layers will concentrate diffusible 'sticky stuff' near the primitive streak rather than laterally. This model can now be used to generate a causal model in which, for example, the 'sticky stuff' receives a molecular interpretation of some kind. As in the previous example, the descriptive sufficient conditions of the initial adequate description of the entire process and portions of it, have served as injunctions to direct attention in particular ways in order to build an hypothesis. And the formation of the hypothesis does indeed seem a rather reasonable affair, open to the judgement of having been done stupidly or well.

As in the previous example, differenet descriptions which are sufficient conditions for the truth of an initial description can yield radically different models. For example, somites may also be described as forming in those mesodermal cells over which Hensen's node has passed. This, coupled with a description of cell movement, is sufficient to deduce the initial description. It leads to the supposition that Hensen's node somehow causes or induces the underlying mesoderm to differentiate into somites. As in the previous example of cell reaggregation, models derived from the two different descriptive sufficient conditions are not isomorphic; each picks out parts and processes in a different way. In the last model, Hensen's node is the crucial part which induces somite formation in the underlying mesoderm; in the former model, Hensen's node is an incidental part of the mechanism having no bearing on the formation of somites. Slightly more advanced forms of the two hypotheses are currently under experimental investigation.

We need now to ask what criteria we utilize to judge whether an initial description is adequate, and what information we use to guide our choice among indefinitely many descriptive sufficient conditions and possible cybernetic and causal models.

In the initial description of the cell reaggregation, and of chick gastrulation, many things which occurred were left undescribed. For example, motions of enzymes within cells, the color of the container in which the system was maintained, and the brownian motion of ions in the medium were not described. Surely, an adequate initial description is already a highly selective description based on our background knowledge about the sort of system and process we are concerned with. The initial description is made in terms of facts and factors we have reason to believe will be the relevant facts and factors, on the basis of our current knowledge and theories. Of course, the initial description may turn out to be inadequate because our suppositions about what is relevant may be incorrect.

There are several interrelated ways in which we use background knowledge to help choose from among diverse descriptive sufficient conditions and cybernetic models those which commend themselves for further serious effort. On the basis of background information, we know a reasonable number of different processes which are, so to speak, stock in trade. For example, cells adhering to one another is such a process. These processes are simple in three senses: (1) They are well known and understood, thus intuitively simple; (2) Many plausible causal mechanisms to realize the process are known or can be supposed, indeed it is just because there are so may causal mechanisms by which cells can adhere to one another that the process 'cells adhere' is very familiar and stock in trade; (3) These processes are simple in the sense that it is by articulating together more than one of these well understood processes that we seek to explain more complex processes. For example, in the hypothetical cell reaggregate, it was permissible to suppose a process 'by which just these cell boundaries form bonds,' for specific bonds between entities is a process which is simple in the specified sense. However, in the chick gastrulation case, the supposition of a process by which cells along the line of the primitive streak, but not laterally, hapened to be the ones to aggregate, would be inacceptable, for this is just the sort of case where we want to see from *other* simple processes, how this complex process might occur.

We therefore choose among the diverse descriptive sufficient conditions those for which we think we can put together a cybernetic model articulating such simple processes. Even if we achieve an articulating cybernetic model, that model might yet be considered implausible for two quite different sorts of reasons. First, the causal mechanisms by which the cybernetic model might be supposed to be realized might be considered implausible in the system under study. Second, even if plausible causal mechanisms might be imaginable, it may be that the cybernetic model demands, in some sense, too much machinery, and it may be implausible to think the system behaves as described initially by such complex means. For example, a descriptive sufficient condition for the initial description of chick gastrulation in terms of a coordinate system is possible, and from it one can create a cybernetic model in which the chick embryo elaborates a coordinate system and cells 'know' where they are and go to the right places. Although causal mechanisms which are plausible might be imagined, the entire construction is likely to be considered an implausibly cumbersome way for a chick to go about gastrulating.

Finally, it must be added that it may not be possible to find a cybernetic model for any of the descriptive sufficient conditions of an initial description which have so far been investigated, for there is no guarantee that, with the currently acceptable set of simple processes, a cybernetic model *can* be formulated. When we are unable to find such a model, one of the most obvious strategies to take is to reexamine the phenomenon we wish to explain and alter our initial description by including new data now hoped to be relevant.

8. Having discussed ways in which we utilize a view of what a system is doing to achieve an articulation of parts decomposition into parts and processes interrelated in particular ways, we now turn to ways in which we utilize information about parts to synthesize new articulation of parts views of what the whole is doing. First, it is clear that new information about a part picked out in an old decomposition of the system can lead to new views of what the system is doing. For example, Harvey's discovery that the blood circulates led to a new view of what the heart, specified on old anatomic grounds, does. The current discoveries that cells and organisms exhibit circadian rhythms, and are, in general, temporally organized, is the focus of debate about the 'significance' of such rhythms. The efforts of this debate are to synthesize a new view of what an organism is doing – essentially to be a timekeeper of some sort. The points to notice about such new syntheses are that, in general, the new synthesis may decompose the system in a new way, cutting across parts specified in an old decomposition. For example, the endocrine system cuts across the old anatomic decompositions; the view of an organism as an open thermodynamic system maintaining low entropy cuts across the decomposition of it as a circadian system.

An important feature of achieving a new synthesis of what a system is doing is that the new view may require no new verification. It may be that true descriptions of behaviors of parts picked out on an old decomposition have accumulated, where these behaviors are irrelevant to the behaviors of the parts which linked them into a system in that old decomposition. It might then be realized that these 'irrelevant' behaviors can be interrelated to achieve a new view of what the system is doing. Since the truth of the descriptions of these irrelevant behaviors is supposed to have been adequately established already, the adequacy of the new composition may not require verification. Of course, the new composition will have *further* consequences which require verification. The 'irrelevant' behavior might have been established in two ways: It may be that, given one description of what the system is doing and a conjugate decomposition into parts and processes, the true descriptions of what the parts are doing, coupled with current theory, may entail the truth of descriptions of other 'irrelevant' causal consequences of the parts. These other causal consequences may be linked together to yield a new synthesis of what the system is doing which would pick out new features of the system and provide an *A* of *P* explanation. The truth of the explanation would be adequately established by the validity of the deductions from the true initial description of the system, and the adequacy of current theory about the stuffs of which the parts were made. On the other hand, we may merely have discovered 'irrelevant' causal consequences of parts on an old decomposition, and established the descriptions of the irrelevant causal consequences by observation, not by deduction from the old descriptions and current theory. These independently established claims about irrelevant behaviors might be linked together to achieve a new composition, or view, of what the system is doing. In either case, the adequacy of the new view does not require verification in the sense in which hypotheses are normally understood to require verification. Of course, normally new compositions often do assert claims about behaviors of parts which are not yet established, and these do require verification. Thus, achievement of a new composition *can* occur without having to reexamine the empirical world; the new composition focuses attention on new features of the system and, presumably, provides a new *A* of *P* explanation of the newly noted behavior. A new behavior is seen as calling for an explanation.

9. We have noted that different views of what a system is doing yield different decompositions of that system into parts and processes. The foregoing discussion has also indicated the ways in which the various true claims made in the diverse decompositions may be related across

decompositions. The claims must, eventually, at least be consistent with one another. However, there is no requirement that the claims made in one decomposition need ever be deducible from those made in another decomposition. That is, the claims must eventually be *compatible* with one another, but not deducible from one another. The set of causal laws sufficient to provide an *A* of *P* explanation of the system seen as doing one thing may not include causal laws necessary for the explanation of a second thing the system may be seen as doing. Thus, the various accounts may proceed independently from one another, with the restriction that they must all, eventually, be non-contradictory.

10. A final point. It may be asked whether there are not preferred views of what an organism is doing, for example, living. I am willing to grant that some views of what an organism does are currently seen as more central to the defined interests of biology, but am not willing to agree that all possible views of what an organism may be seen as doing must, in some sense, be seen as extensions of, versions of, or reducible to those preferred views. To insist, whatever an organism may be seen as doing, that that feature of it is an aspect of a particular view of the organism, say being alive, may be mistaken in each of several ways. It might be the view that only the preferred view of what the organism is doing is to be allowed. It might be the view that whatever happens in an organism is causally necessary for the organism to, e.g. live. Now this claim is almost certainly either false, or an analytic claim masquerading as an empirical one. It might be the view that if the organism is made of just those materials with those particular causal consequences accounting for how it manages to live, then these particular materials have also just the other (indefinitely many) casual consequences. Thus, if the system is to do what it does – e.g. live – then it must do these other things as well. While this version of the claim is true, it does not establish that all views of what a system is doing must be able to be brought under one view of what it is doing and deduced from claims explaining how it does that particular thing. For the truth of the claims on the different accounts need only be compatible, not deducible, from one another. Thus, while the argument above is valid, it does not establish that the diverse account of how the system does the diverse things it can be seen as doing, need be deducible from any one account. The accounts may proceed independently with the restriction that all must eventually be jointly compatible.

Thus, it seems we must admit that many of the features of organisms which we seek to explain are incidental to any particular view of it, even to 'being alive'. Articulation of parts explanations of any way be logically independent, in the sense of not being deducible from many or all the others. There is no reason to restrict the features of organisms we will seek to explain. We can insist that we achieve articulation of parts explanations rather than compilations of true descriptions which fail to articulate. Finally, while we can and do form new inclusive views of what an organism is doing which joins several earlier views in new and rewarding ways, there seems to be no reason whatsoever to insist that all possible views of what an organism is doing and all possible true articulation of parts explanations, need ever be brought under some overarching, ultimate view of what the organism is 'really' doing. For there are, indeed, indefinitely many things which an organism can legitimately be seen as doing, and there is every reason to expect biologists to pursue an ever widening ring of puzzles to explain.

ACKNOWLEDGEMENT

The author wishes to thank Drs. Dudley Shapere and David Hull for helpful criticism.

University of Chicago

III. Equivalence, Analyticity, and In-Principle Confirmability

CLARK GLYMOUR*

THEORETICAL REALISM AND THEORETICAL EQUIVALENCE

> The true correlate of sensibility is not known and cannot be known...
> – KANT

A great many philosophers have thought it impossible that there should exist two distinct theories between which no possible evidence could discriminate. This doctrine, and more rarely its denial, have often been the cynosure of dogmas and disputes about conventions within physical theory, about simplicity, about measurement in quantum mechanics, and most recently, about radical translation. The list is long enough to make it important to know whether or not the doctrine is true. I shall argue that it is, indeed, not true.

The thesis that empirically equivalent theories are synonymous was central to Hans Reichenbach's philosophy of science, and especially to his notion of 'equivalent descriptions' and to his account of simplicity:

> There are cases in which the simplicity of a theory is nothing but a matter of taste or of economy. There are cases in which the theories compared are logically equivalent, i.e., correspond in all observable facts.... For this kind of simplicity which concerns only the description and not the facts co-ordinated to the description, I have proposed the name *descriptive simplicity*. It plays a great role in modern physics in all those places where a choice between definitions is open to us. This is the case in many of Einstein's theorems.... Thus the choice of a system of reference which is to be called the *system in rest* is a matter of descriptive simplicity. It is one of the results of Einstein's ideas that we have to speak here of descriptive simplicity, that there is no difference of truth-character such as Copernicus believed. The question of the definition of simultaneity or of the choice of Euclidean or non-Euclidean geometry are also of this type.[1]

Reichenbach's views on theoretical equivalence have been retained by Professor Putnam, who has suggested that empirically equivalent theories are 'thoroughly intertranslatable',[2] and even more emphatically by Professor Salmon.[3] Adolf Grünbaum,[4] a philosopher who has written a great deal about empirically equivalent geometrics, quotes with approval from the passage of *Experience and Prediction* given above, and claims the correctness of Reichenbach's doctrine of 'equivalent descriptions', which is no other thesis than that empirically equivalent theories always do say the same thing.[5] The view against which I shall argue is, therefore, frequent if not common, and even its few critics have been apostates.[6]

The argument is best begun by giving its point and, in outline, the whole of it. The view I deny is the conjunction of the following theses:

1. The sentences of scientific theories are either true or false; they are not simply 'instruments' for inferring observation statements.

2. If two theories are empirically equivalent, i.e., if any possible piece of evidence for one is likewise evidence for the other, and any possible piece of evidence against one is likewise evidence against the other, then the two theories say the same thing.

There seem to be only two initially plausible ways of defending the conjunction of these claims, and both defenses, I shall argue, must be rejected:

A. The set of all sentences which are statements of possible evidence for a theory form the truth conditions for that theory. Which is to say that a theory is true if and only if all of the observation statements which would be evidence for it are true.

B. The criteria for what is required of a body of sentences in order for it to count as a *theory* are in fact so strong that it is not possible for there to exist two different, empirically equivalent theories.

These are very different lines of defense; the first proposes a novel semantics for the language of science, the second presupposes the usual semantics and, based upon it, a criterion of synonymy for collections of sentences. Against A I shall argue that it makes true what is false and by making logic – in a traditional sense of 'logic' – impossible also makes important aspects of scientific reasoning unintelligible. Against B I shall offer a very weak necessary condition for the synonymy of theories in formalized languages and using it show by counter-examples that reasonable criteria for what is to count as a scientific theory do not suffice to eliminate the possibility of distinct but empirically equivalent theories. Finally I shall mention an example of two geometries which can be interpreted so as to be empirically equivalent but which, I contend, cannot be understood to say the same thing.

What is and what is not a possible piece of evidence for a theory is intolerably fuzzy. The way to keep it clear for our analysis is by dogma rather than by argument. Assume that our scientific theories can be formalized in a first-order language, and that all of the evidence given us by our senses and by our apparatus can be stated using only part of the non-logical vocabulary of the language. It is, for our purposes, of no consequence how this distinction between 'observation terms' and the rest is drawn and justified. Two theories will be empirically equivalent just when every observation statement which is evidence for or against one is, accordingly, evidence for or against the other. We shall assume this is to be the case whenever two theories have the same set of observational consequences, and grasp a palsied justification for this assumption in our ignorance of any organon which would discriminate such theories.

We thus have a crude but usable representation of empirical equivalence. The language of science is a first-order formalized language divided in two by the distinction between observational predicates and the rest. A theory is, at least, a deductively closed collection of sentences in this language, and two theories are empirically equivalent just if they have the same observational consequences. Evidently, this description of empirical equivalence could be given – although not justified – purely syntactically, that is, without using notions of truth, meaning or reference. It is trivial that there can be different deductively closed collections of sentences which contain the same observation sentences, but our obligation is to consider arguments which purport to show that it is nonetheless not true that there are empirically equivalent theories which say different things. Argument A does so by proposing that any two sentences which have the same observational consequences are synonymous, and Argument B reaches the same conclusion by denying that most deductively closed collections of sentences are theories.

The usual semantics for a first-order language gives conditions for the truth of a sentence in a structure for the language, and necessary and sufficient conditions for two sentences of the language to say the same thing. A structure for a language is simply a non-empty set, and for every n-place predicate of the language an n-ary relation on the set. I shall assume your familiarity with what it means for a sentence of the language to be true in a structure, and remind you that two sentences are semantically equivalent just if they are true in exactly the same structures. The semantics 'fits' the logical syntax of the language in a well-known way: a sentence, A, is provable from a sentence, B, just if A is true in every structure in which B is true. It is evident that with the usual semantics there can be empirically equivalent sentences which do not say the same thing.

Reichenbach's remark that empirically equivalent theories are 'logically equivalent' suggests a radical rejection of the usual semantic analysis. It suggests, rather, that we should accept the usual account of truth in a structure only for observation sentences. But, more generally, a sentence will be true in a structure if and only if all of the observation sentences provable from it are true in that structure. Any two sentences, or collections of sentences, from which exactly the same observations sentences are provable will then have the same truth conditions; empirically equivalent theories will be synonymous. This seems a fair account of the tacit semantics in both Reichenbach's and Salmon's accounts of theoretical equivalence, and perhaps also of the semantical views of those philosophers who have been greatly taken with the reaxiomatization theorem of William Craig.

The truth conditions just sketched guarantee the truth of any claim which has no observational consequences other than observational tautologies. This result seems clearly unsatisfactory since there are theories, such as Newton's laws, which we think false but which by themselves have no testable consequences. And there are further difficulties. First order proof theory is a representation of the principles of mathematical reasoning and therefore also of the principles of an important part of scientific reasoning. It is, if you like, an idealized theory of the inference behavior of mathematicians. The importance of the proof theory is evidenced by the natural and enlightening ways in which mathematical theories have been given first-order formalizations. The usual semantics for first-order logic is both an explanation and a partial justification of those principles of inference. If one accepts A and sentence B is provable from A then we accept B, and there is good reason. For if A is true then B is true. Conversely, if B is certainly true when A is true then there is a proof of B from A. This important function of a formal semantics is lost if one takes seriously the new truth-conditions just outlined. The new semantics provides an interpretation of classical logic which is not complete: any formula having only observational tautlogies as consequences is valid but not necessarily provable. Nor is the proposed semantics sound with respect to arguments. Every sentence is provable from a sentence A

and its negation but it is not necessarily the case that every sentence is semantically entailed by such a pair.

Perhaps such considerations only show that classical proof theory is no more tenable than classical semantics. A new account of truth, it might be thought, deserves and requires a new account of proof. Aside from the unlikelihood that a theory of proof agreeing with the proposed semantics would also agree with mathematical inference, there is an important difficulty with such a suggestion. Traditionally, the notion of proof has been required to be effective: there must be a recursive procedure for deciding what strings of formulae are and are not proofs. But the new semantics is not adequate for any proof theory employing an effective notion of proof. A demonstration is given in the appendix.

I shall not pursue argument A any further, for its handicaps seem insuperable. The development of argument B requires conditions for theoretical equivalence which are so framed as to accord with first-order syntax and semantics. For theories expressed within the same language we already have a sufficient condition for theoretical equivalence, namely logical equivalence. But we have heard too much of 'meaning variance' and the like to be comfortable in the assumption that different theories can reasonably be represented in the same language. For generality, then, whenever I mention a pair of theories I will assume their theoretical vocabularies to be disjoint.

The natural necessary condition for the equivalence of two theories framed in different languages is that they be intertranslatable. What that means, at least, is that the two theories have a common definitional extension. For the equivalence of theories T and Q we require a set of definitions D_T of the predicates of Q in terms of open formulas of the language of T, and similarly a set D_Q of definitions of the predicates of T in terms of open formulas of the language of Q. Moreover, the theories and definitions must be such that when D_T is added to T and D_Q to Q the two resultant theories are logically equivalent. Such a criterion guarantees that all and only theorems of T are translated as theorems of Q, and conversely; if T and Q can be axiomatized as single sentences, it is also guaranteed that the translation (into the language of T) of the translation (into the language of Q) of T is logically equivalent to T. Stronger necessary conditions – for example that the composition of translations from the language of T to that of Q and back to T should take every formula into a logically equivalent formula – might conceivably be required, but weaker conditions do not seem likely.

Nonetheless, it is clearly to the advantage of those who like argument B to weaken the necessary conditions for theoretical equivalence; and an argument can be made for weakening the condition just given. Informal number theory is about a specific structure, the natural numbers. But formal number theory has a great many non-standard models; a plethora of statements about the natural numbers which are not provable in formalized number theory are truths of number theory nonetheless. In short, there is more to number theory than its axioms and their models; there is, in addition, an intended model. The same might be said for formalizations of empirical theories; they too may have intended models which are not uniquely characterized by their axioms. But should not two theories with the same intended model be thought equivalent even though they are not intertranslatable?

I grant the objection although I am not sure that I agree with it. Let us take it as sufficient for the equivalence of two theories that they have the same intended model; but to have the same intended model two theories must have at least one model in common, namely the intended one. We must still say what it is for theories in different languages to have a mode in common since the usual notion of a model is relative to a language. Again, however, there is a natural account available. We shall say that two theories, T and Q, have a model in common just if there are models M_T and M_Q of T and Q respectively such that the set D^T of all sentences in the language of T which are true in M_T and the set D^Q of all sentences in the language of Q which are true in M_Q are intertranslatable in the sense given previously. In brief, two theories have a model in common just if they have models whose diagrams have a common definitional extension.[7]

The justification for this account of what it is to have a model in common is straightforward. It follows from the definition that two theories, T and Q, have a model in common just if they have complete consistent extensions which in turn have a common definitional extension, call it TQ. Since TQ is consistent it has a model M, intuitively one of the models which T and Q have in common. The model M is a set with relations corresponding to the predicates in the language of T and relations corresponding to the predicates in the language of Q. The structure M_T, got by dropping the relations corresponding to predicates of Q, is a model of T; similarly, the structure M_Q is a model of Q. But M can be regenerated from either M_T or M_Q by defining the relations which were dropped in terms of the relations which remain.

We have a clear and mathematically useful necessary condition for theoretical equivalence; it is, moreover, a condition which does not beg important questions about the comparability of terms in different theories. Using this condition we have to examine the thesis of argument B, namely that the restrictions on what a collection of sentences must be to be a theory are so strong that all empirically equivalent theories are also equivalent. Since it is trivially possible to give examples of empirically equivalent collections of sentences having no model in common, the claim of B needs arguing. In order for a restriction on what theories are and are not to lend any support to the claim, the restriction must have formal implications. Unless they are of a patently trivial kind, historical, sociological, or psychological restrictions will, of themselves, not limit the variety of structurally different theories which it is possible to construct; neither will material restrictions on the interpretation of the theoretical predicates be of help. It will not advance the case to require, for example, that theories refer only to entities that have a spatio-temporal location, or that for a claim to be a theory it must be possible for someone to believe it, and so on. For such demands, justifiable or not, place no evident restriction on the structure of theories, and if it is to be shown that empirically equivalent theories always have a model in common then some structural restrictions are necessary.

Philosophical papers contain very few formal criteria for a theory, and it is clearly a hopeless task for me to attempt to invent and to study all of the criteria which could conceivably be proposed. What I shall do is to study a criterion which has been explicitly proposed by William Kneale,[8] and tacitly used by a great many others. It is this:

A theory must be finitely axiomatizable, but its collection of observational consequences must not be finitely axiomatizable.

It is not because I think this principle is true that I shall examine it, for it is evidently not true. There are perfectly good theories which appear to have no observational consequences, and there are other, equally good, theories which are not finitely axiomatizable. But the proposed restriction on theories is worth investigating just because, besides being simple, it is too strong in important respects.

The requirement does not suffice to eliminate the possibility of empirically equivalent theories having no models in common. Indeed we have the following result:

If T is any theory in a first-order predicate language (with identity) and every complete, consistent extension of T is decidable, and the collection of observational consequences of T has, up to isomorphism, at most a finite number of finite models, then there is a finitely axiomatizable theory Q which is observationally equivalent to T but has no models in common with T.

A proof of this claim is immediate from work of Kleene, Craig and Vaught and is given in the appendix. Their proofs show, in effect, how to construct such a theory, Q, given the observational consequences of theory T. The theories they construct meet Kneale's criterion whenever the theory T does, but they are perverse theories nonetheless. In effect, the theories which Kleene, Craig and Vaught construct are truth theories for the observation language which say, in addition, that a specified set of observation statements are true. Although Kneale's criterion does not eliminate such systems as theories, it seems entirely reasonable to do so, for they evidently explain nothing. Just what we prohibit by refusing to to regard truth theories as theories is not entirely clear. To show that the requirements still do not eliminate the possibility of empirically equivalent theories, we must, therefore, settle for giving examples where the theories involved are clearly not truth theories. Such examples can easily be given.

Suppose the relevant evidence can all be stated with the identity predicate, and suppose, when all the evidence is in, what it says is that there are an infinite number of objects. Here are two explanations of the evidence which meet all of the required conditions.

T1
$$\forall x \sim L(x, x)$$
$$\forall x, y, z((L(x, y) \,\&\, L(y, z)) \rightarrow (x, z))$$
$$\forall x, y(L(x, y) \vee (x = y) \vee L(y, x))$$
$$\forall x, y \exists z(L(x, y) \rightarrow (L(x, z) \,\&\, L(y, z)))$$
$$\forall y \exists x L(x, y)$$
$$\forall y \exists x L(y, x)$$

T2 $\forall x \sim B(x, x)$

$\forall x \exists y B(x, y)$

$\exists x \forall y (\sim B(y, x) \,\&\, (\forall z \forall s (\sim B(s, z)) \rightarrow z = x))$

$\forall x, y, z ((B(x, y) \,\&\, B(x, z)) \rightarrow z = y)$

$\forall x, y, z ((B(x, z) \,\&\, B(y, z)) \rightarrow x = y)$

*T*1 is the theory of dense order; it is complete and obviously has no finite models.[9] *T*2 is a fragment of the theory of the successor relation;[10] there are no finite models for the theory, and moreover, no finitely axiomatizable extension of *T*2 is complete. It follows immediately that *T*1 and *T*2 have no model in common. Many other examples of a like kind could be given.

I see no possible fault with either of these explanations, aside from the fact that what they explain is rather trivial. Of course, such examples provide no demonstration that empirically equivalent theories having no model in common can always, or even often, be found, but then no such demonstration can be given until philosophers provide a precise account of what is to qualify as a theory and what is not.

Thus far nothing has been said about simplicity, and that is all to the good. Criteria for simplicity are irrelevant to the issue unless they restrict what can be a theory. Thus it might be required that a theory not have eliminable predicates, that is, predicates such that the theory contains an empirically equivalent sub-theory which does not use these predicates. But that is a rather trivial requirement and one which our counter-example already meets. Still, in keeping with Kneale's criterion, one might hope that, for a given non finitely axiomatizable collection of observation statements, there exists a *simplest* explanation. If such a simplest explanation did exist, it might be taken as the one and only *theory* explaining the given body of observation statements. No philosopher has to my knowledge, proposed such a canon of simplicity, nor have I succeeded in divising one. Nonetheless, I shall mention a simplicity criterion which may occur to some, but which does not work.

Let us say that, within a given language, one sentence *A* is simpler than another, *B*, if *A* entails *B* but not conversely. The simplest explanation in a fixed language of a body of observation statements would, then be that sentence entailing the observation statements and entailed by every other sentence entailing the observation statements. Such an account could easily be extended to provide simplicity comparisons of theories with different theoretical vocabularies, e.g., by taking theories to be equi-simple just if they have common definitional extensions. In a fixed language with only one binary predicate besides identity, the simplest explanation of the existence of an infinite number of objects would, then, be that sentence having only infinite models which is entailed by every other sentence having only infinite models. A proof that no such theory exists is given in the appendix. More generally, it is as a reasonable conjecture that given any first order language L, with identity, and any recursively enumerable but not finitely axiomatizable collection Δ of sentences in a proper sub-language of L, there is no sentence A in L such that A entails all sentences in Δ and every sentence B which entails all sentences in Δ entails A.

Kneale's criterion and modifications of it provide no support for the claim of argument B. Neither does there seem any point in working through other possible restrictions on theories one by one, all the while manufacturing toy counter-examples. What is needed in order to lay the *geist* of the doctrine is an example of two *actual* theories which are empirically equivalent but have no model in common. We can lay aside, as well, the dubious premise that there is an observation language. Most will, I hope, admit that physical geometries are actual theories. They are also theories which have been formalized and the metamathematics of which have been studied. Hans Reichenbach has argued, very convincingly I think, that almost any pair of geometries can be interpreted so as to be empirically equivalent.[11] In particular he has argued that, by changing his views about identity, a person who lives in a topologically compact space can consistently maintain a geometry which requires a non-compact topology. The example Reichenbach gives concerns a torus-space interpreted as Euclidean. However, completely analogous arguments can be given for any pair of geometries, one compact the other not, so long as the compact topology is homeomorphic to a quotient topology obtained from the non-compact topology. In particular, Reichenbach's argument can be straightforwardly reproduced for the case of elliptic and Euclidean (or hyperbolic) geometries. The elementary first-order fragments – that is, everything that can be said without set theoretic devices – of elliptic and Euclidean geometries have been formalized and studied. They have no model in common. (See Appendix.)

The argument I have made is not so complete as I should like, nor is its epistemic significance so clear. It is an objection, and a good one, that elementary geometry is only part of geometry not the whole of it. The introduction of set-theoretic devices brings with it, we know, a great explosion in the variety of structures which are models for a geometry. The argument would only be complete if it could be shown that reasonable formalizations of elliptic and Euclidean geometries (or some other pair), based on set theories with geometric points taken as ur-elements, have no model in common. My guess is that the argument can be completed in this way. For those who are realists but not Platonists, the argument is perhaps already good enough.

There are philosophers who find Reichenbach's arguments concerning the empirical equivalence of geometries convincing enough, but only so long as the geometries have the same topology. Professor Grunbaum seems to hold this view, although it is difficult to be sure. For them it would be pleasant to show that elementary Euclidean and hyperbolic geometries have no model in common, but I have no such demonstration. Still, since each of these theories is complete, and therefore has up to elementary equivalence only one model, it would indeed by surprising if they had a model in common.

It may well be that the discussion in this essay presupposes what is false, namely that there is an intelligible notion of empirical equivalence. But if the presupposition is correct, then the admission that there are empirically equivalent theories which are not synonymous seems to entail either that the true theory is sometimes unknowable or that, more simply, even all possible evidence can sometimes have more than one correct explanation.

APPENDIX

1. The claim is that the semantics proposed does not admit a proof theory with an effective notion of proof. To show this it is sufficient to show that the consequence relation generated by the proposed truth conditions is not recursively enumerable (r.e.); thus:

Let L be a first-order predicate language (with identity), V a proper sub-language of L such that there is at least one binary predicate (other than identity) which is in L but not in V. Let **Cn** be a relation on L such that for all sentences A, B in L, **Cn** (A, B) holds if and only if for every sentence C of V, C is provable from A only if C is provable from B. Then **Cn** (x, y) is not r.e.

Proof: Let T be a sentence of L which is undecidable and such that the consequences of T expressible in V alone are complete in V. A sentence Q of L is then refutable from T just if **Cn** $(A \mathbin{\&} \neg A, T \mathbin{\&} Q)$ is true. Again, a sentence Q of L is irrefutable from T just if **Cn** $(T \mathbin{\&} Q, T)$. Assuming that **Cn** is r.e., it follows that both the set of sentences refutable from T and the set of all sentences irrefutable from T are r.e. Since each of these sets is the complement of the other, both are recursive. Hence the set of all sentences refutable from T is recursive so T is decidable, which is a contradiction. This argument is essentially due to Richard Grandy.

2. The claim is that, given an axiomatizable theory T in a first-order language L (with identity and a finite number of predicates) with a proper sub-language V, if all of the completions of T are decidable and the set of consequences of T expressible in V alone has at most a finite number of non-isomorphic finite models, then there is a theory Q, in a language $\mathscr{L}$ having V as a proper sub-language but otherwise disjoint from L, which has the same V-consequences as T but no models in common with T.

Proof: By Craig's reaxiomatization theorem, the V-consequences of T are recursively axiomatizable. Craig and Vaught[12] have established the following: Let Δ be a recursively axiomatizable collection of sentences in a language V having at most a finite number of predicates. Further assume that Δ has at most a finite number of non-isomorphic finite models. Then there is a finitely axiomatizable extension Q of Δ in a language $\mathscr{L}$ got by adding a single binary predicate to V. Moreover Q is essentially undecidable. Let T be as in the claim above and Q the Craig-Vaught extension of the V consequences of T. Then Q and T have no model in common. For, supposing the contrary, then there exists a completion T^* of T and a completion Q^* of Q, having a common definitional extension. The intertranslation provides a recursive mapping $\phi: \mathscr{L} \rightarrow L$ such that for every sentence A in $\mathscr{L} Q^* \vdash A$ if and only if $T^* \vdash \phi(\mathrm{A})$. Since T^* is, by hypothesis, decidable, Q^* is therefore also decidable. But this is impossible since Q is essentially undecidable.

3. The claim is that in a language with identity plus a binary predicate there is no sentence which has only infinite models and which is entailed by every sentence having only infinite models. For suppose there were such a sentence, *A*. Then if an arbitrary sentence *S* had only infinite models we could effectively find that it had only infinite models by searching through proofs from *S* for a proof of *A* from *S*. Hence the set of all sentences having only infinite models would be recursively enumerable. By a theorem of Vaught's[13] this set is not r.e. I am indebted to Andrew Adler for this argument.

4. That elementary elliptic and Euclidean geometry have no model in common is an immediate consequence of a theorem due to R. Robinson.[14] Robinson shows that elementary elliptic geometry can be expressed in a language with a single binary predicate, besides identity, whereas elementary Euclidean and hyperbolic geometry cannot be. Since they are complete, to have a model in common elliptic and Euclidean geometry would have to be intertranslatable. Robinson's result is essentially a proof that they are not.

Princeton University

NOTES

* I am indebted to Dr Andrew Adler, of the Department of Mathematics, University of British Columbia, and to my colleague Dr Richard Grandy, for many hours of discussion on both philosophical and technical aspects of this paper. Professors Wesley Salmon, Carl Hempel, and Dana Scott kindly gave me useful comments on drafts of this essay.

[1] H. Reichenbach, *Experience and Prediction*, University of Chicago Press, 1961, pp. 374–75.

[2] H. Putnam, 'An Examination of Grünbaum's Philosophy of Geometry' in *Philosophy of Science. The Delaware Seminar*, Vol. 2 (ed. by B. Baumrin), 1963.

[3] W. Salmon, 'Verifiability and Logic' in *Mind, Matter and Method* (ed. by P. Feyerabend and G. Maxwell), 1966.

[4] A. Grünbaum, 'Geometry, Chronometry, and Empiricism', in H. Feigl and G. Maxwell, eds., *Minnesota Studies in the Philosophy of Science*, Vol. III (ed. by H. Feigl and G. Maxwell), University of Minnesota Press, 1962, p. 446.

[5] (Added in proof). In an essay forthcoming in *Philosophy of Science* (December, 1970), Professor Grünbaum gives up the doctrine of equivalent descriptions. Grünbaum distinguishes claims which describe 'intrinsic' facts from those which do not. He now holds that empirically adequate, empirically equivalent theories do not necessarily say the same thing but do say the same thing about the 'intrinsic' facts. If it is the case, and I do not know whether it is, that two theories making different claims about the 'intrinsic' states of affairs cannot be empirically equivalent, then Grünbaum's most recent views do not contradict what I shall say in this essay.

[6] Thus, compare W. V. O. Quine's more recent writings with the views on theoretical equivalence expressed in 'Semantics and Abstract Objects', *Proceedings of the American Academy of Arts and Sciences* **80** (1951). Similarly in *Patterns of Discovery*, N. R. Hanson espoused the view I shall criticize. He later gave it up; see his 'Equivalence: The Paradox of Theoretical Analysis' in Feyerabend and Maxwell, *op. cit.*
[7] An equivalent notion is given in K. de Bouvere 'Synonymous Theories', in *The Theory of Models* (ed. by J. Addison *et al.*), 1965.
[8] W. Kneale, *Probability and Induction*, 1949.
[9] See J. Schoenfield, *Mathematical Logic*, 1969, p. 102.
[10] The example and its properties are due to William Hanf, 'Model Theoretic Methods in the Study of Elementary Logic', in *The Theory of Models* (ed. by J. Addison *et al.*), *op. cit.*
[11] H. Reichenbach, *The Philosophy of Space and Time*.
[12] W. Craig and R. Vaught, 'Finite Axiomatizability Using Additional Predicates', *Journal of Symbolic Logic* **23** (1958).
[3] R. Vaught, 'Sentences True in All Constructive Models', *Journal of Symbolic Logic* **25** (1960).
[14] R. Robinson, 'Binary Relations as Primitive Notions', in *The Axiomatic Method* (ed. by L. Henkin *et al.*), 1959.

JOHN A. WINNIE

THEORETICAL ANALYTICITY

The logical analysis of the structure of scientific theory is a task which is central to any comprehensive philosophy of science, and in this sphere no philosopher has contributed more incisively than Rudolf Carnap. The clarity and rigor of his analyses have always made them apt candidates for fruitful discussion and criticism, and Professor Carnap's last proposal dealing with the problem of theoretical analyticity is, I hope to show, no exception to this rule. For my aim here is to take advantage of the rigor of Carnap's formulation by developing its consequences in some detail. The results which emerge from such a development are then employed to overcome some of the more weighty objections against the possibility of drawing an analytic-synthetic distinction for theoretical systems.

The solution which Carnap has offered to the problem of theoretical analyticity derives from his more general account of the logical structure of physical theory (cf. [1] and [4]). According to this account, the descriptive terms of the theory to be reconstructed are initially divided into two mutually exclusive classes: the class of observation terms and the class of theoretical terms. The basis of this division is extra-systematic and, admittedly, not a clear-cut matter. Carnap sometimes speaks of observation terms as those which refer to easily ascertainable properties and relations of middle-sized physical objects, yet he also characterizes an observation term as any term which occurs in sentences of a language 'used by a certain language community', and 'understood by all members of the group in the same sense' ([1], p. 40). Clearly, the two characterizations will not always coincide. Moreover, in his *Philosophical Foundations of Physics*, Carnap declares that observational and theoretical concepts "lie on a continuum", and goes on to suggest that the distinction best be drawn by looking to the linguistic practice of the physicists who employ the theory which is the object of our reconstruction ([4], pp. 225–28).

So although the general framework of Carnap's reconstructions presupposes *some* sort of observational term-theoretical term division, it does not in itself dictate the extra-systematic criteria to be employed in making this division. In particular, the mere adoption of the general distinction does not commit us to drawing this distinction on an *ontological* basis with the all-too-common corollary that theoretical concepts are henceforth ontologically suspect. Moreover, if, like Carnap, we explicitly disavow any distinction drawn along ontological lines, we may thereby acknowledge the strength of the serious objections which have been raised against attempts to *ontologically* ground this distinction (cf. [7]), while at the same time allowing for the fundamental *epistemological* and *semantical* importance of observation statements within scientific theory. But this last thesis will not be argued here.[1] Instead, it will here be assumed that a viable and significant observation term-theoretical term distinction has been suitably drawn in a way which is compatible with the over-all character of Carnap's general reconstruction.

The observation term-theoretical term distinction now permits the following three-fold classification of the sentences of the theory. Those sentences which contain observation terms as their only descriptive signs are called 'observation sentences'; sentences which contain theoretical terms as their only descriptive signs are called 'theoretical sentences'; and sentences which contain both observation and theoretical terms are called 'mixed sentences'. When a theoretical sentence is taken as one of the postulates of a reconstructed theory, it then is called a 'theoretical postulate', and when a mixed sentence is so-taken, it is said to be a 'correspondence postulate' of the theory (cf. [1], pp. 42 ff).

Carnap next goes on to propose that a physical theory be reconstructed in such a way that its postulates become a conjunction of the form $T \cdot C$ (or, more simply, TC), where T and C are each in turn conjunctions of theoretical and correspondence postulates, respectively. Whereas the theoretical postulates (T) merely assert relations between the theory's theoretical concepts, the correspondence postulates relate the theoretical and observation terms of the theory, thus providing the theoretical terms with empirical significance and conferring testability upon the theory as a whole. Hence the correspondence postulates play a dual role: they both lawfully relate the theoretical and the observational and at the same time serve the semantical function of empirically interpreting the theory's theoretical terms. Moreover, once this dual function of the correspondence postulates is recognized, it is but a small step to see that theoretical postulates play this double role as well. For Carnap does not require that there be a correspondence postulate associated with each one of the theory's theoretical terms, and thus those theoretical postulates which involve a term not contained in some correspondence postulate must also serve, however indirectly, to provide these otherwise-dangling theoretical terms with empirical significance.

The fact that Carnap does not require each theoretical term to be associated with a correspondence postulate is generally regarded as one of the virtues of his proposal. Previous proposals which have required either the explicit definability of each theoretical concept in terms of observational concepts alone, or have demanded the introduction of each theoretical term by means of reduction sentences, are now recognized to have been inordinately severe. Nevertheless, it is just the more holistic and liberal character of this framework that provides a basis for serious and reasonable doubts as to the possibility of drawing a sharp analytic-synthetic distinction within this framework. For as Hempel has pointed out ([6], p. 703), the nomological and interpretative functions of the theory's postulates seem to have become so tightly fused as to make their sharp separation impossible. To hold that a statement of the theory is analytic is surely to consider that statement as *devoid* of factual content; but then we must deny that statement any *nomological* function within the theory, and the latter, of course, is in direct conflict with our previous recognition that each of a theory's postulates serves, to some extent, a nomological function.

It is this difficulty which appears to have prompted Carnap's proposal. If I am right in what follows, Carnap's response was successful, and, this response in addition provides the means to cogently rebut Quine's most telling objections to the possibility of drawing an analytic-synthetic distinction within formalized languages. In order to substantiate these claims, let us now consider the details of the Carnap proposal itself.[2]

Consider a physical theory τ whose axioms are expressed as a conjunction of theoretical postulates T and correspondence postulates C. Then we may write the theory as

$$(1) \qquad TC = (\text{---}T_1, \ldots, T_k\text{---}) \cdot (\text{---}O_1, \ldots, O_j;\ T_1, \ldots, T_k\text{---}),$$

where the first conjunct represents its theoretical postulates T, and the second conjunct represents the correspondence postulates C. The Ramsey sentence of the theory is now obtained by existentially quantifying over each of the theoretical terms of TC in succession. Thus the Ramsey sentence of TC becomes, in our notation:

$$(2)\qquad R^{TC} = (Et_1)\ldots(Et_k)\;[(\text{---}t_1, \ldots, t_k\text{---})\cdot(\text{---}O_1, \ldots, O_j;\, t_1, \ldots, t_k\text{---})].$$

As is well-known, the Ramsey sentence of the theory, although not containing any if the theory's theoretical terms, entails just those observation sentences entailed by the theory TC. This result has led some philosophers, Carnap included, to construe the theory's Ramsey sentence as expressing the entire factual content of the theory (see also [10] and [12]).

In keeping with this view, Carnap now proposes that any adequate solution to the problem of analyticity proceed by re-expressing theory TC in the form

$$(3)\qquad F^{TC}\cdot A^{TC},$$

where A^{TC} is to be taken as the theory's meaning postulate. The postulate A^{TC} and all of its logical consequences are now considered to be analytic within that theory.

Any specific proposal for making this division must, however, satisfy the following three adequacy conditions ([3], 965). First, the conjunction $F^{TC}\cdot A^{TC}$ must be logically equivalent to the theory TC; next, the first component, F^{TC}, must entail exactly those observation sentences entailed by TC itself; and finally, the second component, A^{TC}, must have null observational content. These conditions may be summarized as follows:

$$(\text{C1})\qquad F^{TC}\cdot A^{TC} \Leftrightarrow TC,$$

$$(\text{C2})\qquad F^{TC} \Rightarrow (\text{---}Ob\text{---}) \quad \text{if and only if} \quad TC \Rightarrow (\text{---}Ob\text{---}),$$

and

$$(\text{C3})\qquad \text{if} \quad A^{TC} \Rightarrow (\text{---}Ob\text{---}), \quad \text{then} \quad (\text{---}Ob\text{---}) \text{ is logically true}.$$

Carnap now proposes that we take the theory's Ramsey sentence for the first component (F^{TC}), and for the second component (A^{TC}), he suggests the rather curious sentence

$$(4)\qquad C^{TC} = (R^{TC} \rightarrow TC),$$

henceforth to be called 'the Carnap sentence' of the theory. The theory TC is now written as the conjunction of its Ramsey sentence and its Carnap sentence, i.e., in the form

$$(5) \qquad R^{TC} \cdot (R^{TC} \rightarrow TC),$$

with the analytic sentences of the theory now becoming the class of logical consequences of the theory's Carnap sentence. It may now be demonstrated, as Carnap points out, that this proposal satisfies each of the three conditions of adequacy previously laid down (cf. A, Thms. 1 and 2, and Cor. 3).

The above proposal will henceforth be referred to as Carnap's *basic proposal*. For Carnap does go on to remark that should we wish to establish *additional* sentences S of the theory as analytic, we may, if we like, conjoin them to the Carnap sentence provided that the resulting statement $(C^{TC} \cdot S)$ continues to satisfy the third adequacy condition (C3) set down above ([3], 965). Thus suggestion will be referred to as the *extended proposal*, and I shall soon argue that the very features which make for the attractiveness of Carnap's *basic* proposal are those which tell *against* the extended proposal. However, before considering these matters in more detail, let us examine the basic proposal in a preliminary way with a view to dispelling some of the curiosity which surrounds the Carnap sentence itself.

I have previously mentioned the fact that the Ramsey sentence of a theory entails exactly those observation sentences entailed by the theory TC. This fact has led some writers to the view that a physical theory might best be regarded as asserting no more or less than what is asserted by its Ramsey sentence. At times, the motives for such a claim are tied in with an instrumentalistic approach to theoretical entities (cf. [12]), but this need not be the case. Professor Hempel, for example, has written that Ramsey's method "is perhaps the most satisfactory way of conceiving the logical character of a scientific theory...", while at the same time disavowing an instrumentalistic construal of the Ramsey sentence ([5], p. 85; cf. also [7]).

Now although a theory TC logically entails its Ramsey sentence, the converse does not, in general, hold as well. Hence in order to assert the equivalence of a theory and its Ramsey sentence, it suffices to maintain that the Ramsey sentence in some sense entails the theory TC. It is just this that Carnap proposes when he suggests that we consider $R^{TC} \rightarrow TC$ to be analytic. For notice that, by virtue of the fact that the theory TC logically entails its Ramsey sentence, it follows at once that the Carnap sentence is logically equivalent to asserting the equivalence of a theory and its Ramsey sentence, i.e.,

$$(6) \qquad C^{TC} \Leftrightarrow (R^{TC} \equiv TC).$$

Thus Carnap's proposal may aptly be characterized as a formalized version of the philosophical thesis that a theory and its Ramsey sentence are equivalent.

Still another way of viewing Carnap's proposal is to consider its semantical aspects. An analytic sentence ought to hold in all possible worlds i.e., in all possible models of the theory. Since the Carnap sentence is to be taken as the analytic postulate of the theory, we are thereby, in effect, excluding all structures in which the Carnap sentence *fails* to hold from the sphere of possible models. Moreover, since the Carnap sentence is of the form $R^{TC} \rightarrow TC$, the structures thus excluded are those in which the theory's Ramsey sentence holds while the theory TC fails. This is, of course, just what we should expect upon equating the theory to its Ramsey sentence.

This semantical characterization of the basic proposal is reminiscent of Carnap's earlier 'meaning-postulate' approach to analyticity. There, all state-descriptions in which some meaning postulate failed to hold were excluded from the class of possible models of the theory which incorporated these postulates (cf. [2], p. 226). However, these models are so excluded merely on the basis of a list of sentences dubbed 'analytic' within that theoretical reconstruction. Thus, as Professor Quine has pointed out ([8], pp. 32ff) the lack of generality which attaches to this procedure hardly helps to alleviate any qualms about the possibility of drawing a *non-arbitrary* analytic-synthetic distinction in this manner.

Herein lies an important way in which Carnap's last proposal differs from its predecessor. For clearly, a *general* account of theoretical analyticity is now provided which is applicable to any appropriately reconstructed theory, and which yields a clear determination of those sentences which are to be considered as analytic within the theory. Furthermore, unlike the syntactical notions of 'postulate' and 'list of meaning postulates' the class of analytic sentences singled out by the basic proposal is invariant under distinct but logically equivalent axiomatizations of a physical theory τ. For let T_1C_1 and T_2C_2 be logically equivalent axiomatizations of a physical theory which utilize non-equivalent theoretical and correspondence postulates (i.e., $T_1 \not\Leftrightarrow T_2$ and $C_1 \not\Leftrightarrow C_2$). Since T_1C_1 and T_2C_2 are logically equivalent, so are their respective Ramsey sentences, and, as a result, their Carnap sentences are also logically equivalent (i.e., $(R^{T_1C_1} \rightarrow T_1C_1) \Leftrightarrow (R^{T_2C_2} \rightarrow T_2C_2)$). Hence the class of analytic sentences is the same for both axiomatizations, showing that the arbitrariness which attaches to the notion of 'postulate of τ' does not also attach to the notion of 'analytic in τ' (cf. [8], p. 35).

Yet, it must be admitted that a defense of the basic proposal on grounds of generality alone does not go to the heart of Quine's objection. For a proposal which urged us to construe, say, every sentence of a theory as analytic would, for all its generality, be absurd. And it is clear that Quine considers the lack of generality which characterizes the 'list approach' to be a symptom of the underlying *arbitrariness* of any distinction so-drawn. Thus in order to adequately deal with Quine's objections we must provide a *rationale* for taking the *Carnap sentence* and its consequences as analytic, rather than so-taking some *other* set of sentences of the theory.

Herein lies a fundamental difficulty for the *extended* analyticity proposal. Recall that, according to the extended proposal, we may add meaning postulates so long as the result continues to satisfy the third adequacy condition, i.e, so long as the resulting extended meaning postulate entails no contingent observation sentence. However, the following results indicate that the extended proposal is far too liberal, and that its acceptance yields a distinction which is to a great extent arbitrary.

Let us call a sentence *properly analytic* if it is analytic but not merely a logical truth. According to both the extended and basic proposals, all logical truths will be analytic, and thus what is of interest here is how the *properly* analytic sentences compare under each of these proposals. Now it can be shown (cf. A, Cor. 9) that *no* theoretical sentence of a theory is properly analytic under the basic proposal, i.e., only logically true theoretical sentences follow from the Carnap sentence alone.[3] Thus, should we take, in addition to the Carnap sentence, *any* set of the theory's theoretical consequences as meaning postulates, we would be genuinely extending the class of the theory's analytic statements, rather than merely appending sentences already entailed by the Carnap sentence. But it can also be shown that if we were to append *any* conjunction of the theory's theoretical statements to the Carnap sentence, the result will in *all* cases satisfy Carnap's three adequacy conditions (cf. A, Th. 5, Th. 6). Thus, according to the extended proposal, all or none of the theory's theoretical statements may be justifiably construed as analytic. The decision as to just which theoretical sentences to so choose must then remain a matter of arbitrary choice.

While these considerations provide grounds for the rejection of the extended proposal, they do not, of course, afford reasons for the acceptance of the basic proposal, and clearly we must go beyond the three adequacy conditions set down by Carnap in order to provide such a defense. As a preliminary to this task, the notion of an *observationally vacuous* sentence of a theory will now be introduced.

A well-known feature of logical truths is that they occur inessentially or vacuously as premisses in any valid argument. In other words, any correct derivation which utilizes logical truths in its premisses remains a valid argument (although not necessarily a correct derivation) when the logically true premisses are dropped. Now an observationally vacuous sentence of a theory plays a role with respect to the derivation of the theory's *observational* consequences which is analogous to the role played by logical truths in *all* derivations. Observationally vacuous sentences are those which function inessentially in *any* deduction of an observational consequence from the theory. This is what is stated a bit more precisely by the following definition.

(7) *S is observationally vacuous in TC*=df.
$TC \Rightarrow S$, and for any S_1 such that $TC \Rightarrow S_1$, if
$(S_1 \cdot S) \Rightarrow (\text{---}Ob\text{---})$, then $S_1 \Rightarrow (\text{---}Ob\text{---})$. (cf. A, D. 7ff).

Before applying this notion to the analyticity problem, let us first consider its relation to the similar, but weaker, notion of observational uncreativity. Recall that a sentence is said to be observationally uncreative, or to have null observation content, if all its observational consequences are logically true. Carnap's third adequacy condition thus amounts to requiring that the sentence taken as a theory's meaning postulate be an observationally uncreative sentence. As we have seen, the Carnap sentence is observationally uncreative, but so is its conjunction with any theoretical consequence of the theory. Now although any observationally vacuous sentence of a theory is also observationally uncreative, the converse does not hold in general (cf. A, Cor. 10). To see this, note that for a sentence S to be observationally uncreative, it suffices that S *in isolation* yield no contingent observation sentence. But for a sentence S to be observationally vacuous, it is necessary that S, *even in combination with other consequences of the theory*, always play an inessential role in the deduction of contingent observation sentences from the theory.

This observation inutility of such sentences leads at once to the conclusion that, under any reasonable construal of 'confirmation' or 'disconfirmation' sentences which are observationally vacuous in a theory can be neither confirmed nor disconfirmed by observational results. A negative observational outcome cannot be attributed in any way to the incorrectness of an observationally vacuous premiss, since the same prediction results when the observationally vacuous premiss is dropped from the deduction. For the same reason, a positive observational outcome cannot be held to confirm an observationally vacuous premiss.

Since immunity from experimental confirmation or disconfirmation is surely an essential ingredient of the pre-systematic notion of analyticity, it is then reasonable to propose that we supplement Carnap's adequacy conditions by adopting the additional requirement that the meaning postulate of the theory be observationally vacuous, i.e.,

(C4) A^{TC} is observationally vacuous in TC.[4]

At first sight, it might seem that this requirement is too strong, and could result in logical or mathematical truths as being the only qualifiers for analyticity (cf. [6], p. 705). However, it turns out that such fears are unjustified. For it can be shown that (and this is the main result of this paper) *the Carnap sentence of a theory and its logical consequences constitute exactly those statements which are observationally vacuous in that theory*. (cf. A, Cor. 12). Thus a theory's Carnap sentence, and only its Carnap sentence, satisfies the added fourth adequacy condition. Let us now assess the consequences of this result for the general problem of theoretical analyticity.

First of all, the source of the difficulty with Carnap's *extended* proposal is now apparent. By allowing meaning postulates in addition to those entailed by the Carnap sentence, we are thus allowing sentences which are not observationally vacuous to be considered as analytic. But such sentences will function essentially in the deduction of some contingent observation statement from the theory, and could thus be reasonably said to be confirmed or disconfirmed by the appropriate observational outcome.

While this consideration tells against the extended proposal, it stems from a result which provides grounds for maintaining that Carnap's *basic* proposal successfully meets the charge of arbitrariness levelled by Quine against earlier approaches. For the immunity of the Carnap sentence (and its consequences) to both confirmation and disconfirmation provides the needed rationale for holding *just these sentences* to be analytic within the theory.

Finally, we come to the difficulty raised by Hempel concerning the dual role played by the postulates of a theory within the framework of Carnap's general reconstruction. The difficulty, recall, was the following. Since each of a theory's postulates serves *both* a nomological and an interpretative function, no single statement of the theory can serve *merely* an interpretative function, i.e., be an analytic statement.

The way in which Carnap's basic proposal resolves this difficulty is as unexpected as it is simple. For it can be shown that *no* postulate of the theory *TC* is a consequence of the Carnap sentence of *TC*, i.e., each of the postulates of *TC* becomes *synthetic* under Carnap's proposal (cf. A, Cor. 13).[5] Thus the difficulty raised by Hempel is overcome by shifting the locus of the analytic-synthetic distinction from the theory's postulates to its consequence class in general, thereby allowing for the nomological role of these postulates. The analytic ingredients of the postulates emerge at a deductively lower level than the postulates themselves; they are, so to speak, filtered out from the postulates by means of the Carnap sentence.[6] Indeed, it can be shown that the properly analytic sentences of a theory must be mixed sentences of that theory, but mixed sentences which are not correspondence postulates of the theory (cf. A, Th. 2, Cor. 9, Cor. 13).

The above results indicate the ability of Carnap's proposal to deal with objections to the analytic-synthetic distinction which many writers have found extremely compelling, and perhaps conclusive. But this proposal is also of interest by virtue of the light it throws upon the relation of the Ramsey sentence method to the instrumentalism-realism debate. As was mentioned earlier, Carnap's proposal can be seen as a formalized version of the philosophical thesis that a theory and its Ramsey sentence are equivalent. Now it has been rightly pointed out (cf. [5] and [7]) that, by virtue of the existential quantification ingredient in the Ramsey sentence method, we are no less existentially committed to theoretical entities (classes and relations) by the Ramsey sentence than we were at the outset. The above results provide added support for the correctness of this view. For recall that a consequence of Carnap's proposal is that each of the postulates of *TC*, and thus each of the theoretical postulates of *TC*, become synthetic under this account.[7] The factual status thus assigned to each of the theoretical (and correspondence) postulates of a theory *TC* is thus quite in keeping with a realistic interpretation of *TC*, even when the theory *TC* is held to assert no more or less than its Ramsey sentence.

APPENDIX: THEOREMS AND PROOF-SKETCHES

I. *The Language L*

(i) descriptive signs: (a) observation predicates: $O_1, O_2, \ldots, O_j$; (b) theoretical predicates; $T_1, T_2, \ldots, T_k$.

(ii) logical signs: (a) connectives: $\sim, \cdot, \vee, \rightarrow, \equiv$; (b) quantifiers and variables: (1) individual: x, (x), (Ex), x_1, (x_1), (Ex_1), etc.; (2) predicate: t, (t), (Et), t_1, (t_1), (Et_1), etc.

(iii) syntactic variables (a) (---*Th*---), $(\text{---}Th\text{---})_1$, etc. – sentences of *L* containing only *theoretical* predicates. These are called 'theoretical sentences'; (b) (---*Ob*---), $(\text{---}Ob\text{---})_1$ etc. – sentences of *L* containing only *observation* predicates. These are called 'observation sentences'; (c) (---*Ob*, *Th*---), $(\text{---}Ob, Th\text{---})_1$ etc. – sentences of *L* containing *both* theoretical and observation predicates. These are called 'mixed sentences'; (d) S, S_1, S_2, etc. -- sentences of *L*.

II. *L-Concepts of L*

(i) possible models of *L*:

Def. $M = < U, U^1, U^2, U^3, \ldots, U^m, O_1, O_2, \ldots, O_j; T_1, T_2, \ldots, T_k >$ is a *possible model of L* iff:

(i) U is any non-empty class.

(ii) U_i $(1 \leqslant i \leqslant m)$ is the set of all sets of ordered *i*-tuples of members of U, where m is the degree of the predicate of highest degree occurring in *L*.

Intuitively, the U^{i}'s are the range of the predicate variables of L.

(iii) O_i $(1 \leqslant i \leqslant j)$ is a set of ordered n-tuples of elements of U, where n is the degree of the predicate O_i.

(iv) T_i $(1 \leqslant i \leqslant k)$ is a set of ordered n-tuples of elements of U, where n is the degree of the predicate of T_i.

The notion of a sentence S of L being *true in a possible model* is assumed defined in the customary manner. As is usual, a sentence S will be called *L-true* (in symbols, '$\Rightarrow S$') iff S is true in all possible models of L. A sentence S_1 *logically implies* S_2 $(S_1 \Rightarrow S_2)$ if and only if every model of S_1 (i.e., every possible model in which S_1 is true) is also a model of S_2. S_1 and S_2 are said to be L-equivalent $(S_1 \Leftrightarrow S_2)$ if and only if (iff) $S_1 \Rightarrow S_2$ and $S_2 \Rightarrow S_1$. In the following, the standard results involving the logical relations among the sentences of L are freely used. These include the following: (1) $S \Rightarrow S$; (2) if $\Rightarrow S$, then for any S_1, $S_1 \Rightarrow S$; (3) if $S_1 \Rightarrow S_2$ and $S_2 \Rightarrow S_3$, then $S_1 \Rightarrow S_3$; (4) if $S_1 \Rightarrow S_2$ and $\Rightarrow S_1$, then $\Rightarrow S_2$; (5) $(S_1 \vee S_2) \Rightarrow S_3$ off both $S_1 \Rightarrow S_3$ and $S_2 \Rightarrow S_3$, etc.

III. *The Carnap Sentence*

D.1. Where $\tau = Ax(\text{---}O_1, O_2, \ldots, O_j, T_1, T_2, \ldots, T_k; \text{---})$, then $R^{\tau} =$ df. $(Et_1)(Et_2)\ldots(Et_k)\, Ax(\text{---}O_1, O_2, \ldots, O_j; t_1, t_2, \ldots, t_k; \text{---})$.

Th.1. $\tau \Rightarrow (\text{---}Ob\text{---})$ iff $R^{\tau} \Rightarrow (\text{---}Ob\text{---})$.

Proof: Cf. [12], 291ff.

As an immediate corollary of D.1, we also have:

Cor.1. $\tau \Rightarrow R^{\tau}$.

The Carnap Sentence of τ is now defined as follows.

D.2. $C^{\tau} =$ df. $R^{\tau} \rightarrow \tau$.

Cor.2. $C^{\tau} \Leftrightarrow (R^{\tau} \equiv \tau)$.

Proof: Since, by Cor. 1, $\tau \Rightarrow R^{\tau}$, it follows that $\Rightarrow (\tau \rightarrow R^{\tau})$. Hence $(R^{\tau} \rightarrow \tau) \Leftrightarrow (R^{\tau} \rightarrow \tau) \cdot (\tau \rightarrow R^{\tau})$. Thus by D.2., $C^{\tau} \Leftrightarrow [(R^{\tau} \rightarrow \tau) \cdot (\tau \rightarrow R^{\tau})]$, and so $C^{\tau} \Leftrightarrow (R^{\tau} \equiv \tau)$.

Cor.3. $\tau \Leftrightarrow (R^{\tau} \cdot C^{\tau})$.

Proof: (a) by Cor.1, $\tau \Rightarrow R^{\tau}$, and, trivially, $\tau \Rightarrow (R^{\tau} \rightarrow \tau)$, i.e., $\tau \Rightarrow C^{\tau}$. Hence $\tau \Rightarrow (R^{\tau} \cdot C^{\tau})$. (b) By D.2, $(R^{\tau} \cdot C^{\tau}) \Leftrightarrow [R^{\tau} \cdot (R^{\tau} \rightarrow \tau)]$, and thus, by *modus ponens*, $[R^{\tau} \cdot (R^{\tau} \rightarrow \tau)] \Rightarrow \tau$ must obtain. Hence, $(R^{\tau} \cdot C^{\tau}) \Leftrightarrow \tau$.

Cor.4. $\tau \Rightarrow C^{\tau}$.

Proof: Immediately from Cor.3.

D.3. S is analytic (A-true) in $\tau =$ df. $C^{\tau} \Rightarrow S$.

D.4. S is properly analytic in τ=df. S is analytic in τ and S is not L-true.

Cor.5. If $\Rightarrow S$ (S is L-true), then S is A-true in τ.

Proof: follows immediately from D.3 and the fact that any statement entails an L-true statement.

Cor.6. S is A-true in τ iff $\tau \Rightarrow S$ and $\sim R^{\tau} \Rightarrow S$.

Proof: By D.3, S is A-true in τ iff $C^{\tau} \Rightarrow S$. By D.2., $C^{\tau} \Rightarrow S$ iff $(R^{\tau} \rightarrow \tau) \Rightarrow S$, i.e., iff $(\sim R^{\tau} \vee \tau) \Rightarrow S$, i.e., iff $\sim R^{\tau} \Rightarrow S$ and $\tau \Rightarrow S$.

IV. *Observation Statements and the Carnap Sentecne*

Th.2. If S is an observation statement, then S is not properly A-true in any theory τ.

Proof: By D.3 and D.4, this amounts to showing that if $C^{\tau} \Rightarrow (\text{---}Ob\text{---})$, then $\Rightarrow (\text{---}Ob\text{---})$. Assume that $C^{\tau} \Rightarrow (\text{---}Ob\text{---})$. By Cor.4, $\tau \Rightarrow C^{\tau}$, and thus $\tau \Rightarrow (\text{---}Ob\text{---})$. Also, by Th.1, $R^{\tau} \Rightarrow (\text{---}Ob\text{---})$ as well. But since $C^{\tau} \Rightarrow (\text{---}Ob\text{---})$, by Cor. 6, $\sim R^{\tau} \Rightarrow (\text{---}Ob\text{---})$. Thus R^{τ} and $\sim R^{\tau}$ each entail $(\text{---}Ob\text{---})$, and so $\Rightarrow (\text{---}Ob\text{---})$.

D.5. S is Ob – uncreative =df. If $S \Rightarrow (\text{---}Ob\text{---})$, then $\Rightarrow (\text{---}Ob\text{---})$.

Cor.7. If S is A-true in τ, then S is Ob-uncreative.

Proof: Assume that S is A-true in τ and $S \Rightarrow (\text{---}Ob\text{---})$. By D.3, $C^{\tau} \Rightarrow S$, and thus $C^{\tau} \Rightarrow (\text{---}Ob\text{---})$. Hence by D.4 and Th.2, $\Rightarrow (\text{---}Ob\text{---})$.

V. *Theoretical Statements and the Carnap Sentence*

D.6. τ has Ob-content=df. There is a first-order observation statement $(\text{---}Ob\text{---})_1$, such that $\tau \Rightarrow (\text{---}Ob\text{---})_1$ and $\not\Rightarrow (\text{---}Ob\text{---})_1$.

Cor.8. If τ has Ob-content, then $\Rightarrow R^{\tau}$.

Proof: Suppose that τ has Ob-content. Then, by D.6, there is a first-order observation statement, say $(\text{---}Ob\text{---})_1$, such that $\tau \Rightarrow (\text{---}Ob\text{---})_1$ and $\not\Rightarrow (\text{---}Ob\text{---})_1$. By Th.1, $R^{\tau} \Rightarrow (\text{---}Ob\text{---})_1$ also, and since $\not\Rightarrow (\text{---}Ob\text{---})_1$, it follows that $\not\Rightarrow R^{\tau}$.

Th.3. if $(\text{---}Ob\text{---})_1$ and $(\text{---}Th\text{---})_1$ are both satisfiable, first-order sentences, then: (i) if $(\text{---}Ob\text{---})_1 \Rightarrow (\text{---}Th\text{---})$, then $\Rightarrow (\text{---}Th\text{---})$, and (ii) if $(\text{---}Th\text{---})_1 \Rightarrow (\text{---}Ob\text{---})$, then $\Rightarrow (\text{---}Ob\text{---})$.

Proof: This is a standard result of first-order logic which relies upon the fact that the pairs of sentences involved have no descriptive terms in common. For a proof, see [11], 169.

Th.4. If τ is consistent and has *Ob*-content, then if $C^{\tau} \Rightarrow$(---*Th*---), where (---*Th*---) is a first-order sentence, then $\Rightarrow$(---*Th*---).

Proof: Assume that $C^{\tau} \Rightarrow$(---*Th*---). Since τ has *Ob*-content, by D.6, there is a first-order observation statement, $(\text{---}Ob\text{---})_1$, such that $\tau \Rightarrow (\text{---}Ob\text{---})_1$ and $\not\Rightarrow (\text{---}Ob\text{---})_1$. By Th.1, $R^{\tau} \Rightarrow (\text{---}Ob\text{---})_1$ and thus $\sim(\text{---}Ob\text{---})_1 \Rightarrow \sim R^{\tau}$. Now, by Cor.6, $\sim R^{\tau} \Rightarrow$(---*Th*---), and since $\sim(\text{---}Ob\text{---})_1 \Rightarrow \sim R^{\tau}$, we obtain: $\sim(\text{---}Ob\text{---})_1 \Rightarrow$(---*Th*---). Since $\not\Rightarrow (\text{---}Ob\text{---})_1$, it follows that $\sim(\text{---}Ob\text{---})_1$ is satisfiable. Hence, from Th.3 (i) it follows that $\Rightarrow$(---*Th*---).

Cor.9. If τ is consistent and has *Ob*-content, then no theoretical statement is properly *A*-true.

Proof: Immediately from D.4 and Th.4.

Th.5. If τ is consistent, and $\tau \Rightarrow$(---*Th*---), then C^{τ}. (---*Th*---) is *Ob*-uncreative.

Proof: Suppose that $[C^{\tau} \cdot (\text{---}Th\text{---})] \Rightarrow (\text{---}Ob\text{---})$. Then $C^{\tau} \Rightarrow [(\text{---}Th$ $\text{---}) \rightarrow (\text{---}Ob\text{---})]$, and by Cor.6., $\sim R^{\tau} \Rightarrow [(\text{---}Th\text{---}) \rightarrow (\text{---}Ob\text{---})]$. Now by Cor.4, $\tau \Rightarrow C^{\tau}$ and by assumption $\tau \Rightarrow$(---*Th*---). Hence $\tau \Rightarrow (C^{\tau} \cdot$ (---*Th*---)), and since $[C^{\tau} \cdot (\text{---}Th\text{---})] \Rightarrow (\text{---}Ob\text{---})$, $\tau \Rightarrow$(---*Ob*---). By Th.1, it now follows that $R^{\tau} \Rightarrow$(---*Ob*---) also, and thus, trivially, $R^{\tau} \Rightarrow ((\text{---}Th\text{---}) \rightarrow (\text{---}Ob\text{---}))$. But we have already shown that $\sim R^{\tau} \Rightarrow$ $[(\text{---}Th\text{---}) \rightarrow (\text{---}Ob\text{---})]$, and thus $\Rightarrow (\text{---}Th\text{---}) \rightarrow (\text{---}Ob\text{---})$, i.e., (---*Th*---) $\Rightarrow$ (---*Ob*--- Since τ is consistent, and $\tau \Rightarrow$(---*Th*---), (---*Th*---)). is satisfiable. Hence, by Th.3 (ii), $\Rightarrow$(---*Ob*---).

Th.6. If $\tau \Rightarrow$(---*Th*---), then $\tau \Leftrightarrow R^{\tau} \cdot [C^{\tau} \cdot (\text{---}Th\text{---})]$

Proof: As in Cor.3.

VI. Observational Vacuity

D.7. S is observationally vacuous in τ=df. $\tau \Rightarrow S$, and for any S_1, if $\tau \Rightarrow S_1$, and $(S_1 \cdot S) \Rightarrow$(---*Ob*---), then $S_1 \Rightarrow$(---*Ob*---).

Cor.10. If S is observationally vacuous in τ, then S is *Ob*-uncreative.

Proof: If S is observationally vacuous in τ, then from D.7 it follows that, in particular if $(p \vee \sim p) \cdot S \Rightarrow$(---*Ob*---), then $(p \vee \sim p) \Rightarrow$(---*Ob*---), i.e., $\Rightarrow$(---*Ob*---). Hence the result, by D.5.

Th.7. C^{τ} is observationally vacuous in τ.

Proof: By Cor. 4, $\tau \Rightarrow C^{\tau}$. Suppose next that $\tau \Rightarrow S_1$ and $(S_1 \cdot C^{\tau}) \Rightarrow$(--- *Ob*---). Hence, by D.2, $[S_1 \cdot (R^{\tau} \rightarrow \tau)] \Rightarrow$(---*Ob*---), i.e., $[S_1 \cdot (\sim R^{\tau} \vee \tau)]$

$\Rightarrow(\text{---}Ob\text{---})$, i.e., $[(S_1 \cdot \sim R^\tau) \vee (S_1 \cdot \tau)] \Rightarrow (\text{---}Ob\text{---})$. Hence $(S_1 \cdot \sim R^\tau)$ $\Rightarrow(\text{---}Ob\text{---})$.

Since $\tau \Rightarrow C^\tau$ and $\tau \Rightarrow S_1$, $\tau \Rightarrow (C^\tau \cdot S_1)$, and thus (since $(C^\tau \cdot S_1) \Rightarrow$ $(\text{---}Ob\text{---})$) it follows that $\tau \Rightarrow (\text{---}Ob\text{---})$. And from Th.1 we now have that $R^\tau \Rightarrow (\text{---}Ob\text{---})$.

But since $(S_1 \cdot \sim R^\tau) \Rightarrow (\text{---}Ob\text{---})$, and $R^\tau \Rightarrow (\text{---}Ob\text{---})$, we see that $[(S_1 \cdot \sim R^\tau) \vee R^\tau] \Rightarrow (\text{---}Ob\text{---})$, i.e. $[(S_1 \vee R^\tau) \cdot (R^\tau \vee \sim R^\tau)] \Rightarrow (\text{---}Ob\text{---})$ i.e., $(S_1 \vee R^\tau) \Rightarrow (\text{---}Ob\text{---})$. Hence $S_1 \Rightarrow (\text{---}Ob\text{---})$. Thus the result, from D.7.

Cor.11. If S is A-true in τ, then S is observationally vacuous in τ.

Proof: Suppose that S is A-true in τ. Then, by D.3, $C^\tau \Rightarrow S$. Let $\tau \Rightarrow S_1$ and $(S_1 \cdot S) \Rightarrow (\text{---}Ob\text{---})$. Then by D.7, we need now show that $S_1 \Rightarrow (\text{---}Ob\text{---})$. Since $(S_1 \cdot S) \Rightarrow (\text{---}Ob\text{---})$, $S \Rightarrow (S_1 \rightarrow (\text{---}Ob\text{---}))$. Since $C^\tau \Rightarrow S$, we then have that $C^\tau \Rightarrow (S_1 \rightarrow (\text{---}Ob\text{---}))$, i.e., $(C^\tau \cdot S_1) \Rightarrow (\text{---}Ob$ $\text{---})$. Hence, from Th. 7, $S_1 \Rightarrow (\text{---}Ob\text{---})$.

Lemma 1: If $C^\tau \Rightarrow S$ and $\tau \Rightarrow S$, then $\sim R^\tau \Rightarrow S$.

Proof: Immediately from Cor. 6.

Th.8. If $C^\tau \Rightarrow S$, then S is not observationally vacuous in τ.

Proof: Assume that $C^\tau \Rightarrow S$.

Case 1: $\tau \Rightarrow S$. Then the result from D.7 immediately.

Case 2: $\tau \Rightarrow S$. By Cor. 1, $\tau \Rightarrow R^\tau$, and thus $\tau \Rightarrow (S \rightarrow R^\tau)$. Now $S \cdot (S \rightarrow R^\tau) => R^\tau$. Hence by D.7, in order to show that S is not observationally vacuous in τ, it suffices to show that $(S \rightarrow R^\tau) \not\Rightarrow R^\tau$.

Thus suppose that $(S \rightarrow R^\tau) \Rightarrow R^\tau$, i.e., $(\sim S \vee R^\tau) \Rightarrow R^\tau$. Then $\sim S \Rightarrow R^\tau$ and thus $\sim R^\tau \Rightarrow S$. Since $C^\tau \Rightarrow S$ and $\tau \Rightarrow S$, this contradicts Lemma 1.

Cor.12. S is A-true in τ iff S is observationally vacuous in τ.

Proof: Immediately from Cor. 11 and Th. 8.

VII. *Concisely Formulated Theories*

Let τ be expressed as a conjunction of postulates $P_1 \cdot P_2 \cdot \ldots \cdot P_n$.

D.8. $P_i (1 \leqslant i \leqslant n)$ is an Ob-redundant postulate of $\tau =$df. $P_1 \cdot P_2 \cdot \ldots$ $\ldots \cdot P_n \Rightarrow (\text{---}Ob\text{---})$ iff $P_1 \cdot \ldots \cdot P_{i-1} \cdot P_{i+1} \cdot \ldots \cdot P_n \Rightarrow (\text{---}Ob\text{---})$.

D.9. $P_1 \cdot P_2 \cdot \ldots \cdot P_n$ is a concise formulation of $\tau =$df. No postulate $P_i (1 \leqslant i \leqslant n)$ of τ is Ob-redundant.

Th.9. If $P_1 \cdot P_2 \cdot \ldots \cdot P_n$ is a concise formulation of τ, then no postulate of τ is observationally vacuous in τ.

Proof: Let P_i be a postulate of a concise formulation of τ. From D.8

and D.9 it now follows that for some (---*Ob*---), $P_i \cdot [P_1 \cdot \ldots \cdot P_{i-1} \cdot P_{i+1} \cdot \ldots \cdot P_n] \Rightarrow$ (---*Ob*---), yet $[P_1 \cdot \ldots \cdot P_{i-1} \cdot P_{i+1} \cdot \ldots \cdot P_n] \Rightarrow$ (---*Ob*---). The result now follows immediately from D.7.

Cor.13. If $P_1 \cdot P_2 \cdot \ldots \cdot P_n$ is a concise formulation of τ, then no postulate of τ is A-true.

Proof: Immediately from Cor.12 and Th.9.

University of Hawaii

BIBLIOGRAPHY

[1] Carnap, R., 'The Methodological Character of Theoretical Concepts', in *Minnesota Studies in the Philosophy of Science*, Vol. I (ed. by Feigl and Scriven), Minnesota, 1956.
[2] Carnap, R., 'Meaning Postulates', in *Meaning and Necessity*, 2nd. ed., Chicago, 1958.
[3] Carnap, R., 'Replies and Systematic Expositions', in *The Philosophy of Rudolf Carnap* (ed. by P. A. Schilpp), Open Court, 1963.
[4] Carnap, R., *Philosophical Foundations of Physics*, Basic Books, 1966.
[5] Hempel, C. G., 'The Theoretician's Dilemma', in *Minnesota Studies in the Philosophy of Science*, Vol. II (ed. by Feigl, Maxwell and Scriven), Minnesota, 1958.
[6] Hempel, C. G., 'Implications of Carnap's Work for the Philosophy of Science', in Schilpp [3] above.
[7] Maxwell, G., 'The Ontological Status of Theoretical Entities', in *Minnesota Studies in the Philosophy of Science*, Vol. III (ed. by Feigl and Maxwell), Minnesota, 1962.
[8] Quine, W. V. O., 'Two Dogmas of Empiricism', in *From a Logical Point of View*, 2nd. ed. Rev., Harvard, 1961.
[9] Quine, W. V. O., 'Epistemology Naturalized', in *Ontological Relativity and Other Essays*, Columbia Univ., 1969.
[10] Ramsey, F., 'Theories', in his *The Foundations of Mathematics*, 3rd. ed., Kegan Paul, 1954.
[11] Robinson, A., *Introduction to Model Theory and to the Metamathematics of Algebra*, North-Holland Publ. Co., 1968.
[12] Rozeboom, W., 'The Factual Content of Theoretical Concepts', in Feigl and Maxwell, [7] above.

NOTES

[1] For such arguments, see [9], 88ff.

[2] In the body of what follows, the results obtained are stated without proof. Theorems and an outline of their proofs are included in the Appendix, and the references (e.g., '(A, Th. 2a)') in the main body of this paper are to this Appendix. I am indebted to Irving Copi and Alberto Coffa for their comments, criticisms, and suggestions concerning these results. Any errors which may remain are, of course, my own responsibility.

[3] This result, and many of those to follow, is proven under the assumption that the theory *TC* is consistent and has observational content.

[4] Since (C4) implies (C3), we can merely replace (C3) by the stronger (C4) and obtain the same effect.
[5] This result holds on (trivial) condition that *TC* is a *concise* formulation of the theory (cf. A, D.8).
[6] This is more than mere metaphor. For in order that *S* be analytic, the Carnap sentence requires that: (i) $TC \Rightarrow S$; but then *goes on to require* that (ii) $\sim R^{TC} \Rightarrow S$ (cf. A, Cor. 6).
[7] Unless they are logically true or observationally redundant. Indeed, *no* theoretical statements which are consequences of *TC* are properly *A*-true.

EDWARD ERWIN

THE CONFIRMATION MACHINE

It is customary to draw a distinction between statements unconfirmable in practice and statements unconfirmable in principle.[1] It may be impossible, for example, to confirm the statement "The star most distant from us in the universe has recently doubled in size," but this impossibility is not of a logical kind. We could conceive of tests which would either confirm or disconfirm the statement, even if in fact we cannot carry them out. In contrast, the statement 'Everything has recently doubled in size' cannot be confirmed for logical reasons, or so some philosophers have claimed. If everything were to double in size, including the very measuring rods used to detect expansions, it would be logically impossible, so the argument runs, to either confirm or disconfirm the above statement.[2]

Many philosophers, I suspect, would have doubts about citing the above statement about universal expansion as being unconfirmable in principle, but many would agree that one or more of the following do fall in this category:

1. 'Gavagai' means 'rabbit'.
2. Saturday is in bed.
3. $7+5=13$
4. Unnecessary violence is evil.
5. Jones' toothache is exactly like mine.

In addition, many philosophers would be willing to rest important philosophic conclusions on the assumption that a certain statement, or a group of statements, is unconfirmable in principle. Some of these conclusions are listed below:

1. Radical translation is indeterminate.
2. There cannot exist a logically private language.
3. We can never know the mind of another, at least not if Cartesian dualism is true.
4. Statements of either ethics or aesthetics (or both) cannot be known to be true.

Boston Studies in the Philosophy of Science, VIII.

5. Philosophic statements cannot be proved, except by making certain basic assumptions which, in turn, cannot be proved or disproved.

The distinction between unconfirmability-in-principle and unconfirmability-in-practice is in some respects like the distinction between analytic and synthetic statements. Both distinctions are fundamental and are employed in a wide range of philosophic arguments; and both are intuitively plausible, so much so that they have become well-entrenched in the schemes of many philosophers. One important difference, of course, is that the analytic/synthetic distinction has become less well-entrenched in recent years as the result of attacks by Quine and others. I shall now try to eliminate this difference, however, by mounting an attack on the other distinction. What I shall argue is not that the distinction is unclear, as Quine argued in the parallel case, but that it it useless. It is entirely useless, if the strong version of my thesis is correct, because there is *no* statement which is unconfirmable in principle. However, even if there are exceptions, the distinction is of little or no value to philosophy because, to state a weaker thesis, the class of exceptions contains no statement which is philosophically interesting. I shall defend the stronger thesis by describing what I call a 'confirmation machine'. I shall then consider objections to my argument – plus some of its implications, some of which concern the nature of evidence and rationality.

PART 1

A confirmation machine is a black, rectangular box which is six feet high, three feet deep, and four feet wide. The machine has two slots: one on the top and one in the front at its base.

To operate the machine, one drops in the top slot a card with a statement typed on it, and then presses the adjoining button. After a few minutes, the card is returned through the bottom slot with the symbol *T* printed next to the statement if the statement is true or the symbol *F* similarly placed if the statement is false. When the machine functions properly, it prints nothing if the statement is neither true nor false or if no statement has been made.

It is highly improbable that a workable confirmation machine will be constructed on the basis of our present technical knowledge, but it is at least in principle possible to build such a machine. There is no logical or conceptual contradiction contained in the above description. If our knowledge of applied physics and technology develops sufficiently, such a machine may, in fact, be built.

If such a machine is built and does work sufficiently well, then our present methods of testing scientific theories and of confirming or (disconfirming) propositions of common sense may be replaced by use of the machine. Of course, the machine would have to work very well for that to happen, assuming we were rational in choosing our confirmation procedures. Suppose, for example, that we were to feed into the machine cards containing statements we knew to be true and in each case the card was returned marked *T*. Assume further that all cards containing statements known to be false were returned marked *F*. This string of successes would be evidence that the machine does work, but to have knowledge of the successes we would need to know the truth values of the propositions tested and that would require the use of independent testing procedures. We might have additional evidence in support of the machine if, for instance, we had a well confirmed theory explaining how it works. In fact, the theory of the machine might itself be deducible from a higher theory, perhaps a very fundamental theory of physics which has already been extremely well confirmed. We might then be able to use the theory of the machine's operation to predict that a card will be returned marked *T* only if it contains a true statement. This prediction would then be supported not only by the machine's past successes, but also by the mass of observational data which supported both the theory of the machine and the parent theory of physics. Nevertheless, the additional support provided by the theory of the machine would have evidential value only if supported by the mass of observational data. The reliance on more direct methods of confirmation, then, would still be present. Moreover, let us suppose that we have no well confirmed theory of how the machine works. Perhaps the machine was discovered on some distant planet or was developed in secret by one man, who is now dead. In any event, let us assume that we know, or at least have some evidence, that the machine does work, but do not know how it works. It would not be reasonable, then, to jettison immediately all of our usual confirmation procedures in favor of use of the machine. For, it is only through use of the usual methods that we would get evidence in the first place that the machine works, and we would still need the usual procedures to see if the machine continues to work.

Nevertheless, what might be initially unreasonable might be eminently reasonable later. Suppose that the machine is always right; it makes no mistakes at all, no matter what the subject matter of the statement being tested. It works properly, moreover, over a long period of time. How long would be sufficiently long cannot be determined easily. The length of time would depend on many factors, including the kinds of propositions tested. For example, if only trivial and obviously true statements were tested, then the period might have to be a very long one. However, if we tested propositions whose truth value was unknown and subsequently discovered, through an independent check, that the machine was always correct in marking a *T*, then the period might well be shortened. It could be shortened even further if we tested propositions which we believed to be true (or false) on the basis of a large body of evidence, and later found that our belief had been wrong and that the machine had been right. It would help, for example, if we ran through such a statement as "Man has evolved from lower organisms" and later found that the machine had been correct in marking *F*, and that we had been wrong in believing such a statement to be true.

In any event, although we cannot easily know beforehand what length of time will prove sufficient, let us assume, rather arbitrarily, that the machine will be entirely successful for a period of fifty years. Assume, further, that the machine has been used continually by scientists working in very diverse areas (physics, genetics, neurology, psychology, history, etc.) and has been subjected to the most rigid kinds of tests that we can conceive. Not only do we test propositions with uncertain truth values, but we also test propositions which the scientific community, relying on a great deal of observational and theoretical evidence, is unanimously convinced are true. In some of these latter cases, the machine and the consensus of the scientific community conflict; but in every such case, the machine is shown to be correct. It is easy to see how this might happen. Close observers of robberies, football games, street demonstrations, and other kinds of events of ten give inaccurate descriptions of what they see. That such errors occur is well known. However, even more experienced observers, even the most highly trained scientific observers, occasionally go wrong. Perhaps they are guided too much by a current scientific paradigm which conflicts with the data at hand; perhaps they notice certain items, and not others, because of certain personal motivation; and perhaps they are simply careless. But, for whatever reason, such mistakes do occur. They could occur, moreover, perhaps not so often as to lead us to mistrust observation altogether – there may be insurmountable difficulties in defending a total skepticism of the senses – but often enough to mistrust reports of sense observation when they conflict with the issuances of the confirmation machine. If the machine says that *P* is true, then we trust the machine instead of the more direct evidence of our senses which says otherwise. The point here is not simply that some day we might, in fact, trust our use of the confirmation machine over our other methods of confirmation; the point, rather, is that under certain conditions we would be irrational to do otherwise. It would still be reasonable to check the machine's conclusions periodically, especially when they conflict with some of our most well-founded beliefs: for, it is possible for a confirmation machine to break down. But even this procedure might someday be rendered superfluous after it becomes very improbable that the machine will err. There will still be skeptics, of course, who will not trust the judgments of the machine, no matter how often it is right, just as there are now skeptics, for example, who doubt the reliability of even the most reliable scientific observations – merely because it is always logically possible to be in error. Such skeptics of the future will insist on returning to the more antiquated methods of confirmation. In doing so, however, they will be unreasonable; they will be like our present-day skeptics in that they will not have the backing of logic or scientific evidence to support their beliefs. The confirmation machine in any case whatsoever will very probably be right; and the skeptics, if they disagree with the machine, will very probably be wrong.

After the confirmation machine is successful for a sufficiently long period, during which it tests a sufficient number of the right kinds of statements, the very next statement which it marks with a *T*, no matter what the statement might be, will thereby be confirmed. Since it is conceivable that a confirmation machine will be built and will function as described above, and that under such conditions any statement might have a *T* marked next to it by the machine, it follows that any statement can be confirmed in principle.

I am not saying that any statement whatsoever might be true (or false).

Assuming there are any analytic truths at all, '7+5=13' may be a statement which cannot possibly be true and '7+5=12' may be a statement which cannot possibly be false. It might even be the case that we now *know* of certain statements that they could not possibly be true and of others that they could not possibly be false. What I am arguing, rather, is that for any statement whatsoever, there are conceivable conditions under which it could be confirmed (or disconfirmed) to such a degree that we would be rational in believing it to be true (or false). It might be, however, that there are some statements which we would be rational to accept under certain conditions, but which are necessarily false; the machine might mark a *T* next to such a statement, but the machine, unknown to us, would have broken down. Since my argument does not depend on the claim that no statement is known to be analytically true, we can agree with Quine that no statement is immune from revision – if that means that any statement might be rejected on rational grounds – and still maintain that some statements are now known to be analytically true.

I shall now consider some likely objections, and, at the same time, draw out some of the implications my argument has concerning the nature of evidence and rationality.

PART II

Objection 1: It seems that my reasoning proves too much. Perhaps the positivists were wrong in using "confirmability" as a necessary condition of meaningfulness, but surely they were right in thinking it a sufficient condition. If my argument were sound, however, that too would be wrong. Even an obviously meaningless statement, such as 'Saturday is in bed', could, in principle, have *T* marked next to it by the confirmation machine under the conditions I described. If that were sufficient to confirm the statement, even an obviously meaningless statement could be confirmed.

Reply: Suppose, as some writers do, that it is incoherent to speak of meaningless statements, that any *X* which is a genuine statement is necessarily meaningful. On this view, the above objection would fail; for, there would be no meaningless statements and, hence, even if my argument were sound, there would still not be any confirmable meaningless statements. The confirmation machine might break down and mark a *T* next to some *X* which is not a genuine statement, perhaps 'Saturday is in bed', but that would not be an instance of the machine confirming a meaningless statement. We might have good reason to believe that the machine had confirmed a statement, but our belief would be wrong: there would be no statement here to be confirmed.

However, we might accept an alternative position which does allow us to speak of statements as being meaningless.[3] We might say, for example, that we can understand what statement someone is making when he says that Saturday (i.e., the seventh day of the week) is in bed, but what we cannot understand is what it would be like for such a statement to be true. On this view, my conclusion *would* have the implication that meaningless statements can be confirmed. However, that is no more objectionable than saying that self-contradictory statements can be confirmed; and, if my argument is sound, the latter kind of statement can be confirmed by the confirmation machine's printing *T* under suitable conditions. Of course, I cannot know that a statement is meaningless and *at the same time* confirm that it is true; but what now seems meaningless would not, or should not, seem meaningless of the relevant evidence were to change in an appropriate fashion.

Since all meaningless statements (if there be any) can be confirmed in principle, it is a mistake to try to prove that a statement is meaningful by trying to prove that it could be confirmed in principle. Consider, for example, the statement 'There has been time without change.'[4] If we actually confirm this statement, we get evidence that it is true and, hence, that it is meaningful (assuming we realize that all true statements are meaningful). However, if we merely describe conceivable but improbable circumstances under which we would be rational to accept the statement, we would not be providing evidence that the statement is either true or meaningful. Rather, we would be providing evidence that we could conceivably obtain evidence that the statement was true and meaningful; and that is compatible with the statement being meaningless.

Objection 2: We might be rational in trusting the confirmation machine when it rejects some of our most well-founded beliefs, but there are others which we could not rationally relinquish. For example, if the confirmation machine were to mark *F* next to '7+5=12' and *T* next to '7+5=13', the rational thing to do would be to refuse to believe the machine.

Reply: The above objection ignores a crucial point about evidence and rationality. The rationality of accepting any particular belief, no matter how fundamental or seemingly self-evident it might be, depends in part upon our evidence for certain other beliefs. If the evidence for these other beliefs changes, it might no longer be rational to hold on to the particular belief. For example, we believe that most people who have a knowledge of arithmetic do not usually go wrong in doing simple sums, such as '7 + 5' however, the evidence for that belief might change. We might begin by discovering that people were making mistakes in simple calculations which earlier had seldom subject to error. Cashiers are found to be almost always giving the wrong change; clerks and accountants are unable to balance their accounts; and the rest of us have difficulty in keeping our budgets and bank books straight. We might even discover that a certain disease is causing our ineptitude, and that everyone now has the disease. Suppose that the situation grows worse. Even when adding simple sums such as '15 + 18' and '7 + 42', we almost invariably make mistakes. We also have overwhelming evidence that the confirmation machine, which has not gone wrong for the last fifty years, is not likely to make similar mistakes. At some point, if the disease lingered and if our evidence now indicated that people usually go wrong when doing simple sums, we would be irrational if we believed that '7 + 5 = 12' after the confirmation machine marked *F* next to that statement. That this would be so can be seen in a number of ways. For example, if we were to reject the machine's denial of '7 + 5 = 12' how would we explain the machine's error? We might say, at first, that the machine has broken down. That hypothesis, however, could be tested by using the old methods of confirmation to check subsequent answers of the machine, and the hypothesis might be found to be false. We might then propose that only when dealing with a certain kind of statement, say a statement of arithmetic, does the machine err. How ever, that hypothesis might also be disconfirmed. As each proposed hypothesis explaining the machine's failure were tested, disconfirmed and discarded, it would become more and more likely that our belief that the machine had failed was false. In addition, we would already have so much evidence that the machine is unlikely to break down, that the 'failure' hypothesis would become, at some point, very implausible. On the other hand, our belief that we had made a mistake in formerly thinking that 7 + 5 = 12 would not be implausible at all: for, we could now explain such mistakes by pointing to the fact that we have a certain disease. Under these conditions, we would be rational in rejecting our belief that 7 + 5 = 12.

Of course, we might remember that we believed that '7 + 5 = 12' is true even before the outbreak of the disease. However, we might also obtain evidence that the disease has always been with us, even if we have been seriously affected by it only recently. We might even obtain evidence that whenever we seem to remember having accepted certain statements of mathematics, our memory deceives us. We think we remember believing that '7 + 5 = 12' before the discovery of the disease, but we are probably wrong.

Besides our belief about the ability of most people (excluding small children) to do simple sums without continually making errors, there are other beliefs whose disconfirmation would affect our belief in the unshakability of '7 + 5 = 12'. As these other beliefs began to crumble, the seeming solidity of '7 + 5 = 12' would begin to disappear. For example, most of us do not believe, as John Stuart Mill did, that '7 + 5 = 12' is an empirical statement. We may be right, but we could, in principle, find good reasons (in the from of philosophical arguments) for doubting our views on the subject. If that were to happen, we would have less reason for maintaining our present confidence in the impregnability of '7 + 5 = 12' and more reason to accept the overwhelming empirical evidence provided by the confirmation machine that the statement is false.

There are many other background beliefs which affect the firmness of our belief that we cannot be wrong about '7 + 5 = 12'. For example, there is the belief that rational men can consistently agree about the truth values of statements like '7 + 5 = 12' – i.e., statements which are analytically true (or at least are obviously true). However, that belief, too, may someday be disconfirmed. Consider the following sequence of events.

Prior to the discovery of our sudden inability to do arithmetic, we find that rational men begin to disagree about seemingly analytic statements such as 'All bachelors are unmarried' or "All male siblings are brothers.' At first, we assume that those who deny such statements do not understand what is being said. However, we get evidence that is not the case. In addition, we begin describing statements such as 'All bachelors are playboys' and 'People wo show no pain behavior are not in pain' as analytic truths, only later discovering that we were wrong. At some point, we would be right to lose confidence in our ability to recognize analytic truths, and hence would be right to lose confidence in our belief that we could not be mistaken about a statement which seemed to us to be analytically true. However, suppose that I were to retain my confidence despite such evidence. Assume that I, as a Rationalist, believe that some statements can be known to be true with absolute certitude through the use of rational insight. I do not modify my rationalism when I see statements I think to be in this category denied by seemingly rational men; instead, I revise my belief that these men are rational. (Of course, that option might be quickly cancelled if these men clearly demonstrated their rationality in other ways.) However, I would still be faced with the problem of what to say about my own case. Suppose, I continually match wits with the machine and each time I lose. In every case, I say that the machine cannot be right in saying that *P* is false; for *P*, I claim, is analytically true. Yet, after further reflection and further checking, I discover in each case that I was wrong. I can still retain my belief that some statements can be known by rational men with absolute certitude, but it is evident that I can no longer recognize such statements (perhaps because I am not completely rational). Under such conditions, it would not be rational for me to insist that the machine is wrong in marking an *F* next to '$7+5=12$'. I made the same claim earlier, on many occasions, concerning other seemingly analytic statements; but I have always been wrong. In addition, I now have a great deal of new evidence which shows that we can no longer trust our ability to do mathematics. Finally, the success of the machine in the past for so long a time, when dealing with analytic as well as non-analytic statements, would make it quite unlikely that I would be right and the machine wrong.

Objection 3: Some statements could not possibly be confirmed by the confirmation machine because of what they say about confirmation. For example, if the machine marked a *T* next to *S*, where *S* reads "No statement has even been confirmed," we could not trust the machine in this case without undermining our very reasons for such trust; for, we would have reason to trust the machine only because we believed that the machine had in the past confirmed many other statements and accepting *S* would give us reason for rejecting that belief.

Reply: (A) Although I lack the space to do so here, I believe a detailed description can be given of how *S*, and any similar example, might be confirmed. Part of my grounds for that belief are contained in my reply to the previous objection: the rationality of accepting or rejecting any particular belief depends upon the evidence we have concerning related beliefs and such evidence might change in a crucial way. For example, as I described the situation, we would be irrational in accepting the machine's endorsement of *S* so long as we had good reason to believe *R*, where *R* reads "The machine's endorsement of any statement has evidential value only if *S* is false." Although *R* might be true, events might occur which would give us good reasons for rejecting it. For example, *R* might fall into a certain class of beliefs, say the class of epistemological beliefs held be anticonfirmation machine theorists, and we might obtain overwhelming evidence, through a prior use of the machine, that all beliefs in that class were false. After our belief in *R* were undermined in this fashion, the way would be open for our rationally accepting *S* – although of course, *S* would be false.

(B) Even if the above statement, *S*, or any similar one, were unconfirmable, that would not affect my weaker thesis; that the unconfirmability in principle and practice distinction is of little or no value to philosophy. For, the existence of such counter-instances would not show that the class of exceptions contained any philosophically interesting candidates (such as those listed in the beginning of the paper). So too the analytic-synthetic distinction is not very important for philosophy if the only instances of analyticity are trivial examples such as 'All bachelors are unmarried.'

Objection 4: Appealing to the use of a confirmation machine to show that a statement can be confirmed seems like an *ad hoc* way of defending the thesis that all statements are confirmable. Could we not outmaneuver such a procedure by simply revising our claims of unconfirmability? For example, we might say: "Except by using a confirmation machine, such and such a statement cannot be confirmed even in principle."

Reply: Although I did use the notion of a confirmation machine in arguing for my thesis, the notion is dispensable. Consequently, any *ad hoc* maneuver of the kind suggested above will not enable one to escape my argument.

Instead of speaking about a confirmation machine, I could, for example, describe an all knowing man who gives correct answers to varied kinds of questions for a long period of time. Furthermore, the point of using the notion of a confirmation machine, and any similar notion, is merely to show in a single stroke how any statement might be confirmed. Once we understand the facts about confirmation which make that stroke possible, it becomes unnecessary to speak of anything like a confirmation machine. Instead of describing a single situation in which any statement might be confirmed, we might take every statement, one by one, and describe some conceivable set of circumstances for each statement in which it could be confirmed. The demonstration would be much more complex, but the result would be the same.

The relevant facts about the nature of confirmation which make the result inevitable are these. Whether a certain statement, *S*, is confirmed or disconfirmed by a certain event, *E*, is not merely a matter of meaning or logic; it depends as well on the evidence we have concerning the truth or falsity of related beliefs. As the evidence for these related beliefs changes, an *E* which earlier would have confirmed *S* may now fail to do so and an *E* which earlier would not have confirmed *S* may now do so. Given certain conceivable shifts in the evidence for related beliefs, any *S* might be confirmed by any *E*. Consequently, any *S* is confirmable in principle. For example, does the appearance of a white swan tend to confirm the generalization that all swans are white? Whether it does or not is not determined simply by either the logical form or the meaning of 'All swans are white.' Obviously, if we already have a good deal of evidence that *not* all swans are white, the appearance of a white swan does not confirm to any degree our (former) belief that all swans are white. Less obviously, even if our belief in the whiteness of all swans has not yet been disconfirmed, an instance of a white swan would be a confirming instance only given the evidence we have for other well confirmed beliefs. If the evidence for these other beliefs were to change, then the confirmation power of the appearance of a white swan might be zero, or even negative.[5] In general, whether event *E* confirms statement *S* always depends, in a crucial way, on the evidence we already possess concerning both *S* and related beliefs. What, for example, is confirmed by the discovery of a white swan (or the marking of a *T* on a white card by a confirmation machine)? Does the discovery of the swan (or the marking of the *T*) confirm that an electron had a certain position and velocity at a certain time? Or that two lightning strokes occurred with absolute simultaneity? Or that the Absolute is perfect? Very probably, it would confirm none of the above. However, the discovery of the white swan (or the marking by the confirmation machine) could conceivably confirm any or all of these statements if the otality of our evidence were to change in the right way.

Objection 5: In discussing the status of a certain troublesome statement such as 'Unnecessary violence is evil' or 'Smith's toothache feels exactly like the toothache I had yesterday,' it is trivial, and therefore uninteresting, to say that a confirmation machine, which *probably cannot be built*, could confirm the statement under certain conditions *which probably will never hold.* In some science-fiction world, it may be possible to confirm such statements, but that does not help very much if in the actual world confirmation is impossible.

Reply: If we are interested in the practical question of how actually to confirm a certain statement, then talk of a non-existent confirmation machine may be irrelevant. However, if we are interested, instead, in whether it is logically or conceptually possible to confirm a given statement, then it *is* relevant to show that a confirmation machine could, in principle, confirm any statement whatsoever. Of course, once that is demonstrated, it does become trivial to say of any particular statement that confirmation or disconfirmation is in principle possible. That is one of the points I am making: that it is trivial, and of no philosophic interest, to say a particular statement can be confirmed (or disconfirmed) in principle. But what is not trivial, given the claims of many philosophers about the impossibility of confirming certain statements and given the philosophic conclusions which have been erected on these claims (such as those mentioned in Part I), is to demonstrate, in the first place, that all statements are confirmable in principle.

Furthermore, there are other issues at stake here besides the question of the soundness of "unconfirmability in principle" arguments. There is, for example, the widespread view of confirmation as being intimately related to meaning. It is quite natural to accept this view. For, one way to test a statement is to see what the statement *implies* (when combined with other statements) and then to see if, in fact, the implications are true. But that is not the only way, or even the best way; use of a confirmation machine may someday become the best way to test a statement. Hence, many formulations of the view that meaning and confirmation are closely connected are inadequate. For example, according to one formulation, the meaning of a statement is its method of verification. Although most philosophers would now reject this formula,[6] some still believe that a useful test for deciding whether a statement has meaning is to determine whether it can in principle be confirmed or disconfirmed. Others, while rejecting both confirmability and disconfirmability as tests of meaning, believe that disconfirmability in principle is what distinguishes scientific statements from meta-physical ones. Finally, there is the claim that if a statement is true by virtue of the meaning of its constituent terms, then it cannot be empirically disconfirmed. If I have argued correctly, however, the above views of the relation between meaning and confirmation (and science and meta-physics) are defective. All statements can in principle be confirmed in the same way: by use of a confirmation machine: yet, not all statements are equivalent in meaning. Furthermore, no statement whether it is meaningless or meta-physical, can fail to be confirmable and disconfirmable in principle; and even if a statement is actually confirmed, it might still be meaningless. Finally, all statements, including analytic truths and contradictions, can be empirically confirmed or disconfirmed in principle

Consider, too, certain views of evidence and rationality. It has been widely held that all of our rational, non-fundamental beliefs can be justified in terms of other beliefs which ultimately rest on a foundation which either cannot be justified or does not need to be justified. The foundation has been said by different philosophers to consist of different elements: for example, definitions, self-evident truths, sense-datum statements, 'seeming' statements, etc. However, any such foundational view of knowledge is misleading. Some of our beliefs are more fundamental than others; but it is rational to accept even our most fundamental beliefs only so long as the evidence for certain less fundamental beliefs remains constant. There is, therefore, no necessity that our most funfamental beliefs remain fundamental, nor necessity that it always be rational to accept these beliefs. For example, if a confirmation machine is built, and if it works in the manner described earlier (in Section I), then our most well-confirmed belief may someday be the following: that what the confirmation machine indicates is true and what it indicates is not true is not true. If this belief does become central, then it may be rational for us to demote to a place of less importance what had formerly been a very fundamental belief – or to give up the belief altogether. This demotion or rejection might become rationally necessary if the confirmation machine were to provide evidence that either the belief or other beliefs which acted as its support were false.

The above point suggests one final conclusion which concerns confirmation and empiricism. Empiricists have often stressed the primacy of perception in the acquisition of knowledge. We are to trust the evidence of our senses (aided by instruments), except in special circumstances, instead of relying on rational insight, intuition, scripture, the testimony of authorities, the sayings of seers, prophets or witch-doctors, etc. It is seldom stressed, however, that the evidential value of perception depends upon the reasonableness of other beliefs. We believe, if we are empiricists, that perception is primary; but this belief is enmeshed in a complex network of other beliefs: if certain strands in the network were eliminated in order to accommodate new evidence, our belief in the evidential value of perception could be pulled from the center of our conceptual scheme to a location of less importance. Although it is unlikely, the sayings of seers or the issuances of a confirmation machine might someday prove to be of more epistemological importance, at least in certain respects, than the testimony of our senses. Given a certain body of evidence, *E*, concerning the reliability of seers or confirmation machines, it would be rational to trust the seer or the machine when they disagreed with our observation reports. Perception, it is true, would still have a peculiarly important status. We would still need to rely on our senses in order to obtain the evidence produced by the seer or the machine, but the evidential value of perception would nonetheless be significantly reduced. Furthermore, even the meager role left for perception might someday be eliminated. If the body of evidence overwhelmingly supported the continuing reliability, not of the seer nor of the confirmation machine, but of intuition, then perception might then be of no epistemological use at all. If you were to feel intuitively that a certain theory of physics were true or that a certain drug had therapeutic value, then it would be very probable that the theory were true or that the drug had value. Confirming our beliefs by seeing, or hearing, or touching might still be practiced by skeptics or old fashioned empiricists; but, at some point, such practices would be unnecessary if not irrational.

State University of New York, Stony Brook

NOTES

[1] I shall speak of statements as being confirmable or disconfirmable, but the argument would not be affected if for 'statement' we substituted 'assertion', 'belief', or 'sentence'.

[2] See: P. F. Strawson's review of John Passmore's *Philosophical Reasoning* in *Philosophical Books* (1962) and A. Grünbaum, 'Geometry, Chronometry and Empiricism', in *Minnesota Studies in the Philosophy of Science*, Vol. III (ed. by H. Feigl and G. Maxwell), University of Minnesota Press, Minneapolis, 1962, p. 429.

[3] For an argument in favor of the view that speaking of 'meaningless statements' is coherent, see: E. Erwin, *The Concept of Meaninglessness*, The Johns Hopkins Press, Baltimore, 1970, Chapter IV.

[4] For an ingenious attempt to show how the above statement might be confirmed, see: Sydney Shoemaker, 'Time Without Change', *Journal of Philosophy* (1969) 363–38.

[5] For example, we might get evidence that there is an inverse correlation between finding a white specimen at a given time of year and finding the entire species to be white.

[6] Another formulation which is also inadequate is that of A. G. N. Flew: "And anything which would count against the assertion, or which would induce the speaker to withdraw it and to admit that it must be mistaken, must be part of (or the whole of) the meaning of the negation of that assertion." See his, 'Theology and Falsification', reprinted in E. Nagel and R. Brandt, *Meaning and Knowledge*, Harcourt, Brace & World, Inc., New York, 1965.

IV. Probability, Statistics and Acceptance

JAAKKO HINTIKKA

UNKNOWN PROBABILITIES, BAYESIANISM, AND DE FINETTI'S REPRESENTATION THEOREM

In the development of the theory of subjective probability and in discussions about it an important role has been played by de Finetti's representation theorem and by a number of related results presented in de Finetti's classic monograph.[1] This theorem, together with the notion of exchangeability (among its aliases there are the terms equivalence, permutability, and symmetry), was originally put forward by Bruno de Finetti as a solution to a problem to which his subjectivistic interpretation of probability had led him. Although this theorem is mentioned in many expositions of the theory of subjective probability, there are not many satisfactory discussions of its significance available to philosophers. (The best one, and almost the only one, is Braithwaite's paper 'On Unknown Probabilities'.[2]) The purpose of the present paper is not to present any new results but rather to call attention to certain philosophical issues connected with it that apparently are not as fully appreciated as they ought to be.

As de Finetti himself indicates, the situation he is considering is virtually equivalent to the general inductive situation. The only restriction from a logician's point of view is that the set of concepts by means of which the inductive inferences that are considered here are thought of as being accomplished are somewhat restricted. A logician would say that these concepts are restricted to those available in a monadic first-order language. Without any loss of generality, it may be assumed that we have a complete classification of all possible individuals into (say) K pairwise mutually exclusive and collectively exhaustive classes. Let the predicates that accomplish this be $Ct_1, Ct_2, \ldots, Ct_K$, and let us call them Q-predicates. We are also presented with a domain of individuals $a_1, a_2, \ldots$, usually very large, often infinite. It is assumed that nothing is known about their interrelations. (If something were known, this would have to be expressed as an explicit assumption in some suitable language, which presumably would then have to be richer than a monadic one at least in that it contains relational symbols.) For this reason, when different individuals are considered one by one, we may (for the purpose of illustration) think of them as being drawn randomly from our domain of individuals.

Instead of speaking of *individuals*, de Finetti typically considers what he calls successive *events* or *experiments*. (I shall avoid the former term, however, since it is normally used in an entirely different sense in probability theory.) The attributes of individuals will then correspond to the different things we can say of the outcomes of these several experiments. (Of course, if events like experiments are to be considered as individuals[3] – in the logical sense of the word, i.e. as far as their logical type is concerned – there will not be much essential difference between the two points of view.) Anyhow, in the case of experiments, the assumption of ignorance with respect to the interrelations of the individuals will amount to an assumption that one experiment is as good as another, i.e. that we may consider them as repetitions of the same experiment. Probabilistically speaking, the outcomes of the successive experiments, considered alone, will then have the same distribution.

It is very tempting to strengthen this assumption and to assume that the successive experiments are probabilistically *independent*, that is, that we not only have $p(Ct_i(a_j))=p(Ct_i(a_k))$ for all individuals a_j, a_k but also $p(Ct_i(a_j) \mid e)=p(Ct_i(a_k) \mid e)$ for any evidence e consisting of observations of individuals other than a_j and a_k.[4] However, this very idea of independence is a problem – or seems to be – for a consistent subjectivist. For usually the independence assumption is expressed by speaking of the probabilities of the different events involved. An objectivist would explain it by saying that the probability that any given individual a_j has a fixed Q-predicate Ct_i is a certain definite number p_i independent of a_j *and* independent of our evidence, as long as this evidence consists of observations of other individuals. A frequentist would add that p_i equals the relative frequency of individuals with this property in the whole universe.

The trouble is that very often we do not know what these constant probabilities are but have to gather them on the basis of observations of individuals. At least this is what an objectivist would say. A subjectivist, however, is in a quandary even before he gets as far as this, if we are to believe de Finetti. According to him, the very idea of an unknown probability does not make any sense for a subjectivist (like himself). A probability is always someone's betting ratio, but an unknown probability cannot be that. (An unknown betting ratio is an absurdity.) Hence (according to de Finetti) a subjectivist should either give up talking about unknown probabilities or else explain what this talk amounts to in terms of his own assumptions. Now if these unknown probabilities are completely given up, the notion of independence becomes useless. In view of the importance of this concept, a vindication of it is an urgent need for a subjectivist. De Finetti's representation theorem and his notion of exchangeability are designed to accomplish such a vindication.

There are many purposes for which these unknown probabilities are apparently vital. The most important ones are undoubtedly the central results of probability theory known as the law of large numbers and as the central limit theorem. They are often formulated in terms of a sequence of *independent* random variables with the same distribution. The usual formulations of these vital results lose their meaning if the notion of independence is not available.

A brief philosophical excursion may be in order already at this stage. Certain qualifications are needed to what has been said. Often de Finetti's program is presented, as we just did partially, as an attempt to dispense with the notion of independence, which allegedly cannot be accommodated within the subjectivistic framework. This is encouraged by some of de Finetti's own formulations,[5] although it is interesting to see that no reference is made to independence in the crucial second chapter of his monograph. It is also true that de Finetti is (successfully) trying to replace certain limit theorems (laws of large numbers) relying on independence by theorems relying on the more acceptable notion of exchangeability (to be explained later). However, it is amply and trivially clear that there cannot in general be anything wrong with the notion of independence from a subjectivistic point of view.[6] For this notion can be defined in terms of one's probability distribution, two events A and B being independent if and only if $p(A \mid B)=p(A)$. The only plausible reasons a subjectivist can have for being unhappy with this definition are twofold: (i) The events A, B themselves may be suspect in some of the applications of the definition. (ii) The notion of independence so defined, though formally unobjectionable, may fail to perform the tasks it is commonly assumed to perform.

One of the main tasks of this kind is formulated by de Finetti as the question: "Why are we obligated in the majority of problems to evaluate a probability according to the observation of a frequency?" If we are willing to generalize a little bit, we might say that the main task we have to keep an eye on is that of understanding how we learn from experience. (A minimal requirement for such understanding is to be able to give a probabilistic account of the process in question.) The situation of the kind indicated above (monadic case with constant probabilities

$$(1) \qquad p(Ct_j(a_{n+1}) \mid (Ct_{i_1}(a_1) \,\&\, Ct_{i_2}(a_2) \,\&\, \ldots \,\&\, Ct_{i_n}(a_n))$$

independent of n and of the choice of a_{n+1}) is a case in point. If these constant probabilities are interpreted objectivistically, they are in many interesting applications unknown and have to be gathered from observations. A simple example is a biased coin which presumably exhibits a constant but unknown probability of heads and tails. What this probability is must be gathered from experience.

Of course de Finetti formulates the situation by saying that such objective unknown probabilities are conceptually unacceptable. However, we can now see that there exists a perfectly good problem here for anyone quite independently of any particular interpretation of the concept of probability. It is not so much that a subjectivist à la de Finetti cannot accept unknown probabilites. The main problem is to see what we can actually do in cases like that of the biased coin.

The best way of formulating the problem is not to say that it does not make sense of probability distributions which make all the probabilities (1) equal. Surely even an extreme subjectivist understands what such a probability distribution is like. If he confronted a gambler using such a probability distribution, a subjectivist would understand what the gambler were doing, but would put him down as cheating, practicing clairvoyance, or just behaving in an imprudent way, depending on whether the gambler's constant probabilities turn out to match long-range relative frequencies or not. What such a gambler is not doing is to learn from experience. What is bad about making (1) equal is that such a probability distribution, conceived subjectivistically as a set of somebody's betting ratios, will incorporate no learning from experience.

Somewhat similar remarks apply to laws of large numbers. The main point is *not* that these mathematical results do not make sense to a subjectivist, as they perhaps do not. Rather, what is wrong about them is that they do not apply to many typical cases in which we only gather (as an objectivist would say) what certain probabilities are from experience and hence cannot make the kind of independence assumption which these results presuppose for their applicability. Hence there exists on any interpretation of probability a perfectly good problem of offering similar results which do not involve the unrealistic assumption of independence.

There are nevertheless differences between an objectivist and a subjectivist here. By refusing to consider unknown probabilities de Finetti is – at least *prima facie* – depriving himself of certain ways of analyzing the situation. One thing an objectivist can do here is to use the unknown probabilities themselves as random variables. An objectivist can speak of the probability that an unknown probability assumes this or that value. Given a dented coin, he can speculate not only as to what the (unknown) probability is at which it yields heads, but also as to how probable it is that this probability should be between certain limits.

These probabilities are in fact very useful. Considering these probabilities of Type II (as they have been called by Good) a little more closely will in fact help us to find our bearings here. These probabilities often enable us to master the situation in a way we could not otherwise do. For instance, if we know that the unknown probability of heads has two possible values x_1 and x_2 whose probabilities are p_1 and p_2, respectively, then we can calculate our probability for obtaining heads by means of the usual of probability calculus. For instance, in virtue of the law of total probability, the probability of heads in the first throw will be

$$p_1 x_1 + p_2 x_2 .$$

The probability of obtaining n_1 heads and n_2 tails in specified throws among a totality of $n = n_1 + n_2$ throws will be

$$p_1 x_1^{n_1} (1 - x_1)^{n_2} + p_2 x_2^{n_1} (1 - x_2)^{n_2} . \quad (2)$$

If intead of just two values x_1 and x_2 we can have value x between 0 and 1 and if the probability-density is $\bar{p}(x)$ (i.e. if the probability that this value be between x and $x + dx$ is $\bar{p}(x) \cdot dx$), then we have instead of (2)

$$\int_0^1 \bar{p}(x)\, x^{n_1} (1 - x)^{n_2}\, dx , \quad (3)$$

provided $\bar{p}(x)$ is a continuous function or sufficiently like a continuous function for the ordinary integral-sign to make sense here.

More generally, let us consider, in the situation envisaged above, the probability of obtaining a sample of n_i *specified* individuals of the kind Ct_i (where $i=1, 2, ..., K$ and $\sum n_i = n$). Let us assume that the probability distribution function (density function) is $\bar{p}(x_1, x_2, ..., x_{K-1})$ (i.e. that the probability that each of the probabilities p_i is between $x_i + dx_i$ $(i=1, 2, ..., K-1)$ is $\bar{p}(x_1, x_2, ..., x_{K-1})\, dx_1\, dx_2 ... dx_{K-1}$ – whenever this makes sense). Such a function $\bar{p}$ is said to express the *probability density.*

If this function exists and is sufficiently well-behaved, the required probability is

$$Q(n_1, n_2, ..., n_K) = \int_0^1 \int_0^1 \cdots \int_0^1 x_1^{n_1} x_2^{n_2} \ldots x_{K-1}^{n_{K-1}}$$
$$(1 - x_1 - \cdots - x_{K-1})^{n_K}\, d_1 d_2 \ldots d_{K-1} \bar{p}(x_1, x_2. ..., x_{K-1}),$$

where $\bar{p}$ is called the distribution function.

In the discrete case we have

$$Q(n_1, ..., n_K) = \sum_{x_1, x_2, ..., x_{K-1}} x_1^{n_1} x_2^{n_2} \ldots x_{K-1}^{n_{K-1}}$$
$$\times (1 - x_1 - x_2 - \cdots - x_{K-1})^{n_K} \cdot p(x_1, x_2, ..., x_{K-1}),$$

where the sum is taken over all the admissible probability-value combinations $x_1, x_2, ..., x_{K-1}$ and where $p(x_1, x_2, ..., x_{K-1})$ is the probability of such a combination.

Thus in any case knowing the probabilities of the unknown probabilities (whether in the form of the pair p_1, p_2 or in the form of the function $p(x_1, x_2, ..., x_{K-1})$) is precisely what enables us to draw the vital conclusion as to what we may expect to find in observations concerning individuals (singular statements). In brief, on the basis of the unknown probabilities (and their probabilities) we can tell you how to bet on finite sequences of outcomes of our experiment.

This can be spelled out in greater detail. If we have a look at the last two formulas, we see that given the n_i's and the function $p(x_1, ..., x_{K-1})$ we can find the function Q. As was just said, this function Q gives us are probabilities that we can (according to de Finetti) bet on and which a subjectivist can according to him make good sense of. This can be made even more dramatic by expressing Q in terms of conditional probabilities

$$q_i(n_1, n_2, ..., n_K) = \frac{Q(n_1, n_2, ..., n_{i-1}, n_i + 1, n_{i+1}, ..., n_K)}{Q(n_1, n_2, ..., n_i, ..., n_K)}.$$

Thus what q_i gives us the probability that the next individual will be of the kind Ct_i when of the earlier ones n_j has been of the kind Ct_j, $1 \leq j \leq K$. It is readily seen that we have (for instance)

$$(4) \qquad Q(n_1, n_2, \ldots, n_K) = \prod_{i=0}^{n_1 - 1} q_1(i, 0, 0, \ldots, 0) \cdot \prod_{i=0}^{n_2 - 1} q_2(n_1, i, 0, \ldots, 0) \ldots \prod_{i=0}^{n_K - 1} q_K(n_1, n_2, \ldots, n_{K-1}, i)$$

Other representations are of course also possible. The fact that we have this multiplicity of connections between Q and the q_i's reflects the basic feature about this situation. It does not matter in what order the n_1 individuals of kind Ct_1, n_2 individuals of kind Ct_2, etc. come. It is only the sequence of numbers n_i $(i=1, 2, \ldots, K)$ that counts. This is reflected precisely by the fact that Q depends on $n_1, n_2, \ldots, n_{K-1}$ only, which implies the corresponding result for the q_i's. Conversely, if we are given the q_i's and it be required that the q_i's have this property (and the intuitive interpretation given above), then it follows that Q has the same independence property, provided that one more requirement (independence-of-path assumption) is satisfied. This requirement says that all the different product representations like (4) have to give the same result, independently of the order of the individuals.

This basic independence property was seen to be a consequence of the existence of the probability density (distribution) function $\bar{p}$ or the probability function p on the unknown probabilities. In short, it obtains if we are allowed to talk about unknown probabilities, and obviously represents part of the subjectivistic 'cash value' of the idea of unknown probabilities.

Now this independence feature can be taken to be the key to the present situation. That it obtains is what is meant by exchangeability. In other words, exchangeability might be characterized as follows: Suppose that we have assigned a probability to each finite sequence of observations concerning individuals. If the individuals that are observed are, in this order, $a_1, a_2, \ldots$, then the probabilities in question might be said to be the probabilities of the different statements of the form

$$(5) \qquad Ct_{i_1}(a_1) \; \& \; Ct_{i_2}(a_2) \; \& \ldots \& \; Ct_{i_n}(a_n)$$

where n is the length of the sequence we are considering. The exchangeability of our assignment will then mean that these probabilities should depend only on the number of differenet *Ct*-predicates in the sequence $Ct_{i_1}, Ct_{i_2}, \ldots, Ct_{i_n}$ in question, not on the order in which they make their appearance.

It is important to realize that exchangeability is a property that can be defined in terms of actual betting ratios, and hence makes sense for a subjectivist. It simply means certain things as to how one would bet on the truth of statements of form (5). This becomes especially striking in terms of our *q*-functions. For a large part of what it requires of them is that the probability that the (unique) individual to be observed next if of a certain kind should depend only on the numbers of the different kinds of individuals so far observed. What else, we may ask, is there to go on in this situation?

It is to be observed, however, that more has to be required of the functions q_i if they are to yield an exchangeable function Q, than merely that each of them depends on $n_1, n_2, \ldots, n_K$ only. We have to impose also an 'independence-of-the-path' assumption. We have to require that the q_i's yield a unique function Q independently of the order in which Q is 'built up of' the successive values of the different q_i's, as in (4).

Plausible though this 'independence-of-the-path' assumption is, it is not devoid of interest. A simple observation which illustrates this interest pertains to certain 'counter-inductive policies' which are sometimes considered in the philosophical literature on the subject. For instance, when $K=2$, an especially simple class of such counter-inductive policies is obtained by putting

$$q_1 = \frac{n - n_1 + \lambda/2}{n + \lambda} = \frac{n_2 + \lambda/2}{n + \lambda}$$

$$q_2 = \frac{n - n_2 + \lambda/2}{n + \lambda} = \frac{n_1 + \lambda/2}{n + \lambda}.$$

These functions are perhaps not completely foreign to certain (rather 'irrational') modes of thinking and of behaving under which 'luck' is conceived of as a commodity which can 'run out'. The more consistently experience has favored one alternative, the more certain it is to change over to the other one, this line of thought suggests. (The longer a run of heads one has obtained, the stronger one believes that the next throw will yield tails.) This of course is essentially what the counter-inductive policy just mentioned amounts to.

However, this counter-inductive policy is ruled out by the independence-of-the-path requirement, as one can see by means of simple examples.

Since exchangeability is thus a notion that pertains to Q or to the q_i's only, and since these allows an easy interpretation in terms of betting ratios, exchangeability is a notion that a subjectivist can feel happy about (*apud* de Finetti).

De Finetti's basic result is that an exchangeable Q (with the original interpretation as giving the probability that a particular sequence of Ct_i-predicates is observed) determines uniquely the probability distribution function p, which exists if Q is exchangeable. Mathematically, to speak of p is no better and no worse than to speak of Q: they determine each other completely. (This is not quite strictly true. However, the mathematical niceties which my statement disregards are not relevant here.) Hence, de Finetti in effect says, a subjectivist can speak freely of unknown probabilities and of probability distributions on them. These are mere mathematical fictions whose real meaning (for a subjectivist) can always be expressed by speaking of the associated Q-function that is immediately explainable as a system of betting ratios subject to certain conditions (which define exchangeability). The same goes for the notion of probabilistic independence, which is thus vindicated.

Since Q and the q_i's determine each other, we can say equally that the functions q_i determine uniquely the probability distribution which p or $\bar{p}$ expresses. In short, if we know how to bet on whatever *next* individual we can conceivably encounter, we know how to bet on unknown probabilities. Bets on the latter are nothing but systems of bets on the former.

What is there to be said about this result (which is of course *not* proved here)? It is in many ways an interesting and important result; in fact, I think its importance extends much beyond the meaning de Finetti himself gives to it. However, I suspect that its main function is *not* to get rid of unknown probabilities. What is wrong, anyway, for a subjectivist, in the idea of unknown probabilities? It is not their being unknown, it seems to me, that is relevant here. There is a sense in which betting ratios can be unknown. Examples are easily devised to bring home this truth. Suppose the number of dots on the different sides of a fair dice are determined by some chance device, say by throwing another fair dice which is completely normal. (All there is to the idea of fairness for a subjectivist is of course that he is willing to bet 1/6 for any given side to come up.) Now that is there more to the idea of *unknown* probabilities than the *hypothetical* betting ratios I would use if I *knew* which results the second dice had yielded?

This can obviously be generalized. Anyone who considers different objectively or subjectively interpreted probabilities and a probability distribution on them, may be taken to be considering an experiment with several different possible outcomes such that these outcomes determine the former probabilities. These unknown first-order probabilities can then be interpreted as betting ratios conditional on the different outcomes of the first experiment. *Unknown* probabilities become simply *conditional* probabilities.

So where is the difficulty? The difficulty, it seems to me, is due simply and solely to the fact that the outcome of the first experiment cannot always be directly verified. To see this, let us be even bolder than de Finetti and reduce the whole frequentistic theory at once to a subjectivistic one. Is this difficult? By no means. All we have to say is that the objective probability of an event, say of the next individual's having the predicate Ct_1, is the betting ratio I would use if I knew the relative frequency of individuals with this predicate in the whole universe, or perhaps its limit in successively more and more extensive observations. This betting ratio is unknown, it is true, but it is unknown merely because it is hypothetical. Why should this betting ratio fare worse than those we considered earlier? There is a difficulty here, it seems to me, only if a further assumption is made. Recalling the subjectivistic idea of probabilities as betting ratios, this assumption can be expressed by requiring that the question as to who wins a bet be *decidable on the basis of the observation of individuals* (singular events).[7] This requirement amounts to saying that what causes the problem here is not just de Finetti's *subjectivism* but as much what might be called his *positivism*. (This point has been made earlier in different terms by Braithwaite.) For what is there wrong about betting on the limit of an infinite sequence of relative frequencies? Nothing. There is no law, natural legal, or moral, against doing so. What causes difficulties here is solely the fact that we have to wait infinitely long in order to find out who has won such a bet. Consequently, the only thing that prevents a subjectivist from betting on probabilities as limits on relative frequencies, or from considering bets conditional on these limits having certain values, is the positivistic requirement that bets be decidable on the basis of observations of individuals. This is a difficulty that is now largely independent on the basic assumptions of the subjectivistic position; it is a difficulty (a perfectly good one) into which anyone will run who tries to use the concept of betting as applied to objectivistically interpreted probability-statements, of for that matter to any similar statements (e.g. general laws). It is a problem that also confronts a frequentist if he tries to use his own unknown objective probabilities as a guide to rational betting.

In a psychological experiment, subjects might easily be induced to bet on general statements. What might cause difficulties is merely the potential inquisitiveness of the subjects; they might start asking unpleasant questions as to how the experimenter decides who wins or loses a bet of this kind.

A partial answer is that *if* we had an omniscient oracle whom we could always consult concerning the truth of generalizations (or who could conceivably give such an answer after a finite length of time), even a positivist could make sense on bets on general statements. (In the experiments envisaged above, the experimenter has to play the role of such an oracle.) However, part of what a positivist disbelieves is precisely the possibility of making these decisions in a finite amount of time.

Thus de Finetti has not really explained what an objective probability (more specifically, a probability-statement interpreted as a statement of relative frequency) is in terms of subjective probabilities. Rather, he has explained what it means to bet on such objective probability-statements, i.e. what it means to associate probabilities to them subjectively. Although this is something a subjectivist might want to do, it is a problem that a frequentist might also be interested in, e.g. when he tries to find out what certain unknown probabilities actually are. (We might almost say: in so far as he is positivistically inclined.)

De Finetti's basic concepts can be further illustrated by relating them to those of Carnap. In the finite case, Carnap calls a state-description a statement of the form

$$(6) \qquad Ct_{i_1}(a_1) \,\&\, Ct_{i_2}(a_2) \,\&\, \ldots \,\&\, Ct_{i_N}(a_N),$$

where $a_1, a_2, \ldots, a_N$ are all the individuals of our universe. (It is assumed that the order of the conjuncts does matter.) A probability-distribution on state-descriptions determines the probabilities of *all* our statements. Such probability-distribution is called *symmetric* by Carnap if and only if all isomorphic state-descriptions (i.e. state-descriptions with the same numbers of respective Q-predicates occurring in (6), i.e. state-descriptions transformable into each other by permuting the names $a_1, a_2, \ldots, a_N$) receive the same probability.

A disjunction of all state-descriptions isomorphic to each other is called by Carnap a structure-description. The requirement of symmetry may thus be expressed by saying that all the disjuncts of a structure-description must be equally probably. It is readily seen that if the distribution of probabilities to our basic statements $Ct_i(a_j)$ satisfies the requirement of exchangeability, the underlying distribution of probabilities to state-descriptions is symmetric, and *vice versa*.

These observations relate in the first place to the case of a finite universe. However, by means of limit considerations we can extend them to infinite universes.

More important than any relation to Carnap's notion is the possibility of using the concept of exchangeability to do essentially the same job as independence traditionally does in results like the law of large numbers. De Finetti establishes some such results in his classical monograph on 'Foresight'. They are among the most important results of his work. (Results in the same direction are briefly mentioned by Carnap, too). They may be said to show that the notion of exchangeability in fact can serve the same purposes as the concept of independence in probability theory.

Adopting the point of view which was sketched above when the law of large number was first mentioned, the gist of these results may be expressed (very roughly) by saying that when the number of observed individuals grows, the probabilities q_i that the *next* individual is of the kind Ct_i will converge to the *actual relative frequencies* f_i of these kinds of individuals in the universe, provided that these exist. This is clearly a most interesting type of result. So far we have been thinking of the probabilities q_i subjectively as someone's betting ratios, assuming symmetry. Now we see that the assumption of symmetry (*or* exchangeability, for they are equivalent) may be interpreted as giving an asymptotic objective justification to these betting rations: they are seen to converge toward actual frequencies.[8] Hence they may be taken to be approximations of some sort toward these (limits of) relative frequencies.

Thus what de Finetti has in fact done is in effect much more than merely to enable a subjectivist to talk of probabilities attributed to objectively interpreted probability-statements. He has shown what a subjectivist has to do in order to make his subjective betting ratios to converge towards the (objectively true) relative frequencies of different kinds of things in our universe. I submit that it would be misleading to say that de Finetti has shown how to reduce objective probabilities to subjective ones. Rather, he has shown how to adapt subjective probabilities to objective probabilities (relative frequencies). Concessions are here being made as much by the subjectivist as by the objectivist.

Perhaps the best description of the significance of de Finetti's results is thus that they enable one to combine objective and subjective probabilities. On one hand, they enable us to conceive of subjective probabilities as approximations towards objective relative frequencies. On the other hand, they show a frequentist how to bet on (objective) probability statements and thereby to use subjective probability concepts in certain important contexts.

It is of course true that an objectivist cannot use directly his unknown probabilities for the purpose of betting on individual events, and that his betting ratios will therefore in the beginning be partly arbitrary (or at least subjective). But saying this is tautological. (We cannot use *directly* what is unknown.) What is important is the way in which one should allow experience to change one's betting ratios; and concerning this de Finetti and an objectivist are likely to agree. Indeed, the great general interest of de Finetti's results is due precisely to the fact that it throws some sharp light as to how one can rationally learn from experience. This was seen to be a question one can raise independently of one's interpretation of probability-statements, and hence de Finetti's insights has importance that extends way beyond any special difficulties of the subjectivistic theory of probability.[9]

This point can be elaborated. One thing de Finetti has criticized in the kind of situation we have considered is the idea of probabilistic independence. If all the probabilities (1) are equal, then the consecutive observations of individuals are independent of each other.

This notion de Finetti finds useless, primarily (it seems to me) because it turns on the use of unknown probabilities. As we have seen, he operates instead with the concept of exchangeability, which for him is a notion with clear-cut meaning.

Tables can be turned on de Finetti here, it seems to me. His own results show that we can deal with the unknown probabilities in the same way a Bayesian would deal with any unknown quantities. We assume some probability distribution on their different possible values, and we rely on experience (conditionalization with respect to observation-statements) to take us closer and closer to the actual values of these quantities. The exchangeability assumption may be thought of as being calculated to secure convergence to the right limit. For, according to de Finetti's theorem, different systems of exchangeable bets on individual events are tantamount to so many distributions of *a priori* probabilities on the unknown probabilities. Moreover, from the forms of the law of large numbers which de Finetti proves it follows that these probability distributions are subject to the same kind of justification as a Bayesian would normally strive to give to his methods in any case. What they show is (roughly speaking) that, whatever unhappy features the choice on one's *a priori* probabilities might have, they are in the long run cancelled by the effects of experience. Thus there is a sense in which once can say that de Finetti's results mark a gain in insight not so much for a subjectivist as for a Bayesian.[10]

It is rather interesting to see that in the first place we motivated the assumption of exchangeability in terms which were not tied to long-range success or any other sort of objective idea but rather to a kind of symmetry principle or principle of insufficient reason. We asked: apart from the numbers of observed individuals of the different kinds, what else is there to go on when we make predictions about the next individual? Now it turns out that the assumption so introduced has a kind of objectivistic justification too.

One way of illustrating the philosophical significance of de Finetti's line of thought is to apply it to a case in which the question of so-called unknown probabilities does not arise at all. Instead of the question as to what can be understood by bets on probability-statements one might ask what can be understood by bets on universal (general) statements. In finite domains, they are obtained as special cases of probability-statements by putting certain probabilities equal to zero. In so far as infinite domains are treated as limiting cases of large finite domains, the same applies to them. The result de Finetti's theorem gives may be expressed by saying that to bet on a universal statement is to bet in a certain specified manner on singular events. The fact that a subjectivistic interpretation of probability enables one to do this greatly enhances the value of this interpretation for a student of inductive probability.

Thus one of the results that de Finetti's theorem in effect has is to make it possible for a subjectivist to talk about the probabilities of general statements. This seems to me an important insight. It has been suggested by several philosophers that betting ratios offer the clearest and most concrete method of approaching probability concepts. Some of them have gone further and asserted that "we never bet on general laws" and that it is therefore inappropriate to attach probabilities to general statements.[11] To these views, de Finetti's line of thought gives a definitive answer. Presupposing the exchangeability assumption, it implies that a bet or a general law is simply to make an infinite system of bets of a certain specifiable kind on singular statements.

A similar conclusion also emerges from the recent 'foundational' work in probability logic (Gaifman, Scott & Krauss, Fenstad). Already Gaifman showed[12] that there is a certain systematic way in which the probabilities of singular statements determine the probabilities of quantified statements. In fact, $p((Ex)\, F(x))$ will be the least upper bound on the probabilities of finite disjunctions $p(F(a_1) \vee F(a_2) \vee \ldots \vee F(a_n))$ and dually for $p((x)\, F(x))$.

A partial perspective is thus also gained on the question whether induction is 'essentially' concerned with generalizations or with inferences from particulars to other particulars. Perhaps the most important fact here is that it cannot help being concerned with both. Since to determine the functions q_i is to determine one distribution of prior probabilities, one's inductive policies are completely fixed by the determination on how one bets on the *next* individual (on all possible finite bodies of evidence concerning fully observed particulars). Philosophers like Mill and Carnap therefore have a good conceptual reason for emphasizing inferences from particulars to particulars.

What is often overlooked, however, is that saying this by no means rules out assigning non-zero probabilities to generalizations. Even that decision is inevitably reflected in one's methods of betting on unobserved particulars. The best justification I can think of to Carnap's and Mill's concern with inferences from particulars to particulars thus turns their philosophical claims into a tautology.

A further general perspective suggested by these considerations, albeit somewhat indirectly, is the versatility of prior probability distributions, i.e. how flexibly and how completely we can sum up conceptually interesting features of one's inductive behavior by specifying certain features of one's prior probabilities. The 'behavioristic' notion of exchangeability goes together with the independence assumptions (which pertain to the prior distribution). Our remarks on general laws suggests that the important form of inductive behavior that characterizes one's belief in the lawlikeness of certain generalizations (strong belief or propensity to bet on relatively slim evidence that so far unobserved cases conform to the law) will inevitably – and fully – be reflected by one's prior probabilities. In another context, it will be shown in some simple cases that this is indeed so.

This power of prior distributions in codifying various important assumptions further suggests that the choice of these prior should be kept as flexible as possible, so as to enable us to express by their means whatever we on different occasions want to express. It seems to me that this possibility is very strongly Bayesian in spirit, and perhaps not too far removed from the spirit of de Finetti's work, either.[13]

Academy of Finland and Stanford University

NOTES

[1] Bruno de Finetti, 'La prévision: ses lois logiques, ses sources subjectives', in *Annales de l'institut Henri Poincaré*, Vol. 7 (1937). Translated by Henry E. Kyburg under the title 'Foresight: Its Logical Laws, Its Subjective Sources', in *Studies in Subjective Probability* (ed. by H. E. Kyburg and H. E. Smokler), John Wiley & Sons, New York, 1964, pp. 93–158.

[2] R. B. Braithwaite, 'On Unknown Probabilities', in *Observation and Interpretation: A Symposium of Philosophers and Physicists* (ed. by S. Körner), Butterworth, London, 1957, pp. 3–11.

[3] As some philosophers have urged – see e.g. Donald Davidson, 'The Logical Form of Action Sentences', in *The Logic of Action and Preference* (ed. by Nicholas Rescher), Pittsburgh University Press, Pittsburgh, 1967.

[4] In other words, for any e of the form

$$Ct_{i_1}(a_{i_1}) \,\&\, Ct_{i_2}(a_{i_2}) \,\&\, \ldots$$

where a_j, a_k are different from all the a_i's.

[5] For instance, he speaks of "the nebulous and unsatisfactory definition of 'independent events with fixed but unknown probability'" (Kyburg and Smokler, p. 142).

[6] In his article, 'Probability: Philosophy and Interpretation', in the *International Encyclopedia of Social Sciences*, de Finetti says that for a subjectivist "'independent'

means 'devoid of influence on my opinion'". Is this not a fairly clear and not at all nebulous notion?

[7] In de Finetti's own exposition (see Kyburg and Smokler, pp. 141–142), we find a contrast between biased coin and random drawing of balls from an urn of an unknown composition. What makes the difference between the two cases seems to be that the frequency of different kinds of balls in a (finite?) urn is "an objective fact which can be directly verified".

[8] As de Finetti himself expresses the point (Kyburg and Smokler, p. 142): "A rich enough experience leads us always to consider as probable future frequencies or distributions close to those which have been observed".

[9] Some caution is needed here, however. Although the notion of exchangeability together with the limit theorems that can be established in terms of it, throw some extremely interesting lights on the idea of learning from experience, the precise limits of the set of those probability distributions which can be said to represent (rational) learning from experience remains somewhat unclear.

[10] Notice especially how the switch from independence to exchangeability makes a Bayesian treatment possible here. This point is not always appreciated. In his paper, 'A New Approach to a Classical Statistical Decision Problem', in *Induction: Some Current Issues* (ed. by H. E. Kyburg and E. Nagel), Wesleyan University Press, Middletown, 1961, pp. 101–10; Herbert Robbins has discussed situations closely similar to de Finetti's examples. He compares his own recommendations with two others which he labels the Bayesian decision functions (p. 108) and the minimax decision functions (p. 107). I shall not discuss the first or the third here. As to a Bayesian treatment, Robbins restricts it to a case in which the random variables analogous to our properties *Ct* are *independent*. This, however, represents precisely the kind of refusal to learn from experience which de Finetti has rightly rejected and which to me seems antithetical to the spirit of Bayesianism.

[11] In his interesting book, *Logic of Statistical Inference*, Cambridge University Press, Cambridge 1965, pp. 215–16, Ian Hacking denies that it makes sense to bet on hypotheses like general laws. He appeals in so many words to de Finetti for support. However, it seems fairer to say that de Finetti's very results show precisely what it means for a subjectivist to bet on generalizations. The admissibility of generalizations to one's probabilistic discussion is shown by de Finetti's result to be a problem independent of whether one uses "betting rates as the central tool in subjective statistics".

[12] H. Gaifman, 'Concerning Measures on First-Order Calculi', *Israel Journal of Mathematics* **2** (1964) 1–18.

[13] In working on this paper, I have greatly profited from discussions with Dr. William K. Goosens. Among many other things, footnote 10 above is essentially due to him.

WILLIAM W. ROZEBOOM

NEW DIMENSIONS OF CONFIRMATION THEORY II: THE STRUCTURE OF UNCERTAINTY

You are, I am sure, just as aware as I am that the operational nodes of a complex problem, the points at which it can be split open to yield nuggets of new insight or achieve lasting advances, often lie in tediously technical details perhaps incomprehensible to all but specialists in the matter and anyways totally lacking in the romance and easy excitement which attract the topic's dilettantes. I would like you to hold fast to this appreciation, for the concerns I shall raise here are technical indeed. They are, however, *seminal* technicalities which, as we seek to fathom them, should fundamentally reshape our comprehension of the logic of confirmation, causal connectedness, and the foundations of statistical inference. In brief, it will be seen that certain powerful everyday intuitions concerning which propositions are confirmationally irrelevant to which others are exceedingly difficult to justify. When the roots of these intuitions are laid bare, an inherent intimacy emerges between the structure of rational uncertainty and presuppositions about nomic order.

I

Throughout what follows I shall make one important working assumption which, though far from apodictic, is not so controversial as to degrade the value of the sharp focus it makes possible. This is that in a system of rational beliefs, degrees of credibility can be expressed by a measure Pr(___ | ...) – read "the probability (credibility) that ___, given that..." – which obeys the axioms of the conditional probability calculus under the latter's standard propositional interpretation. For reasons not unlike certain points raised by Hacking (1967), I am far from convinced that this is an entirely appropriate basis on which to deal with all major issues of confirmation theory. Even so, it may well hold under sufficiently ideal conditions, such as the believer's being aware of all logical relations among the propositions in his belief system, and is in any event the only quantitative credibility model which currently enjoys a modicum of

Boston Studies in the Philosophy of Science, VIII.

provisional philosophic accord. The distinctive problems and part-solutions here developed in its terms should, I think, persist relatively unmodified into whatever more realistic account of credibility relations may eventually supersede it.

The basic theorems of this model are sufficiently familiar to require no review here. I will, however, remind you that for any background information k and propositions p and q,

(1) $$\Pr(p \cdot q \mid k) = \Pr(q \mid k) \times \Pr(p \mid q \cdot k),$$

of which a simple corollary is that

(2) $$\frac{\Pr(p \mid q \cdot k)}{\Pr(p \mid k)} = \frac{\Pr(p \cdot q \mid k)}{\Pr(p \mid k) \times \Pr(q \mid k)} = \frac{\Pr(q \mid p \cdot k)}{\Pr(q \mid k)}.$$

If we introduce the 'confirmation ratio', CR, by the definition

(3) $$\mathrm{CR}(p, q \mid k) =_{\mathrm{def}} \frac{\Pr(p \mid q \cdot k)}{\Pr(p \mid k)}$$

as a measure of the degree to which p confirms q relative to background k, (2) may be notationally simplified to

(2a) $$\mathrm{CR}(p, q \mid k) = \frac{\Pr(p \cdot q \mid k)}{\Pr(p \mid k) \times \Pr(q \mid k)} = \mathrm{CR}(q, p \mid k),$$

which says that under any background information, p confirms q to exactly the same degree that q confirms p. In particular, insomuch as q confirms, is confirmationally independent of (i.e. is indifferent to), or disconfirms p given background k according to whether $\mathrm{CR}(p, q \mid k)$ is respectively greater than, equal to, or less than unity,[1] (2) entails that whether one proposition confirms, is indifferent to, or disconfirms another is symmetric in the two propositions.

A second principle which will soon be needed is

(4) $$\frac{\Pr(p \mid q \cdot r \cdot k)}{\Pr(p \mid k)} = \frac{\Pr(p \mid q \cdot k)}{\Pr(p \mid k)} \times \frac{\Pr(p \mid q \cdot r \cdot k)}{\Pr(p \mid q \cdot k)},$$

the sense of which can perhaps best be grasped from its confirmation-ratio equivalent

(4a) $$\mathrm{CR}(p, q \cdot r \mid k) = \mathrm{CR}(p, q \mid k) \times \mathrm{CR}(p, r \mid q \cdot k).$$

In fact, to save later distraction we may as well establish now some more complicated CR-relationships which will eventually prove useful. (The reader who is willing to take on faith my proofs of Theorems 1 and 2, p. 355 below, may omit the remainder of this section.) The proofs of these had best be prefaced with a word about extreme probabilities. Many otherwise straightforward conditional-probability theorems require boundary restrictions to fend off degeneracies which sometimes arise from zero probabilities. (E.g., basic principle $\Pr(p \mid k) = \Pr(p \cdot q \mid k) + \Pr(p \cdot \sim q \mid k)$ fails when k is inconsistent, while the otherwise exceptionless equivalence $\Pr(p \mid q) = \Pr(p \cdot q)/\Pr(q)$ is unreliable when $\Pr(q) = 0$ insomuch as its right-hand side is then ill-defined even though its left-hand side may be perfectly determinate.) To minimize conceptual and visual affront, I will list the nonzero probabilities upon which each result here is conditional in a parenthesized addendum wherein 'p' and 'p/k' abbreviate '$\Pr(p)$' and '$\Pr(p \mid k)$', respectively. Thus '(Nonzero: p, p/k)' following a lemma or theorem advises that it presupposes $\Pr(p) > 0$ and $\Pr(p \mid k) > 0$. Some of the listed boundary restrictions are not strictly necessary, but they can be relaxed only with difficulty and we shall have no need for the increased generality this would afford. In a couple of places, the proofs as here given presuppose additional nonzero probabilities which are neither required to be so by these theorems nor are listed among their nonzero boundary conditions. Modification of the proofs to accommodate zero values of these probabilities should be obvious and will be left to the reader.

For principles (5) and (6), let $r_1 \ldots, r_n$ be a set of mutally exclusive and jointly exhaustive propositions, i.e. $\sum_{i=1}^{n} \Pr(r_i) = 1$ and $\Pr(r_i \cdot r_j) = 0$ for $i \neq j$. Then,

$$(5) \qquad \sum_{i=1}^{n} \Pr(r_i \mid k) \times \mathrm{CR}(p, r_i \mid k) = 1. \quad \text{(Nonzero: } p/k\text{)}.$$

(PROOF: Since $\sum_{i=1}^{n} \Pr(p \cdot r_i \mid k) = \Pr(p \mid k)$ while $\Pr(p \cdot r_i \mid k) = \Pr(p \mid k) \times \times \Pr(r_i \mid k) \times \mathrm{CR}(p, r_i \mid k)$, dividing through by $\Pr(p \mid k)$ establishes the theorem.)

$$(6) \qquad \mathrm{CR}(p, q) = \sum_{i=1}^{n} \Pr(r_i) \times \mathrm{CR}(p, r_i) \times \mathrm{CR}(q, r_i) \times \mathrm{CR}(p, q \mid r_i) \quad \text{(Nonzero: } p, q\text{)},$$

or equivalently,

(6a) $$\mathrm{CR}(p, q) = \sum_{i=1}^{n} \Pr(r_i \mid p) \times \mathrm{CR}(q, r_i) \times \mathrm{CR}(p, q \mid r_i).$$

(Nonzero: p, q).

(PROOF: $\Pr(p \cdot q \cdot r_i) = \Pr(r_i) \times \Pr(p \cdot q \mid r_i) = \Pr(r_i) \times \Pr(p \mid r_i) \times \Pr(q \mid r_i) \times [\Pr(p \cdot q \mid r_i) / \Pr(p \mid r_i) \times \Pr(q \mid r_i)] = \Pr(r_i) \times \Pr(p) \times \Pr(q) \times \mathrm{CR}(p, r_i) \times \mathrm{CR}(q, r_i) \times \mathrm{CR}(p, q \mid r_i)$. Substituting into $\Pr(p \cdot q) = \sum_{i=1}^{n} \Pr(p \cdot q \cdot r_i)$ then yields (6), while (6a) follows in turn by noting that $\Pr(r_i) \times \mathrm{CR}(p, r_i) = \Pr(p \cdot r_i) / \Pr(p) = \Pr(r_i \mid p)$. Theorems (6) and (6a) also hold, of course, relative to any consistent background information k.)

Equations (5) and (6) deserve interpretive comment. By letting $\{r_i\}$ be the pair $r, \sim r$, (5) may be seen to justify our intuition that p confirms a proposition iff $\sim p$ disconfirms it. (6) is more surprising, for it shows that $\mathrm{CR}(p, q)$ is guaranteed to lie between $\mathrm{CR}(p, q \mid r)$ and $\mathrm{CR}(p, q \mid \sim r)$ only if r is confirmationally independent either of p or of q. Otherwise, it is possible e.g. for p to confirm q both given r and given $\sim r$, yet to disconfirm it unconditionally.

LEMMA 1: If $\mathrm{CR}(p, q \mid k) = 1$, then $\mathrm{CR}(p, \sim q \mid k) = 1$. (Nonzero: p/k, $\sim q/k$).

LEMMA 1′: If $\Pr(q \mid k) = 0$, then $\mathrm{CR}(p, \sim q \mid k) = 1$. (Nonzero: p/k).

(PROOF: Let $w = \Pr(q \mid k)$, $a = \mathrm{CR}(p, q \mid k)$, $b = \mathrm{CR}(p, \sim q \mid k)$. Then from (5) we have $wa + (1 - w)\, b = 1$, under which $a = 1$ entails $b = 1$ so long as $w < 1$. If $w = 0$, then $b = 1$ unconditionally.)

LEMMA 2: If $\mathrm{CR}(p, q \mid r) = \mathrm{CR}(p, q \mid \sim r) = 1$, then $\mathrm{CR}(p, q) = 1$ only if either $\mathrm{CR}(p, r) = 1$ or $\mathrm{CR}(q, r) = 1$. (Nonzero: p, q, r).

LEMMA 2′: If $\Pr(q \mid r) = 0$ and $\mathrm{CR}(p, q \mid \sim r) = 1$, then $\mathrm{CR}(p, q) = 1$ only if $\mathrm{CR}(p, r) = 1$. (Nonzero: p, q, r).

(PROOF: Let $v = \Pr(r \mid p)$, $w = \Pr(r)$, $a = \mathrm{CR}(q, r)$, $b = \mathrm{CR}(q \mid \sim r)$. Then by (6a) under the lemma's assumptions, $\mathrm{CR}(p, q) = 1$ entails $va + (1 - v)\, b = 1$. But by (5) we also have $wa + (1 - w)\, b = 1$. Combining these equations and collecting terms yields $(v-w)(a-b)=0$, which holds only if either $v=w$ or $a=b$. But $v=w$ entails $\mathrm{CR}(p, r)=1$ when $w>0$, while $a=b$ when $wa+(1-w)b=1$ only if a, i.e. $\mathrm{CR}(q, r)$, equals 1 – which establishes Lemma 2. For Lemma 2′, note from (4a) and (6a) that $\mathrm{Pr}(r \mid q)=0$ entails $\mathrm{CR}(p, q)=\mathrm{CR}(p, \sim r)\times\mathrm{CR}(p, q \mid \sim r)$, from which Lemma 2′ is then obvious.)

LEMMA 3: If $\mathrm{CR}(p, q \mid r)=\mathrm{CR}(p, q \mid \sim r)=1$, then $\mathrm{CR}(p, q)=1$ only if either $\mathrm{CR}(q, r)=1$ or $\mathrm{Pr}(p \mid q\cdot r)=\mathrm{Pr}(p \mid q\cdot r)$. (Nonzero: $p, q, r, \sim r, p/r, \sim q/r$).

LEMMA 3′: If $\mathrm{Pr}(q\cdot r)=0$ and $\mathrm{CR}(p, q \mid \sim r)=1$, then $\mathrm{CR}(p, q)=1$ only if $\mathrm{Pr}(p \mid q\cdot\sim r)=\mathrm{Pr}(p \mid \sim q\cdot r)$. (Nonzero: $p, q, r, \sim r, p/r$).

(PROOF: By Lemma 1, given the stipulated nonzero probabilities, $\mathrm{CR}(p, r)=1$ and $\mathrm{CR}(p, q \mid r)=1$ respectively entail $\mathrm{CR}(p, \sim r)=1$ and $\mathrm{CR}(p, \sim q \mid r)=1$. Hence by (4a), $\mathrm{CR}(p, r)=1$ and Lemma 3′s assumptions jointly entail $\mathrm{CR}(p, q\cdot\sim r)=\mathrm{CR}(p, \sim r)\times\mathrm{CR}(p, q \mid \sim r)=1$ and $\mathrm{CR}(p, \sim q\cdot r)=\mathrm{CR}(p, r)\times\mathrm{CR}(p, \sim q \mid r)=1$, whence $\mathrm{Pr}(p \mid q\cdot\sim r)=\mathrm{Pr}(p)=\mathrm{Pr}(p \mid \sim q\cdot r)$. Reference to Lemma 2 completes the proof of Lemma 3. Lemma 3′ follows similarly from Lemma 2′ by noting that when $\mathrm{Pr}(r)>0$, $\mathrm{Pr}(q\cdot r)=0$ entails $\mathrm{Pr}(q \mid r)=0$ and hence, by Lemma 1′, $\mathrm{CR}(p, \sim q \mid r)=1$.)

LEMMA 4: If $\mathrm{CR}(p, r \mid s)=\mathrm{CR}(p, r \mid \sim s)=\mathrm{CR}(q, s \mid r)=\mathrm{CR}(q, s \mid \sim r)=1$, and p and r jointly entail q while q and s jointly entail p, then both $\mathrm{CR}(p, r)=1$ and $\mathrm{CR}(q, s)=1$ only if either $\mathrm{CR}(r, s)=1$ or $\mathrm{Pr}(p\equiv q \mid r\cdot\sim s)=\mathrm{Pr}(p\equiv q \mid \sim r\cdot s)=1$. (Nonzero: $p, q, r, s, \sim r, \sim s, p/r, q/s, \sim s/r, \sim r/s$).

LEMMA 4′: If $\mathrm{Pr}(r\cdot s)=0$, $\mathrm{CR}(p, r \mid \sim s)=\mathrm{CR}(q, s \mid \sim r)=1$, and p and r jointly entail q while q and s jointly entail p, then both $\mathrm{CR}(p, r)=1$ and $\mathrm{CR}(q, s)=1$ only if $\mathrm{Pr}(p\equiv q \mid r\cdot\sim s)=\mathrm{Pr}(p\equiv q \mid \sim r\cdot s)$. (Nonzero: $p, q, r, s, \sim r, \sim s, p/r, q/s$).

(PROOF: By Lemma 3, Lemma 4's CR-assumptions imply that $\mathrm{CR}(p, r)=\mathrm{CR}(q, s)=1$ only if either $\mathrm{CR}(r, s)=1$ or both $\mathrm{Pr}(p \mid r\cdot\sim s)=\mathrm{Pr}(p \mid$

$\sim r \cdot s$) and $\Pr(q \mid r \cdot \sim s) = \Pr(q \mid \sim r \cdot s)$. Similarly, since $\Pr(r \cdot s) = 0$ entails $\Pr(r \mid s) = \Pr(s \mid r) = 0$ when $\Pr(r)$ and $\Pr(s)$ are nonzero, Lemma 3′ and the assumptions of Lemma 4′ imply that $\mathrm{CR}(p, r) = \mathrm{CR}(q, s) = 1$ only if both $\Pr(p \mid r \cdot \sim s) = \Pr(p \mid \sim r \cdot s)$ and $\Pr(q \mid r \cdot \sim s) = \Pr(q \mid \sim r \cdot s)$. To complete the proof, let h_1 and h_2 abbreviate $r \cdot \sim s$ and $\sim r \cdot s$, respectively. Then if $p \cdot r$ entails q while $q \cdot s$ entails p, we have $\Pr(p \cdot \sim q \mid h_1) = \Pr(\sim p \cdot q \mid h_2) = 0$ while $\Pr(p \mid h_1) = \Pr(p \cdot q \mid h_1)$ and $\Pr(q \mid h_2) = \Pr(p \cdot q \mid h_2)$. The latter equations, together with the already established $\Pr(p \mid h_1) = \Pr(p \mid h_2)$ and $\Pr(q \mid h_1) = \Pr(q \mid h_2)$, yield $\Pr(p \cdot q \mid h_1) = \Pr(p \cdot q \mid h_2) + \Pr(p \cdot \sim q \mid h_2)$ and $\Pr(p \cdot q \mid h_1) + \Pr(\sim p \cdot q \mid h_1) = \Pr(p \cdot q \mid h_2)$, which can hold only if $\Pr(\sim p \cdot q \mid h_1) + \Pr(p \cdot \sim q \mid h_2) = 0$, i.e. only if $\Pr(\sim p \cdot q \mid h_1) = 0$ and $\Pr(p \cdot \sim q \mid h_2) = 0$. Then $\Pr(p \equiv q \mid h_1) = 1 - [\Pr(p \cdot \sim q \mid h_1) + \Pr(\sim p \cdot q \mid h_1)] = 1$ while similarly $\Pr(p \equiv q \mid h_2) = 1$.)

II

Let us begin, as one so often does in the philosophy of confirmation, with Hempel's classic 'paradox of confirmation' concerning empirical support for conditional generalities of form

(7) All *A*s are *B*s.

Since it is intuitively obvious – or so it seems – that (7) is confirmed by observing an *A* which has property *B*, i.e. by evidence of form $Aa \cdot Ba$, while by the same logic, evidence of form $\sim Ba \cdot \sim Aa$ confirms

(8) All non-*B*s are non-*A*s,

why is it that observing some non-*B* to be a non-*A* feels confirmationally irrelevant to (7) when (7) and (8) are prima facie logically equivalent? I have previously explored this puzzle – which is a paradox of intuition only, not of logic – at some length (Rozeboom, 1968); but with the aid of principle (4a) its heart can be bared in a few sentences.

Consider first the alleged confirmation of (7) by 'positive instance' $Aa \cdot Ba$. Whatever may be the most appropriate technical reading of 'All *A*s are *B*s' – and as we shall see, this is not nearly so univocal as has often been assumed – it is analytically clear that (7) and the hypothesis that some particular object *a* is an *A* jointly entail that *a* is a *B*; hence given any background information k,

$$(9) \qquad \mathrm{CR}((7), Ba \mid Aa \cdot k) = \frac{\Pr(Ba \mid Aa \cdot (7) \cdot k)}{\Pr(Ba \mid Aa \cdot k)} = \frac{1}{\Pr(Ba \mid Aa \cdot k)} \geq 1,$$

in which the inequality is strict so long as *Ba* is not already certain given just *Aa* and *k*. Thus for natural *k* (specifically, which does not pre-empt the force of (7) for the observation in question), *Ba* is guaranteed to confirm (7) given *k* and *Aa*. This, however, is not quite the conclusion wanted here, for the new evidence with which observation of favorable object *a* augments our background information *k* is that *a* is both an *A* and a *B*. To justify our confirmational intuition in this case it needs be shown that 'All *A*s are *B*s' is confirmed under *k* by conjunctive datum $Aa \cdot Ba$, not just by datum *Ba* once *Aa* is also established.[2]

Happily, principle (4a) makes completely clear the nature of this distinction. For by direct substitution we have

$$(10) \qquad \mathrm{CR}((7), Aa \cdot Ba \mid k) = \mathrm{CR}((7), Aa \mid k) \times \mathrm{CR}((7), Ba \mid Aa \cdot k).$$

Hence if datum *Aa* is by itself confirmationally irrelevant, given *k*, to whether all *A*s are *B*s – i.e. so long as $\mathrm{CR}((7), Aa \mid k) = 1$ – joint observation *Aa* and *Ba* confirms this generality to exactly the same degree, given *k*, as *Ba* confirms it given *Aa* as well as *k*. However, a greater-than-unity value for the second confirmation ratio on the right in (10) cannot prevent a sufficiently low value of the other from dragging their product below unity. Thus despite confirmation of (7) by *Ba* given *Aa* and *k*, the total epistemic import of $Aa \cdot Ba$ for (7) given *k* can indeed be disconfirmatory if *Aa* by itself disconfirms (7) given *k*.

But how, you ask, could observing just that something is an *A* possibly make any difference for whether all *A*s are *B*s – at least under natural background information which does not include some philosophical contrivance such as 'If anything is an *A*, then all *A*s are *B*s'? For we know from (2a) that *Aa* confirms (7) if and only if (7) confirms *Aa*, and surely the hypothesis that all *A*s are *B*s implies nothing about whether some object of yet-unidentified character will prove to be an *A*!? Your surprise nicely illustrates how profoundly we have allowed unexamined intuition to dominate our thinking on this matter. I do not suggest that this intuition is baseless and should be ignored. Quite the opposite: My contention here, as before (Rozeboom, 1968), is that it is an extrusion of our beliefs about nomic order which promises to yield important new analytical leverage on the latter. Even so, it is an intuition which turns out to be generally in error. This point, and its possible significance, is the main concern of this essay; but since I have already aroused the beast of Hempel's paradox, which others have so often sought to slay, I will pause long enough to pull its fangs before proceeding to the treasure which lies beyond

Regardless of whether generalities (7) and (8) are entirely equivalent, the bite of Hempel's paradox lies in our intuition that finding a non-*B* which is not an *A* – a 'positive instance' of (8) – ought not to matter for whether all *A*'s are *B*'s. However, by principle (4a), the degree of (7)'s confirmation by datum $\sim Aa \cdot \sim Ba$ given background *k* factors as

$$(11) \qquad \mathrm{CR}((7), \sim Aa \cdot \sim Ba \mid k) \\ = \mathrm{CR}((7), \sim Aa \mid k) \times \mathrm{CR}((7), \sim Ba \mid \sim Aa \cdot k).$$

If intuition is right to insist, as it does for natural *k*, that once something is known to be *not* an *A* its *B*-state has no further relevance to whether or not all *A*s are *B*s, then the last term in (11) is unity and observing that *a* is neither an *A* nor a *B* confirms (7) to the very same degree as does datum $\sim Aa$ alone – which, in turn, confirms or disconfirms (7) just in case *Aa* disconfirms or confirms it, respectively. Consequently, once it becomes clear that generalized conditionals are *not* as a rule confirmationally independent of data concerning just their antecedents, then it will no longer seem strange that $\sim Aa \cdot \sim Ba$ might confirm (7), or, alternatively, that $\sim Ba$ may be so disconfirmatory of all non-*B*s being non-*A*s that joint observation $\sim Ba \cdot \sim Aa$ does *not* confirm the latter and hence not that all *A*s are *B*s, either.

III

Why should I expect you to take seriously the patently silly proposal that whether or not all *A*s are *B*s is generally relevant to whether some particular thing is an *A*? Because once one stops to investigate the matter it soon becomes evident that this must be so. The deeper question is why intuition should be so insistent to the contrary and what special circumstances, if any, would make this intuition correct?

Consider first of all that insomuch as (7) and *Aa* jointly entail *Ba* for any object a, (7) by itself entails the extensional $Aa \supset Ba$. Now, having a false antecedent suffices – as is well known and often rued – to make an extensional conditional true. Since $\sim Aa$ thus confirms $Aa \supset Ba$, we are assured by principle (5) that Aa disconfirms it. But from principle (6a), since $\Pr(\sim[Aa \supset Ba] \mid (7) \cdot k) = 0$, (or from (4a), since (7) is equivalent to $(7) \cdot [Aa \supset Ba]$),

(12) $$\mathrm{CR}((7), Aa \mid k) = \mathrm{CR}(Aa \supset Ba, Aa \mid k) \times \mathrm{CR}((7), Aa \mid [Aa \supset Ba] \cdot k),$$

in which the first term on the right is, as just observed, less than 1, while the last term therein is unity if the relevance of (7) for Aa given k is mediated entirely by $Aa \supset Ba$. Hence, *unless* 'All *A*s are *B*s' *has some bearing on a's being an A over and above the import of* $Aa \supset Ba$ *for this, datum Aa disconfirms that all As are Bs via its disconfirmation of* $Aa \supset Ba$. Moreover, under the classic extensional reading of 'All *A*s are *B*s' as

(13) $(x)(Ax \supset Bx)$,

it becomes extraordinarily difficult to see what relevance this generality *could* have, under natural k, for a's prospects on B except through its instantiation $Aa \supset Ba$ for a.

However, we need not insist that (13) captures the full ordinary-language force of 'All *A*s are *Bs*'. Indeed, a good reason for *not* assuming this is simply that we do *not* normally consider data of form Aa to confirm this conditional generality. There are, in fact, at least two other prime candidates for the interpretation of all/are statements. One is to paraphrase (7) as

(14) If anything is an A, then it is a B,

which emphasizes universal quantification over a propositional connective and may hence be formalized as

(14a) $(x)(Ax \rightarrow Bx)$,

and then to construe the if/then connective $\rightarrow$ as stronger than material implication. Alternatively, (7) may be understood as a statistical assertion of form

(15) The proportion of *A*s which are *B*s is r,

or

(15′) The probability of *B*-ness, given *A*-hood, is *r*,

both of which may be formalized as

(15*a*) $\mathrm{fr}(B \mid A) = r$

so long as the distinction between statistical probability and de facto relative frequency is not crucial for the purpose at hand. Schemata (14a) and (15a) differ in two important technical details. In the first place, (15a) contains a numerical parameter *r* whose unitary value in the statistical interpretation of (7) is merely one of alternative *r*-possibilities whose non-extreme cases we understand fully as well as we do $r=1$; whereas to give (14a) comparable scope we would have to introduce a whole family of implicational connectives $\overset{r}{\rightarrow}$ such that for $r<1$, *Aa* and $Aa\overset{r}{\rightarrow}Ba$ only make *Ba* likely to a certain degree. Secondly, generalizations of form (14a) entail stronger-than-extensional conditionalities between particular events, namely, that for every object *a*, the event of *a*'s *A*-ing produces a *B*-ing by *a*.[3] In contrast, (15a) expresses a second-level relation between properties *A* and *B* which logically entails no conclusion about any particular object *a* except when $r=1$, in which case (15a) implies only the extensional conditional $Aa \supset Ba$.[4] The differences among (13), (14a), and (15a) are sufficiently great that no conclusion about the confirmational behavior of one is certain to hold for the others as well. And if either (14a) or (15a) can be shown to be confirmationally independent of *Aa*, that one should clearly replace (13) as our top-drawer explication of (7) if we wish to preserve the latter's full intuitive force.

That (14a) may well have the confirmational property we seek is made plausible by recalling the sustained failure of philosophy of science to come up with a creditable extensional analysis for subjunctive and counterfactual conditionals. It is essential to the meaning of such statements as

> If the bullet had struck a quarter-inch lower, Jones would have been killed instantly,

and

> If Dull Victory wins the Derby this year, his stud fees will triple,

that falsity of their antecedents not be a sufficient condition for their truth as a whole. But if refuting the antecedents of such conditionals is analytically forbidden to verify them, can the truth state of their antecedents make any confirmational difference for them at all? Certainly commonsense is adamant that knowing how Dull Victory's stud fees will be affected by the outcome of this year's Derby is totally useless for judicious betting on the race itself. When we assert 'If p then q' in this sense, we mean that if p is/had been the case, then q is/would have been the case *because* of p. In particular, when this conditionality is thought to be causal,[5] 'If p then q' envisions p as bringing q about – i.e. *forcing* q to occur – by some principle of inter-event action.[6]

It might thus seem to be analytically true of the if/then connective in subjunctive and counterfactual conditionals that for any propositions p and q,

$$(16) \qquad \Pr(q \mid p\cdot[p \rightarrow q]\cdot k) = 1$$

for every k with which p and $p \rightarrow q$ are compatible, and

$$(17) \qquad \Pr(p \mid [p \rightarrow q]\cdot k) = \Pr(p \mid k)$$

for most natural k, including in particular the null-background case where k is tautological. Indeed, this seemed reasonable enough to me previously (Rozeboom, 1968) to warrant defining a concept of 'modalic entailment' in terms of these properties. But unhappily, (17) proves to be far too tough a condition to hold in any significant generality – as I will now show not just for the conjectured propositional connective $\rightarrow$ but for any construction which has the confirmational properties ascribed to $\rightarrow$ by (16, 17).

Let H be a propositional schema, two different completions of which entail propositions H_{pq} and H_{qp}, respectively, which has if/then force in that for any two propositions r and s, r and H_{rs} jointly entail s. For example, if H is the schema '___ $\rightarrow$ ...', then H_{pq} and H_{qp} are '$p \rightarrow q$' and '$q \rightarrow p$', respectively, while if H is the schema 'All ___s are ...s' or some proposed explication thereof, H_{pq} and H_{qp} might be 'If *Aa* then *Ba*' and 'If *Ba* then *Aa*' for object a.[7] (Heuristically, H_{pq} may be read as 'If p then q' or 'p brings q', but particular versions of H may endow H_{pq} with a much richer meaning than just this.) Our problem is now to assess how general may be the backgrounds k under which

$$(18) \qquad \Pr(p \mid H_{pq}\cdot k) = \Pr(p \mid k), \quad \Pr(q \mid H_{qp}\cdot k) = \Pr(q \mid k),$$

or equivalently,

(18a) $\quad CR(p, H_{pq} \mid k) = 1, \quad CR(q, H_{qp} \mid k) = 1,$

obtain jointly. The point of the question of course is that if propositions p and q are essentially alike in their logical relations to k, then if k is a natural background under which p's truth state is intuitively irrelevant to whether or not p brings q, q should likewise be irrelevant under k to whether q brings p. This is especially so if k is null (i.e. tautological) while p and q have the same logical form.

Unfortunately, however, (18) holds only under constraints far too strong, even when k is null, to demand of a rational belief system in any generality. To begin, note that (18) directs us to consider jointly the two pairs of possibilities, that p may or may not bring q and that q may or may not bring p. Together, these yield four mutually exclusive and jointly exhaustive alternatives, $H_{pq} \cdot H_{qp}$, $H_{pq} \cdot \sim H_{qp}$, $\sim H_{pq} \cdot H_{qp}$, and $\sim H_{pq} \cdot \sim H_{qp}$, the first of which is apt to be analytically impossible if H embodies a strong sense of conditionality. (E.g., whereas $p \supset q$ and $q \supset p$ are clearly compatible for most p and q, p's being causally responsible for q presumably precludes q's being a cause of p.) Now: Does information about possibility H_{qp} upset the intuitive irrelevance of p to H_{pq}? Surely not. Yet if not, we are impaled on the dilemma which follows from Lemma 4 by substituting H_{pq} and H_{qp} for r and s, respectively – or worse, if H_{pq} and H_{qp} are incompatible, are spitted upon the single shaft which similarly protrudes from Lemma 4′:

THEOREM 1: If $CR(p, H_{pq} \mid k) = 1$ and $CR(q, H_{qp} \mid k') = 1$ when k is variously null, H_{qp}, and $\sim H_{qp}$, and k' is variously null, H_{pq}, and $\sim H_{pq}$, then either $CR(H_{pq}, H_{qp}) = 1$ or $\Pr(p \equiv q \mid H_{pq} \cdot \sim H_{qp}) = \Pr(p \equiv q \mid \sim H_{pq} \cdot H_{qp}) = 1$. (Nonzero: $p, q, H_{pq}, H_{qp}, \sim H_{pq}, \sim H_{qp}, p/H_{pq}, q/H_{qp}, \sim H_{qp}/H_{pq}, \sim H_{pq}/H_{qp}$).

THEOREM 2: If H_{pq} and H_{qp} are incompatible while $CR(p, H_{pq} \mid k) = 1$ and $CR(q, H_{qp} \mid k') = 1$ when k is variously null and $\sim H_{qp}$, and k' is variously null and $\sim H_{pq}$, then $\Pr(p \equiv q \mid H_{pq} \cdot \sim H_{qp}) = \Pr(p \equiv q \mid \sim H_{pq} \cdot H_{qp}) = 1$. (Nonzero: $p, q, H_{pq}, H_{qp}, \sim H_{pq}, \sim H_{qp}, p/H_{pq}, q/H_{qp}$).

Theorems 1 and 2 tell us that intuitive principle $CR(p, H_{pq} \mid k) = 1$ can be had for the choices of k listed therein only at a price – or, if H_{pq} and H_{qp} are not mutually exclusive, at one of two prices. The price $\Pr(p \equiv q \mid H_{pq} \cdot \sim H_{qp}) = \Pr(p \equiv q \mid \sim H_{pq} \cdot H_{qp}) = 1$, which is our only option for strong senses of conditionality, is totally unacceptable, for it says that given if-p-then-q but not if-q-then-p or conversely, then p if and only if q. This would be self-contradictory were the if/then force of H merely extensional implication, and is in any event an absurd requirement to impose on a belief system. But even for weak if/thens which allow the alternative option in Theorem 1, $\mathrm{CR}(H_{pq}, H_{qp}) = 1$ does not seem like a reasonable stricture either; for even if H_{pq} and H_{qp} are compatible, why should information about the one *inherently* make no difference for the credibility of the other? It may be concluded, then, that intuitive principle $\mathrm{CR}(p, H_{pq} \mid k) = 1$ can be defended even for null k only by abandoning it when k is H_{qp} or $\sim H_{qp}$.

But is this a reasonable retrenchment? For while p should be more likely given H_{qp} than given $\sim H_{qp}$, it is extremely difficult to see how H_{pq} could have any more confirmational bearing on p once H_{qp} or $\sim H_{qp}$ is given than it does in the absence of the later information. This is especially true of $\sim H_{qp}$, which by Theorem 2 is all that matters for strong if/thens. If knowledge that p brings q tells nothing about whether p is the case, how does information that q does not bring p disturb that irrelevance? But if intuition can be this wrong about $\mathrm{CR}(p, H_{pq} \mid k)$ when k is $\sim H_{qp}$, dare we trust it at all on this point?

Moreover, we cannot restore (18) to confirmation-theoretic health just by relinquishing the intuitive indifference of p to H_{pq} given H_{qp} or $\sim H_{qp}$, for even then (18) remains incompatible with many credibility combinations which we have no present reason to consider irrational. To show this in an extreme case, suppose that $\Pr(H_{pq} \cdot H_{qp}) = \Pr(\sim H_{pq} \cdot \sim H_{qp}) = 0$, i.e. that H_{pq} and H_{qp} are thought to be jointly exhaustive as well as mutually exclusive. Then $\Pr(p \mid H_{pq}) = \Pr(p)$ and $\Pr(q \mid H_{qp}) = \Pr(q)$ jointly imply $\Pr(p \mid H_{pq}) = \Pr(p \mid H_{qp})$ and $\Pr(q \mid H_{pq}) = \Pr(q \mid H_{qp})$, from which in turn, since $\Pr(p \cdot \sim q \mid H_{pq}) = \Pr(\sim p \cdot q \mid H_{qp}) = 0$, it follows (cf. proof of Lemma 4) that $\Pr(p \equiv q \mid H_{pq} \cdot \sim H_{qp}) = \Pr(p \equiv q \mid \sim H_{pq} \cdot H_{qp}) = \Pr(p \equiv q) = 1$. Yet why should one who is convinced that either H_{pq}-or-H_{qp} is the case be required to deny all possibility of $p \cdot \sim q$ or $q \cdot \sim p$ when neither of these two precluded alternatives is at all incompatible with $H_{pq} \cdot \sim H_{qp} \vee \sim H_{pq} \cdot H_{qp}$?

One can idle away many a cheerful hour in search of unusually outrageous constraints which follow from (18), especially if still another intuition almost as strong as (18), namely, that H_{pq} should not affect the credibility of q given not-p, is also thrown into the game; and I will not totally foreswear a future return to this matter. Meanwhile, until we learn much more about what, over and above axiomatic coherence (i.e. agreement with the axioms of the probability calculus) determines the rationality of a belief system, hoping to find some sense of conditionality with built-in confirmational indifference to its antecedent is like yearning for the gold at rainbow's end. A proposition which implies that if-p-then-q *may* be confirmationally irrelevant to p in particular instances, but only because certain other credibilities in the system, notably those involving the parallel possibility that if-q-then-p, oddly happen to hit upon just the right numerical values. In general, then, we must assume that $\Pr(p \mid k$ and if-p-then-$q)$ is *not* equal to $\Pr(p \mid k)$, not even for null k, nor can this lack of indifference be expunged by a sufficiently ingenious contrivance of new conditionality concepts.

But what, then, are we to make of the intuition which insists so strongly that q's being conditional upon p has no relevance to whether p is the case? Is it just a witless blunder which deserves no further heed? I think not. It is at the very least a philosophically profound (*contra* silly) error which, sedated against naive overexuberance, may yet be usefully rehabilitated.

There are perhaps two reasons, one a superficial reflection of the other, why it is so tempting to consider if-p-then-q indifferent to p. Superficially, this is a failure to distinguish logical independence from informational irrelevance. Unlike the relation between an extensional conditional and its antecedent, whose negations are incompatible, the possibility envisoned by a subjunctive if-p-then-q has no logical relevance to whether p is the case. Hence it is no easier for commonsense to appreciate that if-p-then-q might inform about p than it can see how learning that John has red hair could affect the credibility that John loves Mary – one's natural inclination in both cases being to assume that there is no confirmational relevance at all. However, logical and confirmational independence are *not* the same, neither analytically nor extensionally. It is quite appropriate for our learning of John's redheadedness to modify the credence we give to his loving Mary if e.g. we suspect that redheads are more passionate than other people; and knowing that p brings q would indeed be evidence for

p were it believed that whatever establishes subjunctive dependencies also tends to actualize their antecedents.

But I am being unfair to intuition here. My examples point out that there may well be natural background information under which logically independent propositions are confirmationally related. But commonsense does not dispute that; rather, the confirmational-independence intuition becomes ascendent only with an increasing empoverishment of background information under which the logical possibilities for one state of affairs become increasingly symmetric in their relations to the possibilities for another. In the limiting case of *no* background information, then, perhaps logical and confirmational independence do coincide after all, at least extensionally. I venture that the deep reason why (18) seems so plausible is simply that for null k it is an instance of

The Fundamental Indifference Intuition [*FII*]. If propositions p and q are logically independent, then $\mathrm{CR}(p, q \mid k)=1$ if k is null.

However, even apart from qualms about whether credibility relations are really well-defined at all relative to no background whatsoever, it is important to realize that *FII* is dramatically untenable so long as 'logical independence' is understood in its usual sense whereby two or more propositions are logically independent (of one another) if all their truth-combinations are logically possible. Let p and q be any two logically independent propositions while $r =_{\mathrm{def}} p \cdot q \vee \sim p \cdot \sim q$. Then any two of the three propositions p, q, r are logically independent of one another (since for p and r, $p \cdot r$, $p \cdot \sim r$, $\sim p \cdot r$, and $\sim p \cdot \sim r$ are logically equivalent to $p \cdot q$, $p \cdot \sim q$, $\sim p \cdot q$, and $\sim p \cdot q$, respectively, and similarly for r's logical independence of q), so by *FII*,

$$\text{(19)} \qquad \Pr(p \cdot q) = \Pr(p) \times \Pr(q), \quad \Pr(p \cdot r) = \Pr(p) \times \Pr(r), \quad \Pr(q \cdot r) = \Pr(q) \times \Pr(r).$$

From (19), letting $w = \Pr(p \cdot q)$, $x = \Pr(p \cdot \sim q)$, $y = \Pr(\sim p \cdot q)$, $z = \Pr(\sim p \cdot \sim q)$, we have

$$\text{(20)} \qquad w = (w + x)(w + y), \quad w = (w + x)(w + z), \quad w = (w + y)(w + z),$$

which a little algebra shows to obtain if and only if either w, x, y, or z is unity or $w = x = y = z$, i.e. only if either (*i*) $\Pr(p)$ and $\Pr(q)$ are both extreme (i.e. zero or unity) or (*ii*) $\Pr(p) = \Pr(q) = 0.5$. But a credibility system which satisfies either (*i*) or (*ii*) for every pair of logically independent propositions is stunningly degenerate. (The degeneracy of (*i*) is obvious; while for a system containing three logically independent propositions p, q, and r, since r is then also logically independent of $p \cdot q$ and $p \cdot \sim q$, (*ii*)'s holding for all of these would require $\Pr(p) = \Pr(p \cdot q) = \Pr(p \cdot \sim q) = 0.5$, which is impossible.) Hence in general, $\mathrm{CR}(p, q) \neq 1$ even when p and q are logically independent, and not merely does our intuitive skill at recognizing confirmational relevance fail in relatively esoteric cases like (18), it is demonstrably rotten at the root.

Or is it? May there not exist some reading of 'logical independence', stronger than its technically standard sense, under which *FII* is tenable after all? (For example, the conceptual linkage between p and $p \cdot q \vee \sim p \cdot \sim \sim q$ in my disproof of *FII* might seem to violate the spirit if not the letter of 'logical independence', though to be sure if *FII*'s interpretation of this were to allow no conceptual overlap at all between logically independent propositions it would no longer explain the intuitive confirmational independence between a subjunctive conditional and its antecedent.) Indeed, unless it is arbitrary which propositions in a belief system are confirmationally independent under null background information, this must be so. For if there is some *basis* for propositions p and q being confirmationally independent under null background while p and r are not, even though there are no more logical exclusions on truth-combinations of p and q than of p and r, then a strong sense of 'logical independence' which yields *FII* can be defined on this basis. Contrapunctally, if a person considers p and q, but not p and r, to be confirmationally independent under null background, then, unless such opinions are rationally a matter of whim, that person has in effect committed himself to some possibility for how things are, namely, to whatever would make his particular allocation of confirmational independencies an appropriate one.

In short, the situation is this: Given a boolian algebra $\{p_i\}$ of propositions, there are infinitely many different unconditional credibility distributions $\{\Pr(p_i)\}$ over these which are 'formally admissible' in that they satisfy the restrictions of the probability calculus. Are all these admissible credibility allocations equally rational, or do some have greater epistemic merit than others? Nothing in the confirmation-theoretic literature, whether by philosophers, statisticians, or practicing scientists, has yet established any convincing grounds on which to prefer one admissible distribution of unconditional credibilities to another.[8] Yet were this completely arbitrary, then so would be the conclusion of any ampliative argument; for if premise p neither entails nor is incompatible with prospective conclusion q given background information k, then for any nonnegative real number n there exists an admissible unconditional credibility distribution under which $\mathrm{CR}(p, q \mid k) = n$. Accordingly, if the unconditional credibilities in a person's belief system were entirely free parameters, normatively constrained only by formal admissibility, then apart from logical entailments the degree to which any given proposition confirms another would likewise be arbitrary. Yet surely, when I think that p confirms q given k while you think that p disconfirms it, we have a genuine cognitive dispute rather than just a mismatch of personal tastes in free parameters. Even if your credibility system agrees with mine in rating $\mathrm{Pr}(p \cdot q \cdot k) \times \mathrm{Pr}(k)$ higher than $\mathrm{Pr}(p \cdot k) \times \mathrm{Pr}(q \cdot k)$, so that for both of us $\mathrm{CR}(p, q \mid k) > 1$, I may still wonder if we have *correctly* construed p to confirm q given k. Is it then a mistake for me to feel normative qualms about whether the structure of uncertainty in my belief system – i.e. what evidence confirms what conclusions under what background information – is exactly as it *should* be even though my credences are in perfect accord with the constraints of the probability calculus? Is there not *something* which, could I but become cognizant of it, would reassure me/confirm my suspicions that my uncertainty structure is satisfactory/misguided?

I do not pose these questions rhetorically, for the proper directions in which to answer them seem far from clear to me. I do, however, insist that the questions themselves are extraordinarily important ones whose full depths and complexities will unfold only gradually as we tease apart the issues which coil within them. Previously, I have ventured that placements of confirmational independence within a person's belief system are intimately tied to his suppositions about the world's responsibility order (Rozeboom, 1968), and the final consideration to be explored here will provide powerful reinforcement for that conjecture along lines rather different from my earlier argument.

IV

We have yet to probe the confirmational import of datum *Aa* for generalization 'All *A*s are *B*s' when this is interpreted statistically. Section III has already established (contrary to my intimations in Rozeboom, 1968) that the credibility of *Aa* cannot be generically indifferent to information about the frequency of *B*-ness among things which are *A*s, but it will nonetheless prove instructive to examine the specifics of this case.

To expedite the discussion, I had best review a few technical concepts which will be needed. A scientific *variable* **X** over a domain or 'population' *P* of objects is a function from *P* into a set $\{X_i\}$ of attributes (or, if **X** is numerically scaled, into numbers which represent attributes) such that for each object *a* in *P* exactly one attribute in $\{X_i\}$ applies to *a*, this attribute being what is meant by the *value* of variable **X** for argument *a*. (For details, see Rozeboom, 1961, 1966.) For conceptual simplicity – and technically, this is an *enormous* simplification which, happily, is entirely harmless for present purposes – I shall speak as though a variable has only a finite number of alternative values.

The statistical *distribution* of variable **X** in population *P* is a function $\mathrm{fr}(\mathbf{X})$ which maps each value X_i of **X** into the relative frequency or statistical probability $\mathrm{fr}(X_i)$ of attribute X_i in *P*, while more generally the joint distribution of two (or, similarly, more) variables **X** and **Y** in *P* is a function $\mathrm{fr}(\mathbf{XY})$ mapping each combination of a value X_i of **X** with a value Y_j of **Y** into the relative frequency or statistical probability $\mathrm{fr}(X_iY_j)$ in *P* of the conjunction of attributes X_i and Y_j. The *conditional* distribution of **Y** given **X** in *P* is a function $\mathrm{fr}(\mathbf{Y} \mid \mathbf{X})$ mapping each pair of values X_i of **X** and Y_j of **Y** into the relative frequency or statistical probability $\mathrm{fr}(Y_j \mid X_i)$ $[=\mathrm{fr}(X_i Y_j)/\mathrm{fr}(X_i)]$ of Y_j among just those members of *P* which have attribute X_i, while the conditional distribution of **Y** in *P* given a particular value X_i of **X** is $\mathrm{fr}(\mathbf{Y} \mid \mathbf{X})$ restricted to X_i.

Very often what is of statistical interest about a given distribution is not so much its unanalyzed entirety as certain of its parametric features such as means, variances, correlations, etc. If *D* is the family of alternative distributions logically possible for a given set of variables in a certain population, then a *parameter* of *D* is any many-one function from *D* into numbers (or sometimes more complex abstract entities), while a *parameterization* of distribution family *D* is a set of parameters such that each distribution in *D* is uniquely identified by the totality of its values on these parameters. (Henceforth, when I speak of a distribution 'family' I shall always mean the set of alternative distribution possibilities for a fixed set of variables in a particular population.) There are many alternative ways to parameterize a given distribution family; however, if $\langle \xi, \Omega, \ldots \rangle$ and $\langle \xi', \Omega', \ldots \rangle$ are two different parameterizations of D then each parameter in the one set is completely specified by the parameters in the other, i.e. for the first parameter in the second set (and similarly for the others) there exists a function ϕ such that $\xi'(d) = \phi\ [\xi(d), \Omega(d), \ldots]$ for each d in D. Parameters $\xi, \Omega, \ldots$ of distribution family D are all *logically independent* (of one another) iff any logically possible value on any one of these parameters is logically compatible with all logically possible combinations of values on the others.

Finally, a notational ellipsis: When discussing our uncertainty about the numerical values of a given distribution's parameters, it will be convenient to let 'The value of parameter ξ for distribution d is ξ_i', be abbreviated simply as 'ξ_i'. Then $\Pr(\xi_j \mid \Omega_i)$ is the credibility that the joint distribution at issue has value ξ_j of parameter ξ given that its value of parameter Ω is Ω_i.

When 'All As are Bs' is interpreted statistically as the claim that $\mathrm{fr}(B \mid A) = 1$, the theory of its confirmation by instances falls under the more general problem of determining the value of a statistical parameter from sample data. In our present case, the distribution at issue is that of two dichotomous variables **A** and **B**, whose values are A, $\sim A$ and B, $\sim B$, respectively, in some population which may here be described noncommittally as the class T of 'things'. Because these variables are both dichotomies, their joint distribution in T can be exhaustively specified by just three numbers, two such parameterizations being

$$(21) \qquad \alpha =_{\mathrm{def}} \mathrm{fr}(A), \quad \beta =_{\mathrm{def}} \mathrm{fr}(B \mid A), \quad \gamma =_{\mathrm{def}} \mathrm{fr}(B \mid \sim A),$$

and

$$(22) \qquad \alpha' =_{\mathrm{def}} \mathrm{fr}(B), \quad \beta' =_{\mathrm{def}} \mathrm{fr}(A \mid B), \quad \gamma' =_{\mathrm{def}} \mathrm{fr}(A \mid \sim B).$$

(In what follows, 'α', 'β', etc. will be used specifically as defined in (21) and (22).) Each parameter in (22) is a determinate function of the parameters in (21), namely,

$$(23) \qquad \alpha' = \alpha\beta + (1 - \alpha)\gamma, \quad \beta' = \frac{\alpha\beta}{\alpha\beta + (1 - \alpha)\gamma},$$

$$\gamma' = \frac{\alpha(1 - \beta)}{(1 - \beta) + (1 - \alpha)(1 - \gamma)}$$

and similarly with sets (21) and (22) interchanged. *Within* each set (21) or (22), however, the parameters are all logically independent of one another. (Still another parameterization of this distribution is

$$\mathrm{fr}(A \cdot B), \quad \mathrm{fr}(A \cdot \sim B), \quad \mathrm{fr}(\sim A \cdot B).$$

These last are not logically independent, however, insomuch as 1 less the value of any one is an upper bound on the value of any other.)

What does observing that something is an A tell us about the joint distribution of **A** and **B** in T? Or rather, to emphasize our present special interest, when is a parameter ξ of fr(**AB**) *not* informed about by datum Aa – i.e. under what circumstances is it the case that for every value ξ_i of ξ, $\mathrm{CR}(\xi_i, Aa \mid k)=1$? (Henceforth, unless there is need for explicit mention of background information, I will let k be null.) One parameter to which Aa clearly ought to be confirmationally relevant is α, i.e. fr(A), and indeed, under the orthodox assumption that $\Pr(Aa \mid \alpha_i)=\alpha_i$, it can easily be shown that

$$\mathrm{CR}(Aa, \alpha_i) = \frac{\alpha_i}{\mathrm{Exp}(\alpha)},$$

where $\mathrm{Exp}(\alpha)$ is the mean of the credibility distribution over possible values of α. More useful for present purposes, however, is an assumption which will let us get at Aa's significance for the rest of fr(**AB**), namely, that the relevance of datum Aa for any parameter ξ of fr(**AB**) is mediated by whatever Aa tells about the frequency of As. Specifically, I shall assume that for all values α_i of α and ξ_j of ξ, Aa adds noting to what α_i tells about ξ_j, i.e. that $\Pr(\xi_j \mid \alpha_i \cdot Aa)=\Pr(\xi_j \mid \alpha_i)$ or, equivalently,

$$\text{(23)} \qquad \mathrm{CR}(\xi_j, Aa \mid \alpha_i) = 1. \quad \text{(assumed)}$$

(Adoption of principle (23) – which would have to be hedged against certain quirky special cases were it to be defended in complete generality – is strictly provisional here, but as a working idealization it seems reasonable in ways which can be verbalized, and it is hard to think of an alternative which makes any intuitive sense.) From (23) by (6a) we then have

$$\text{(24)} \qquad \mathrm{CR}(\xi_j, Aa) = \sum^{i} \Pr(\alpha_i \mid Aa) \times \mathrm{CR}(\xi_j, \alpha_i),$$

in which summation is over all possible values of α. Equation (24) says that the degree to which datum Aa confirms that fr(**AB**) has a particular value ξ_j of parameter ξ is a weighted average of the degrees to which the various possible values of α respectively confirm ξ_j. If parameters α and ξ are confirmationally independent of one another – i.e. if $\mathrm{CR}(\xi_j.\ \alpha_i)=1$ for all α_i and ξ_j – then $\mathrm{CR}(\xi_j, Aa)=1$. If α and ξ are *not* confirmationally independent, it is still possible for the right-hand side of (24) to equal unity, but only under special allocations of credibility which, so far as we have any reason to suspect, obtain in only a vanishingly small proportion of rational credibility systems.[9] Hence,

THEOREM 3. A parameter ξ of the joint distribution of variables **A** and **B** in population T is confirmationally independent of datum Aa if and, for all practical purposes, only if ξ and $\mathrm{fr}(A)$ are confirmationally independent of each other.

In particular, letting arbitrary parameter ξ be β, we have

COROLLARY. $\mathrm{fr}(B \mid A)$ is confirmationally independent of Aa if and, for all practical purposes, only if it is confirmationally independent of $\mathrm{fr}(A)$.

Under assumption (23), whether observing an A has any relevance to whether statistically all As are Bs thus reduces to whether information about the frequency of A-hood makes any confirmational difference for the conditional frequency of B-ness among things which are As. This, in turn, falls under the more general question of which statistical parameters are confirmationally independent of which others. So by rights, to achieve resolution on this point it should suffice to consult what advanced statisticians have had to say about it. However, extant technical doctrine on credibility relations among statistical parameters can be summarized with shocking brevity: *There is none.* Or rather, the theory of this exists only implicitly in real-life statistical practices which are as prevalent as they are unreasoned, namely,

The Statistical Independence Presupposition [*SIP*]. If parameters ξ, Ω, ... of a given distribution family are logically independent under background information k, then $\xi, \Omega, \ldots$ are also confirmationally independent given k.

It is not profitable to attempt documenting the universality of this presupposition here,[10] but readers familiar with the practice of partitioning sample data as a conjunction of 'sufficient' statistics, each of whose likelihood function depends only upon a proper subset of the unknown parameters, will also recall that sample values of these sufficient statistics are construed to inform *only* about their own specific likelihood parameters. As a less sophisticated illustration, a researcher who searches his journals for some indication of the correlation between certain variables would feel totally unenlightened if he could find reports only of their means and variances.

Yet no matter how attractively *SIP* may intuit in particular cases, it is altogether untenable as a general principle. For it can easily be proved that if $\xi, \Omega, \ldots$ and $\xi', \Omega', \ldots$ are nontrivial[11] alternative logically independent parameterizations of the same distribution family, and parameters $\xi, \Omega, \ldots$ are all confirmationally independent of one another, then this is *not* generally also true of $\xi', \Omega', \ldots$. In particular, this holds for alternative parameterizations (21) and (22) of fr(**AB**). To show this by means of an extreme example, suppose that our belief system gives nonzero and, say, equal credibility to only two values, 0.2 and 0.8, of α, and that the same is also true of β and γ. Then there are eight nonzero-credibility alternatives for fr(**AB**), one for each row in Table I, all having the same probability of 0.125 if α, β, and γ are confirmationally independent. If the confirmation ratios $\mathrm{CR}(\alpha'_i, \beta'_j)$, $\mathrm{CR}(\alpha'_i, \gamma'_k)$, and $\mathrm{CR}(\beta'_j, \gamma'_k)$ are worked out for all values of α', β', and γ' which appear in the last three columns of Table I, these are seen to be either zero (complete disconfirmation), 4, or 8 (both

TABLE I

Values of parameters (21) and (22) under eight different possibilities for fr(**AB**)

α	β	γ	α'	β'	γ'
0.2	0.2	0.2	0.20	0.20	0.20
0.2	0.2	0.8	0.68	0.06	0.50
0.2	0.8	0.2	0.32	0.50	0.06
0.2	0.8	0.8	0.80	0.25	0.25
0.8	0.2	0.2	0.20	0.75	0.75
0.8	0.2	0.8	0.32	0.50	0.94
0.8	0.8	0.2	0.68	0.94	0.50
0.8	0.8	0.8	0.80	0.80	0.80

strong and in some cases complete confirmation.) When the marginal credibility distributions for α, β, and γ are more realistically continuous, the confirmational relations among α', β', and γ' under confirmational independence of α, β, and γ are too complex (at least for me) to evaluate analytically. However, computer-assisted finitary approximations indicate that even when the marginal credibilities over α, β, and γ are perfectly flat (i.e. all their possible values are equally likely), the average deviation of the $\mathrm{CR}(\alpha'_i, \beta'_j)$, $\mathrm{CR}(\alpha'_i, \gamma'_k)$, and $\mathrm{CR}(\beta'_j, \gamma'_k)$ from unity is about 0.2, while if two of the three parameters in (21) are known fairly precisely when the third is rather uncertain, the confirmational relations among α', β', and γ' become quite strong.[12] It is, in fact, a nice question whether *any* nondegenerate distribution of credibilities over the possibilities for $\mathrm{fr}(\mathbf{AB})$ permits confirmational independence among parameters (21) to be combined with the same for parameters (22).

"Very well", I can hear you mutter, "so logically independent statistical parameters are not in general also confirmationally independent. What of it?" Don't turn off too quickly, for there is indeed much to be made of this. First and least important, it substantiates my earlier claim that the $\mathrm{fr}(B \mid A) = 1$ reading of 'All *A*s are *B*s' will not yield systematic confirmational indifference of conditional generalities to instances just of their antecedents. Secondly, it proposes a contribution to real-life statistical technology: Since in practice we always act as though estimates of one statistical parameter tell us nothing about the others which interest us when these are logically independent of the first under our background assumptions, we have not been reaping the full statistical harvest provided by our sample data and should revise our statistical models to exploit the parameter correlations which, we now see, must lurk somewhere within our belief system if it is formally admissible. Conceivably, we could even parlay these improved statistical methods into more powerful or more economical research designs. (It would, for example, save an *awful* lot of experimental effort if the regression of one variable upon another could be reliably estimated from the distribution of the latter alone.) However, I must in all honesty confess that any hopes I might promulgate for so enhancing our statistical engineering prowess would be essentially bogus. There are reasons to doubt that confirmational dependencies among logically independent statistical parameters are often large enough for practical significance except under conditions to which applied statisticians already accommodate intuitively through the background assumptions of the models they adopt on particular occasions.[13] Even so, it should be of considerable value for advanced statistics to work out the likely magnitude of these dependencies under various idealized and realistic circumstances, and more generally to think through the *logic* of confirmationally correlated statistical parameters for whatever new wisdom this may bring to the theory and practice of inductive inference.

An essential part of the aforementioned logic will be some normative standards concerning which parameters of what distributions *should* be confirmationally independent under a given informational background. For if I take the first of two different parameterizations of the same distribution family to be confirmationally independent under background k while for you this seems true of the second set, how is our disagreement to be appraised? Is it just a misalignment of arbitrary whims, or are we not each making some cognitive committment in which at least one of us is *wrong*? Obviously much needs to be said on this matter, a great deal of it requiring considerable sophistication in statistics and experimental design, and needing first of all to grope out some sense of direction in those many sectors of the problem where even professionally sensitized intuition still flounders perplexedly. Even so, regarding what is probably the most basic of all ways to partition a joint statistical distribution, educated intuition does seem to me to speak with some assurance even if it takes a moderately trained ear to discern its message. This is

The Causal Structure Criterion [*CSC*]. With some qualifications, every parameter of $\text{fr}(\mathbf{X})$ in population P is confirmationally independent of every parameter of $\text{fr}(\mathbf{Y} \mid \mathbf{X})$ in P under background information k if and, for all practical purposes, only if k entails that variable **X** has complete causal precedence over variable **Y** in P. By '**X** has complete causal precedence over **Y** in P' is meant that any causal antecedents common to **X** and **Y** affect the variance of **Y** in P only through the mediation of **X**, i.e. that any causal coupling between **X** and **Y** not held constant in P is directed from **X** to **Y** rather than from **Y** to **X** or to **X** and **Y** jointly from a mutual source.

In particular, for our two dichotomous variables **A** and **B**, **A** has complete causal precedence over **B** iff our uncertainty about the frequency of *A* is independent of our uncertainty about the frequency of *B* both among things that are *A*s and things that are non-*A*s.

I am not going to discuss *CSC* in any detail here, much less try to justify it, for I voice it not as any sort of *conclusion* but merely as a scientifically realistic point of departure for what needs to be an extended series of probes and exchanges on possible relations between the confirmational grain of our statistical beliefs and our suppositions about nomic order. I describe *CSC* as 'scientifically realistic' because it embodies two powerful undercurrents in professional thinking on multivariate analysis and research design, namely, (i) that some parameterizations of a statistical distribution align more closely with its causal structure than do others, and (ii) that features of the world sharing no common determinants whatsoever are independent in some fashion stronger than mere logical independence. From (i), we have that fr $(\mathbf{Y} \mid \mathbf{X})$ is a pure reflection of **Y**'s causal dependence upon **X** (in both its character and degree) only if **Y** is not, conversely, a source of **X** and no additional sources of **Y** unmediated by **X** also influence **X**. Otherwise, the statistical conditionality of **Y** upon **X** conflates **X**'s causal import for **Y**, if any, with **Y**'s noncausal predictive regression on **X** in ways that, for linear dependencies at least, can be formulated in considerable quantitative detail. Secondly, if the sources of **X** and the sources of **Y**'s statistical dependence on **X** have nothing in common – as presumably obtains if fr $(\mathbf{Y} \mid \mathbf{X})$ reflects purely **X**'s causal influence on **Y** – then undercurrent (ii) presses us to construe fr $(\mathbf{X})$ and fr $(\mathbf{Y} \mid \mathbf{X})$ as strongly independent in some sense, the most promising candidate for this being confirmational independence.[14]

In short, *CSC* is what results if we attempt to salvage statistical intuition *SIP* by limiting it to just those statistical parameters which align without interpretive contamination with the causal structure of the distribution they describe. Just how robustly *CSC* can be maintained I have no present idea. Certainly it needs some restrictions on the background information over which it generalizes, insomuch as for any background k under which two statistical parameters ξ and Ω are confirmationally independent, additional information h can always be conceived such that ξ and Ω are no longer confirmationally independent given both k and h. Were this the only problem here, it might suffice to restrict *CSC* to just those k under which the parameters at issue are logically independent, since that will block the construction I have in mind while preserving the spirit of *SIP*; however, I can also think of other ways in which *CSC* may be overextended as given. But *CSC*'s boundary limitations are not all that crucial just now. More important is for us to begin tuning our thoughts to the qualitative possibilities of principles such as this, to speculate whether *something* like it may not be an inevitably basic feature of any complex rational belief system.

SUMMARY AND PROSPECTUS

There is no good place to stop in a domain of new ideas so fragmentarily explored as this one, but I have already said more than enough for easy comprehension at one sitting. It is time to pull together what has been accomplished so far and to sight ahead in the direction of its thrust.

We began by noting a major supposition of commonsense confirmation theory, the importance of which lies not in its support (or even recognition) by extant philosophic models of rational inference but in its existentialistic operation as a rule we *live* by. This is that information about the truth of the if-clause in an ordinary-language if /then conditional, or about instances just of the antecedent in an all/are generality, tells nothing relative to natural background information about whether the conjectured dependency itself obtains. Now, until it is normatively clear what propositions *should* be confirmationally relevant/irrelevant to one another under what background information, we can ill afford as philosophers and methodologists of science to disrespect our practical intuitions about this; rather, any discernable consistencies within the latter warrant careful identification and analysis in order that we may be instructed by the principles which govern them. Even if these intuitive principles prove normatively defective in some respects, they may well nonetheless provide (as commonsense so often does) an essential first-approximation to what needs be said by a more technically adequate account of the matter, without whose inspiration the latter would never get off the mark at all.

My primary objective in this essay, then, has been to see what can be made of this commonsensical confirmatory indifference of subjunctive conditionals and conditional generalities to data on their antecedents; specifically, to inquire whether it supposes anything of philosophical significance about thought or reality, whether in the large it is epistemologically defensible, and, if its naive version fails to achieve cognitive coherence, whether there may not be *something* in it which technical reconstructions of rational inference need to preserve. This question's philosophic importance is all the greater in that the ordinary-language sense of conditionality at issue here is the one wherein 'If p then q' has roughly the force of saying that q would be *due to p* were the latter the case, that p would *bring about*, or *cause*, or be *nomically responsible* for q. These italicized notions have been notoriously refractory to philosophic comprehension, but the possibility now arises that an essential part of their meaning may reside in or at least become accessible through their behavior in confirmational relationships.

My first excursion into this territory (Rozeboom, 1968) achieved, I think, a useful sighting on heretofore unsuspected wilds of confirmation-theoretic *problems*, but retrieved little in the way of conclusions. Now, however, we have arrived at two results solid enough to build upon. The first is that, commonsense to the contrary notwithstanding, if-p-then-q is *not* in general confirmationally independent of p, nor is all-As-are-Bs of Aa, under natural or impoverished background information in *any* sense of conditionality. Hence in particular, this cannot be an intrinsic credibility property of nomic conditionals. Even so – and this is our second main result, though I have not yet quite finished putting it together – there is a statistical counterpart to this commonsense coupling of causal-order suppositions with our uncertainty structure which not only appears logically coherent when suitably qualified but which captures remarkably well the spirit of the original intuition albeit at the price of greater technical complexity. This is that if something's being or not being an A is assumed to have complete causal precedence over its being or not being a B – i.e. given that any causal connection between these variables goes exclusively from A to B – while our remaining background information is sufficiently natural, then our uncertainty about $\mathrm{fr}(A)$ is independent of our uncertainty about $\mathrm{fr}(B \mid A)$, with the result that whether or not a particular object a is an A is irrelevant to whether or not statistically all As are Bs. It is worth making clear just how this statistical revision of the nomic indifference principle improves upon the original. In both cases, the driving idea is that our construing one event, or one kind of event, to be responsible for another is mirrored somehow by confirmational independence between such conditionalities and their antecedents. In the original intuition, however, this nomic force is projected into the if/then and all/are content of the statements to which the corresponding if-antecedents are then thought irrelevant. In contrast, since statements about conditional frequencies are neutral regarding which properties have responsibility for bringing about which others, the intuitive principle's statistical emendation withholds nomic force from the conditionality expressed by 'All *A*s are *B*s' and its implicate 'If *Aa* then *Ba*', but puts it into the background under which the latter are deemed confirmationally independent of *Aa*.

Where do we go from here? I'm sure that you are sceptical that principle *CSC* can survive much hostile examination, and I must confess to my own qualms about this. Yet there is a right way and a wrong way to go about criticizing *CSC*. The wrong one would be to ignore it altogether as philosophically outlandish (i.e. disquietingly unorthodox) or dismiss it on grounds that your own causal concepts just don't work that way. Such an argument, however, would presume that your present intuitive grasp of causation is so unproblematically clear that any proposed explication of nomic responsibility which disagrees with it must perforce be misguided. If the whole sorry history of philosophic fumbling with causal concepts proves anything at all, it is surely that we do *not* understand these at all well. Certain attitudes of ordinary-language philosophy to the contrary notwithstanding, our everyday working concepts never achieve the heights of clarity, precision, and consistency which ideally they should enjoy, but are under constant pressure to evolve as the individuals and cultures which employ them mature in experience and intellectual sophistication. Accordingly, it might be argued that *CSC* expresses not so much an incontrovertibly analytic property of nomic structure as we now conceive of this as it is a persistent theme therein which offers the best foundation for a technically superior reworking of this still-obscure notion. That is, if some version of *CSC* is not now true of your de facto worldview, perhaps it *ought* to be.

The most auspicious way to appraise *CSC*, it seems to me, is through more general inquiry into the significance of confirmational independence in statistical inference. If it is not rationally arbitrary which parameters of a given distribution family are confirmationally independent given information k – and if this *is* arbitrary the logic of induction is in deep, deep trouble – then it is not at all implausible that whatever k implies about causal order has major bearing on this. If so, then deciphering the nature of that import will in passing also critique *CSC*, which, meanwhile, will usefully serve to model how information about nomic structure *might* control the structure of statistical credibilities. However, *CSC* by no means delimns the breadth or horizons of this front of inquiry. Ultimately at issue is a much more basic possibility: When how we manage our beliefs has pragmatic repercussions foreseeably serious enough to evoke anxiety whether we are doing this *correctly*, then may it not be that we are inexorably driven to create conceptions of states of affairs which, were they to obtain, would *justify* – i.e. insure the correctness of – our managing our beliefs one way rather than another? The notion of statistical probability may well have had such an origin, namely, as justification for degrees of subjective confidence;[15] the intuitive credibility decoupling of lawlike conditionals from their antecedents urges that our still-formative ideas about nomic responsibility have a similar basis; and it is not too much to hope that still others of the modal connectives and operators which have for so long frustrated analytic philosophers will reveal hitherto unsuspected transparencies when viewed from this perspective.

Meanwhile, the prospect directly at hand is that not only do causal suppositions determine the structure of our uncertainties *if* we happen to believe in such (as viewed by some) philosophically dubious event-couplings, but conversely that without some such commitment we have no rational basis on which to allocate confirmational dependencies – specifically, that whatever can be cited to justify one confirmational structure over another will also serve to define what it is for one event, or event type, to be nomically responsible for another. How such definitions can best be shaped will depend upon just what logical connections between the two structures, causal and confirmational, prove to be most analytically fundamental; but we may anticipate constructions *roughly* to the effect that if S is the totality of confirmational properties necessarily possessed by a rational system of conditional credibilities given causal hypothesis H, while $\{C_i\}$ is a suitably comprehensive set of conditions less philosophically problematic than H such that each C_i is sufficient reason for a credibility system to have properties S, then H may be explicated as what is common to these C_i.

I conclude, then, with a revitalized hope and an operational directive for the philosophy of causation. The hope arises from discovery that the closed circuit of blind alleys within which this concept's analysis has for so long been pursued opens upon heretofore unknown corridors which reach in promisingly new directions. It would be naive to expect these to lead us directly into the naked light of total comprehension, but at the very least they should merge our currently disparate confusions on several major philosophic issues into a single integrated perplexity which substantially reduces our total uncertainty over these matters even if the marginal uncertainty in any one is not greatly diminished. As for the directive, this is not merely that we search out the confirmational implications of causal hypotheses as we now intuit these, but that especial effort be made to diagnose the conditions under which two propositions approach confirmational independence under real-life background information. For roundabout as this might seem, detailed study of *approximate* irrelevance in statistical inference is perhaps the most powerful research press now at our avail for cracking the causal conundrum. This is because if ξ and Ω are two parameters of some unknown statistical distribution, then just as our uncertainty about ξ or Ω given k_i may converge under a series of empirical data $k_1, k_2, \ldots, k_i, \ldots$ (notably, where k_i describes observed frequencies in a size-i sample of events) to asymptotic conviction in a particular value determined by k_i regardless of our unconditional credibility distribution for this parameter, so may statistical intuition also insist that for some such k-series, $\mathrm{CR}(\xi, \Omega \mid k_i)$ converges to unity with increasing i. If so, k_i stands revealed as de facto *empirical evidence* for whatever causal conjecture is required to justify treating ξ and Ω as perfectly independent, where the larger is i the more conclusively does k_i support this causal inference. My concluding proposal, in short, is that nomic structure may well relate to the limit of certain data aggregates in much the same way as statistical probabilities relate to limiting relative frequencies. Even if statistical probability cannot convincingly be *identified* with relative frequency, it is nonetheless illuminating to know that these are asymptotically equivalent numerically; and if it can be established that special kinds of empirical evidence similarly converge to causal conclusions, we shall finally have achieved our first significant domestication of this indispensable but still-feral concept.

University of Alberta

BIBLIOGRAPHY

Blalock, H. M., *Causal Inferences in Nonexperimental Research*, The University of North Carolina Press, Chapel Hill, 1961.

Hacking, I., 'Slightly More Realistic Personal Probability', *Philosophy of Science* **34** (1967) 311–325.

Raiffa, H. and Schaiffer, R., *Applied Statistical Decision Theory*, Harvard University, Graduate School of Business Administration, Division of Research, Boston, 1961.

Rozeboom, W. W., 'Ontological Induction and the Logical Typology of Scientific Variables', *Philosophy of Science* **28** (1961) 337–77.

Rozeboom, W. W., 'Scaling Theory and the Nature of Measurement', *Synthese* **16** (1966) 170–233.

Rozeboom, W. W., 'New Dimensions in Confirmation Theory', *Philosophy of Science* **35** (1968) 134–55.

Rozeboom, W. W., 'New Mysteries for Old: The Transfiguration of Miller's Paradox', *British Journal for the Philosophy of Science* **19** (1969) 345–53.

Whittle, P., 'Curve and Periodogram Smoothing', *Journal of the Royal Statistical Society, Series B* **19** (1957) 1615–27.

Whittle, P., 'On the Smoothing of Probability Density Functions', *Journal of the Royal Statistical Society, Series B* **20** (1958) 334–43.

NOTES

[1] I trust that it is essentially uncontroversial by now to explicate confirmation/disconfirmation as an increase/decrease in credibility relative to the background information. But if not, the evident appropriateness of this may perhaps be enhanced by noting that since $\Pr(p \mid k) = \Pr(q \mid k) \times \Pr(p \mid q \cdot k) + \Pr(\sim q \mid k) \times \Pr(p \mid \sim q \cdot k)$, $0 < \Pr(q \mid k) < 1$ implies that $\mathrm{CR}(p, q \mid k)$ is greater or less than unity if and only if $\Pr(p \mid q \cdot k)$ is respectively greater or less than $\Pr(p \mid \sim q \cdot k)$. That is, so long as q is uncertain given k, q confirms p under k by the CR-criterion iff p's credibility is greater given q and k than given $\sim q$ and k.

[2] This point is well worth emphasis, for much of the literature on Hempel's paradox in fact reeks of failure to distinguish confirmation of (7) by *Aa*-and-*Ba* given *k* from its confirmation by *Ba* given *Aa*-and-*k*.

[3] Assertion by '$Aa \rightarrow Ba$' that an objective relation $\rightarrow$ holds between the events *Aa* and *Ba* is relatively straightforward so long as *Aa* is in fact the case, but how might '$Aa \rightarrow Ba$' then be counterfactually true (as we often think it is) when there is no such event as *a*'s *A*-ing? Perhaps the ontologically safest way to interpret (14a) is as an ellipsis for $(x)[Ax \supset (Ax \rightarrow Bx)]$.

[4] Even this isn't strict deductive entailment unless *fr* is relative frequency (*contra* statistical probability) in a finite class, since otherwise $\mathrm{fr}(B \mid A) = 1$ is technically compatible with the existence of a finite number of *A*s which are not *B*s.

[5] As noted in Rozeboom (1968), the concept of 'because' subsumes logical dependencies as well as causal ones.

[6] Any reader who thinks that the notion of one event *forcing* another to occur is a primitive superstition long abandoned by modern science will find it edifying to browse through Chapter 1 of Blalock (1961).

[7] Note that propositions *r* and *s* are not required to be literally contained in H_{rs}.

Thus when p and q are 'If *Aa* then *Ba*' and 'If *Ba* then *Aa*', respectively, H_{pq} and H_{qp} might respectively be just 'All *A*s are *B*s' and 'All *B*s are *A*s'. The important technical point here is that the present argument applies to the confirmational behavior of universal and particular conditionals alike.

[8] There are at least two important partial exceptions to this which may be expected to play an increasingly major role in confirmation theory but are still afflicted with internal difficulties. One is that symmetry properties of one sort or another have often seemed to be a reasonable demand on unconditional credibilities – i.e. that if p and q are formally alike in certain ways, then $\Pr(p)$ ought to equal $\Pr(q)$. Unfortunately, symmetry stipulations tend to entail inconsistencies unless carefully restricted (cf. the vicissitudes of the classic 'principle of insufficient reason'), and just what symmetry demands can be successfully defended under what boundary conditions is still very much an open question. Secondly, applications of statistical generalities to inferences about particular events usually presuppose that some event credibilities conditional upon statistical information are analytically determinate, thereby placing constraints beyond formal admissibility on which unconditional credibilities are rationally acceptable. For example, it is usually assumed that the credibility of a particular object a's being an A, given that the relative frequency or statistical probability of A-hood is r, is r; whence it follows that $\Pr[Aa\cdot \text{fr}(A)=r]=r\times\Pr[\text{fr}(A)=r]$. But this principle, too, is considerably more problematic than it intuitively appears (cf. Rozeboom, 1969).

[9] Namely, when there is zero covariance between the two series of quantities $\{\text{Cr}(\xi_j, \alpha_i)\}$ and $\{\Pr(\alpha_i \mid Aa)-\Pr(\alpha_i)\}$ for $i=1,\ldots$. This is mathematically possible, but has virtually no chance of being exactly true unless one of these two series has zero variance – while moreover the variance of the second series is zero only if parameter α is already completely known, i.e. if $\Pr(\alpha_i)$ equals 1 for one value of α and 0 for the rest.

[10] The only significant exceptions to it of which I know are Raiffa and Schlaiffer (1961), who at least recognize the possibility of confirmationally related parameters in their abstract development even if they say nothing about it subsequently, and two important applications of correlated-parameter notions by Whittle (1957, 1958) to which Professor L. J. Savage has graciously called my attention. I think it can safely be said (i) that only Bayesian statisticians have shown awareness that *SIP* is not a statistical truism, while (ii) even Bayesians have remained at a loss how to replace it with some more defensible normative principle.

[11] I.e., the parameters in the one set do not have a one-one equivalence to those in the other, as would be the case e.g. if $\xi'=\xi$, $\Omega'=\Omega$, etc.

[12] Average CR deviation from unity, though easy to comprehend, is not actually a very good measure of confirmational relatedness. Much superior to it, technically, are the Information-theoretic measure of transmitted information and the Correlation Ratio for goodness of curvilinear regression. In terms of the latter measures, my finitary approximation shows the predictability of β' or γ' from α' and conversely to be negligible when the credibilities over α, β, and γ are flat (a correlation ratio of only 0.16, or 2 % variance reduction), and to remain generally low even when α, β, and/or γ become known more precisely. In contrast, the correlation ratios for predicting β' from γ' and conversely are about 0.36 (a variance reduction of 13 %) even when α, β, and γ have flat marginal credibilities, and become quite respectable in size when one or more of these marginal distributions is sharp.

[13] Notably, it is virtually always assumed that the to-be-estimated distribution is of a particular restricted form, given which a small subset of what would otherwise be logically independent parameters precisely determines the remainder. Thus if a multi-

variate distribution is assumed to be Normal, all its parameters can be computed from just its first and second moments, while the totality of a Poisson distribution is specified by its mean alone.

[14] (ii)'s most common embodiment is in the classic and still widespread assumption that source variables which themselves have no deeper sources in common are statistically independent of one another. Statistical independence is not applicable to the present case, however, insomuch as the dependence of Y upon X in *P* is a parametrically constant *feature of* the joint distribution in *P* of X, Y, and other variables, not a variable *within* this system.

[15] Why should I feel rather sure, considering the appearance of the sky, that it will rain tonight? Because when the sky looks like that the chance of rain is about 70 %.

ALEX C. MICHALOS

COST-BENEFIT VERSUS EXPECTED UTILITY ACCEPTANCE RULES*

ABSTRACT. A rule for the acceptance of scientific hypotheses called 'the principle of cost-benefit dominance' is shown to be more effective and efficient than the well-known principle of the maximization of expected (epistemic) utility. Harvey's defense of his theory of the circulation of blood in animals is examined as a historical paradigm case of a successful defense of a scientific hypothesis and as an implicit application of the cost-benefit dominance rule advocated here. Finally, various concepts of 'dominance' are considered by means of which the effectiveness of our rule may be increased.

1. THE PROBLEM

One of the fundamental problems of life in general and the philosophy of science in particular is the specification of reliable criteria for the determination acceptable hypotheses (in the broad sense of ordinary sentences, laws and theories) and courses of action. While no one has been able to provide a set of criteria that could be regarded as necessary and sufficient for all sorts of hypotheses and circumstances, a number of more or less plausible rules specifying sufficient conditions of acceptability *given* certain data, purposes and attitudes have obtained fairly wide acceptance. These include such familiar principles as Gauss's least squares rule of estimation, Fisher's method of maximum likelihood, Wald's minimax loss rule, Savage's minimax regret rule, Bernoulli-Bayes's rule for the maximization of expected utility, and so on.[1] The last rule in this list is of particular importance for this paper.

A number of influential philosophers have recommended the Bernoulli-Bayes rule or some variation of it as a first approximation or step in the right direction toward a solution of the problem of providing a criterion, principle or rule for determining the acceptability of scientific hypotheses (e.g., [11, 25, 27, 37, 40]). But, so far as I know, no one has suggested that some sort of benefits-less-costs rule might be more advantageous, and it is roughly this idea that I wish to explore and ultimately vindicate. More precisely, I shall attempt to prove the *normative* claim that a cost and benefit dominance principle of acceptance ought to be preferred to any sort of Bernoulli-Bayesian principle because right now and for the foreseeable future the former performs better and cannot perform worse than the latter (in a sense of 'perform' that will be elucidated below).[2]

Although most of the paper consists of a detailed analysis and comparison of the two relevant principles, their requirements and applications, I shall begin with a brief outline of the basic elements of each in order to provide a general orientation and more or less common background for our discussion.

2. Maximization of expected utility

Proponents of the rule enjoining the maximization of expected utility, which we shall hereafter abbreviate as MEU, imagine a decision-maker confronted with a set of (practically speaking) mutually exclusive and exhaustive possible courses of action from which one that is optimal must be adopted. The decision-maker knows that the payoff or *utility* (in some sense of this word which will be explained later) that he obtains from his choice will be partially determined by events which are (practically speaking) mutually exclusive, exhaustive and beyond his control. If he has objective probability values (i.e. relative frequencies, propensities, physical range measures, etc.) for the occurrence of these events then he is operating under conditions of risk. If he does not have such values then he is in a situation of uncertainty, but he will transform it into a situation of risk by determining appropriate subjective probability values (i.e., betting quotients, degrees of belief, etc.) for the events. Given all this data he is ready to use MEU, which, as a normative principle, prescribes the acceptance of that course of action whose sum of probability-weighted utilities is larger than that of any of its alternatives. If 'U_i' and 'p_i' represent the utility and probability values, respectively, of the ith of n payoffs obtainable by adopting some hypothesis, then

$$(1) \qquad \sum p_i U_i \quad (i = 1, 2, \ldots, n)$$

or

$$(2) \qquad p_1 U_1 + p_2 U_2 + \cdots + p_n U_n$$

represents the expected utility of accepting that hypothesis. Clearly, when there is no risk involved then formulae (1) and (2) shrink to the single utility value whose procurement is certain (has a probability value of unity) given the acceptance of that hypothesis. Following MEU then, one would simply accept that hypothesis whose expected-utility (or utility in the limiting case) was greater than that of its alternatives. If two or more hypotheses have the same expected-utility and it is higher than those of the alternatives, then each of the former should be regarded as equally acceptable.

For our purposes we may think of the decision-maker described above as a practicing theoretical or applied scientist, and we may assume that his possible courses of action are the adoption of certain hypotheses as bases for further action. Although one *can,* as Isaac Levi has shown us [39], imagine a person accepting a hypothesis not as a basis for action but as somehow suitable for admission into his total corpus of knowledge, our concern here is with the provision of hypotheses more or less directly related to action (cf. [12]). If we regard a hypothesis as acceptable then, at the very least, it merits the investment of further resources such as research facilities and activity, time, money, energy, and so on. Note, however, that our concern with the acceptability of hypotheses as bases for action does not alter any of the formal or logical aspects of the problem. Only the content of the utility function or the arguments to be included in that function would be different for Levi's decision-makers and mine.

3. Cost-benefit dominance

Our exposition of a variation of a benefits-less-costs rule, which we will call 'the principle of cost-benefit dominance' and abbreviate CBD, may begin with a decision-maker in roughly the same situation we described for MEU. He is confronted with a similar set of (practically speaking) mutually exclusive and exhaustive hypotheses and events. He may be able to assign some sort of probability value to the occurrence of each event and he may not. In any case, such values are not required. Similarly, utility values are not required. Instead of assigning a utility value to the payoffs he will receive as a result of this or that combination of accepted hypothesis and turn of events, he merely determines the "raw forms" of the benefits and costs attached to each combination. For example, instead of noting that a hypothesis adequately accounts for a certain phenomenon provided that certain events take place rather than some others, coheres with a well-established theory in another domain given those events and has a Reichenbachian weight of .7, *and* that all of this gives it a utility value of .8, he merely lists its benefits in their "raw form" (i.e., it coheres provided that such and such is the case, etc.). Similarly, he lists its costs and the benefits and costs for its alternatives given the various contingencies. It must be assumed, of course, that he is able to weakly order the "raw form" data *within* each attribute and the corresponding preferences. For example, he must be able to determine whether two hypotheses are equally explanatory or one explains more than the other; he must be able to rank order any three distinct levels of explanatory power transitively; and he must recognize that his preferences ought to be perfectly positively correlated with the "raw form" data, e.g., he ought to prefer a hypothesis that explains more phenomena to one that explains less (if all other things are equal). It does not have to be assumed that he is able to weakly order the "raw form" data across attributes, e.g., he never has to be able to rank order different levels of explanatory power, coherence, testability, etc. on a single scale of some sort. In other words, he is obliged to make intra-attribute comparisons but not inter-attribute comparisons.

The net result of this analysis is a battery of matrices (i.e., a different matrix for each attribute) which constitute a comparative "profile" of each hypothesis with respect to its alternatives. Schematically, the matrices in such a battery for the attributes of, say, simplicity, explanatory power, precision, coherence, testability, etc. would each look like this.

Possible Events (States of Nature)

$$\begin{array}{llll} & E_1 & E_2, & \ldots, E_n \\ H_1 & p_1B_1^1 & p_2B_2^1, & \ldots, p_nB_n^1 \end{array}$$

Hypotheses

$$\begin{array}{llll} H_2 & p_1B_1^2 & p_2B_2^2, & \ldots, p_nB_n^2 \\ \vdots & \vdots & \vdots & \vdots \\ H_n & p_1B_1^n & p_2B_2^n, & \ldots, p_nB_n^n \end{array}$$

The entry for row H_1 column E_2, for example, would tell us that if the hypothesis represented by 'H_1' is accepted and the event(s) represented by 'E_2' occur, then we will obtain a benefit represented by 'B_2^1' with a probability represented by 'p_2'. Since 'p_2' and 'B_2^1' need not represent numerical values of any sort, the juxtaposition of these two signs must not be taken to mean multiplication (as in formulae (1) and (2)). The superscript on 'B_2^1' indicates the hypothesis 'H_1' and the subscript indicates the event(s) 'E_2'. The 'B' is short for 'benefits'. They might be a high degree of explanatory power, simplicity, coherence with other theories, precision, etc. While various philosophers and scientists (e.g., [1, 6–8, 19, 20, 25–28, 32, 36, 40, 46–48, 53, 58, 69, 76, and 83]) have made recommendations as to which attributes ought to be included in an optimal set, all that is assumed here is that such a set would contain, say, more than a couple and less than a couple dozen members, none of which would have to be entirely independent (in any sense) of the others. In matrices for such costs as required set-up time, computational effort, special facilities, technical assistance, money, operationalization, etc. the 'B' would be replaced by a 'C' for costs'.[3] As with formulae (1) and (2), when there is no risk involved then such matrices shrink to a single item, namely, a column indicating the various benefits (costs) that will be obtained (borne) with certainty given the acceptance of a particular hypothesis.

If for every possible contingency and for every attribute, the benefits and costs of accepting one hypothesis are preferable (i.e., ought to be preferred) to those of accepting another then the former *strongly dominates* the latter. If for some contingency and for some attribute, the benefits and costs of accepting one hypothesis are preferable to those of another *and* for all of the remaining contingencies and attributes the benefits and costs of accepting the latter are not preferable to those of the former (i.e., some are exactly alike and others are less preferable), then the former dominates the latter.[4] Hence, if one hypothesis strongly dominates another then the former also dominates the latter, but the converse is not true. According to CBD then, the hypothesis that ought to be accepted is the one which dominates all of its alternatives. If two or more hypotheses have the same benefits and costs but dominate all others then they should be regarded as equally acceptable. In general, depending on the particular benefits and costs involved, research in a given problem-area should continue until some hypothesis emerges as dominant over all of its alternatives.

4. Preferability and Superior Performance

I am assuming that one principle of acceptance is *preferable* to another provided that the former may perform better and cannot perform worse than the latter. Moreover, one principle *performs better* than another if and only if it is more effective and more efficient than the other. An acceptance principle is *effective* exactly insofar as it is possible in every sense of this term to isolate, identify or select acceptable hypotheses by applying it. If it were impossible in any sense to apply a principle then it could not be applied and, consequently, could not identify anything. So it would be completely ineffective. The *efficiency* of an acceptance principle may be defined and measured by the ratio of its effectiveness to the number of assumptions and the degree of sophistication of the kinds of information required for its application.[5] The *degree of sophistication* of a particular piece of information may be determined by the sorts of scales (nominal, ordinal, interval or ratio) or concepts (qualitative, comparative or quantitative) required to accurately express that information, e.g., 'this is hot' may be regarded as a less sophisticated piece of information than 'this is hotter than that' which is less sophisticated than 'this has a temperature of 90 °F'. Hence, if it could be shown that CBD and MEU are equally effective or that CBD is more effective than MEU *but* that CBD requires fewer assumptions and/or less sophisticated information than MEU, then the superior efficiency of CBD would be established. That, of course, would establish its superiority of performance and, therefore, its preferability over MEU, which is my central theses. In the next two sections I shall attempt to establish this thesis roughly as follows. In Section 5 it will be shown that MEU requires information of a more sophisticated sort and, consequently, more assumptions than CBD. Some of this information is not now nor will it be in the foreseeable future available. It follows then, that MEU is now and will be for some time to come *completely ineffective*. In Section 6 we review Harvey's defense of his theory of the motion of the heart and blood against that of Galen as a paradigm case study of an undoubtedly successful defense of a scientific hypothesis *and* of an implicit application of CBD. Any adequate acceptance rule would have to disclose the superiority of Harvey's view over Galen's and, as a matter of fact, it is fairly apparent that something like CBD was behind Harvey's presentation of the evidence for his view. By means of this historical example then, the *effectiveness* of CBD is established. Thus, because CBD is more effective and must have a higher efficiency ratio than MEU, the former performs better and cannot perform worse than MEU for the present and foreseeable future. Hence, the superiority and preferability of CBD over MEU is established.

5. MEU VERSUS CBD: COMPARISON OF REQUIREMENTS

The requirements and general prospects of our two principles may be thoroughly compared in five respects.

5.1. In the first place it is apparent that MEU does but CBD does not require numerical probability values. With MEU it is not enough to be able to determine that a certain event and concomitant attribute value is probable, very probable, improbable, more probable than some other, as probable as some other, etc. Such frequently useful qualitative and comparative probabilistic judgments are worthless for MEU, because the latter can only "process" quantitative judgments; e.g., judgments of the form 'The degree of probability of obtaining a value of x for attribute A in the event that E is r'. Thus, one who uses MEU must assume that all of the notoriously difficult philosophic problems involved in obtaining initial numerical probability values (for the particular procedures employed) have been satisfactorily solved. (See, for example, [5, 31, 33, 35, 49, 50, 52, 54, 56, 69, and 75].) Moreover, granted that a given position is philosophically unobjectionable in itself, it must still be assumed that every contingency that might be relevant to the acceptance of any hypothesis can be meaningfully assigned a numerical value by that procedure, i.e., that in one way or another it it always meaningful to transform conditions of uncertainty into conditions of risk. As Ellsberg has shown [15], however, proponents of minimax, maximin and maximax decision rules cannot and would not accept this assumption. (See also [10, 33 and 64].) on the other hand, both the assumption and its denial are irrelevant to CBD, for the latter does not require probability values of any sort but it can always use them when they are available.

5.2. Just as CBD can but MEU cannot get along without numerical probability values, the former can and the latter cannot proceed without numerical utility values. There seem to be two *prima facie* possible ways to produce these values, the first of which will be shown to be abortive and the second of which is largely wishful thinking. The former will be considered in this subsection and the latter in the next.

To begin, it should be noted that the numerical utility values required are cardinal and not merely ordinal. The former are necessary because they are the only ones that can be meaningfully added or multiplied, and both of these operations must be performed on the utility values used by MEU. So, some means of obtaining at least an interval scale of numerical utility values must be found.[6]

One way to tackle the problem of securing an interval scale of utility values is to try to generate them from a simple rank ordering of attribute values. The most popular and manageable means to this end is the standard-gamble technique of von Neumann and Morgenstern [65]. Unfortunately (for defenders of MEU), while this technique can (in principle though not always in fact [51]) be used to produce an interval scale of utility, the utility involved is not the sort that is of interest to philosophers of science. The latter are concerned with "epistemic utilities". In Hempel's words,

> *epistemic utilities* ... represent "gains" and "losses" as judged by reference to the objectives of "pure" or "basic" scientific research; in contradistinction to ... *pragmatic utilities*, which would represent gains or losses in income, prestige, intellectual or moral satisfaction ... [26, p. 156].

And Levi writes

> when an investigator declares himself to be engaged in an effort to replace agnosticism by true belief ... there is no need to ascertain his "true feelings" ... [40, p. 76].

Presumably the fundamental logical distinction between a scale of "epistemic utility" and "pragmatic (including psychological) utility" is that only the former could have some normative force for a scientist *as* a scientist.[7] Moreover, it might be thought that by making certain adjustments in the von Neumann-Morgenstern technique, a scale with normative force could be constructed. Indeed, this idea seems to be behind Levi's efforts in *Gambling With Truth* (p. 50). It is easy to demonstrate, however, that the von Neumann-Morgenstern technique cannot yield the required scale. According to that technique[8], a decision-maker begins by rank ordering attribute values. To keep things simple, suppose he has a single attribute, say, explanatory power, and can distinguish in a publically observable or intersubjectively testable fashion, three ranks, low, average and high. The latter may be represented by 'L', 'M' and 'H', respectively. Clearly, he prefers

$$H \text{ to } M \text{ to } L.$$

He then assigns a value of 0 to the lowest rank and 1 to the highest rank. To determine the numerical value of the remaining rank, he considers various choices (gambles) that might be put to him having the form

$$M \text{ versus } pL + (1 - p)\,H,$$

where 'pL' is short for 'obtaining a low ranking hypothesis with a probability value of p'. Clearly, if $p=0$ then the right side of the gamble would be preferred (because the choice is then between M and H, and H is preferred to M), and if $p=1$ then the left side is preferred (because M is preferred to L). By varying the value of p appropriately, the decision-maker can (often if not always) reach a point where both sides seem equally attractive. At that point he merely computes the numerical value of the right side and that tells him the value of the left side, i.e., the value of the remaining rank.

Now, in order to put this whole procedure to work *normatively*, one would have to have some criterion, rule or principle that prescribes the appropriate probability values. For the case before us, it might have the form: Given a gamble between 'M' and '$pL+(1-p)\,H$' a decision-maker (with his eye on the "gains" and "losses" to "pure" science) ought to reach a state of indifference when and only when $p=r$. The only way to justify such a prescription, however, is to show that M must have a certain "epistemic utility" value, and that can only be shown *after* the interval scale of "epistemic utility" values for explanatory power has been constructed. That, of course, is much too late for anyone who thought the standard-gamble technique would be useful. It is *not* useful, because one must already possess the very information it is supposed to, but can only redundantly provide. Hence, it is impossible to use this technique to construct a scale of "epistemic utility" values which would have normative force for a scientist *as* a scientist. Therefore, the first *prima facie* possible way of producing the scale required by MEU has been shown to be abortive (cf. [72, p. 5]).

5.3. The other *prima facie* plausible way to obtain the required scale is to construct interval scales of measurement for *every* relevant attribute and then transform the attribute values into utility values. Mathematically, at least, the transformations would be relatively straightforward. As a matter of fact, both Hempel and Levi have begun their constructions of "epistemic utility" measures from approximately this point. (See [26, p. 154; and 40, p. 71]). Both of them have worked primarily *from* a "content measure" which yields values unique up to a linear transformation *to* a measure of "epistemic utility" which yields similar values. While there is nothing objectionable about this procedure in itself, it must be emphasized that it is at best a first step on a very long journey. Even if we had an acceptable measure of content, we would still almost certainly need measures of simplicity, explanatory power, precision of predictions, coherence with theories in other domains, probability, etc. in order to apply MEU. Needless to say, such measures are not available now and, judging from the history and present status of the unresolved issues surrounding the measurement of probability, they will not be available in the foreseeable future either. Again, however, such measures (interval scales) are not necessary for CBD, which may be applied as soon as one is able to construct a simple rank ordering of attribute values.

5.4. As has been suggested, the transformation of one interval scale into another is mathematically straightforward. If 'u' and 'a' represent the "epistemic utility" and attribute values, respectively, of accepting a certain hypothesis then

$$u = f(a),$$

where 'f' represents any transformation function up to and including the linear one

$$u = ma + n,$$

where 'm' and 'n' represent constants and $m \neq 0$. The extremely difficult problems before proponents of MEU with respect to these transformations are not, therefore, mathematical or merely technological. They are plainly philosophical. In particular, they involve the appropriate selection of 'f', or, in the linear case, the values of the constants 'm' and 'n'. In the latter case, for example, given interval scales of measurement for every relevant attribute, in order to apply MEU one must be able to systematically choose the appropriate constants and to present a plausible justification of his choices. The burden of proof then, that such and such a value for a certain attribute is worth this or that much 'epistemic utility' falls squarely upon MEU's adherents, and it is a burden which can hardly fail to create interminable haggling. After all, what principles could one invoke to show that, say, a hypothesis with a degree of simplicity of .8 ought to be assigned an "epistemic utility" of .8 or .6 or anything else? What principles could be used to prove that, say, a coherence value of .8 ought to be worth more or less than a simplicity value of .8? What could we use as a basis of comparing the significance of various attributes in order to transform their values into "epistemic utility" values? Clearly some sort of interattribute comparisons will be required to justify the transformation functions, but there is no basis for such comparisons. Of course, if we already knew the "epistemic utility" values corresponding to every attribute value for all attributes, then such comparisons would be self-evident. But that is beside the point. It is precisely those utility values that we are unable to obtain, because the required principles and bases for comparison do not exist. Indeed, even the idea that they (not to mention their justifications) might be forthcoming in the near future seems farfetched to say the least. Thus, we have another good reason for expecting perennial ineffectiveness from MEU and for turning to CBD instead.

5.5. Supposing, for the sake of argument, that all of the aforementioned problems were satisfactorily solved, proponents of MEU would still be faced with an amalgamation problem i.e., with a problem of combining all of the individual "epistemic utility" values into a single most representative or appropriate value. Since they were forced beyond ordinal to cardinal values from the very beginning, they need not be troubled by Arrow's paradox [4, 60]. Nevertheless, some rule of combination must be constructed and its plausibility defended, and this just creates more unnecessary work in view of the availability of CBD. Because a number of amalgamation rules have already been developed, however, this final step may be the least troublesome of all. (See, e.g., [22, 29, 73, 74].) But it is still a piece of excess baggage.

To summarize the arguments in this section, I have been attempting to establish the ineffectiveness of MEU on the ground that it requires a more sophisticated sort of information than is now or will be in the foreseeable future available. To obtain this information, a number of problematic assumptions have yet to be made and substantiated. CBD, on the other hand, does not require such sophisticated information or, consequently, its attendant assumptions. Thus, a *prima facie* case for CBD over MEU has been established. But this is not enough. It is one thing to show that CBD does not have the infelicities of MEU and another to show that CBD is effective. It is to the latter task that we shall now turn. If it can be shown that CBD is at all effective now or can be expected to be effective in the foreseeable future, then both its superior effectiveness and efficiency over MEU will be established.

6. Harvey's implicit use of CBD

In Chapter 14 of his classic *Anatomical Disquisition on the Motion of the Heart and Blood in Animals* [23], Harvey summarizes his position as follows.

> Since *all things*, both argument and ocular demonstration, show that the blood passes through the lungs, and heart by the force of the ventricles, and is sent for distribution to all parts of the body, where it makes its way into the veins and porosities of the flesh, and then flows by the veins from the circumference on every side to the centre, from the lesser to the greater veins, and is by them finally discharged into the vena cava and right auricle of the heart, and this in such a quantity or in such a flux and reflux thither by the arteries, hither by the veins, *as cannot possibly be supplied by the ingesta* and *is much greater than can be required for mere purposes of nutrition*; it is absolutely necessary to conclude that the blood in the animal body is impelled in a circle, and is in a state of ceaseless motion; that this is the act or function which the heart performs by means of its pulse; and that it is the sole and only end of the motion and contraction of the heart (p. 93, italics added).

This famous paragraph is instructive in a number of ways, four of which are relevant to our discussion.

First, the fact that Harvey insists that "all things" show that his position is sound reveals an implicit acceptance of the methodological rule we have been advocating, namely, CBD. He obviously assumes that the sort of complete dominance that his hypothesis (theory) has over its rivals is a sufficient reason for accepting it. Some of the details of that dominance will be described below.

Second, the long conjunction from "show" to the italicized phrases

Stomach and intestines
Through portal vein
Liver
Vena cava
Right ventricle
Some through septum
Most through pulmonary artery to lungs
Air
Pulmonary vein
Sooty waste
Left ventricle
Aorta
Rite mirabile at base of brain
Lower body
Sense organs by hollow nerves
GALEN'S VIEW
Vena cava
Body
Right auricle
Aorta
Right ventricle
Left ventricle
Pulmonary artery
Left auricle
Lungs
Air
Pulmonary vein
HARVEY'S VIEW
⟹ Venous blood
⟶ Arterial blood

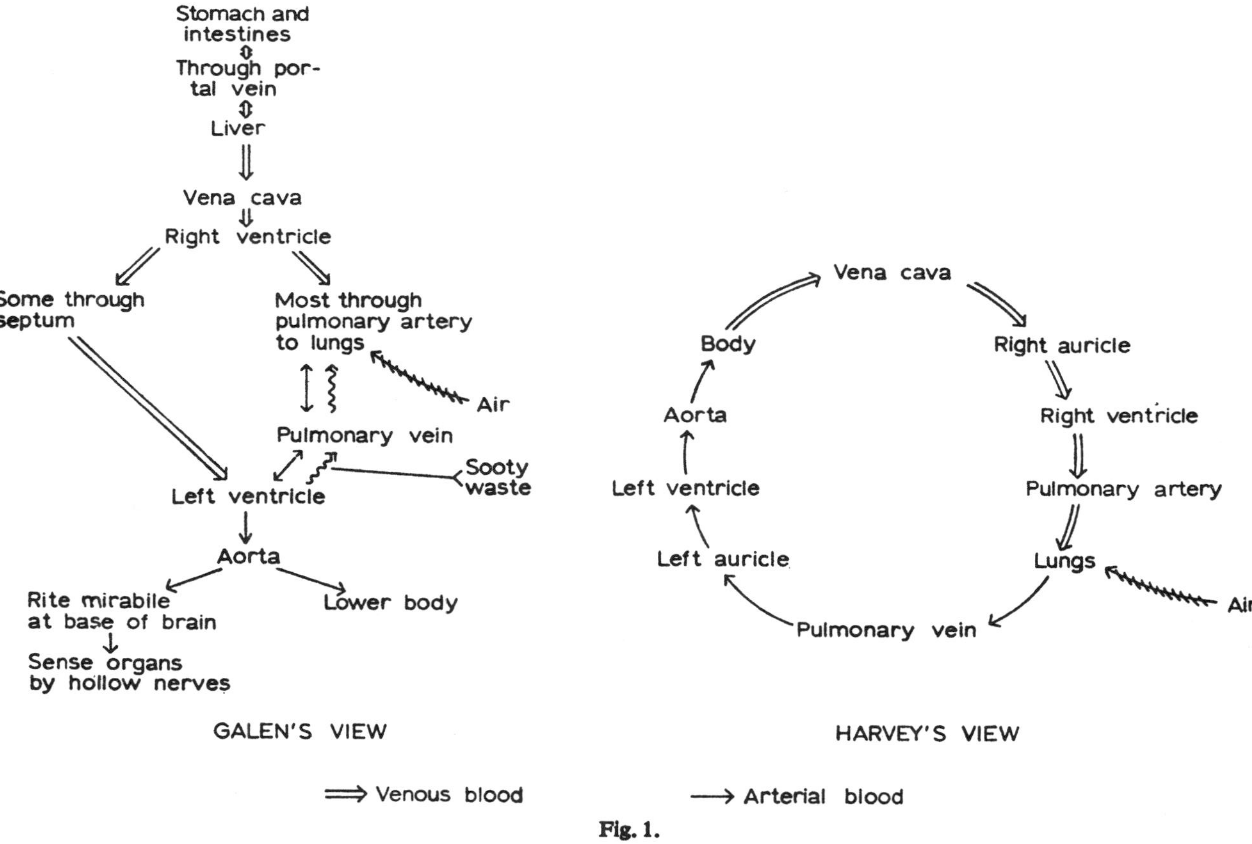

Fig. 1.

outlines the circular path taken by the blood through a body. This is illustrated in Figure 1 beside its most widely held alternative which was developed by Galen.[9] We shall have more to say about these diagrams below, but for now it is enough to notice that beginning with the vena cava, the arrows indicating the direction of flow in Harvey's view form a circle while those in Galen's view proceed along two straight paths with some "back-up". This illustrates the fundamental discrepancy between the view that Harvey is advocating and the alternative that he is rejecting.

Third, the italicized phrases refer to Galen's theory that the blood is produced from ingesta and dispersed outward continuously from the liver to the rest of the body for nutrition. Although Cesalpino [13, II, pp. 226–227] had somewhat vaguely and inconsistently advocated the theory that there was a daily ebbing and flowing (like the tides) between the heart and the veins and arteries, and although there was some two-way movement in Galen's view certainly in the pulmonary and portal veins and possibly throughout the system as a result of his theory of "attractive" and "expulsive faculties" (explained below), Harvey's primary target is the idea that the blood flows more or less continuously from the liver. It is this hypothesis especially that his continuous circulation theory is to replace.

Fourth, the remarks following the italicized phrases emphasize the main aspects of his theory, namely, that the blood in animals (not just men) is moved continuously in a circle by the contraction of the heart.

A careful examination of Harvey's text suggests that Harvey considered the following attributes as especially relevant to the comparison of his theory with Galen's: explanatory power, analogies with accepted theories in other domains, simplicity and logical (internal) consistency. A priori this set of attributes has no more to recommend it than a number of others one might suggest. But for our purposes it is not necessary to reach an agreement on the optimal set of appraising attributes. All that is required here is a generally satisfactory set, or as Herbert Simon would say, a 'satisficing' set [78]. Our primary goal is to show that with respect to each of these attributes, Harvey's theory is superior to Galen's.

A. *Explanatory power*

Although Galen's theory could not account for any observable phenomena for which Harvey's theory had no explanation, the latter could but the former could not account for the facts that: (1) The amount of blood that passed into the aorta in an hour weighed much more than the total amount found in an animal [23, pp. 74–78, 87]. (2) This amount of blood did not drain all the veins or rupture the arteries [23, p. 94]. (3) It could not be produced from the juices of ingested aliment or absorbed as nutriment [23, pp. 75–78, 80, 87]. (4) When the vena cava is closed the heart becomes pale and smaller, and when the aorta is closed the heart becomes deep purple and larger [23, pp. 46–47, 79]. (5) A middling ligature closing a vein can cause a limb to swell and a tight ligature closing an artery can cause it to turn pale [23, pp. 81, 83]. (6) Dissected bodies have much more blood in their veins than in their arteries and much more in the right ventricle than the left [23, pp. 76, 111]. (7) The valves are situated in order to prevent the passage of blood from the large to the small veins which would cause swelling and rupture [23, pp. 89–90]. (8) Tumefaction follows a blow to the temple [23, p. 85]. (9) In phlebotomy a ligature must be applied above the puncture [23, p. 85]. (10) When a patient undergoing a phlebotomy becomes weaker the blood drains more slowly [23, p. 87]. (11) A whole system may become contaminated although the originally infected part is apparently sound [23, p. 96]. (12) Medicine applied externally influences internal organs [23, p. 97].

As if all of this were not enough, Harvey includes the following general remark near the end of his penultimate chapter.

> Finally, reflecting on every part of medicine, physiology, pathology, semeiotics and therapeutics, when I see how many questions can be answered, how many doubts resolved, how much obscurity illustrated by the truth we have declared,... I see a field of such vast extent in which I might proceed so far, and expatiate so widely, that this my tractate would not only swell out into a volume,... but my whole life, perchance, would not suffice for its completion [23, pp. 99–100].

B. *External analogies*

While the sort of movement envisaged by Galen's theory suggested little more than a perpetually flowing stream, the cyclic motion of Harvey's theory was analogous to and fit together admirably with a number of ideas and theories that he accepted in other domains. Indeed, immediately following his description of the path followed by the blood, he cites four analogies that would have been familiar to most of his readers.

This motion we may be allowed to call circular, in the same way as Aristotle says that the air and the rain emulate [a] the circular motion of the superior bodies; [b] for the moist earth, warmed by the sun, evaporates; the vapours drawn upwards are condensed, and descending in the form of rain, moisten the earth again. [c] By this arrangement are generations of living things produced; [d] and in like manner are tempests and meteors engendered by the circular motion, and by the approach and recession of the sun [23, p. 71].[10]

Seven chapters later we find him again arguing in peripatetic fashion that since "under all circumstances" motion generates and preserves "heat and spirits" which are necessary for life, a body must have its "particular seat and fountain, a kind of home and hearth, where ... the original of the native fire, is stored and preserved". This "original" in animals was none other than their pulsating hearts [23, p. 94; 68, p. 181].

C. *Simplicity*

Harvey's theory was simpler than Galen's in the sense that the former required fewer basic assumptions and ad hoc hypotheses than the latter. In the first place, Galen regarded the liver as the center of the venous system and the heart as the center of the arterial system, with anastomosis between the two systems and "communication between the cavities of the heart" through "the tiny pores which appear above all toward the middle of the partition between the cavities ..." [23, pp. 34, 38–39; 18, p. 321]. Harvey's theory of course had a single center, the heart, and as for "the tiny pores" in the septum, he exclaimed "By Hercules! no such pores can be demonstrated, nor in fact do any such exist" [23, p. 42]. Furthermore, the ad hoc assumption of anything passing through the septum raised more questions than it answered.

... how could one of the ventricles extract anything from the other ... when we see that both ventricles contract and dilate simultaneously? Why should we not rather believe that the right took spirits from the left, than that the left obtained blood from the right ventricle ...? But it is certainly mysterious and incongruous that blood should be supposed to be most commodiously drawn through a set of obscure and invisible ducts, and air through perfectly open passages, at one and the same moment. And why ... is recourse had to secret and invisible porosities, to uncertain and obscure channels, to explain the passage of the blood to the left ventricle, when there is so open a way through the pulmonary veins [23, p. 42].

It is perhaps worthwhile to notice here that although the capillaries required by Harvey's theory to permit blood to pass from arteries to veins were as "obscure and invisible" as the "uncertain and obscure channels" through the septum required by Galen's theory[11], the assumption of the existence of the former did not create more problems than it solved. Indeed, it seems to be primarily this aspect of the hypothesis of invisible capillaries which makes it decidedly not ad hoc.

Second, Galen's view of the *causes* of the movement of "material" to and from the heart was enormously more complicated than Harvey's. The latter's view was mentioned in the fourth point that was cited above following his summary of Chapter 14. In Galen's view "almost all parts of the animal" possessed an "attractive faculty" by means of which they obtained their "proper juice", a "retentive faculty", which was responsible for the retention of whatever was of "some benefit", an "alterative faculty", which accounted for the conversion of attracted material into "nourishment", and an "expulsive faculty" that explained the elimination of whatever was not of "some benefit" [18, pp. 223–225, 247–249, 307]. For example, with respect to the stomach he explains:

> ... the attractive faculty in connection with swallowing, the retentive with digestion, the expulsive with vomiting and with the descent of digested food into the small intestine – and digestion itself we have shown to be a process of alteration [18, p. 275].

He also distinguishes "*two kinds of attraction*, that by which a vacuum becomes refilled and that caused by appropriateness of quality" [18, p. 319]. All "hollow organs" such as the heart and arteries display both kinds of attraction during diastole, with the former "always attracting lighter matter first" and perhaps from some distance, while the latter "acts frequently ... on what is heavier" and usually nearby [18, pp. 317–319, 325].

> The arteries draw into themselves on every side; those arteries which reach the skin draw in the outer air ... those which pass up from the heart into the neck, and that which lies along the spine ... draw mostly from the heart itself; and those which are further from the heart and skin necessarily draw the lightest part of the blood out of the veins [18, p. 317].

This should be enough for our purposes. Rather than providing a general causal explanation of the movement of "material" to and from the heart as described in his *On the Functions of Parts of the Human Body* and outlined in Figure 1, Galen has given us reasons to expect *not* that but another kind of movement. For with this explanatory scheme "almost all" of the single arrows in Figure 1 should be replaced by double arrows.[12]

Furthermore, considering the fact that he only uses the term 'faculty' "so long as we are ignorant of the true essence of the cause which is operating" [18, p. 17], even if his view were internally consistent, it would not be very informative. And finally, to return to our original point, even if the scheme worked, it was much more complicated and contained many more loose ends than Harvey's.

D. *Internal consistency*

Galen was fully aware that the uncoordinated activity of the "attractive" and "expulsive faculties" could lead to paralysis or chaos. But he thought that the whole system could run smoothly if the faculties operated "consecutively" like inhaling and exhaling [18, pp. 303–307]. Harvey knew that that explanation was inadequate. If, as noted above, the "ventricles contract and dilate simultaneously" and if they only attract with dilation and repel with contraction, then they could not be exchanging anything "consecutively" [23, p. 42]. So either there was no exchange or the "faculty" scheme was faulty, or, as Harvey claimed, both. Similarly, Harvey claimed that on Galen's view the "spirits"[13] in the aorta (which were necessary to the life of the heart as well as every other organ) should have been drawn into the left ventricle as a result of its "attractive faculty", but somehow they always escaped [23, p. 40]. This is merely a special case of the general point made above, namely, that "almost all" of the arrows on Galen's view in Figure 1 turn out to be both double and single at the same time, which is impossible. Again, Harvey saw that the idea that the mitral valve[14] should allow the "spirituous blood" to pass from the left ventricle to the lungs while at the same time it prevented the "thinner" air from retrogressing through the same channel was plainly inconsistent [23, p. 40]. And finally, he noted a similar infelicity in the alleged "cooling and cleaning system" operating between the left ventricle and lungs by means of the pulmonary vein. If the mitral valve prevented the "cooling" air from escaping once it arrived in the left ventricle then it could not fail to prevent the "fuliginous vapours" from escaping also, in which case there would be no "cleansing" activity [23, p. 40].

This completes my review of Harvey's comparison of his theory with that of Galen's on the movement of the blood and the function of the heart with respect to the four attributes of explanatory power, analogies with other theories, simplicity and internal consistency. The implicit application of CBD with these attributes was *effective*. It produced a decision in favor of Harvey's view over Galen's. Harvey was certainly not "all right", especially in his selection of acceptable scientific theories in other domains, which is quite understandable.[15] But his errors are only relevant to the content of his argument, not to its logical form. From the logical or methodological point of view, his argument was perfectly non-demonstratively valid. His theory dominated its alternatives and was, therefore, more acceptable. Fortunately (for Harvey at least), unlike most theories, his theory has continued to dominate its alternatives. In fact, by the time he and his contemporaries had passed away, serious alternatives were no longer put forward [21, pp. 172–176]. Hence, today it is appropriate to describe his theory not merely as more acceptable than its alternatives, but as acceptable.

Considering the results of Sections 5 and 6 together now, I take it that the effectiveness and efficiency, and therefore, the superiority and preferability of CBD over MEU has been established. Before closing this investigation, however, two more general topics merit our attention. The first pertains to a certain infelicity shared by both MEU and CBD, and will be discussed in the next section. The second concerns various strategies that could be used to increase the effectiveness of CBD and, therefore, strengthen our case for it. These issues are taken up in Section 8.

7. AN INFELICITY OF MEU AND CBD

The major drawback of both of these principles is that they do not provide any built-in evaluation for the variety of evidence for or against a hypothesis. If, for example, one hypothesis has a utility of .2 on the basis of a single attribute and a probability of .6 of obtaining its full value, while another hypothesis has a utility value of .2 for each of three attributes (two plus the one on which the other hypothesis is superior) with probabilities of .2 each, then the expected utility of each hypothesis is the same .12. Similarly, neither hypothesis dominates the other. However, the hypothesis with a greater variety of support might plausibly be regarded as warranting a higher assessment. What can be said about this discrepancy?

It seems to me that this problem of assessing variety may be treated in much the same way that voting theorists treat the problem of "no election". That is, we may introduce the attribute of 'variety' into our analysis just as voting theorists introduce the option 'no election' along with the list of candidates. (See, e.g., [14].) Then, just as a voter is allowed to judge the merits of each candidate in the presence of the option to have the whole election rescinded, our decision-makers are allowed to judge the variety of support for or against a hypothesis and assign it some appropriate value. Whether or not this strategy would work as well for our decision-makers as its analogue works for voters, a priori it certainly seems that it would.

8. Increasing the Effectiveness of CBD

There are a number of ways to increase the effectiveness of CBD, some of which put more severe demands on the number of assumptions and kinds of information required than proponents of CBD would be willing to satisfy. In the remaining paragraphs of this section I shall introduce five general tactics and indicate their peculiar costs.

8.1. You recall that one hypothesis was said to dominate another if *for all* attributes and contingencies the benefits and costs associated with the latter are not preferable to those of the former and for some attribute and contingency the benefits and costs associated with the former are preferable to those of the latter. The italicized phrase 'for all' may be regarded as an abbreviation of the longer locution 'for all $n(n \geqslant 1)$ relevant attributes and $m(m \geqslant 1)$ contingencies'. As long as n is sufficiently large, we may weaken the notion of dominance by degrees, by replacing 'for all n' by 'for all $n-1$', 'for all $n-2$', and so on up to 'for all $n-(n-1)$'. Thus while strong dominance and dominance place certain requirements on all attributes, $(n-1)$-dominance puts requirements on all but one attribute, $(n-2)$-dominance on all but two, etc. For example, a hypothesis whose benefits and costs were preferable to those of another with a single exception in which the latter's benefits (or costs) were preferable to the former might be said to $(n-1)$-dominate the latter, although it could not dominate the latter. Clearly, the chances of obtaining a single acceptable hypothesis with CBD increase as the degrees of dominance decrease from n. Moreover, no new information is required. On the other hand, it must be assumed that the data from some *prima facie* relevant attribute(s) may be safely ignored. That, of course, may be difficult to justify, especially in the completely general fashion proposed. For notice that, say, $(n-1)$-dominance does not specify any particular attribute to be ignored. It merely permits a reversal in any attribute whatsoever, and one may be reluctant to grant such sweeping permission.[16]

Just as one can reduce n to $n-1$, etc. one can reduce m to $m-1$, etc. when there are a sufficient number of contingencies. This would produce the same advantages and disadvantages as reductions in n.

Finally, it should be mentioned that one could again use the analogy between this investigation and voting theory, and consider such familiar notions of dominance as simple majority dominance, absolute majority dominance, $\frac{2}{3}$ dominance, and so on.[17] For specific n's and m's, all of these phrases could be translated into the 'n minus something' terminology.

8.2. Having considered the apriori elimination of attributes and contingencies, there are two fairly natural tacks to take. One may consider the elimination of specially selected contingencies or of specially selected attributes. I shall discuss the former in this subsection and the latter in the next.

It is a familiar fact that the decision rules known as 'minimax loss', 'minimax regret', 'maximin gain', 'maximax gain' and 'Hurwicz's rule' focus a decision-maker's attention on only some of the contingencies before him. For example, minimax loss tells one to merely review the maximum losses (costs) possible as a result of accepting any hypothesis given each contingency, and to act so as to guarantee the smallest of the maximum losses possible. Similarly, one might eliminate all of the data on benefits from one's analysis and define a concept of minimax-dominance. To determine which hypothesis minimax-dominated which, one would review the maximum costs attached to each hypothesis for every contingency and regard that one as minimax-dominant which insured the smallest maximum possible cost. Concepts of minimax regret-dominance, maximax-dominance, etc. could be constructed analogously.

All of these qualified types of dominance would be easier to obtain than unqualified dominance. So their use would increase the effectiveness of CBD. Furthermore, they do not require any more information. Indeed, some of them require less information, because they completely disregard either benefits or costs. However, this demands the rather bold assumption that such data *and more* can be safely ignored, and a priori there seems to be no justification for this assumption.

8.3. As Miller and others have shown [3, 24, 62, 67, 77], there is a general tendency for decision-makers to unwittingly let one or two of many relevant attributes determine their final judgment. Moreover, as Mac-Crimmon and Raiffa [43, 72] have recently emphasized, for one reason or another, a decision-maker may *choose* to regard one attribute as more important than all of the others together. In the latter case then, concepts of specific attribute dominance might be defined such as 'probability-dominance', 'explanatory power-dominance', etc. to be used with CBD. while such an approach presupposes interattribute comparisons of importance and, therefore, additional evaluative criteria, "weighing" devices, assumptions and justifications, it is still less demanding than MEU. Hence, with this modification, CBD would probably (depending primarily on the number of vitally important attributes selected) still be more effective than MEU.

8.4. If the total expulsion of some attributes and/or contingencies from the set of relevant evidence seems unjustifiable, a less drastic procedure may seem attractive. It has already been noted that evaluative criteria and "weighing" devices would have to be developed in order to select the one or two supremely important attributes mentioned in the previous paragraph. *If* every attribute could be assigned a numerical value indicating its weight of importance *and* all attribute values could also be expressed numerically, then each hypothesis could be assigned a numerical value equal to the sum of its weighted attribute values and the hypothesis with the largest sum could be regarded as the most acceptable. By requiring all weights of importance to be real numbers greater than zero, one could be certain that every relevant attribute had *some* influence on the total evaluation sum for each hypothesis. Neat as it sounds, such a procedure could not be practicable, because it demands even more information and assumptions, and could not be more effective than MEU.

8.5. As Simon [78] and Ellsberg [15] have insisted, it is sometimes easier to determine that something is unsatisfactory than it is to determine just how satisfactory something is. It is usually easier to decide which shirts, suits, socks, or ties "just won't do" than it is to decide which of a couple fairly decent ones one should buy. Following Simon, we may say that an attribute value is satisficing if and only if hypotheses with such a value could in every sense of this term be acceptable. It follows then, that any hypothesis with a non-satisficing value for some attribute in some contingency cannot be acceptable. Hence, such hypotheses may be immediately eliminated from consideration, with CBD applied to the remainder. By providing a good reason for rejecting some hypotheses that might otherwise remain in the set of live options, a review of attribute values from a satisficing point of view could increase the a priori chance of obtaining a single acceptable hypothesis with CBD. The apparent additional information required is the minimum satisficing attribute value for every attribute. Even if such comprehensive data was not available, however, it might still be worthwhile (i.e., increase the effectiveness of CBD) to know *some* minimum satisficing values. Naturally the identification of such values presupposes assumptions and justifications for the evaluative criteria employed.

8.6. Finally, it should be noted that one could combine some of the tactics described in 8.1–8.5.[18] For examples, one could use a satisficing review of attribute values (8.5) along with a weaker concept of dominance (8.1); a contingency eliminating rule (8.2) with a weaker concept of dominance; (8.1), (8.2) and (8.5) together; and so on. What must always be remembered, of course, is that increases in the number of tactics employed create corresponding increases in the number of assumptions and justifications required. Furthermore, the very reason CBD has been recommended here in the first place is that it is supposed to reduce the latter without excessive costs.

9. Conclusion

Since I have summarized the argument for my central thesis at the end of Section 4, it is not necessary to repeat it here. All that remains to be said now is that what is required at this point is a strong defense of a particular set of relevant attributes or, in Bunge's words [8], "assaying criteria" for the evaluation of all hypotheses. Perhaps no single set will do for all kinds of hypotheses. What we require of acceptable laws may be different from what we require of acceptable theories. Attributes that are relevant for the determination of the acceptability of ordinary sentences (rather than laws or theories) may well be something else again. At any rate, if my case for CBD over MEU has been argued persuasively, then it is clearly the relevant attributes or "assaying criteria" that should be the focus of our attention now.

University of Guelph, Ontario

REFERENCES

[1] Ackermann, R., 'Inductive Simplicity', *Philosophy of Science* **28** (1961) 152–161.
[2] Ackoff, R. L., *Scientific Method*, New York 1962.
[3] Archer, E. J., Bourne, L. E., and Brown, F. G., 'Concept Identification as a Function of Irrelevant Information and Instructions', *Journal of Experimental Psychology* **49** (1955) 153–164.
[4] Arrow, K. J., *Social Choice and Individual Values*, New York 1951.
[5] Barker, S. F., *Induction and Hypothesis*, Ithaca 1957.
[6] Buchdahl, G., *Metaphysics and the Philosophy of Science*, Oxford 1969.
[7] Bunge, M., *Metascientific Queries*, Springfield 1959.
[8] Bunge, M., 'The Weight of Simplicity in the Construction and Assaying of Scientific Theories', *Philosophy of Science* **28** (1961) 120–149.
[9] Bunge, M., *Scientific Research*, Vol. II, Berlin 1967.
[10] Burks, A. W., 'The Pragmatic-Humean Theory of Probability and Lewis' Theory', in *The Philosophy of C. I. Lewis* (ed. by P. A. Schilpp), LaSalle 1968, pp. 415–464.
[11] Carnap, R., *Logical Foundations of Probability*, Chicago 1950.
[12] Chisholm, R. M., 'Lewis' Ethics of Belief', in *The Philosophy of C. I. Lewis* (ed. by P. A. Schilpp), LaSalle 1968, pp. 223–242.
[13] Crombie, A. C., *Medieval and Early Modern Science*, Volume I and II, Garden City 1959.
[14] Dodgson, C. L. (Lewis Carroll), 'A Discussion of the Various Methods of Procedure in Conducting Elections', reprinted in D. Black, *The Theory of Committees and Elections*, Cambridge 1963, pp. 214–222.
[15] Ellsberg, D., Risk, Ambiguity, and the Savage Axioms, P-2173, The RAND Corporation, Santa Monica, 1961.
[16] Fishburn, P. C., *Decision and Value Theory*, New York 1964.
[17] Fleming, D., 'Galen on the Motions of the Blood in the Heart and Lungs', *Isis* **46** (1955) 14–21.
[18] Galen, *On the Natural Faculties* (trans. by A. J. Brock), London 1916.
[19] Good, I. J., 'Corroboration, Explanation, Evolving Probability, Simplicity and a Sharpened Razor', *British Journal for the Philosophy of Science* **19** (1968) 123–143.
[20] Goodman, N., 'Recent Developments in the Theory of Simplicity', *Philosophy and Phenomenological Research* **19** (1959) 429–446.
[21] Graubard, M., *Circulation and Respiration*, New York 1964.
[22] Harsanyi, J. C., 'Cardinal Welfare, Individualistic Ethics and Interpersonal Comparisons of Utility', *Journal of Political Economy* **63** (1955) 309–321.

[23] Harvey, W., *An Anatomical Disquisition on the Motion of the Heart and Blood in Animals*, Willis's translation revised and edited by A. Bowie, London, 1889 and reprinted in *Classics of Medicine and Surgery* (ed. by C. M. B. Camac), New York 1959.
[24] Hayes, J. R., 'Human Data Processing Limits in Decision Making', *Electronics System Division Report*, ESD-TDR-62-48, 1962.
[25] Hempel, C. G., 'Inductive Inconsistencies', *Synthese* **12** (1960) 439–469.
[26] Hempel, C. G., 'Deductive-Nomological Versus Statistical Explanation', in *Minnesota Studies in the Philosophy of Science*, Vol. III (ed. by H. Feigl and G. Maxwell), Minneapolis, 1962, pp. 98–169.
[27] Hempel, C. G., 'Recent Problems of Induction', in *Mind and Cosmos* (ed. by R. G. Colodny), Pittsburgh 1966, pp. 112–134.
[28] Hertz, H., *The Principles of Mechanics*, New York 1956.
[29] Hildreth, C., 'Alternative Conditions for Social Orderings', *Econometrica* **21** (1953) 81–91.
[30] Hinrichs, H. H. and Taylor, G. M. (eds.), *Program Budgeting and Benefit-Cost Analysis*, Pacific Palisades, Calif. 1969.
[31] Hintikka, J. and Suppes, P., *Aspects of Inductive Logic*, Amsterdam 1966.
[32] Jeffreys, H., *Scientific Inference*, Cambridge 1957.
[33] Keynes, J. M., *A Treatise on Probability*, London 1921.
[34] Kuhn, A., *The Study of Society*, Homewood, Ill. 1963.
[35] Lakatos, I., 'Changes in the Problem of Inductive Logic', in *The Problem of Inductive Logic* (ed. by I. Lakatos), Amsterdam 1968.
[36] Laudan, L., 'Theories of Scientific Method from Plato to Mach', *History of Science* **6** (1968), 1–63.
[37] Leach, J., 'Explanation and Value Neutrality', *British Journal for the Philosophy of Science* **19** (1968) 93–108.
[38] Leinfellner, W., 'Generalization of Classical Decision Theory', in *Risk and Uncertainty* (ed. by K. Borch and J. Mossin), London 1968, pp. 196–210.
[39] Levi, I., 'On the Seriousness of Mistakes', *Philosophy of Science* **29** (1962) 47–65.
[40] Levi, I., *Gambling with Truth*, New York, 1967.
[41] Little, I. M. D., *A Critique of Welfare Economics*, Oxford 1950.
[42] Luce, R. D. and Raiffa, H., *Games and Decisions*, New York 1957.
[43] MacCrimmon, K. R., Decisionmaking Among Multiple-Attribute Alternatives: A Survey and Consolidated Approach, Memorandum RM-4823-ARPA, The RAND Corporation, Santa Monica, 1968.
[44] Mackenzie, W. J. M., *Free Elections*, London 1967.
[45] Manheim, M. L. and Hall, F. L., 'Abstract Representation of Goals', P-67-24, Department of Civil Engineering, M.I.T., 1968.
[46] Margenau, H., *The Nature of Physical Reality*, New York, 1950.
[47] McLaughlin, A., 'Science, Reason and Value', *Theory and Decision* **2** (1970), to be published.
[48] Michalos, A. C., *Probability and Degree of Confirmation:* A Study of the Disagreement Between Karl Popper and Rudolf Carnap from 1934 to 1964. Doctoral dissertation, University of Chicago, 1965.
[49] Michalos, A. C., 'Two Theorems of Degree of Confirmation', *Ratio* **7** (1965) 196–198.
[50] Michalos, A. C., 'Estimated Utility and Corroboration', *British Journal for the Philosophy of Science* **16** (1966), 327–331.

[51] Michalos, A. C., 'Postulates of Rational Preference', *Philosophy of Science* **34** (1967) 18–22.
[52] Michalos, A. C., 'Descriptive Completeness and Linguistic Variance', *Dialogue* **6** (1967) 224–228.
[53] Michalos, A. C., 'An Alleged Condition of Evidential Support', *Mind* **78** (1969) 440–441.
[54] Michalos, A. C., *Principles of Logic*, Englewood Cliffs 1969.
[55] Michalos, A. C., 'A Theory of Decision-Making Evaluation', paper read at the Annual Meeting of the Eastern Division of the American Philosophical Association, 1969.
[56] Michalos, A. C., 'Analytic and Other "Dumb" Guides of Life', *Analysis*, to be published.
[57] Michalos, A. C., 'Decision-Making in Committees', *American Philosophical Quarterly* **7** (1970) 91–106.
[58] Michalos, A. C., 'Positivism Versus the Hermeneutic-Dialectic School', *Theoria* **35** (1969) Part 3, 267–278.
[59] Michalos, A. C., 'The Costs of Decision-Making', *Public Choice*, to be published.
[60] Michalos, A. C., 'The Impossibility of an Ordinal Measure of Acceptability', unpublished manuscript.
[61] Michalos, A. C., 'Efficiency and Morality', paper read at the Annual Meeting of the Western Division of the American Philosophical Association, 1970.
[62] Miller, G. A., 'The Magical Number Seven, Plus or Minus Two', *Psychological Review* **63** (1956) 81–97.
[63] Miller, D. W. and Starr, M. K., *Executive Decisions and Operations Research*, Englewood Cliffs 1960.
[64] Milnor, J., 'Games Against Nature', in *Decision Processes* (ed. by R. M. Thrall, C. H. Coombs, and R. L. Davis), New York 1954, pp. 49–60.
[65] von Neumann, J. and Morgenstern, O., *Theory of Games and Economic Behavior*, Princeton 1947.
[66] Newman, P., *The Theory of Exchange*, Englewood Cliffs 1965.
[67] Osgood, C. S., Suci, G. J., and Tannenbaum, P. H., *The Measurement of Meaning*, Urbana 1957.
[68] Pagel, W., 'The Position of Harvey and van Helmont in the History of European Thought', in *Toward Modern Science*, Vol. II (ed. by R. M. Palter), New York 1961, pp. 175–191.
[69] Popper, K. R., *The Logic of Scientific Discovery*, New York 1959.
[70] Prest, A. R. and Turvey, R., 'Cost-Benefit Analysis: a Survey', *The Economic Journal* **75** (1965) 683–735.
[71] Pruzan, P. M., 'Is Cost-Benefit Analysis Consistent with the Maximization of Expected Utility?', in *Operational Research and the Social Sciences* (ed. by J. R. Lawrence), London (1966) pp. 319–336.
[72] Raiffa, H., Preferences for Multi-Attributed Alternatives, Memorandum RM-5868-DOT/RC, The RAND Corporation, Santa Monica, 1969.
[73] Rescher, N., *Introduction to Value Theory*, Englewood Cliffs 1969.
[74] Rothenberg, J., *The Measurement of Social Welfare*, Englewood Cliffs, 1961.
[75] Salmon, W. C., *The Foundations of Scientific Inference*, Pittsburgh 1966.
[76] Schlesinger, G., *Method in the Physical Sciences*, London 1963.
[77] Shepard, R. N., 'On Subjectively Optimum Selection Among Multi-Attribute

Alternatives', *Human Judgements and Optimality* (ed. by M. W. Shelly and G.L. Bryan), New York 1964, pp. 257–281.
[78] Simon, H. A. and March, J. G., *Organizations*, New York 1958.
[79] Singer, C., *A Short History of Anatomy and Physiology from The Greeks to Harvey*, New York 1957.
[80] Stedry, A. C. and Charnes, A., 'The Attainment of Organization Goals Through Appropriate Selection of Subunit Goals', in *Operational Research and the Social Sciences* (ed. by J. R. Lawrence), London 1966, pp. 147–164.
[81] Tullock, G. and Buchanan, J. M., *The Calculus of Consent*, Ann Arbor 1962.
[82] Wilkie, J. S., 'Harvey's Immediate Debt to Aristotle and to Galen', *History of Science* **4** (1965) 103–124.
[83] Williams, P. M., 'The Structure of Acceptance and Its Evidential Basis', *British Journal for the Philosophy of Science* **19** (1969) 325–344.
[84] Wilson, C. Z. and Alexis, M., 'Basic Frameworks for Decisions', *Journal of the Academy of Management* **5** (1962), 151–164.

NOTES

* The number of friends who have kindly given me suggestions and encouragement is almost embarrassingly large, but I would like to express my gratitude to Myles Brand, Cliff Hooker, David Hull, Scott Kleiner, Hugh Lehman, Werner Leinfellner, Andrew McLaughlin and Tom W. Settle.
[1] All but the last of these rules are reviewed in [54].
[2] [43] and [71] contain less thorough comparisons of these and similar rules, with respect to different applications. [72] contains a defense of the Bernoulli-Bayes principle for "multi-attribute" problems. [30] and [70] contain excellent surveys of recent work on cost-benefit analysis.
[3] A fairly thorough analysis of decision-making costs may be found in [59].
[4] These two definitions, of course, are merely special applications of the famous Pareto Principle that has been used widely by economists since Vilfredo Pareto's *Cours d'Économie politique*, 1877, e.g., [4, 22, 29, 42, 66, 74, 81].
[5] 'Effectiveness' and 'efficiency' are analyzed in greater detail in [61].
[6] On the problem of scales and their transformations see [2, 16, 42, and 74].
[7] A vast amount of literature has been produced by proponents and opponents of "pragmatic utility", and it is doubtful that I could contribute anything novel to the discussion here. Interested readers may find critiques of the concept in [4, 15, 34, 43, 45, 48, 50, 51, 74 and 80].
[8] A more thorough analysis of the technique may be found in [63].
[9] The path outlined here for Galen's view has been put together from excerpts from *On the Functions of Parts of the Human Body* in [13, 17, 21, and 79] and from the remarks of the historians themselves.
[10] Both Crombie [13, II, pp. 235–237] and Pagel [68, pp. 177–182] regard these analogies as highly influential on Harvey's thinking.
[11] The flow of blood through capillaries was not observed until 1661 by Marcello Malpighi [13, II].
[12] Crombie [13, I, pp. 164–165] and Fleming [17] completely missed this point, and criticized historians who had referred to a general ebbing and flowing in the whole venous system. See also Subsection D below.

[13] Galen imagined that the vital functions were produced by the activity of three kinds of "spirits", namely, the "vital spirit" of the heart, the "natural spirit" of the liver and the "animal spirit" of the brain. The first "accounted for" the "vital faculty" or "principle of animal life", the second for the "vegetative faculty" or "principle of nutrition and growth" and the third for the "psychic faculty" or "spiritual principle of life" [13, I, pp. 163–167].

[14] The mitral valves are located between the auricles and ventricles in the mitral orifaces on both sides of the heart. The one referred to here is on the left side and known as the 'bicuspid valve' because it has two flaps or doors.

[15] According to the first quotation in subsection B above, he evidently accepted a geocentric theory of the planetary system.

[16] Most of the case histories cited in [8] seem to have admitted some reversals, although a more careful analysis might reveal a different picture.

[17] From a formal or logico-mathematical point of view, voting theory and the theory of multi-attribute decision-making are virtually indistinguishable. See, for example, [44, 55, 57, 59 and 60].

[18] This is also suggested by MacCrimmon [43].

BEN ROGERS

MATERIAL CONDITIONS ON TESTS OF STATISTICAL HYPOTHESES

Orthodox statistical theory, which is the theory used by most practicing statisticians, physicists, biologists, and social scientists, assumes there are objective probabilities which are properties of physical systems. Statistical hypothesis testing is a branch of statistical theory concerned with a theory for accepting or rejecting statements which ascribe these objective probabilities to particular physical systems. Since use of the theory of statistical hypothesis testing leads to the acceptance or rejection of a kind of empirical statement, I shall assume that this theory is a part of inductive logic. Recent philosophical inquiries have largely ignored orthodox statistical theory, and personalistic (or Bayesian) statisticians have attacked it. Since the theory is so widely used, it is important to reexamine its foundations and its place in the canon of inductive logic. In particular I shall be directly concerned with the Neyman-Pearson theory of statistical hypothesis testing, because most of orthodox hypothesis testing is based on the considerations which underlie this theory.

It has been universally maintained that the application of the Neyman-Pearson theory requires that the user make empirical assumptions. I agree. What has not been made clear is why these material assumptions are necessary. An important part of the clarification of the foundations of the theory is the task of showing why these conditions are necessary and what consequences the requirement of material conditions has for the place of the theory in the canon of inductive logic.

With regard to the Neyman-Pearson theory, Neyman has maintained that "... the practical applicability of the solution reached within the model depends solely upon the adequacy of the model."[1] In the context of his discussion, he is saying that unless some material assumptions peculiar to the particular application hold, then the test arrived at from the use of the theory is not practically applicable. But nowhere does he discuss why some material assumptions must obtain for the theory to be applicable.

Similarly Reichenbach, who also held that statistical inference was concerned with objective empirical probabilities, indicated that the Neyman-Pearson theory was a legitimate part of inductive logic but that it was applicable only in contexts where empirical knowledge already existed. He distinguished between rules of inductive inference which are appropriate where there is no previous knowledge available and rules which are appropriate only where previous knowledge is available. The former he called contexts of primitive knowledge, and the latter he called contexts of advanced knowledge.[2] Only the rule of simple enumeration, he maintained, is applicable in the context of primitive knowledge. From the point of view of *his* system of inductive inference, which assigns probabilities to hypotheses, the Neyman-Pearson theory must require the knowledge of some empirical statements for its application. The Neyman-Pearson theory does not assign probabilities to hypotheses, but licenses their acceptance or rejection. Empirical statements asserted without a weight attached (derived from probabilities) Reichenbach calls anticipative or blind posits. He then maintains that the Neyman-Pearson theory leads to anticipative posits. Then of the Neyman-Pearson theory and of other similar methods, he says:

> The justification of the method of anticipative posits in advanced knowledge differs from the one concerning primitive knowledge so far as it cannot be proved that the method must lead to success if success is attainable. It can be proved only that success is probable if it is attainable. The limitation to a probability originates from the merely probable character of the inductive assumptions entering into the method. It may be false that the distribution belongs to the class considered; then there might still exist a limiting distribution, but it would not be found through the method. In primitive knowledge the method of anticipative posits is free from this limitation.[3]

Leaving aside the question of whether the Neyman-Pearson theory can be justified in the manner Reichenbach indicates, he makes a fundamentally important point. If knowledge is necessary to restrict the class of hypotheses from among which the asserted hypothesis will come, then the restriction may be incorrect for the true hypothesis may be outside the class from which the asserted hypothesis is chosen. In such a case, the hypothesis which is asserted will be false. I shall return to this point later. What is important here is that Reichenbach does not analyze why the application of the Neyman-Pearson theory requires that material conditions be satisfied. He merely shows that from the point of view of *his* theory it *should* require empirical assumptions, but not that there is something about the theory itself which requires some conditions hold.

I shall follow Ian Hacking in calling the physical property described by objective probabilities 'chance' to distinguish it from other uses of 'probability'.[4] A physical system which has this kind of property is a chance set-up. A kind of trial on a chance set-up is defined when the possible outcomes of a trial of that kind are specified. A statistical hypothesis is a statement about the distribution of chances among the possible outcomes on a trial of a particular kind. A *simple hypothesis* completely specifies the distribution of chances among the possible outcomes of some kind of trial on a chance set-up. Any statistical hypothesis which is not simple is *composite* and consists of a set of simple hypotheses.

The basic idea of statistical hypothesis testing is to accept or reject a statistical hypothesis, H, on the basis of the observed outcome of a specific trial of the kind which the hypothesis is about. A *test* of the hypothesis is specified when a subclass, R, of the possible outcomes, Ω, is selected. The class R is the *rejection class* for the hypothesis H. If the observed outcome E on the trial T of kind K is a member of R, then H is rejected; if E is not a member of R, H is accepted.

There are two kinds of error possible when H is tested on the basis of E.

Type I Error: H is rejected, when H is true.
Type II Error: H is accepted, when H is false.

Since H is a statistical hypothesis, it is logically compatible with any observed outcome, except in extreme cases. But it may be possible to determine the chance of Type I error. Let the chance of E on trial T of kind K as stated by H be denoted by '$P(E/H)$'.

The chance of Type I error, α, is the chance that E is a member of R if H is true, i.e.:

$$P(E \in R/H) = \alpha .$$

In line with the traditional idea that a hypothesis should be rejected when the observed outcome is 'unlikely' if the hypothesis were true, Neyman and Pearson recommend that α be made small; that is, $P(E \in R/H) = \alpha$, where α is small. α is called the *size* of a test; it is also called, somewhat misleadingly, the significance level.

A difficulty arises at this point. In general, for a particular hypothesis there are a large number of rejection classes, R, for each of which it is true that their size is α. Thus, controlling Type I error is not enough; the correct rejection class is not uniquely determined. But what about Type II error? Type II error is committed if H is accepted and H is false. H is accepted if E is not a member of R. Let $\bar{R}$ be the complement of R with respect to Ω. H is accepted if E is a member of $\bar{R}$; i.e., an error of Type II is committed when $E \in \bar{R}$ and H is false. Then the *chance of an error of Type II* is:

(1) $P(E \in \bar{R}/H \text{ is false})$.

If H is false, then some other hypothesis is true. Suppose for the moment that for some reason it is known that the true hypothesis in a given situation is either H or H'. Then substituting in (1), the chance of Type II error, β, is:

(2) $P(E \in \bar{R}/H') = \beta$, and the power of R, $1 - \beta$, is:

(3) $P(E \in R/H') = 1 - \beta$, because $\bar{R}$ is the complement of R with respect to Ω.

Now, the criterion for choosing a rejection class according to the Neyman-Pearson theory can be given. Suppose that both H and H' are simple hypotheses. (a) Choose a small α. This choice determines that the chance of Type I error is less than or equal to α. (b) Having chosen α, from the class of possible rejection classes determined by α, choose that one which minimizes the chance of Type II error, β, (or, equivalently, choose the class which maximizes the power). Note that this criterion essentially involves the consideration of an alternative hypothesis to the hypothesis under test.

The hypothesis for which Type I error is defined is called the *test hypothesis*, and the hypothesis against which the test hypothesis is tested is called the *alternative hypothesis*. If both the test and alternative hypotheses are simple hypotheses, there always exists a test of the chosen size which maximizes the power. This important theorem is called the Fundamental Lemma of Neyman and Pearson. When either the test or the alternative hypothesis is composite, no rejection class can be chosen using the criteria stated above. In such cases, additional criteria are needed. For the purposes of this inquiry, the exact nature of these criteria need not be specified. Use of these new criteria result in tests in which an upper limit is placed on the chance of Type I error for each simple hypothesis in the test hypothesis and a lower limit is placed on the power for each simple hypothesis in the test hypothesis.

Now let us return to the original question: Why does a test conforming to the Neyman-Pearson criteria require that some material condition hold for the test to be applicable? The question is important because the criteria themselves nowhere mention explicitly that some material condition must hold. So why is a material condition necessary for the application of the test?

Let J be a hypothesis to be tested. Set the Type I error by choosing the size of the test. For J, Type II error is:

(1) accepting J when J is false, or

(2) $E \in \bar{R}$, when J is false.

So the chance of Type II error is:

(3) $P(E \in \bar{R}$, when J is false$)$.

But (3) cannot be calculated except on the basis of some hypothesis or other; i.e.,

(4) $P(E \in \bar{R})$,

is not defined, is not a well-formed expression. Let K be another hypothesis. Then,

(5) $P(E \in \bar{R}$, when K is true$)$

is a well-formed expression. But is it the chance of Type II error defined with respect to J? The latter is defined in (3). (5) is the chance of Type II error with respect to J *only if K is true when J is false.* For (5) to be the chance of Type II error with respect to J, either J or K must be true.

If J is composite, $P(E \in R/J)$ cannot be calculated unless the distribution of probabilities that each simple J_i is true is known, and similarly for K. But the Neyman-Pearson theory is independent of these prior 'probabilities'. When J is composite, the tests resulting from use of the Neyman-Pearson criteria do not allow the calculation of $P(E \in R/J)$ but instead set an upper limit on it. But for these upper limits to be defined, it is still required that either J or K contain the true hypothesis.

The preceding argument proved the need for the following condition on the use of the Neyman-Pearson theory:

The Neyman-Pearson theory can be used to license a legitimate inference only if the true hypothesis is included either in the test hypothesis or in the alternative hypothesis.

I shall call this condition the criteria condition, for it is required only because of the logical relations between the definitions of chances of Type I and Type II error. If this condition is not satisfied, it is not the case that both the chance of Type I error is limited to the size of the test and the chance of Type II error is minimized for a test of that size. It has not been shown that the criteria condition is a material condition or that it can be satisfied only if some further material condition is satisfied.

Suppose that the criteria condition is satisfied. Before the theory can be applied a further condition must be satisfied:

The Neyman-Pearson theory can be used to license a legitimate inference only if there exists a rejection class for the test and alternative hypotheses which satisfies the appropriate Neyman-Pearson criteria.

This is the *existence condition* for the application of the theory. For given test and alternative hypotheses it is a logical question as to whether there exists a rejection class meeting the criteria. It is conceivable that for a given pair of hypotheses there is no such rejection class. Hence, the condition is necessary. If there always is such a rejection class, the condition is merely redundant.

The criteria condition may not be a material condition. I shall call a statistical hypothesis an LP-hypothesis if, given the specification of a definite kind of trial on some chance set-up, the hypothesis is a statement of a distribution which is logically possible for a trial of that kind. Consider the set of all LP-hypotheses about a specified kind of trial. Suppose that this set is partitioned into two, and one of the partitions is arbitrarily labeled the test hypothesis while the other is labelled the alternative hypothesis. Since the set of all LP-hypotheses is the set of all statements about possible distributions on trials of the specified kind, it must contain the statement of the true distribution. Thus either the test hypothesis or the alternative hypothesis will contain the true hypothesis, and the criteria condition is satisfied. Must there be, in this case, a rejection class; i.e., must the existence condition be fulfilled? If it can be guaranteed that there is a rejection class in this case, then for this kind of trial the criteria condition does not constitute a material condition. I know of no conclusive argument that there can be no rejection class in these circumstances, although I suspect there cannot. I shall attack piecemeal the problem of showing that, in these circumstances, there can be no rejection class meeting Neyman-Pearson criteria. Some attention will be paid to findings from the history of statistics which are relevant to my position.

The applications to which the Neyman-Pearson theory is put in ordinary statistical practice require material assumptions; that is, there is a material condition on the legitimacy of each of these practical arguments. If a test is concerned with a proper subset, H, of the set of all LP-hypotheses, T, for a particular kind of trial, then the criteria condition requires that H contain the true hypothesis. There must have been some empirical reason for rejecting the complement of H with respect to T, i.e., for claiming that every hypothesis in $\bar{H}$ was false. Thus there are many tests in actual use which are legitimate only if a material condition holds. The question is, however, whether every Neyman-Pearson test requires a material condition. The history of statistics for the last 45 years can reasonably be seen as an attempt to discover tests which progressively weakened the material conditions on the tests or to discover tests to establish ordinary material assumptions of other kinds of tests. Historically, it can be noticed that as material conditions are weakened it becomes increasingly more difficult to prove the existence of rejection classes meeting the Neyman-Pearson criteria.

I conjecture that it is logically necessary that there be material conditions on the legitimacy of any argument based on the Neyman-Pearson criteria. That there must be material conditions follows from three factors: the criteria condition, the existence condition, and the nature of the set of all LP-hypotheses. In order for there to be a Neyman-Pearson test which had no material condition on it, it would be required that the union of the set of hypotheses constituting the test hypothesis and of the set constituting the alternative hypothesis be the set of all LP-hypotheses, that is, the set of all possible distributions of the kind of trial specified. That this is necessary follows from the criteria condition, which requires the true hypothesis to be either in the test hypothesis or in the alternative.

The conjecture requires proof of the following statement: No rejection class exists for any twofold partition of the set of LP-hypotheses. For every such partition of the set of LP-hypotheses satisfies the criteria condition; and if there exists a rejection class for any of these partitions, there exists a test without material conditions. Thus the proof of the conjecture reduces to proving that for no twofold partition of the set of all LP-hypotheses does there exist a rejection class.

What kind of reason is there to believe that there exist no rejection classes for any twofold partition of the set of all LP-hypotheses? The reason lies in lack of structure of the set of LP-hypotheses. In order to construct a rejection class where composite hypotheses are concerned, the set of hypothese consisting of the test and the alternative hypotheses must have enough structure to allow an upper limit to be fixed on the chance of Type I error while at the same time a lower limit is fixed on the power. But because the set of all LP-hypotheses is very nonhomogeneous there is no way to divide it into only two sets (the twofold partition) so the resulting sets are ordered enough to allow the necessary limits to be set and the resulting rejection class calculated. The criteria proposed by the Neyman-Pearson theory for testing composite hypotheses are essentially criteria that, in certain circumstances, separate out a set of hypotheses which have an ordering of the kind described.

I shall restate the conclusion. No twofold partition of the set of LP-hypotheses for a trial of a given kind will allow the calculation of a Neyman-Pearson test. In order to apply the Neyman-Pearson theory, the set of hypotheses considered in a particular case must be a subset of the set of all LP-hypotheses. This means that some subset of the set of all LP-hypotheses must be treated as false so that the true hypothesis is in the subset considered in the construction of the test. Thus, some material condition must be imposed on the set of hypotheses considered so that a rejection class may be found which meets the Neyman-Pearson criteria. How does this conclusion affect the position of the Neyman-Pearson theory in the canon of inductive logic?

First, I wish to emphasize that the requirement for a material condition on inferences licensed by the Neyman-Pearson theory is a logical requirement, not a methodological requirement. It is a logical consequence of the criteria proposed to govern tests and of the logical structure of the set of all LP-hypotheses on a given kind of trial. This logical requirement has a methodological counterpart. In order for there to be adequate grounds for believing, acting on, etc. the conclusions of an argument licensed by the Neyman-Pearson theory, there must be adequate grounds for believing, knowing, etc. the material conditions imposed on the particular inference. The 'practical applicability' of which Neyman talked is a version of the methodological counterpart of the material conditions on legitimate inference.

Secondly, the material assumptions on many arguments licensed by the Neyman-Pearson theory can themselves be tested by tests selected according to the criteria furnished by that theory itself. For example, one test may assume that the test and alternative hypotheses are both of normal form, while another test can be constructed to test whether the form is normal or not. So some material conditions can be tested within the theory. But if my conjecture is true, then not every material condition can be tested within the theory without assuming some additional, new, material condition. In Reichenbach's terminology, the Neyman-Pearson theory is an inductive method applicable only in the context of advanced knowledge. In particular, the knowledge that must be known is knowledge of some statistical hypothesis or other. So the Neyman-Pearson theory is a kind of secondary induction even within the restricted class of empirical hypotheses consisting only of statistical hypotheses.

Third, a more general conclusion also follows from my argument that the Neyman-Pearson theory requires a material condition. The theory is essentially a theory of eliminative induction. If one thinks of using it in a sequential fashion, first narrowing the set of LP-hypotheses, then using another test to narrow the resulting set still further, and so on, then what is obtained is a theory of induction which proceeds by eliminating sets of hypotheses as false and retaining other sets as true. Used in this fashion, the theory would be a method of elimination and an inductive method because the hypotheses eliminated were logically compatible with the observational evidence. Now any theory of eliminative induction which claimed to be a method of induction applicable in the context of primitive knowledge would be required to show how it would *systematically* eliminate parts of the set of all LP-hypotheses for a given kind of trial. It seems that this would require the set of all LP-hypotheses be ordered in some way connected with chance of the various outcomes on each of the hypotheses in the set. But the gist of my argument that the application of the Neyman-Pearson theory requires material conditions is that in general no such ordering exists. Hence, any method of eliminative induction which requires such ordering will require, in general, some material conditions on its application.

Wichita State University

NOTES

[1] Neyman, J., 'The Problem of Inductive Inference', *Communications on Pure and Applied Mathematics* **8** (1955) 19.
[2] Reichenbach, H., *The Theory of Probability*, University of California Press, Berkeley, 1949, p. 364.
[3] *Ibid.*, p. 454.
[4] Hacking, I., *Logic of Statistical Inference*, Cambridge University Press, Cambridge, 1965. My exposition of the Neyman-Pearson theory follows closely that given there by Hacking and also that of M. G. Kendall and A. Stuart, *The Advanced Theory of Statistics*, Vol. 2, Hafner, New York, 1967.
[5] The most commonly invoked criteria lead to unbiased and invariant tests. See Hacking, *op. cit.*, pp. 97–99, and Kendall and Stuart, *op. cit.*, pp. 200–6, 256–57. A more complete discussion of statistical hypothesis testing is given in E. L. Lehmann's *Testing Statistical Hypotheses*, John Wiley and Sons, New York, 1959.

V. Problems in Quantum Physics; Genetic Epistemology

PAUL FITZGERALD

TACHYONS, BACKWARDS CAUSATION, AND FREEDOM

I

'Tachyons' are hypothetical faster-than-light particles.[1] Their existence has been suspected and sought,[2] on the ground that they appear to be compatible with the laws of nature, particularly with Special Relativity, and so there is at least some likelihood that they exist. That last mad inferential leap is justified by the experience of particle physicists, summed up in 'Gell-Mann's totalitarian principle' that "whatever is not forbidden is compulsory." Experimental searches made so far have failed to find them.

The classical reason for ruling out faster-than-light particles has been that according to Special Relativity an infinite force would be required to accelerate a particle up to and beyond the speed of light. For if the particle's rest mass m_0 is greater than zero, then its relativistic mass, given by the formula

$$m_0/\sqrt{(1 - v^2/c^2)},$$

approaches infinity as the particle's velocity v approaches c, the speed of light. But this objection is dodged if the particle does not have to be accelerated up to and beyond the speed of light. Suppose that it has a 'superluminal' velocity at all points on its world-line, and there is no need to invoke infinite forces. There *is* a need to reckon with that imaginary-valued denominator in the above formula for the relativistic mass. Since the relativistic mass should be represented by a real number if it is to be an experimentally measurable quantity, we can posit that the rest mass m_0 in the numerator of the formula is also imaginary. Introducing the real-valued quantity 'meta-mass' by the formula $m_0 = i\mu$, and multiplying numerator and denominator of the formula for relativistic mass by $i = \sqrt{-1}$, we get the new formula for relativistic mass:

$$m = \mu/\sqrt{(v^2/c^2 - 1)}$$

Boston Studies in the Philosophy of Science, VIII.

Here all of the quantities which appear are reassuringly real-valued. Tachyons are also assigned imaginary proper length and proper time by the relativistic formulas. Once again, those who are spooked by this and need a comforting word can introduce a real-valued 'meta-length' and 'meta-time' by analogy with metamass. Photons have no such saving grace; as they move *in vacuo* their proper lengths and proper times remain obstinately zero under this sort of fiddling.

It is assumed that the total energy E of a tachyon is related to its relativistic mass m by the usual $E = mc^2$. We also assume Lorentz invariance of the energy-momentum four-vector associated with a point on a tachyon's world-line, and thus covariance of

$$E^2 = p^2c^2 = m_0^2c^4$$

where p is the momentum. This hyperbola is graphed in Figure 1 for the simple special case where the total momentum is along the x-direction for the whole class of inertial coordinate-systems considered. Each point on the hyperbola represents a pair of values, for energy and momentum respectively, assigned to the tachyon at the place-time by one of these coordinate systems. Compare the points S and S' to see that one man's positive-energy tachyon is assigned negative energy by another.

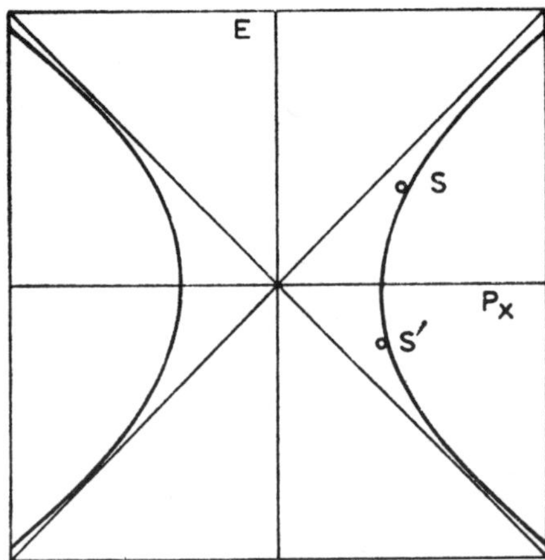

Fig. 1. Diagram from 'Particles Beyond the Light Barrier' by O. M. Bilaniuk and E. C. G. Sudarshan, *Physics Today*, May 1969.

Assignment of negative energies is related to another curious phenomenon; namely, that pairs of events along a tachyon's world-line do not have Lorentz-invariant temporal order. This follows from the fact that those world-lines are spacelike rather than timelike. This means that if you pick any section of such a world-line along which the time, t, relative to a given coordinate system, increases monotonically, then the 'length' of that section between its endpoints t_0 and t_1 is an imaginary number given by $\int_{t_0}^{t_1} \sqrt{(1 - v^2/c^2)}\, dt$. (Note that v is in general a variable which must be expressed in each case as a specific function of t.)

To see the significance of this, take the case of a tachyon whose velocity relative to a chosen coordinate-system S is constant throughout its history. If, relative to that coordinate system, O is the earliest event in the tachyon's life and P the latest, then another coordinate system S' can be found for which P is the earliest and O the latest. If O can be described as a partial cause of P, let us say, then relative to S' this cause is later than its effect. For instance, O might be the pulling of the trigger of a 'tachyon gun', which fires a tachyon, and P might be the reception of that tachyon in a tachyon camera. Conceive of the particle as travelling from the gun to the camera and you will regard the particle as travelling backward in time, if you use coordinate system S'. This simply means that you date the firing as later than the reception of the tachyon at the camera. Figure 2 illustrates the situation.

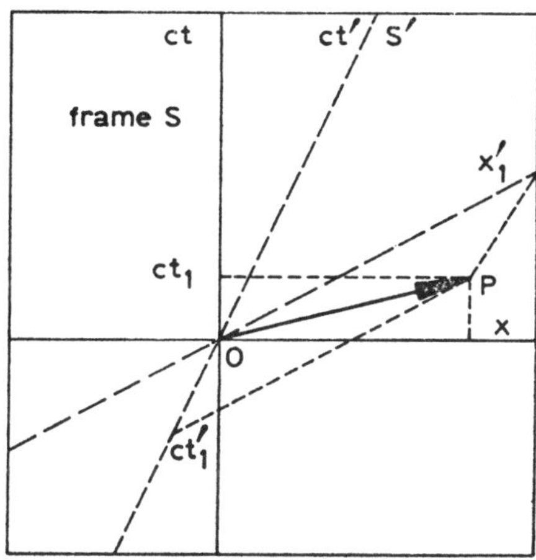

Fig. 2.

II

The quantum mechanical treatments[3] have revealed more interesting features of tachyons. They have imaginary rest mass. There are no finite-dimensional representations of the Lorentz group other than the one-dimensional, so tachyons have either zero spin, or else an infinite number of polarization states. Assuming zero spin, the tachyon field must be a scalar field obeying the Klein-Gordon equation. The solutions are restricted to cases where the absolute value of the momentum $|k|$ is greater than or equal to the meta-mass μ. That ensures that the energy E, which is equal to $\sqrt{(k^2-\mu^2)}$, will be real-valued, and the solutions will be oscillatory in time. This is required so that a superposition of such solutions can be construed as the wave function of a particle with real energy. But the restriction entails that no superposition of the allowed set of solutions can be made into $\delta^3(x)$, which in turn means that tachyons can not be sharply localized.

In accord with the usual method of second quantization the field variable $\emptyset$ satisfying the Klein-Gordon equation is an operator, and the coefficients $a^+(k)$ and $a(k)$ of its Fourier expansion are creation and annihilation operators. In Gerald Feinberg's field theory of non-interacting tachyons, the creation and annihilation operators satisfy anti-commutation relations and obey Fermi-Dirac statistics, despite their zero-spin character. This departure from the usual connection between spin and statistics does not violate the usual theorems in which the connection is proven, and is not surprising in a theory with unorthodox features. The vacuum state is not Lorentz-invariant, for instance, and neither is the particle-number.

Feinberg had been led to anti-commutation relations for the creation and annihilation operators by following out what was implied by the Lorentz-invariance of the field theory. Lorentz-invariance entails the existence of unitary operators $L(\lambda, a)$ associated with Lorentz transformations $x'=\lambda x+a$, where x' is the Lorentz-transform of the position four-vector x and the transformations are in general inhomogeneous. The operators $L(\lambda, a)$ must satisfy the operator equation

$$L(\lambda, a)\, O(x) L^{-1}(\lambda, a) = O(\lambda x + a)$$

The energy-momentum operator P_μ generates the space-time translations.

It is then discovered that annihilation operators are transformed by the operator L into creation operators, which is inconsistent with the canonical commutation relations for them and forces anti-commutation relations.

Arons and Sudarshan purport to show that Feinberg's theory is not Lorentz-invariant on the grounds that the energy-momentum operator P_μ has a Lorentz-transform which differs from the usual one by an infinite constant, and that the vacuum state Lorentz-transforms into a state with an infinite number of particles. The former fact 'strongly implies' and the latter fact shows the non-existence of the required unitary operators $L(\lambda, a)$. Arons and Sudarshan then construct an alternative free-field theory which introduces creation and destruction operators for both positive and negative energy tachyons, which can now satisfy Bose-Einstein statistics. Processes involving negative energy tachyons are reconstrued as ones involving only positive energy particles, with the roles of emission and absorption interchanged. Subsequently, this theory was extended by Dhar and Sudarshan to cover interactions.

III

A number of proponents of tachyons, including Bilaniuk, Deshpande, Sudarshan, and Feinberg[4] have tried to avoid the suggestion that tachyons would involve negative energies, travel backward through time, and causes later than their effects ('retrocausality') by introducing a 'reinterpretation principle'. As Bilaniuk and Sudarshan state it, the reinterpretation principle is the "... interpretation of negative-energy tachyons propagating backward in time as positive-energy tachyons propagating forward in time. This reinterpretation invalidates causality objections to the possibility of existence of faster-than-light signals and permits construction of a consistent theory of tachyons." The principle bids us to interpret tachyon world-lines that relative to our chosen coordinate-system the tachyons are viewed as having velocities less than or equal to infinity, and therefore as carrying positive energy. So in the case described in Section I and illustrated in Figure 2, frame S construes event O as the tachyon's emission and P as its absorption, whereas relative to frame S' event P is emission and O is absorption. Both regard the tachyon as travelling 'forward in time' with positive energy, but the spatial direction of travel is construed oppositely. Thus we are thought to have our tachyons without retrocausality and negative energies.

You might think that the only kind of retrocausality that threatens when tachyons are involved and which the reinterpretation principle is designed to block is that mild strain in which the causal influence is propagated between topologically simultaneous place-times, such as O and P, whose time-order varies from one coordinate-system to another. This isn't so. Given the milder strain, the more virulent kind of retrocausality strikes too, namely, that in which the cause is later than the effect relative to all coordinate-systems. Tolman had pointed this out long ago,[5] and the reinterpretation principle was crafted to avoid his charge that faster-than-light particles would give rise to the strong causal anomaly just described.

To see how that anomaly arises, assume for simplicity that we can send tachyons with infinite velocity. Person *alpha* at place-time P_2 sends a burst of tachyons, at infinite velocity relative to his coordinate-system, to person *beta*, who is receding from him with velocity W. *Beta*, on receiving these particles at place-time P_1, immediately fires back a burst toward *alpha*. These travel at infinite velocity relative to *beta*'s coordinate-system. As the accompanying Minkowski figure (Figure 3) shows, they are received by *alpha* at a place-time P_0 which is *absolutely* earlier than the place-time P_2 at which he sends the burst of tachyons to *beta*.

Bilaniuk and Sudarshan tried to avoid this retrocausal interpretation

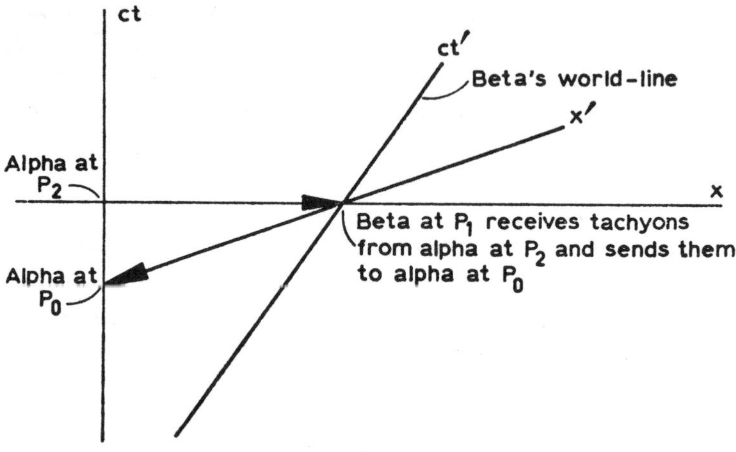

Fig. 3.

of the situation by invoking the reinterpretation principle. They point out, quite correctly, that person *beta* regards his own 'reception' at P_1 of *alpha*'s signal as earlier than *alpha*'s emission of it, at P_2. So by their principle *beta* must regard the signal as travelling from P_1 to P_2. Moreover, *alpha* regards his 'reception' of the signal, at P_0, as earlier than *beta*'s sending of it at P_1. Bilaniuk and Sudarshan then press the reinterpretation principle into service and say 'there is no more back and forth exchange of signals. In each case the observer believes that he is sending out both signals."[6]

This seems implausible, on the face of it, if the signals sent by me carry information which only I am in a position to know, and the same *mutatis mutandis* for the other person in the story. Shoichi Yoshikawa and Bryce DeWitt have in fact objected, correctly I believe, that the reinterpretation principle won't do the job of enabling us to avoid regarding the situation as retrocausal. For if person *alpha* at P_2 could modulate the beam of tachyons which he emits then he could transmit information to *beta* at P_1. *Beta* at P_1 could then presumably retransmit the information to *alpha* at P_0. The fact that we have information transmitted by deliberate action, albeit by a negative energy beam, just makes it implausible to use the reinterpretation principle or to deny that retrocausality is involved. More fundamentally, there are criteria other than temporal order for picking out which of a pair of events causes the other.[7] And these criteria take precedence over temporal priority.

Moreover, there is another reason why the reinterpretation principle does not enable us to dodge retrocausal situations. Clever thought-experiments[8] have been contrived in which signals are sent in a chain from observer to observer and eventually arrive back at the observer who starts the signalling process.

Each pair of observers agrees on which of them sent the tachyon signal to the other, without there being any occasion for them to invoke the reinterpretation principle. Yet the observer who starts the process receives the final signal in the chain, ... *before* he sends the starting signal! Nowhere is there any occasion to use the reinterpretation principle to dodge this anomaly.

DeWitt's variant of this thought-experiment goes as follows. Let *alpha* at P_2 send information by tachyon to *beta* at P_1, but have *alpha* and *beta* at rest relative to one another, so that both agree that *alpha sent* the message, even if they accept the reinterpretation principle. *Beta* now sends the same information by photon to observer *gamma*, who is in the immediate neighborhood receding from *alpha* and *beta* with velocity *w*. Everyone agrees about the direction of transmission here, too.

Receiving the information from *beta*, *gamma* immediately retransmits it by tachyon beam to a confederate, *delta*, at rest relative to *gamma*. Let *delta* receive gamma's signal right near place-time P_0 (which is occupied by *alpha*). *Delta* and *gamma*, the only direct participants in this last transaction, agree that it was *gamma* who sent and *delta* who received the information. On receiving the massage, *delta* immediately sends it by photon to *alpha* at place-time P_0, absolutely earlier than place-time P_2 whence *alpha* sends the initiating message. The entire situation is pictured in Figures 4 and 5. Pirani has worked out a similar thought-experiment in which only tachyons are used to transmit the messages.[9]

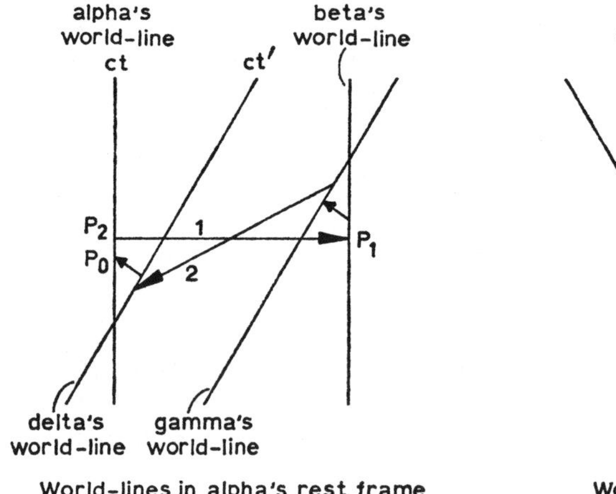

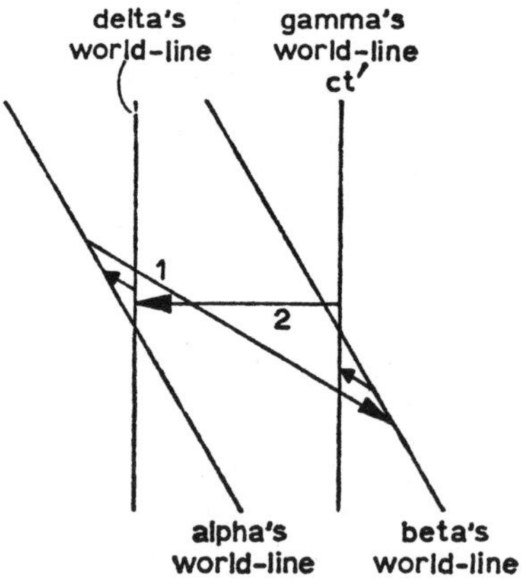

Figs. 4 and 5. The Dewitt Gambit.

We can conceive of cases in which the reinterpretation principle can plausibly be invoked. But those cases are ones in which the 'nomic relation' between the events in question is not of the *asymmetric* causal type anyhow, so no highly anomalous retrocausal situation threatens. Where only symmetric nomic relations are involved the reinterpretation principle does enable us to avoid saying that a tachyon "travels backward in time."

But it does not enable us to avoid saying that the cause is later than the effect, in the cases where having to say that is particularly bothersome, namely, where asymmetric causal relations are involved. Where the principle can be used, it is not needed for this purpose, since we're dealing only with time-symmetric cause effect pairs anyhow, not with full-fledged asymmetric causes and effects.

Maybe tachyons, if they exist, are essentially Janus-faced or time-symmetric particles in the sense that given any pair of events, *A* and *B*, on a tachyon's world-line it is no more plausible to say that *A* causes *B* than that *B* causes *A*, unless time-order is invoked. If so, the reinterpretation principle would always be usable, and either endpoint of a tachyon's world-line could be regarded as its birth, the choice being determined solely by the date which your coordinate-system assigns to the endpoints. But if tachyons were like that they could not be relied on to transmit signals, since their production and modulation would not vary sensitively enough with external manipulable conditions. Where tachyons can be used to transmit information we would probably have grounds other than dating to distinguish emission from reception, at least in some cases, and where we do, the reinterpretation principle loses its plausibility.

To see what those other grounds can be like, let's take a case of tachyon interaction which doesn't directly involve human action, though it does involve artifacts. The reader can readily adapt it to purely natural objects, such as pulsars emitting a varying tachyon beam in synchrony with their electromagnetic emission.

Suppose that we have at place P_1 a roulette wheel which stops randomly on red or black once a minute and is automatically respun by an attached device. Each time the wheel stops a scanner records whether it has stopped on red or black, and sends a signal recording that fact to a machine named 'Alonzo'. Alonzo either sends or receives tachyons; we don't want to prejudge the issue by saying which. At another place P_2 there is a similar tachyon machine, Bertram. Attached to Bertram is a moving tape on which is punched, once a minute, either a '0' or a '2'.

We discover that once each minute the distance separating the two tachyon machines is traversed by either a single tachyon or by a pair travelling together. Our coordinate-system is so chosen that the tachyons move with infinite velocity between the two machines. If at time t_0 the roulette wheel lands on red then very shortly thereafter, at t_1, a pair of tachyons traverses the distance separating the two machines. After another short interval, a '2' is punched on Bertram's moving tape. But if at t_0 the roulette wheel lands on black, then at t_1 a single tachyon travels between the two machines, and at t_2 a '1' is punched on Bertram's tape. This happens regularly; an unfailing correlation between colors and numbers.

Since we have chosen a coordinate system relative to which the tachyons move infinitely fast, the reinterpretation principle doesn't tell us which machine we should regard as emitting the tachyons and which as receiving them. We'll have to use our heads. Should we construe Alonzo as emitting and Bertram as receiving, or the other way round? (See Figure 6).

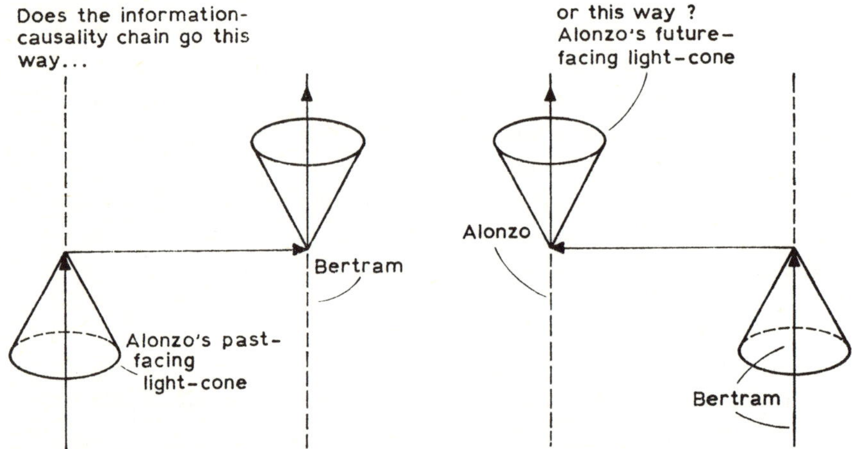

Fig. 6. Who's sending the Tachyons.

If we conceive of Alonzo as emitting and Bertram as receiving, then we get an explanation of why the *n*th outcome at the roulette wheel matches the *n*th symbol punched on Bertram's tape (slightly later, according to our coordinate-system.) View Alonzo as receiving and Bertram as emitting and we have no equally plausible account of this amazing coincidence. This would look even less plausible if one disconnecting the roulette wheel from Alonzo we find either that tachyon action stops or the correlation between roulette wheel and punched symbol breaks down. That would confirm the hypothesis that information went from roulette wheel to Alonzo to Bertram to his tape. No principle of reinterpretation should lead us to revise this opinion just because we switch to a different co-ordinate-system in which the outcome at the roulette wheel is later than the appearance on tape of the correlated symbol.

As mentioned before, the presence of artifacts such as Alonzo and Bertram is inessential, for a similar thought-experiment could be constructed involving purely natural objects, such as tachyon-emitting pulsars. If the intensity of a tachyon beam between earth and a pulsar shows the same pattern of variation as the intensity of an electromagnetic beam, then it's reasonable to assume that the pulsar is talking to us in tachyonese as well as in the usual way, rather than that we're sending it a tachyon message which it is echoing back to us by photons. This is particularly true if the tachyon talk dies to a whisper in our vicinity when a mass of tachyon absorbing material appears between us and the pulsar.

IV

There is an idea abroad that tachyons, and retrocausal influences generally, would involve an intolerable kind of causal anomaly or paradox, perhaps even a self-contradiction.[10] In the philosophical literature several papers have sprouted purporting to show by detailed argument that it is *a priori* impossible for a cause to be later than its effect. The literature of physics has seen thought-experiments in which retrocausal tachyons are claimed to involve paradox or self-contradictory situations.[11] And there is an undercurrent of suspicion in both fields that retrocausal influences acting on us would involve a peculiar kind of limitation on our freedom of decision or action. In this section I'll consider a typical thought-experiment in which tachyons are supposed to involve a paradox not relating primarily to freedom, and in the next section I'll face the freedom issue.

Of course, there is the anomaly involved in the bare fact of retrocausality; we're not used to the idea of causes being later than their effects. But that by itself is not a very strong argument against tachyons, because even if they exist it's not surprising that we'are unaccustomed to retrocausal transactions ... after all, they've passed unobserved up to now, since we've hardly begun to look for them!

A more powerful argument to show that retrocausal tachyons involve an intolerable conceptual difficulty is illustrated by the Case of the

Logically Pernicious Self-Inhibitor. This is a device consisting of a tachyon transmitter and a tachyon reflector. It is so constructed that a tachyon message is sent from the transmitter, *A*, at place-time P_2 to the reflector, *B*, at place-time P_1 only if no tachyon has impinged on the transmitter during the 'sensitive period', which is the preceding five minutes of proper time. If the message is sent from *A* then it succeeds in reaching *B* at P_1, because the whole device is carefully sealed off from outside interference and the path through space-time is cleared of obstructions. As Figure 7 shows, reflector *B* is designed to reflect any tachyon impinging

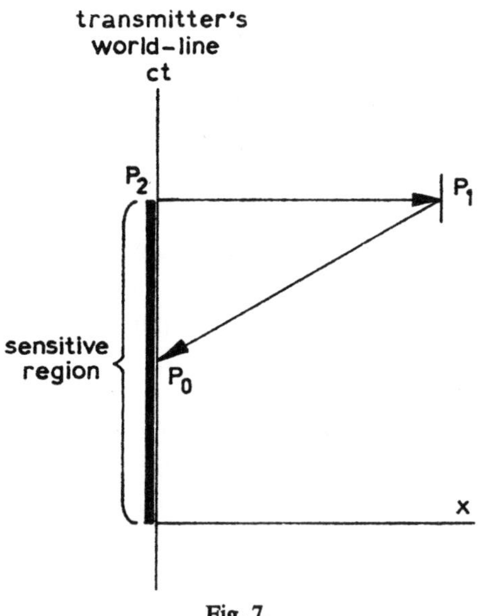

Fig. 7.

on it at P_1 to *A* at P_0, which is two minutes earlier than P_2 (proper time again.) Here too, the space-time airways are clear so that a tachyon sent from *B* at P_1 is guaranteed to hit *A* at P_0.

But this means that the transmitter will have been hit by a tachyon during the sensitive period, and thus won't transmit from place-time P_2. In short, *A* transmits if and only if it is not hit during the sensitive period, but it transmits only if the transmission is reflected from *B* at P_1 and hits *A* during the sensitive period. (See Figure 7.)

Of two things we can be sure. First, no such device will function according to plan, for the plan involves self-contradiction. Second, the existence of tachyons does not by itself entail or render probable the existence of a logically pernicious self-inhibitor, so we can't say that the existence of controllable tachyons *entails* self-contradiction. We would have to say simply that they cannot be controlled in all of the ways required for the operation of the self-inhibitor. The existence of completely uncontrollable tachyons is left unscathed by this argument, however dubious the uncontrollability may be on other ground.

Not that the defender of tachyons or 'tachyonite', has a right to sit back and breathe easy. For paradox machines like the one just sketched do pose a threat. Though falling short of proving that tachyons involve self-contradiction, they may provide good reason to cease looking for them. For granted that paradox machines like the self-inhibitor will never function successfully, tachyons or no tachyons, the tachyonites should give an account of why, if there are tachyons, such machines will nonetheless fail. There are two possibilities. The first is that there are general empirical reasons why such machines fail to function. Perhaps tachyons are uncontrollable. Or maybe the strong electromagnetic fields needed to direct charged tachyons plays hob with the reflectors. Or maybe it's impossible to seal off tachyons from entering the device, with the result that the tachyon messages are swamped by intruders from outer space, and the sensitivity region of any machine is soaked in a wave of these cosmic marauders.

The second possibility is that there are no generally applicable empirical reasons for the malfunctioning of paradox machines; we simply have a fortuitous series of accidents which avert the embarrassment of self-contradiction every time. The first self-inhibitor fails to function because a screw comes loose; the second because an earthquake destroys it, the third because... on and on come the tedious details. It seems to me that this sort of appeal to fortuitous accidents, while is does show that tachyons are not intrinsically self-contradictory, fails to blunt the thrust of the thought-experiment, which is that it is empirically unlikely that tachyons exist.[12] A similar appeal to providential accident could be made in the face of those thermodynamical thought-experiments which run along some such lines as 'If such-and-such were possible, then we could construct a perpetual motion machine of the second kind ... but we can't, so such-and-such is not possible." It would be a poor defense against this sort of argument to hold that even given such-and-such we might persistently fail, through an unrelated series of accidents, in attempts to use such-and-such to make the perpetual motion machine. I think that the reason why such appeals to the fortuitous are unsuccessful defenses is this. To sustain the appeal, we would have to regard it as a lawlike regularity that unrelated accidents will always conspire to render impossible the successful functioning of a machine which should be constructible if the suggested phenomenon (e.g. tachyons) obtains. We are not willing to accept such a principle as a law since it lacks direct empirical support, has no indirect support from nor organic interplay with accepted laws and theories, and doesn't even particularly resemble other things which we accept as lawlike regularities. So we surrender instead our belief in the empirical possibility of tachyons, or what have you.

So what the tachyonite needs is a plausible general empirical reason why paradox machines can't be constructed though tachyons exist. If tachyons can be produced, modulated and detected in a fairly reliable way then it's hard to see how the paradox machines can be ruled out on general empirical grounds. It won't do to say, for example, that individual tachyons can't be produced at will. For we might still be able to alter the statistics of large numbers of them in a way suitable for signal transmission. Alväger and Kreisler (1968) tried to generate a beam of charged tachyons by sending photons into lead plate, hoping thereby to pair-produce enough of them to get visible Cerenkov radiation. Even if tachyons are uncontrollable they might still be detectable, say by the 'missing mass' method (cf. e.g. Feinberg, 1970). But they'd be treading a tightrope; interacting with matter just often enough to make detection possible but not often enough to make control and signalling feasible.

Bilaniuk and Sudarshan (1969b, p. 51), following up a suggestion of DeWitt (ibid.), have suggested a means of escape from paradox machines which I think is not successful. It assumes the reinterpretation principle, but that's not the point of my objection to it. Bilaniuk and Sudarshan propose that cosmological boundary conditions are such that relative to an observer P_0 at rest in a standard frame S_0 "the flux of tachyons coming from distant regions of the universe is finite;" we can say for simplicity and without loss of generality that it is zero. 'For another observer P_1 moving with a velocity $w<c$ relative to the standard frame this situation

implies the impossibility of his emitting tachyons with a velocity greater than a certain threshold velocity $u_1 = c^2/w$." (Bilaniuk and Sudarshan, 1969b, p. 51). The point is that in virtue of the cosmological boundary conditions only those signals which carry information and energy in the same (futureward) direction as seen in the standard frame are possible. "Under such circumstances *no causal loops could arise* and no 'anti-telephone' (a paradox machine), such as proposed by Gregory Benford, David Book and William Newcomb, could be built." (Ibid. p. 51; italics theirs, parenthetical remark mine.)

I so admire the ingenuity which the tachyonites have shown in elaborating the physics of superluminosity that I want them to emerge triumphant over the dragons of paradox. But this defense, if I understand it aright, seems to fail. To succeed it would have to involve positing a non-ad *hoc*, independently plausible empirical reason why retrocausal paradoxes don't appear when tachyons do. Now the appeal to cosmological boundary conditions looks at first blush like such a principle. We all know about the regularities governing the flux density of cosmic radiation entering earth's atmosphere, the cosmic three degree Kelvin black-body radiation, and so on. Why shouldn't similar lawlike regularities pertain to tachyon flux density?

Here's the rub. If we take the cosmological boundary condition in a sense which makes it similar to accepted principles of the kind mentioned then it won't help us dodge the retro-causal paradoxes. And if we take it in the sense required for avoidance of the retrocausal paradoxes, it looks implausible and *ad hoc*; almost as bad as positing the above-mentioned rash of redeeming accidents.

The cosmological conditions posited are plausible if they are dependable regularities about the tachyon flux from *the cosmological background*, and it is understood that occasional departures from the rule are permitted, *as well as* superimposed signals from foreground sources, such as nearby observers signalling to us. After all, we don't suppose that our rules concerning cosmic 3 K black-body radiation entail the impossibility of sending a rocket into orbit which will transmit a "3 K black-body radiation signal" superimposed on the cosmic background. Let our observer P_1 be a foreground source sending retrocausal signals to P_0 and we're faced with the paradox machines and retrocausal anomalies.

To prevent these from rearing their ugly heads, we have to specify that the 'cosmological boundary condition' is a very strong one to the effect that *no* tachyons, from the cosmical background or foreground sources, can possibly be made to impinge on the standard frame P_0. No matter how hard one tries to send tachyons to P_0, even from a frame at rest with respect to it, one *must* fail. This is particularly paradoxical because "an arbitrary number of tachyons can be emitted by P_0" (ibid., p. 51). Is it then by a fortuitous series of accidents that other observers at rest relative, to P_0 must always fail to get tachyons through to him? Why are they so accident-prone and he so accident-free?

Other ways of avoiding the paradoxes may be more promising, such as DeWitt's suggestions that tachyons exist only as virtual particles, or that "the universe as a whole is so finely tuned (for example, by quantum mechanical interference effects) that whenever information is sent into the past,... it is always wiped from the receivers memory in time to prevent paradoxes from occurring." (DeWitt, 1969, p. 50).

V

A second sort sort of thought-experiment is sometimes cited to show that retrocausal influences and signals from the future involve either an intolerable conceptual difficulty and/or a peculiar, *sui generis* kind of limit on our freedom. Let's illustrate it with "The Case of the Potential Murderer."

You have a pistol in your hand, and your mind idly wanders to the idea of shooting and killing Rameses, who is standing before you. Suddenly you receive through your earphones a tachyon-transmitted message from the future, telling you that Rameses is alive and well two days hence. You might distrust the message. But that might be irrational; for maybe you've learned by extensive experience that messages from that tachyon station are always reliable. (This assumes that epistemological Luddites have not thwarted all attempts to send information regularly by tachyon-transmitter, so we have been able to establish the reliability of certain transmitting stations.) In fact the evidence for the reliability of the tachyon messages might be so strong that they can give us knowledge of what's going to happen.

So you come to believe, in fact to know, that Rameses will be alive two days from now, and therefore that you won't kill him now. So you are not in the familiar situation of deliberating about what act to perform and resolving your uncertainty only by deciding. Is your freedom or your ability to shoot and kill Rameses thereby curtailed?

Not necessarily!

The argument that your freedom is curtailed would presumably run this way. The tachyon message gives you enough information to figure out that you will not shoot Rameses. This information does not include anything about your present intentions. So you can say, "Obviously it's not in my power to kill Rameses. For if I were to kill him then he'd be dead two days from now, in which case I wouldn't have received this tachyon message telling me that he's alive then. Since the message is reliable, it's not in my power to kill Rameses now." There's the limitation of freedom; now for the paradox.

The paradox arises when you add this line of reasoning. 'Nothing is preventing me from killing Rameses. He's in plain view; the gun is in working order; I'm a crack shot and never miss a target that large at such short range. So obviously I can shoot and kill Rameses. It's up to me."

In brief, one line of reasoning says, "I *can't* kill Rameses, for there is a reliable message from the future telling me that he will be alive." The other line of reasoning says, "I *can* kill Rameses, for I have the general ability to pull off feats like that and nothing is preventing me in this particular case." Well, can you or can't you?

Let's grant to begin with that the tachyonite must find a reason for believing that tachyons cannot be used in this way, not so much because of any peculiar feature of human freedom but simply because the situation invites introduction of non-human, non-free, pre-programmed paradox machines.

But waiving that, we can ask whether "I can kill Rameses" here contradicts "I can't kill Rameses," and if so, whether any limitation on human freedom is involved. Some think that the contradiction obtains, and that it is false that you can kill Rameses. So tachyon messages informing us of what we will or won't do involve a limitation on our power to act, and perhaps on our freedom or moral responsibility as well. Others suggest that the two statements, correctly interpreted, are mutually compatible. For "I can't kill Rameses" is true only if the 'can't' is read as expressing epistemic impossibility. It is epistemically impossible that I be going to kill Rameses, since I know that he's alive two days from now. But "I can kill Rameses" is true in the sense that it is in my power to do so; whether or not I kill Rameses depends on me. Call this the "in my power" sense of 'can'. Now what must be shown is that this latter sense of 'can' does not entail epistemic possibility, or any other condition which is not satisfied in the present case. We have to show that it is in my power to kill Rameses even though it is epistemically impossible that I kill him.

Unfortunately, there is no generally accepted analysis of that sense of 'in my power' in which it is relevant to free will and moral responsibility. I'm going to assume that these are compatible with universal causal determinism, and *a fortiori*, with the bare truth of statements about what the prospective agent will and will not do. (This last claim is defended in Fitzgerald (1971.) In what follows I'll try to dispel the apparent paradox presented above, and then fill out the case to get 'Situation I', in which no restriction on the prospective agent's freedom or power is involved. Then two other variants of the case at hand, Situations II and III, will be developed, in which there is some limitation on the agent's freedom, but not of a paradoxical sort or one peculiar to retrocausal situations.

To resolve the paradox, let's see what's special about the case as described so far. It differs from ordinary cases in which we say "It's in my power to do *X* or not, as I choose". The difference is that ordinarily when we say this, we don't know what we're going to do. Or if we do know that we're going to do *X*, we know it only in virtue of the fact that we intend to do *X*. Call this the 'Ignorance Condition,' which is satisfied by a person *P* at time *t* if and only if *P* at *t* either does not know what he is going to do, or does know it, but only in virtue of knowing that his intention is to do *X* rather than *Y*. In the case in question the Ignorance Condition is not satisfied, for you know in virtue of the tachyon message that you're not going to kill Rameses. Does this mean that it's not in your power to kill Rameses?

I think not. For the Ignorance Condition should not be regarded as part of the analysis of such locutions as "It's in my power to do *X* or not, as I choose." It is just a condition which happens to be satisfied in cases familiar to us.

But there are two reasons why someone might think that the Ignorance Condition is essential to the 'in my power' sense of 'can'. First, unless the Ignorance Condition is satisfied, the prospective agent is not in the position of having to deliberate, weigh alternatives, and then decide what to do. But in analysing the notions of free will, free decision, voluntary action, and so forth, we often concentrate on those paradigmatic cases of decision-making in which the agent dispels his ignorance of what he'll do by deliberating and deciding. Here the Ignorance Condition is satisfied, so it seems plausible to include it in analyses of such notions as *free action* and *voluntary action.*

Second, where the Ignorance Condition is not satisfied, the agent can figure out, independently of knowing his intention, what he's going to do. He may express his reasoning this way: "Since Rameses will be alive two days hence, it can't be that I'm going to kill him today." This is naturally, though erroneously, assimilated to "I can't kill him today, even if I choose to," and "It's not in my power to kill him today."

But look more closely at the reasoning, and see how it might be continued to give us Situation I. "Since Rameses will be alive two days from now, it can't be that I'm going to kill him today. That is, it is epistemically impossible.[13] It does not follow that I can't kill him today even if I choose to. All that follows is that I *wont t* kill him, perhaps because I won't choose to, or won't try. In fact, my not killing Rameses is perfectly voluntary, for I don't want to kill him. It is in my power to kill Rameses if I choose. For my killing or not killing Rameses depends crucially on what I am inclined to do. Had I wanted to kill Rameses, then I would succeed in killing him. Of course, in that case I would not have this reliable message from the future telling me that he is alive and well. So, I know that I won't kill Rameses. And my knowledge of this does not depend essentially on the evidence provided by any knowledge I may have of my present intention in the matter. But nevertheless, my not killing him does depend on my not wanting to."

The paradox is resolved. Nothing is preventing you from killing Rameses; you have the ability, the opportunity, and the wherewithal. If you want to and try, you'll succeed. The fact that you know by tachyon message that you won't kill him does not by itself testify to any limitation of your freedom to kill him.

Of course, had it been the case (Situation II) that you would not succeed in killing Rameses even if you had tried, then there is some limitation on your freedom or power to kill Rameses. This would be like the case in which it's not within a man's power to leave a room if he tries, since the only door is strongly bolted on the outside. But that sort of limitation is not entailed by messages from the future telling you what you'll do.

It might be thought that a limitation on freedom peculiar to retrocausal messages arises in the following sort of case (Situation III). You want very much to kill Rameses. And if you were to try you would succeed. The only reason you don't try is that you believe the tachyon message from the future. But that message is true only because you don't kill Rameses, since you don't try. And you don't try because you know that the message is true.

I think that in this case the message does involve a limitation on your freedom. For it is the of your not doing something which you want to do, and would have succeeded in doing had it not been present. Anything which prevents a person from being, doing, or having what he wants to be, do or have, or what he ought to be, do, or have, or what it behooves him to be, do, or have, constitutes *some* limitation on his freedom. That is not to say that they prevent his acts from being free in the sense required for moral responsibility. There are some limitations on freedom which are compatible with moral responsibility, e.g., a bad character.

But cases analogous to the limitation on freedom just mentioned can arise without messages from the future. Suppose that on Friday I want to leave my island for the weekend to visit the mainland. I hear a local radio announcement saying "The only bridge joining us to the mainland has been washed away, so we're stranded here until Tuesday." The message is true and I believe it. For that reason alone I do not try to get to the mainland.

But had I tried, I would have succeeded. For unknown to me, a group of army engineers are on the island to build a new bridge. They don't plan to start until Monday. But their commander is an old college chum of mine, who would have them build the bridge immediately if he knew how badly I wanted to get to the mainland this weekend. So the only reason why I don't get to the mainland is that I believe a certain message to be true. And the message is entirely true only because I don't try to reach the mainland. This limitation on my freedom is analogous to the previously described one in involving retro tachyon messages from the future.

University of Pennsylvania

BIBLIOGRAPHY

Alväger, T. and Kreisler, M., *Physical Review* **171** (1968) 1357.

Arons, M. and Sudarshan, E. C. G., 'Lorentz Invariance, Local Field Theory, and Faster-than-Light Particles', *Physical Review* **173** (1968) 1622.

Baltay, C., Feinberg, G., Weh, N., and Linsker, R., US AEC Report NYO-1932 (2)-148 (1969).

Benford, G. A., Book, D. L., and Newcomb, W. A., 'The Tachyonic Antitelephone', *Physical Review* **D2** (1970) 263.

Bilaniuk, O. M. and Sudarshan, E. C. G., 'Particles Beyond the Light Barrier', *Physics Today*, May 1969.

Bilaniuk, O. M., Deshpande, V. K., and Sudarshan, E. C. G., *American Journal of Physics* **30** (1962) 718.

Bilaniuk, O. M., Brown, S., DeWitt, B., Newcomb, W., Sachs, S., Sudarshan, E. C. G., and Yoshikawa, S., 'More about Tachyons', *Physics Today*, Dec. 1969.

Black, Max, 'Why Cannot an Effect Precede Its Cause?, *Analysis* **16**, 1956.

Csonka, P. L., 'Advanced Effects in Particle Physics', *Physical Review* **180** (1969) 1266.

Dhar, J. and Sudarshan, E. C. G., 'Quantum Field Theory of Interacting Tachyons', *Physical Review* **174** (1969) 5.

Dummett, M., 'Can an Effect Precede Its Cause?' *Aristotelian Society Supp.* **28**, 1954.

Dummett, M., 'Bringing about the Past', *Philosophical Review* **73**, 1964.

Feinberg, G., 'Possibility of Faster-Than-Light Particles', *Physical Review* **159** (1967) 1089.

Feinberg, G., 'Particles that Go Faster Than Light', *Scientific American*, Feb. 1970.

Fitzgerald, P., 'Review of *Fate, Logic and Time*, by Steven Cahn', *Philosophy of Science* (Mar. 1971), in print.

Fitzgerald, P., 'On Retrocausality', to be published.

Gale, R., 'Why a Cause Cannot Be Later than Its Effect', *Review of Metaphysics* **19**, 1965.

Gale, R., *The Language of Time*, Routledge & Kegan Paul, London, 1968.

Gorovitz, S., 'Leaving the Past Alone', *Philosophical Review* **73**, 1964.

Flew, A., 'Can an Effect Precede its Cause?' *Aristotelian Society Suppl.* **28**, 1954.

Flew, A., 'Effects Before Their Causes – Addenda and Corrigenda', *Analysis* **16**, 1956.

Hacking, Ian, 'Possibility', *Philosophical Review* **76**, April 1967.

Newton, R. G., 'Causality Effects of Particles that Travel Faster than Light', *Physical Review* **162** (1967) 1274.

Pirani, F. A. E., 'Noncausal Behavior of Classical Tachyons', *Physical Review* **D1** (1970) 3224.

Rolnick, W. B., 'Implications of Causality for Faster-Than-Light Matter', *Physical Review* **183** (1969) 1105.

Root, R. G. and Tregil, J. S., 'An Amusing Paradox Involving Tachyons', *Lettere Al Nuove Cimento* **3**, 13, 412.

Swinburne, R. G., 'Affecting the Past', *Philosophical Quarterly* **16**, 1966.

Tanaka, S., *Progress in Theoretical Physics (Japan)* **24** (1960) 177.

Terletskii, Y. P., *Paradoxes in the Theory of Relativity*, Plenum Press, New York, 1968.

Tolman, R. C., *The Theory of Relativity of Motion*, University of California Press, Berkeley, 1917.

NOTES

[1] Investigation of tachyons was pioneered by Tanaka (1960); Terletskii (1968); and Bilaniuk *et al.* (1962).

[2] For the experimental work see Alväger and Kreisler (1968), Baltay *et al.* (1969), and Feinberg (1970) for a popular account.

[3] Cf. Feinberg (1967); Arons and Sudarshan (1968); Dhar and Sudarshan (1968).

[4] Bilaniuk *et al.* (1962); Bilaniuk and Sudarshan (1969); Feinberg (1967, 1970).

[5] Cf. Tolman (1917), pp. 54–5.

[6] Bilaniuk and Sudarshan, May 1969.

[7] This point is also made by Newton (1967) and Benford *et al.* (1970).

[8] See Yoshikawa and DeWitt in Bilaniuk *et al.* (Dec. 1969).

[9] See Pirani (1970).

[10] Several attempts have been made to show that it is *a priori* impossible for a cause to be later than its effect, among them Flew (1954, 1956); Black (1956); Gorovitz (1964); Swinburne (1966) and Gale (1965, 1968). I believe that all of these attempts are unsuccessful and have argued against all except Swinburne's in 'On Retrocausality', to be published. Also treated there is Dummett's claim (1954, 1964), that backwards causation is *a priori* possible but that to believe in it we would have to surrender some strongly held epistemological principle about the knowability of the past. Here I proceed on the assumption that backwards causation is at least *a priori* possible.

[11] See for example Rolnick (1969), Benford *et al.* (1970).

[12] My thanks to John Stachel for illuminating comment on this point.

[13] Roughly, a putative state of affairs is epistemically possible relative to a person *P* at time *t* if and only if *P* at *t* does not know that the state of affairs does not obtain, and the information and techniques available to him then do not enable him to discover very readily that the state of affairs does not obtain. In such a case *P* can say truly "It is possible that *S* (the putative state of affairs)," where epistemic possibility is meant. So epistemic possibility is relative to a person and his situation at a time, and perhaps even to the audience he is addressing. For if *P* asserts the possibility of *S* to someone who knows that *S* does not obtain, that person can contradict *P* in saying truly, "No, it's not possible that *S*, for... (reasons)" There is a masterful discussion of epistemic possibility in Hacking (1967).

GEN-ICHIRO NAGASAKA

THE EINSTEIN-PODOLSKY-ROSEN PARADOX REEXAMINED

The Einstein-Podolsky-Rosen paradox is an old problem which started back in 1935 with an article in *Physical Review* titled 'Does Quantum Mechanics Describe Physical Reality Completely?' [1] These authors assert that quantum mechanics, being inconsistent, cannot describe physical reality completely. In the present paper the nature of the inconsistency will be analyzed and it will be shown that the E.P.R. paradox is essentially of formal nature and that the apparent formal inconsistency may be regarded as a consequence of unjustifiable presupposition.

According to quantum mechanical theory the states of a microscopic system may be represented by state vectors and in the absence of any disturbance from outside the state vectors will change continuously with time according to Schrodinger's equation. If a measurement is made on the system, however, the state vector will change discontinuously and a reduction of the wave packet will take place, and this projection postulate, as is often called, has been the subject of much discussion.

Suppose we have a combined system consisting of two sub-systems which, having interacted for a certain period of time, have separated. Now if we measure a property of one system, a reduction of the wave packet will occur in that system, but ordinarily the other system will be left in a pure state. But suppose that the state function can be expressed as a series something like

$$\text{(A)} \qquad \Psi(1,2) = \sum_{v} \psi_v(1)\, \varphi_v(2)$$

where ψ_v, φ_v are eigenfunctions respectively belonging to the first and the second systems corresponding to the eigenvalues of certain observables. Then if the measurement on the system may be regarded as a measurement on the combined system, a reduction of the wave packet will reduce the state function of the entire system to just a single term in the Expansion (A), say $\psi_R(1)\, \varphi_R(2)$, thus leaving the second system in a definite state the measurement of which would yield a definite value. Since it is assumed that no interaction exists between the two systems, the measurement on the first system should affect in no way the state of the second system, and therefore the eigenvalue λ_R which belongs to the eigenstate expressed by φ_R should *really* belong to the system. Now in the paradox of Einstein-Podolsky-Rosen it is presumed that in the absence of internal interaction such an expansion is not only possible but the same state function may be expanded in terms of different sets of eigenfunctions corresponding to non-commuting observables. Thus they claim "it is possible to assign two different wave functions to the same reality" and "Starting then with the assumption that the wave function does give a complete description of the physical reality, we arrived at the conclusion that two physical quantities, with non-commuting operators, can have simultaneous reality," which is apparently in contradiction with ordinary assertions of quantum mechanics. This is the so-called Einstein-Podolsky-Rosen paradox. Though a very problematic notion of reality is involved in the argument above, at least it was a part of the conclusions by these authors that the second system should possess independently real properties, and it is easy to see that it reveals that the argument involves a contradiction. And it should be noticed that the paradox is based on that conclusion. If it were true, by parity of reasoning, the first system too must possess independently real properties and therefore the measurement performed on the first system would give rise to no reduction of the wave packet, for it must be in a definite state immediately after the interaction. In order to reach the conclusion, however, the reduction of the wave packet has had to be assumed at the outset. Thus the whole argument is a case of *reductio ad absurdum* and it shows simply that the argument based on the reduction of wave packet should lead to a contradictory conclusion. The question is, therefore, whether or not the argument really refutes the theory that a quantum mechanical measurement should cause a reduction of the wave packet, so the paradox may be considered as an issue in the theory of measurement. It must be remembered that quantum mechanics with statistical interpretation does not prohibit to assign simultaneously definite values of non-commuting observables to a microscopic system. The commutation relation asserts only that the product of dispersions of these values obtained cannot be made infinitely small. The notion of uncontrollable disturbance due to measurements is not inherent in the theoretical system of quantum mechanics. It should be considered rather as an epistemological interpretation which is not exactly a part of rules of correspondence. Therefore the well-known criticism by Bohr [2] which is more of epistemological nature does not resolve the paradox.

Curiously enough, Einstein, Podolsky, and Rosen seem to have believed in the validity of the projection postulates, and for that reason the conclusion was believed to show that quantum mechanics cannot describe reality. As the argument above shows, the problem is essentially of formal nature. It must be remembered that the assertion that the system should possess independently real properties is based entirely on the validity of quantum mechanics, especially of the projection postulate. Furthermore, the states of the two systems before the interaction are assumed to be known, but according to the ordinary interpretation of quantum mechanics this knowledge is to be obtained only through measurements on the systems. Therefore, even if their argument should be accepted, in order to predict actually the values of non-commuting observables with respect to the second system, precise measurements of those variables would have to be carried out on the first system simultaneously and, of course, this is incompatible with quantum mechanics. In short the argument forwarded by Einstein, Podolsky, and Rosen is devoid of the considerations of the relevant semantic rules. So we may be permitted to dismiss the epistemological issues alleged in the argument. By the semantic rules here I mean the coordinations existing between theoretical statements and experimental conditions including preparation of states.

Now if the postulate of reduction of the wave packet leads to a contradiction, apparently the contradiction may be avoided by abandoning the postulate. In fact this is eventually the course taken by Margenau in his theory of measurement [3]. But it seems that his denunciation of the projection postulate does not necessarily mean the rejection of the reduction of wave packet itself. If to carry out a measurement it is necessary to prepare the state, a reduction of the wave packet does occur in the process of the state preparation which involves an interaction between the system and the apparatus for the state preparation.

An obvious alternative is to suppose that when the systems have emerged from the interaction it is already in the state of mixture. But any microscopic system to be measured has always been prepared in some way which should involve some kind of interaction. And since the superposition principle is one of the basic postulates of quantum mechanics the validity of which has been confirmed by the experiments with respect to the systems thus prepared, it is too convenient to suppose that an interaction between the systems always prepares the states in a definite way, which was pointed out by Furry [4]. Furthermore, it has been shown by Wigner in his paper on measurement in 1963 that "it is not compatible with the equations of motion of quantum mechanics to assume that the state of object-plus-apparatus is, after a measurement, a mixture of states each with one definite position of the pointer [5]." Since a measurement is regarded here as an interaction between the two quantum-mechanical systems, namely the object and the measuring apparatus, this statement is evidently extended to the case of any interaction. So it is feasible to assume that the two systems can never be in sharp states after an interaction. Thus we may say that an interaction always gives rise to uncertainty with respect to the states of the systems, and since any measurement involves interactions the same may be said of every interaction process. So the assumption of mixture would not resolve the contradiction and the paradox stays.

Let us then assume that the reduction of the wave packet in the first system is actually accompanied by a corresponding reduction in the second system. Now that the interaction has ceased and the two systems have separated sufficiently apart, the two systems are supposed not to overlap. Unless we are willing to admit the existence of mysterious action at distance, we would be forced to assume that a causal connection between the systems exists ever after the interaction. Thus Bohm and Aharonov state, "the quantum theory of the many-body problem implies the possibility of a rather strange kind of correlation in the properties of distant things," which is a realization of the potentialities of the second system [6]. This statement implies an assertion that the problem should be more than a mere matter of formal consistency. Thus these authors claim that the experiment performed by Wu and Shaknov [7] on the correlation of the polarizations of positronium annihilation photons may be regarded as empirical evidence for the assumption of distant correlation and against the hypothesis of mixture states after the interaction. If the measurement on the first system should be asserted to provide a prediction with respect to the second system, a subsequent measurement on the second system will have to be carried out to confirm the prediction. Now if the prediction should turn out to be true, then it would be a job of the physicist to devise a suitable model which would provide an adequate explanation for the process.

It is clear that the Expansion (A) and the projection postulate furnish the basis of such a model for the process proposed by Bohm and Aharonov. If the result of experiment should confirm the theoretical prediction, then the model may be thought adequate and as long as we are satisfied with instrumentalistic function of quantum mechanics no further explanation for the process would be required, thus no need for such a notion as distant correlation. Therefore the argument by Bohm and Ahronov is irrelevant to the problem of formal consistency of quantum mechanics, but rather it concerns with epistemological interpretation of quantum mechanics. In the original formulation of the E.P.R paradox it must be noticed that no direct measurement should even be performed on the second system so that the data obtained with respect to the second system may be viewed to reveal an element of physical reality. And if any measurement should be made on the second system, it would bring in an uncontrollable disturbance and we must be satisfied with statistical confirmation of the prediction.

The feasibility of the Bohm-Ahronov interpretation of the experiment has been questioned and some suggestions have been made about the possible experimental tests of the assumption [8]. However, even if those experiments should yield results which are in favor of the assumption, I do not think the results would possibly furnish a convincing evidence for its plausibility. It is a basic semantic rule of quantum mechanics that if two systems do not overlap, they may be regarded as completely independent systems. The assumption of distant correlation is apparently incompatible with the basic rule of interpreting experimental findings. In a scattering experiment, for instance, each particle which constitutes the incoming beam should be regarded completely independent of other particles, at least in approximation, and in quantum mechanics the beam can be treated as a single body represented by place waves and never as a many-body problem. In reality the systems which have separated can never be isolated. There would be numerous particles and quanta between the systems and it is absurd to assert that a hidden correlation should exist only between the systems which have interacted previously.

As for the examples of experimental test for the assumption which were cited by those authors, I think they simply overlooked the fact that "a measurement creates a state" as was expressed by Bohr. It must be noticed that such experiments always involves measurements on both systems which are presumed to be independent. Therefore the results of the experiment will be conditioned by the apparatus employed and thus the apparatus itself would help establish a certain connection between the measurements on the two systems, and, if the projection postulate should be correct, reduction of the wave packets would take place in both systems as the consequence of the measurements. Take, for example, the Stern-Gerlach experiment carried out on such systems. The magnetic field employed will apparently condition the results of experiment and establish a certain connection between the results of measurements on the two systems. If a measurement should be performed on the second system after one on the first system, then there should be no way of knowing whether the second system has already been in a definite state due to distant correlation between the systems or a reduction of the wave packet has occurred as the consequence of the measurement on the second system. If the former should be the case, the very existence of the measurement on the first system should, in general, give results of the measurement on the second system, and this would happen only when a measurement on the first system has been made previous to the measurement on the second system. Therefore, if the result of Wu-Shaknov experiment, which is a statistical consequence of approximately simultaneous measurements on both systems, is found to agree with the result of theoretical calculation where the wave function of the combined system is so constructed as invariant under spatial rotation, it shows at most that the assumption may be statistically tenable that the pair of photons emitted in the positronium annihilation are polarized at right angles to each other.

Even if coincidence measurements on each pair of emitted photons are carried out, and it is shown that the temporal order of the first measurement with respect to the second makes a difference in the result of the measurement on the second system, this will not give any support to the plausibility of the distant correlation hypothesis. It must be recognized that the quantum mechanical laws only furnish probability connections between results of measurements and as long as the experiment described in the above involves measurements on quantum-mechanical systems, and each measurement should not be free from the uncertainty due to interactions, the result of a single measurement will not have a significant influence with respect to the question whether quantum mechanics should be furnished with further theoretical hypothesis which is not compatible with the semantic rules of quantum mechanics now accepted.

Suppose that the two systems become completely independent after they have separated, and are in pure states. Then each system is expressed by a superposition of eigenvectors with certain coefficients, and if measurement should be performed on the system, theoretical calculations will give probability distribution depending on the coefficients. Informations about the magnitudes of these coefficients will be obtained by comparing the result of theoretical calculation with that of experiment. Now if this approach is correct, then the two independent systems should not be regarded as forming a single combined system, and as a consequence we cannot expect that certain characteristics like invariance under spatial rotation should be conserved. It is a postulate of quantum mechanics that if an observable commutes with the Hamiltonian of the system, at all times it is a constant of motion, but this is usually possible only if the Hamiltonian itself is constant. Apparently this is not the case with the systems which have separated after interaction. In the theoretical treatment of the process different Hamiltonians must be employed to represent the process of interaction and that after the systems have separated.

It is always possible, however, to use the same Hamiltonian for the entire process and in that case the interaction between the systems will never cease to exist, and this is the argument defended by Sharp [9]. If this is correct, then the state function of the combined system cannot be expanded in the form (A), and consequently the argumentation by Einstein, Podolsky and Rosen will lose its ground. But we cannot assert that this is the exact treatment of the problem, and that the ordinary formalism of isolated systems is an approximation. I think it is true that the notion of completely isolated systems could be regarded as an approximation. The claim that the interaction between the two systems could be made to vanish only approximately involves a tacit assumption that the combined system could be regarded as an isolated system. As far as this argumentation is concerned it is apparent that there is no exact treatment of the problem. The assumption of isolated systems constitutes a model and the notion of non-vanishing interaction another model. So it is simply the question of the feasibility of models, and if the problem is viewed in this way, the model of Sharp is not a plausible one, for if two systems are quite apart the interaction between the two would reasonably be regarded as negligible in comparison with the probable influences from numerous systems which might exist surrounding the systems in question. Furthermore, we must remember quantum mechanics has been constructed on the model of isolated systems, and therefore we may reasonably say that this forms a part of the basic semantic rules of quantum mechanics. In other words the interpretation of quantum mechanics is conditioned by the semantic rule which would determine the range where quantum mechanics could be applied validly.

If the assumption of isolated systems is accepted, then the expansion in the form of (A) will not be possible. If the combined system has been transformed into two isolated systems after interaction, it was proved by Araki and Yanase that the Expansion (A) is only approximately possible if the dimension of one system is sufficiently large [10]. Therefore it is clear that the paradox of Einstein-Podolsky-Rosen has its root not in the postulate of reduction of wave packet but in the unjustifiable assumption of the Expansion (A). These authors overlooked the fact that a state function could receive realistic interpretations only if the state function is furnished with appropriate boundary conditions and that the state function of the isolated systems should be subject to the boundary conditions which express the effect of the interaction.

That the assumption of isolated systems is fundamental in the interpretation of quantum mechanics may be seen from the fact that a large area of researches in modern physics is dominated by a single versatile tool, that of collision experiment. Almost all our knowledge of atomic nucleus and elementary particles has been obtained through the observations of the type. In the theoretical treatment of collision experiment two kinds of interactions are always involved, namely, the one at the place where a scattering takes place and the one which occurs when the outgoing particles are recorded on the measuring apparatus. Unless the outgoing particles are regarded as isolated systems between the two interactions and accordingly the two interactions are assumed to be completely incoherent, the theoretical treatment of collision experiments will not be possible. Quantum mechanics, like any other scientific theories, is a theory constructed out of data thus obtained with the aid of appropriate semantic rules. Since the assumptions of isolated systems and incoherent interactions furnish the basic semantic rules of quantum mechanics, we may say that quantum mechanics is primarily concerned with interactions between microscopic systems and that it may be validly applied only to, say, microscopic regions of space-time. Therefore the behavior of the system between the interactions must be such as may be described by classical physics, and in order to obtain useful informations concerning the nature of the interaction, the experimental arrangement must be so designed that the state of the system may be approximated to a mixture of states, and that the approximation and uncertainty involved in the second interaction may not affect the consequences of inference drawn from the results of measurement significantly. This is the reason why collision experiments should play so predominant roles in the investigation of microscopic systems. It may indicate also the validity of the Copenhagen Interpretation which includes an assertion that any experiment is to be described by the language of classical physics and yet the application of the classical concepts is limited by the uncertainty relations. And if the outgoing particle should be regarded as a part of measuring apparatus, we may understand the implications of the conclusions obtained by Wigner, Araki and Yanase with respect to the quantum-mechanical theory of measurement. Uncertainty in momentum and energy will arise from the limited space-time range of the interaction, and since this uncertainty should be carried by the outgoing systems, the conservation of energy and momentum will hold only approximately. Thus simple considerations from uncertainty principle also conform with the foregoing arguments and this indicates the consistency of the argumentation.

Nanzan University'
Nagoya

BIBLIOGRAPHY

[1] Einstein, A., Podolsky, B., and Rosen, N., *Phys. Rev.* **47** (1935) 777.
[2] Bohr, N., *Phys. Rev.* **48** (1935) 696.
[3] Margenau, H., *Philosophy of Science* **25** (1958) 23.
[4] Furry, W. H., *Phys. Rev.* **49** (1936) 393.
[5] Wigner, E. P., *Am. J. Phys.* **31** (1936) 6.
[6] Bohm, D. and Aharonov, Y., *Phys. Rev.* **108** (1957) 1070.
[7] Wu, C. S. and Shaknov, I., *Phys. Rev.* **77** (1950) 136.
[8] Peres, A. and Singer, P., *Nuovo Cimento* **15** (1960) 907; Inglis, D. R., *Rev. Mod. Phys.* **33** (1961) 1.
[9] Sharp, D., *Philosophy of Science* **28** (1961) 225.
[10] Araki, H. and Yanase, M., *Phys. Rev.* **120** (1961) 622.

MILIČ ČAPEK

THE SIGNIFICANCE OF PIAGET'S RESEARCHES ON THE PSYCHOGENESIS OF ATOMISM

What is remarkable about the history of atomism is the fact that it can be traced back to the very beginning of human reflection on nature. It is needless to recall the Pythagorean monadism and the names of Leucippus, Democritus and Lucretius: atomism is clearly as old as the first scientific and even proto-scientific explanations of nature. What is even more remarkable is that human thought was able so early and without the aid of microscope and measuring devices to anticipate one of the main findings of modern science. The significance of this anticipation is not diminished by the fact that only the most general features of modern atomism were present in the thought of its ancient ancestors: it was still quite an achievement in the fifth century B.C. to hold the view that space is infinite, that matter is homogeneous and discontinuous in its structure and that all diversity and all changes in nature are reducible to the configurations and displacements of homogeneous, permanent units. The question why and how such a successful anticipation of modern physical science occurred, has been and still is largely ignored; and when it is not ignored, the explanations suggested are hardly satisfactory. These explanations fall roughly into two distinct groups.

The first kind of explanation, apparently still widely accepted, is that the so-called anticipation of modern atomism was a sheer coincidence: Leucippus was just lucky in his guess. This view is held by the majority of professional physicists and by not a small number of philosophers and philosophers of science. It is hardly any answer to the question raised above; furthermore, it is based on the false assumption that there is a radical difference between ancient speculative atomism and the modern empirically based atomistic theories. Such view is historically incorrect; for, as Kurd Lasswitz showed in his classical *Geschichte der Atomistik*, the tradition of ancient atomism survived through the Middle Ages, despite the official rule of the Aristotelian philosophy of nature, to be revived by Gassendi and others in the seventeenth century. But precisely Gassendi's atomism represented an intermediate stage between the Greek speculative

Boston Studies in the Philosophy of Science, VIII.

form and its modern empirical form which began with John Dalton. In truth, even ancient atomism cannot be called 'purely speculative'; any attentive reader of Lucretius' *De rerum natura* knows about the wealth of empirical observations, no matter how elementary, to which the author refers. On the other hand, philosophical themes were present not only in the atomism of Newton (cf. the concluding pages of his *Optics*), but in that of Dalton as well.

The second answer to the same question was along the Kantian or neo-Kantian lines: the preference for atomistic explanations is an inherent feature of human reason. It is true that Kant in his Transcendental Dialectics regarded atomism as one thesis of the second antinomy which can be proved with equal cogency as its antithesis, i.e. the continuity of matter. But if we look more attentively at Kant's antithesis, we see that it is not as antithetical to the thesis as Kant presents it.[1] The so called 'continuity' of matter is nothing but its infinite divisibility which is entirely compatible with the concept of *material point* which, unlike the material atom, is strictly dimensionless. This was the view which Kant held in the earlier stage of his thought and which is implicitly present also in his later work *The Metaphysical Foundations of Natural Science*; this was also the view of Boscovich before Kant and, after Kant, that of Ampère, Cauchy, Fechner, George Cantor and Russell, at least in his earlier stage. But the material points are as external to each other as the corpuscles of a finite radius; thus the so called 'continuity' of matter is nothing but 'discontinuity infinitely repeated.' In this light the second antinomy is hardly an antinomy at all since both the thesis and antithesis affirm the discontinuity and indivisibility of the first elements of matter; they disagree only about their *number* and their *size*. The thesis postulates the finite number of the atoms of a finite size; the antithesis an infinite number of the dimensionless material points. It was thus logical when the above mentioned Kurd Lasswitz, who besides being a historian of atomism was also a disciple of Kant, made explicit what was merely implicit in Kant: our preference for atomistic explanations is a part of the *a priori* structure of human mind.[2] The views of Emile Meyerson are not basically different, although he, unlike Lasswitz, avoided the Kantian terminology: according to him, the atomistic explanations of change and diversity in the terms of the permanent self-identical units is one of those identifying patterns which, in his view, underlie all rational explanations.

But how does the assumption of the *a priori* character of the atomistic schemes explain their successful application to experience? Orthodox Kantianism claimed that all *a priori* categories are simply *imposed* on the amorphous sensory experience and it is known how obsolete and unsatisfactory this view proved to be. Meyerson's view was more critical; he was well aware that experience is not amorphous and that it resists the identifying patterns which our reason is imposing on it. This surd element in experience was called by him 'l'irrationnel'; this accounts for the fact that the identifying patterns do not apply strictly to experience since neither change nor diversity can be entirely eliminated. But Meyerson still failed to explain an enormously successful application of these identifying schemes, in particular of the conservation laws and of atomism: why sensory experience yields to such extent to these preconceived patterns? The only explanation – and hardly a plausible one – would be to assume that there is a kind of 'pre-established harmony' between the structure of our reason and the objective reality. But it is precisely this 'harmony', this agreement between what Kant called 'form' and 'material' of experience, which requires an explanation and not a mere statement.

There is thus a need for another, more satisfactory theory which, in my view, is provided by that which, for brevity sake, can be called a 'biological theory of knowledge.' The central thesis of this theory (of which there are some varieties) is the claim that the structure of our reason, instead of being ready-made and immutable, is the result of the long process by which our mind gradually adjusted itself to that sector of reality which is biologically important for us. This was nothing but an application of the general theory of evolution to the cognitive functions of man; for these functions, as parts of the psychophysical organism, cannot be exempt from the evolutionary process of which the organism itself is a result. This accounts for the fact that this theory began to appear approximately at the time when Darwin and Wallace proposed their theory of biological evolution. In truth, the first edition of Herbert Spencer's *The Principles of Psychology*, in which the evolutionary and genetic approach to epistemological problems was for the first time consistently applied, appeared in 1855, – four years prior to *The Origin of Species*. Spencer was followed by a number of outstanding and influential psychologists who shared with him his philosophical interest as well as his positivistic orientation. Helmholtz, Mach and Avenarius in Germany, Ribot, Fouillé and, in some aspects of his thought at least, Henri Poincaré in France, – to name at least those who were the most influential.

The biological theory of knowledge corrected Kant's apriorism in the following way: "What is *a priori* for the individual is *a posteriori* for the species" (Spencer) or, in Helmholtz' words, what Kant regarded as *a priori* forms of knowledge corresponds to the inborn psycho-physiological patterns which themselves are – like any other organic features – the results of a long phylogenetic development. There were two main reasons why the interest in this theory began to decline in this century. First, there was a powerful reaction after 1900 against classical positivism and against psychologism in general. Not only the opponents of positivism, like Husserl, but also the neopositivists of the Vienna School, Bertrand Russell and others, especially those interested in the problems of logic and language, dismissed every genetic and evolutionary approach to the problems of epistemology as irrelevant, as being guilty of the 'genetic fallacy'; according to them, the question of the truth of certain beliefs is independent of the question how these beliefs were formed, which biological and psychological processes were involved in their formation. This attitude was forcibly expressed by Wittgenstein in his *Tractatus* (4.1122): "Darwin's theory has no more to do with philosophy than any other hypothesis in natural science."

This reaction obviously went too far; it is difficult to see how epistemology can ignore *any* area of experience, in particular how it can ignore such crucially important facts as those of genetic psychology and evolutionary biology. Fortunately, the pendulum now starts moving in the opposite direction. It was Professor Quine, one of the foremost American logicians, who concedes that the applicability of our logic to reality can be explained only in the evolutionary way. Professor Donald T. Campbell is now detecting the elements of the biological theory of knowledge in the thought of Karl Popper. I, myself, tried to point out the same elements in the thought of two so widely different thinkers – Henri Bergson and Hans Reichenbach.[3] But the most significant contributions to genetic epistemology in the last three decades have been made by Professor Jean Piaget. His researches on the genesis of the concepts of object, of space, of number, and, in particular, his investigations on the psychogenesis of atomism and of the conservation laws shed, in my view, a new light on the question why certain seemingly *a priori* categories of thought apply to the macroscopic level of experience and also why they fail beyond its limits.

Piaget starts from the sound principle that "every response, whether it be an act directed to the outside world or an act internalized as thought, takes the form of an adaptation or, better, of a readaptation." He thus consistently regards intelligence as "a generic form of organization or equilibrium of cognitive structures." In other words:

It (i.e. intelligence) is the most highly developed form of mental adaptation, that is to say, the indispensable instrument for interaction between the subject and the universe when the scope of this interaction goes beyond immediate and momentary contacts to achieve far reaching and stable relations. But, on the other hand, this use of the term precludes our determining where intelligence starts; it is an ultimate goal, and its origins are indistinguishable from those of sensori-motor adaptation in general or even from those of biological adaptation itself.[4]

For our purpose the most important of Piaget's investigations are those dealing with the formation of the notion of permanent object and those dealing with the psychogenesis of atomism and the laws of conservation. It is clear that the latter presuppose the former; for the idea of atom as an invisible permanent object can be grasped only after the notion of permanent object in general is formed. In his book *La construction du réel chez l'enfant*, Piaget[5] showed how the belief in permanent objects gradually develops in a child's mind under the pressure of experience and how the elaboration of the belief in the reality of space takes place concomitantly. There is no place here to go into all details of Piaget's investigations; let us recall that in the first phase the child's universe consists of images which are capable of recognition, but still lack substantial permanence as well as spatial organization; that only in the fourth phase, the child begins to search for the objects that disappeared from the sensory field without taking into account their displacements; that in the fifth phase, the displacements are taken into account, but only as long as they are perceived; and that only in the sixth phase (16–18 mo.) is the belief in the permanence of the objects existing outside the sensory field established, even when their displacements are not perceived. But, as Piaget stressed again in his Columbia lecture *Genetic Epistemology* (Columbia University Press, 1970) the notion of identity is not equivalent to that of quantitative conservation. The identity which the child's mind grasps is of a qualitative kind; it is the ability to recognize the same things in its different positions without realizing that its substance, weight and volume remained the same. This is only natural since certain phenomena seemingly suggest the absence of quantitative conservation. Piaget's monograph *Le développement des quantités physiques chez l'enfant: Conservation et atomisme*[6] is extremely instructive in this respect. He observed the intellectual reactions of the children of 6–12 yr old observing the dissolution of sugar in a glass of water. This looks as a sheer annihilation of a piece of sugar which seemingly literally disappears in water – and it is so interpreted by children at first. The next stage, when the attention was drawn to the fact that water remains sweetened, the process of dissolution is interpreted as a sort of qualitative change: water + sugar = sweet water. The child seems to be in the Aristotelian stage when the components in the mixtures and solutions are regarded as *ceasing* to exist actually; the resulting solution is a *new* stuff, homogeneous in all its parts. Only when attention is drawn to the fact that the weight of the glass of water remains *increased* by adding a piece of sugar to it and that the total volume also remains slightly larger, does the child begin to imagine that a piece of sugar was divided into tiny indestructible particles scattered through the water; only then the conservation of the mass and finally that of the volume is firmly grasped. (This takes place in the age 10–12 yr.) There is no place here to go into all the details and all the stages of these experimentation; what was given above is a very simplified sketch. It is exceedingly interesting to see how the insight into the conservation of matter is correlated with the notion of corpuscular structure both in the individual human development and in the collective intellectual development of mankind and how early the permanence of the object, apparently contradicted by sensory appearances, is attributed to tiny material particles.

In the same book Piaget showed how the logico-arithmetical operations develop concomitantly with the operations performed with physical solids; in other words, as Piaget says in his more recent work, the operations executed originally materially, with the solid bodies, become 'internalized', that is, performed in thought.[7] This was already clear in his book *La genèse du nombre chez l'enfant* when he showed how the belief in the elementary commutation law $a+b=b+a$ is an internalization of the operation originally performed with material objects; this belief was impossible as long as the permanence of the objects in the aggregate symbolizing their sum was not explicitly recognized. The reversibility of arithmetical operations is thus based on the reversibility of physical operations. Helmholtz explicitly anticipated Piaget's conclusion when he wrote in *Zahlen und Messen* that the countability of objects presupposed their immutability; during the process of counting "they must neither disappear nor merge nor to be split nor any new one be created."[8] Thus, there seems to be a close correlation between atomism and arithmetization of reality.

Yet, there is one essential difference between Piaget's genetic theory of knowledge and the classical biological epistemology of which Helmholtz was one representative. They both agree that the structure of our reason is a result of the gradual adjustment of our psycho-physical organism to the structure of reality. But Helmholtz believed – like Spencer, like Mach, like even Poincaré – that this process is more or less complete; this confident belief was characteristic of the culminating period of classical physics. If Piaget knows bettei, it is because the twentieth century physics taught us to be more cautious and more modest. Needless to repeat what has been said so many times before about the failure of intuitive corpuscular models on the microscopic scale. Physical atomism has been transformed beyond recognition by its very triumph; for if one thing is certain, it is the fact that not a single feature of the classical concept of particle survived the present revolution in physics. The 'particles' of modern physics are neither immutable, nor permanent, that is, neither indestructible nor uncreatable; their 'motions' cannot be traced along continous trajectories nor can be even localized precisely. In truth, the very usage of the term 'particle' or 'corpuscle' is nothing but a mere inertia of the traditional language. In his article 'The Child and Modern Physics' (*Scientific American*, March, 1957) Piaget made a following pertinent remark:

Contemporary physicists abandoned some old intuitions about the nature of the physical world. They have, for instance, renounced the concept of the permanence of objects in submicroscopic realm: a particle does not exist unless it can be localized; if it cannot be located at a particular position, it loses its title as an object and must be described in other terms. Now by an extremely curious coincidence it was found that a very young baby acts with regard to objects rather like a physicist.

This sounds paradoxical, but what appears strange becomes completely intelligible within the framework of the new genetic epistemology. There is no reason to be afraid that modern physics invites the physicist to re-descend to the mental, pre-linguistic and pre-operational level of the infant. The babies are hopelessly wrong about our macroscopic environment. But Piaget's investigation of their mentality showed that the structure of human mind is far more flexible than we assumed. Apparently, it does not possess any *a priori* structure; or, if it does, it is certainly not that which Kant had believed he found in analyzing the cognitive functions of man. For what Kant analyzed was not the *a priori* structure of mind but its particular modification which is imposed on it by our continuous intercourse with the solid bodies of our macroscopic experience which eventually was systematized in the Euclid-Newtonian conceptual framework. And it is this structure which proved to be inadequate beyond what Reichenbach called "the realm of the middle dimensions," located between the realm of quanta and the realm of the fleeting galaxies. It fails beyond these limits for the same reason why it triumphs within the realm of the middle dimensions; for it is itself by its own nature the result of the adaptation to this middle region. Our most important and most widely applied category – that of the permanent, identifiable object – is out of place both on the microscopic and megacosmic (extragalactic) scale. In the second volume of his *L'Introduction à l'épistémologie génétique*, Piaget pointed out that we still continue to speak of the universe as a whole, either finite or infinite, without realizing that we are illegitimately transferring our category of object which is a diaphanous replica of the macroscopic solid body beyond the realm of its applicability.[9] He could have added one powerful argument based on the relativity theory: because of the nonexistence of the objective cosmic 'Now', it is meaningless to treat the universe at large as an aggregate of simultaneously existing objects or particles. Our logic seems to be indeed, the logic of solid bodies and the very term 'object' seems to be nothing but a macroscopic prejudice.[10]

But the crisis of the concept of object or 'thing' may go even deeper and imply more than a departure from the Newton-Euclidian corpuscular model of nature. Its full significance is suggested by Reichenbach who after pointing out that "the experience offered by atomic phenomena makes it necessary to abandon the idea of corporeal substance," concludes:

> With the corporeal substance goes two-valued character of our language, and even the fundamentals of logic are shown to be the product of an adaptation to the simple environment into which human beings are born.[11]

Does it mean that not only Euclidian geometry and Newtonian mechanics, but also the two-valued logic is a result of environmental conditioning, more specifically, of the adaptation to the realm of middle dimensions? This view will be probably always resented by logicians, especially those who are oriented toward some kind of Platonism. But, as we have seen, even Professor Quine concedes that the applicability of our logic to experience is the result of natural selection. But he probably assumes with the classical biological theory of knowledge that this adjustment is complete, that is, that what he calls the "immemorial doctrine of ordinary enduring middlesized physical objects" does not require any basic revision and that our two-valued logic truly 'mirrors' the reality in its *whole* extent. But these views are hardly compatible not only with the new genetic epistemology which stresses the incomplete character of our cognitive adjustment, but with the facts of contemporary physics as well.

This is now being more and more recognized. Hilary Putnam raised the question whether logic is also not, after all, an empirical science. David Finkelstein in his recent paper 'Matter, Space and Logic' raised the question whether the present upheaval in physics will not require a similar far reaching *widening* or *generalization* of logic, as revolutionary as the generalization of classical mechanics and of Euclidian geometry was.[12] Within such generalized logic the traditional two-valued logic would still retain its legitimate place, since it would remain applicable to our macroscopic experience. Yet, the traditional logic would lose its privileged, absolutist status since, like the geometry of Euclidean and Newtonian determinism, it would be applicable only to a very large, yet still limited region of our experience.

Boston University

NOTES

[1] Cf. N. Kemp Smith *A Commentary to Kant's Critique of Pure Reason*, Humanities Press, New York, 1962, pp. 489–91. The correlation between extensionless points and mathematical continuity of space can be seen especially clearly in Bertrand Russell's *Principles of Mathematics*, The Norton Library, 1964.

[2] K. Lasswitz, *Atomistik und Kriticismus*, Braunschweig 1878, in particular Chapter VII, 'Das Apriori in der Physik'.

[3] Cf. W. Van Orman Quine, *Ontological Relativity and Other Essays*, Columbia University Press, New York, 1969, pp. 126–7; Donald T. Campbell, 'Evolutionary Epistemology', to appear in: *The Philosophy of Karl R. Popper*, The Library of Living Philosophers, The Open Court Publishing Company, La Salle, Illinois; M. Capek, 'The Development of Reichenbach's Epistemology', *The Review of Metaphysics* **11** (1957) 42–67; 'La théorie biologique de la connaissance chez Bergson et sa signification actuelle', *Revue de métaphysique et de morale* **44** (1959) 194–211.

[4] Jean Piaget, *Psychology of Intelligence*, Littlefield, Adams & Co., Patterson 1960, p. 7.

[5] *La construction du réel chez l'enfant*, Neuchatel 1937, pp. 11–96

[6] *Le développement des quantités physiques)chez l'enfant: conservation et atomisme*, 2nd ed., Neuchatel 1962, esp. pp. 81–140.
[7] *Ibid.*, p. 279.
[8] *La genèse du nombre chez l'enfant*, Neuchatel 1950; 'Zahlen und Messen', in *Hermann von Helmholtz, Schriften zur Erkenntnistheorie* (ed. and annotated by P. Hertz and M. Schlick), Berlin 1921, p. 82.
[9] *L'Introduction à la épistemologie génétique*, Paris 1951, II, pp. 212–13.
[10] The term used by Ferdinand Gonseth in *Les mathématiques et la réalité*, Paris, 1936, p. 158.
[11] *The Rise of Scientific Philosophy*, University of California Press, Berkeley, 1951, pp. 189–90.
[12] H. Putnam, 'Is Logic Empirical'? in *Boston Studies in the Philosophy of Science*, Vol. V, D. Reidel and Humanities Press, Dordrecht and New York, 1969, pp. 216–41; David Finkelstein, 'Matter, Space and Logic', the same volume, pp. 215. Cf. in the same volume my article 'Ernst Mach's Biological Theory of Knowledge', pp. 400–20. Putnam's radical empiricism in his essay is hardly compatible with his insistence on the timelessness ('tenselessness') of reality, including that of the future, in his article 'Time and Physical Geometry' *The Journal of Philosophy* **64** (1967) 240–7, where he upholds a position very close to the most extreme forms of traditional rationalism.

VI. Theoretical Pluralism; Understanding; Methodological Agreement

ANDREW MCLAUGHLIN

METHOD AND FACTUAL AGREEMENT IN SCIENCE

One important view of the nature of facts in science sees the validity of factual claims as resting upon intersubjective agreement through observation, yet the conditions underlying such intersubjectivity are rarely explored.[1] In this paper I should like to inquire into the role of shared method in arriving at intersubjective agreement concerning fact. One useful way of exploring this issue is to examine the methodological element in factual disputes in science, since such controversy may indicate important aspects of factual agreement.[2]

There are times in science when an appeal to factual evidence is successful in the choice between hypotheses. In fact it is clear that, at least on some occasions, there is agreement among scientists as to "what the facts are." Moreover, such factual agreement can be found between scientists who maintain quite different theories. This seems to imply that those who claim that facts in science are always 'dependent upon' theory in any important sense are mistaken – but the issue is more complicated than that. What factual agreement between proponents of differing theories *does* show is that there are things in scientific practice which I shall call *theory neutral facts.* This concept minimally describes what happens in science when theoretical disputes are resolved by appeals to evidence. If proponents of differing theories have resolved their dispute through appeals to facts, then they must have been able to point to certain evidence which each will *accept* as the relevant court of appeal. Such a procedure does involve intersubjective agreement as to the facts and is descriptive of one important facet of scientific practice. This evidence accepted by proponents of two (or more) differing theories is what I shall call 'theory neutral facts'.

But the fact that some theoretical disputes in science are resolved by appeals to evidence does not, by itself, imply that those evidential claims occupy some privileged position sufficient to justify an important epistemological distinction between fact and theory. Rather, there is still room for the claim that facts in science depend upon theory. All that the existence of theory neutral facts implies is that there are some 'facts' which are taken as neutral between *some* theories, not that they are neutral between *all* theories. One important question, then, in discussing the epistemological relationship between fact and theory is whether there are some facts which are neutral between *all* theories. Such facts, if there are any, might be called *theory free facts*. The question of the presence or absence of such theory free facts is one of considerable interest in current disputes about the relation between theory and fact. This question, however, is not a simple one, since some care must be taken with the meanings given to terms such as 'theory' and 'theory dependence'. As both Achinstein[3] and Spector[4] have shown, there is no single distinction to be drawn between theory and observation. Rather there are several distinctions which may be drawn and these various distinctions do not seem to be equivalent. Consequently, the claim that there are (or are not) theory free facts in science must be accompanied with a specification of the meaning of 'theory'. For some possible meanings of this term there surely are theory free facts. Thus, if we take 'theory' to mean 'a particular low level scientific theory' – such as Darwinian evolutionary theory – then there surely are theory free facts. That is, there are relevant facts which, in no important sense, 'depend upon' Darwinian theory. Since in a paper of this length it is quite impossible to investigate the whole range of meanings which might be attached to 'theory', I would like to distinguish one important sense of this term and show that when 'theory' is understood in this way there are no theory free facts in science.

It is possible to distinguish between substantive scientific theories about "the way the world is" and methodological theories which are about the proper methods to be used in gaining scientific knowledge about the world. I do not mean to imply that these two types of theories are, in the final analysis, wholly independent, for what we take to be adequate methods probably depends upon what we believe to be the nature of the world. But in the early stages of analysis these two types of theories can be distinguished and discussed independently. There are, in all scientific inquiries, implicit or explicit beliefs concerning the proper procedures for obtaining knowledge about the world. Every method for gaining knowledge is a rule governed activity (at least implicitly), and the rules specify what are reliable ways of ascertaining fact. Churchman has put the general point this way:

Why are certain beliefs and estimates much more reliable than others? The only conceivable answer is because experience has shown them to be. But then the answer, if sensible, implies that the reliability of a certain kind of information is based on a *theory*, namely that the information has had and will have a certain degree of reliability over time and a class of 'normal' observers.[5]

What I want to suggest is that agreement over theory, as that term applies to methods of inquiry, is a condition of agreement over facts. If that is true, then in this sense of 'theory', there are no theory free facts in science. There may, of course, be other senses of 'theory', some of which do allow theory free facts and some of which do not.

One way in which the dependence of factual agreement upon theoretical agreement (in the sense specified above) can be illuminated is to look at a factual dispute in science which involves disagreement over method. It is possible in science to deny any evidential claim the status of theory neutrality by criticizing the method by which it was obtained. This move is always possible because all scientific evidence is the product of an inquiry which uses one or another method. Moreover, any concrete application of a method involves utilizing specific procedures which may also be critized. We can talk very generally about the scientific method, less generally about the experimental method and even less generally about rather specific modes of inquiry, i.e. the non-directive interview. Controversy over the methods used in a particular inquiry can be simply about specific experimental designs or they can be about the appropriateness of the experimental method. For the purposes of the argument in this paper it is not necessary to carefully examine the various levels of generality which might be meant by the term 'method', although it is important to note the existence of different possible levels of methodological dispute and that the use of 'method' in this paper is rather wider than usual.

The claim that proposed evidence can be contested by criticizing the method which led to it is not entirely familiar, so it may be of value to illustrate the type of dispute to which I want to call attention. Since the experimental method is regarded by some as the paradigm of the scientific method, it is of value to look at an example which utilizes rigorous experimental procedures. Verbal learning theory in psychology uses such a method and an interesting controversy over the basic nature of learning has developed in this area. The dispute concerns whether the learning of particular items occurs gradually or all at once. The general belief among learning theorists has been that learning is incremental. This means that even if a subject does not respond correctly to an item in a particular reinforced trial, the subject has nevertheless learned an association to some extent. Thus, it is claimed, the probability of the subject correctly responding to the stimulus in the future has increased. According to the incrementalist position, the repetition of an item does increase learning, and sufficient repetition may lead to correct response. The other side of this dispute maintains that learning is an event that either happens – in which case the subject responds correctly – or else the association is not learned. Thus, if learning is not incremental, but rather is all or none, then the repetition of an item does *not* promote learning, except insofar as repetition provides more instances when the item might be learned. On the all or none view, either the association is correctly learned or it is not – there is no middle ground. Controversy over this issue can be traced back in one form or another at least to the 1930's, but for our purposes we may date the issue in its contemporary form from an important study by Irwin Rock in 1957.[6] Rock presented evidence that when the learning task is to link up two items, such as nonsense syllables or letters and numbers, it does not make any difference how often a pair of items are presented as to the probability that the subject will learn their association. Essentially, what Rock did was to replace items missed on each trial with new items not previously presented. He found that this dropping out of items and their replacement did not affect the rate the paired items were learned. There is not time, nor is it necessary for my point, to go into the details of Rock's work and subsequent evidence concerning one trial learning. Rather, what is important is to examine one way of attempting to deny the status of theory neutrality to Rock's evidence through methodological criticism.

Since Rock's claim, if granted, involves considerable change in the theoretical beliefs of most learning theorists, resistance was to be expected. In this light it is interesting to note one theorist's initial resistance and subsequent acceptance of all or none learning. Estes recognized that the traditional design of learning experiments would obscure all or none learning, if that is how learning occurs.[7] The traditional experimental design in learning theory averages the performances of individual subjects and constructs learning curves on the basis of aggregate performance. Such a practice is reasonable for the primary interest is in learning, not the idiosyncratic behavior of individual subjects. But such a procedure *would* mask all or none learning, *if* that is how learning occurs. This is so because it is the individual subject and not the aggregate of those subjects which learns either incrementally or all or none. Consequently, Estes developed some new experimental designs which allowed him to examine how individual subjects performed. Since Estes had set out to falsify Rock's position he was surprised to find that his results indicated that learning was all or none.[8] One of the conclusions which he draws from his experience is of particular interest: "The findings we have considered in this paper suggest that in point of fact some of the most firmly entrenched concepts and principles of learning theory may be in a sense artifacts of a conventionalized methodology."[9] Thus, Estes concludes that much of what has been taken as theory neutral fact is actually methodologically contaminated, so that it is not neutral evidence between incremental and all or none theories of learning. The evidence which has been appealed to in the past is an 'artifact' of conventional experimental designs. The method (meaning here the traditional design of learning experiments) has become the subject of controversy, removing the evidence obtained by that method from theory neutral status.

But just as the traditional design of the learning experiment can be questioned, so can the designs used by Rock and Estes.[10] Thus, Leo Postman claims that the evidence for all or none theories is contaminated by the methods used and that the incrementalist position can still be maintained.[11] According to Postman, Rock's evidence can be discounted because of two factors. In the first place, some of his results were due to 'artifacts inherent' in his procedure, for he used a 'slow' rate of presenting items to be learned.[12] This slow rate of presentation allowed subjects to rehearse what they were learning, and Postman cites evidence to show that faster rates of presenting items prevent such a rehearsal strategy.

Secondly, there was "a source of bias which is inherent in... [Rock's] experimental design," for the method of replacing missed items leads to the selection of more easily learned items.[13] Further, according to Postman, "important differences in materials and procedures" may explain the differences between Estes and Postman's findings, making it "clear that the empirical inadequacy of the incremental model is still far from being a generally established fact."[14] What is important to note in Postman's reaction to evidence taken by some as telling against the incremental position is that he wishes to deny the status of theory neutrality to that evidence by criticizing the procedures used in obtaining it.

Another position taken in this controversy is of interest here. Instead of taking Rock's evidence based upon 'slow' rates of presenting items as an artifact, one might take this evidence to be of crucial importance. This is Miller's reaction.[15] He notes that one theory of learning utilizing computer simulation techniques is based upon a conceptualization of the learning process closely related to the all or none position. If one adopts this approach, then the full importance of Rock's experiments may not have been fully appreciated. According to Miller:

> The Feigenbaum and Simon theory assumes that subjects will rehearse a few items and let the others go unnoticed on each trial. Postman's experimental demonstration that if you interfere with a rehearsal strategy, you destroy the Rock effect is strong evidence that many people do follow an EPAM I strategy of memorization. Postman refers to this rehearsal strategy as an artifact, but if Feigenbaum and Simon are on the right track it may well be the most important fact of all.[16]

A few points must be made about this too brief presentation of a complicated dispute. In the first place, I do not mean to suggest that the claims and counter-claims about results being artifacts or resulting from biases inherent in experimental designs is simply an arbitrary procedure. Several sorts of reasons are offered by each writer to substantiate his particular claims about what is fact and what is artifact. Yet it should be apparent that the claim that certain evidence can be discounted because of the method by which it was obtained is *not* a simple factual claim. Rather, it involves invoking a methodological theory about the proper methods for gaining knowledge. If calling the considerations invoked in the controversy discussed above 'theoretical' seems inflated, it is well to remember that the dispute is relatively 'low level,' concerning particular experimental designs and carried out within the context of general acceptance of the experimental method. More general controversies, such as the question of the general appropriateness of the experimental method for the study of man, do have a distinct 'theoretical' aroma to them.

It is, moreover, worth noting that what is claimed to be a methodological artifact is typically related to the substantive theoretical beliefs of the inquirer. The relationship between what is claimed to be artifact and substantive theories of the learning process is clearest in the attempt to assess the importance of 'slow' and 'fast' rates of item presentation and rehearsal strategies of the subjects. Such a relationship between claims of evidence being artifactual and substantive theories does not appear to be necessary, however, as Estes' experience led him to significantly revise his theoretical position. Finally, it should be noted that both the all or none and the incrementalist theories became increasingly complicated in the course of this controversy.[17]

By far the most important result, for the purposes of this paper, is to note that the claim that certain results are artifacts of the method by which they were obtained, and therefore not suitable candidates for theory neutral facts, is one which is *always* possible. This is so because all scientific evidence is obtained by some method and any method can become the subject of controversy. Such controversies involve theories about the proper methods of gaining knowledge about 'the way the world is.' Facts in science depend both upon the nature of the world *and* the methods used to investigate the world. If this is true, then methodological agreement is a condition of factual agreement and there are no theory free facts in science. Although such methodological agreement is *one* of the bases of intersubjectivity in science, there are probably quite a few other conditions of such intersubjectivity.

It could be asked at this point whether agreement concerning methods is either a necessary or a sufficient condition for intersubjectivity about facts. It is fairly clear that such agreement is not a sufficient condition for factual agreement, since agreement over method does not require agreement about fact. But it is also not at all clear that such agreement is a necessary condition, since disagreement about method does not require disagreement about fact. Two inquirers could agree about some particular factual claim, yet disagree about the proper method for ascertaining that factual claim. Hence the sense in which shared theories about methods is a condition for agreement over factual claims is that such theoretical agreement makes intersubjective agreement concerning fact more likely.

It is true that some agreements of this type could be only apparent, and the verbal agreement could hide some deeper disagreement. Thus, if the claim is that an object has a certain mass, then if different theories about the methods of ascertaining mass are held, it is possible that the term 'mass' has importantly different meanings. The verbal agreement as to the mass of an object, in such a case, could obscure the fact that the two inquirers might mean rather different things in their verbally similar statements. While such agreement may be only apparent in some cases, this does not mean that such agreement must *always* be only apparent.

Rather, it seems that agreement over factual claims arrived at through differing methods constitutes stronger evidence for that claim than agreement involving only one method. Such would be the case, of course, only if there is reason to have confidence in both the methods used. One of the results of recognizing the influence of method upon factual agreement is the prescription that we ought to seek a plurality of methods, since factual agreement which results from the application of various methods seems to be more thoroughly grounded than agreement arising from the use of only one method.[18]

It is worth emphasizing that the range of possible methodological dispute extends well beyond the sort of disagreement concerning particular experimental designs which have been considered above. It is simply not the case that there is only one 'scientific method' which is used in all scientific inquiries. This is surely true, at least, if one seeks a characterization of methods which are reasonably descriptive of the way inquiries are actually conducted in science. Perhaps this is clearest in the social sciences where there are highly visible methodological disagreements over the best means to gain knowledge. Some inquirers think that the experimental method is best, while others adopt survey research techniques, or use mathematical models, or computer simulation; still others insist that more 'subjective' methods such as participant observation are the proper ways to study man. Such methodological disagreements stem partly from divergent beliefs about the types of knowledge which ought to be sought,[19] but such disagreements also involve differing theories about the proper methods of gaining reliable knowledge. It is perhaps more difficult to see what methodological controversy would look like in the natural sciences. Clearly there are examples of low level methodological disputes in the natural sciences (e.g. over proper measurement techniques) but, it could be asked, what would be the force of questioning the general experimental method in the natural sciences. This objection has some force, but two points make it less compelling. It is not clear that the variety of concrete methods utilized in the natural sciences are usefully as all being examples of one single method – namely the experimental method. For example, are inquiries stemming from mathematical physics best understood as 'experimental', or rather might there be many types of experiments which exemplify different methods of inquiry. It might turn out that if we pay closer attention to what natural scientists actually do, there may be such diversity of procedures that the use of a single term to describe these various methods is misleading. But even if there is a certain general uniformity of procedure which can be successfully discerned in the natural sciences, it doesn't follow that such uniformity is desirable. Rather such uniformity may mask the influence that the standard method has upon the substantive results of inquiry. And it is simply not true that the experimental method is the only possible method for the natural sciences. To make only one suggestion as to alternatives, it is possible that the method of simulation could be used in the natural sciences and the use of such an alternative might provide a useful check on the influence of the experimental method on the substantive results achieved.

At this point a distinction which has been tacitly presupposed in my argument should be made explicit. There is, I think, an important difference between what I have called theory neutral facts and what might be called data. What functions as evidence in science are theory neutral facts, not data. Data are what are proposed as neutral fact but not all are accepted as such. The existence in scientific inquiry of the difference between data and fact can be most easily illustrated by noting the difference between denying data the status of fact and denying that data are truly data at all. We have already examined the process of denying data the status of fact in examining the dispute in learning theory. Contrast that type of dispute with a much rarer sort of dispute which involves accusing a scientist of deliberately misreporting the results of his inquiry. Such an accusation is much more serious than the claim that some evidence is the artifact of the method employed and is treated quite differently by the scientific community. Distinguishing between data and fact captures this difference nicely, for it makes room for the important difference between impugning the honesty of an inquirer and merely denying his data the status of theory neutrality.

A final point I want to make involves indicating a fruitful area for further inquiry. Noting the methodological aspect of intersubjectivity in science suggests that such intersubjectivity is neither a brute fact nor does it just fortuitously happen. Rather it is an achievement. There is reason to believe that such intersubjectivity as can be attained has a basis in shared theories about methods, but the process of achieving intersubjectivity is much more complex than that. After a study of the problems involved in reaching intersubjective agreement in a particular interdisciplinary research project, Stewart E. Perry concludes that: "The process of making public what is our private experience is the social-cultural-psychological act in science. The necessity to share and to reach consensus implies an engagement in [social] interaction."[20] Further investigation into the conditions underlying intersubjective agreement should involve recognizing the psychological and social dimensions of such agreement. Inquiry in these directions may prove quite fruitful in illuminating the nature of fact in science.

Lehman College,
City University of New York

NOTES

[1] The following passage brings out the idea of intersubjective agreement as the basis of fact in science, and it also shows how intersubjectivity can be taken as 'given': "The condition thus imposed upon the observational vocabulary of science is of a pragmatic character: it demands that each term included in that vocabulary be of such a kind that under suitable conditions, different observers can, by means of direct observation, arrive at a high degree of agreement on whether the term applies to a given situation.... That human beings are capable of developing observational vocabularies that satisfy the given requirement is a fortunate circumstance: without it, science as an intersubjective enterprise would be impossible." (C. Hempel, *Aspects of Scientific Explanation*, Free Press, N.Y., 1965, p. 127.)
[2] A similar strategy for exploring a different condition of intersubjectivity was adopted by Hanson. See his *Patterns of Discovery*, Cambridge University Press, 1958, p. 18.
[3] P. Achinstein, 'The Problem of Theoretical Terms', *American Philosophical Quarterly* **2**, 1965.
[4] M. Spector, 'Theory and Observation, I and II', *British Journal for the Philosophy of Science* **17**, No. 1 and 2.
[5] C. West Churchman, *Prediction and Optimal Decision*, Prentice-Hall, Englewood Cliffs, N.J., 1961, p. 80, italics in original.
[6] I. Rock, 'The Role of Repetition in Associative Learning', *American Journal of Psychology* **70**, 1957; see also I. Rock and W. Heimer, 'Further Evidence of One Trial Associative Learning', *American Journal of Psychology* **72**, 1959.
[7] W. K. Estes, 'Learning Theory and the New "Mental Chemistry"', *The Psychological Review* **67**, 1960.
[8] *Ibid.*, p. 215.
[9] *Ibid.*, p. 221.
[10] Estes seems to neglect this possibility and thinks he is at least on the way to experimental designs which are beyond dispute and which will lead *directly* to the truth about learning theory: "While I would not for a moment deprecate the role of imagination in science, I suspect that it will begin to serve us effectively in learning theory only as

we begin to accumulate reliable determinations of the effects of single variables upon single learning trials in individual organisms. If by continual simplification of our experimental analyses and refinement of our mensurational procedures we can achieve these determinations, we may find that the long sought laws of association may be not merely 'instigated', or even 'suggested', but literally dictated in form by empirical data." (*Ibid.*, p. 222.)

[11] L. Postman, 'One Trial Learning', in *Verbal Behavior* (ed. by C. N. Cofer and B. Musgrave), McGraw Hill, N.Y., 1963.

[12] *Ibid.*, p. 302.

[13] *Ibid.*, p. 303.

[14] *Ibid.*, p. 319.

[15] G. A. Miller, 'Comments of Professor Postman's Paper', on Cofer and Musgrave, *op. cit.*

[16] *Ibid.*, p. 326.

[17] For example, see R. C. Atkinson and R. C. Calfe, 'Mathematical Learning Theory', in *Scientific Psychology* (ed. by B. B. Wolman), Basic Books, N.Y., 1965.

[18] I take this to be an application of Feyerabend's prescription concerning theoretical pluralism. See, for example, his 'How to Be a Good Empiricist – A Plea for Tolerance in Matters Epistemological', in *Philosophy of Science; The Delaware Seminar*, Vol. 2, (ed. by B. Baumrin), Wiley, N.Y., 1963.

[19] See my paper 'Science, Reason and Value', *Theory and Decision*, forthcoming.

[20] S. E. Perry, *The Human Nature of Science*, Free Press, N.Y., 1966, p. 236.

VII. Induction and Reduction

JAMES H. FETZER

DISPOSITIONAL PROBABILITIES

The propensity interpretation poses an intriguing alternative to the frequency definition for the explication of probability as a physical magnitude. It is intended to provide an explicitly dispositional account of this concept within the context of statistical laws. First systematically advocated by Karl Popper, it has been endorsed – in one form or another – by Ian Hacking and D. H. Mellor, among others. The purpose of this paper is, first, to distinguish two rather different formulations of the propensity construct (which we shall refer to as the 'long run' and 'single case' concepts); second, to explain away some of the objections that, *prima facie*, might be thought to undermine an explication of this kind; and, third, to analyze the inadequacies of a single case dispositional account that fails to take seriously the concept of probability as a statistical disposition.

I

According to Karl Popper, probabilities are "unobservable dispositional properties of the physical world."[1] These properties are envisioned as belonging to specifiable sets of conditions, which may be referred to as 'experimental arrangements' or as 'chance set-ups'. They are dispositional in the sense of being tendencies to display particular patterns of response under appropriate test conditions. As Popper explains it,

> Every experimental arrangement is *liable to produce*, if we repeat the experiment very often, a sequence with frequencies which depend upon this particular experimental arrangement. These virtual frequencies (themselves) may be called probabilities. But since these probabilities turn out to depend upon the experimental arrangement, they may be looked upon as *properties of this arrangement. They characterize the disposition, or the propensity*, of the experimental arrangement to give rise to certain characteristic frequencies *when the experiment is often repeated.*[2]

This passage thus suggests that probabilities should be conceived as dispositional tendencies for experimental arrangements to generate certain outcomes with characteristic relative frequencies during long runs of experiments.

Boston Studies in the Philosophy of Science, VIII.

Other passages, however, suggest another possibility, namely: that probabilities should instead be regarded as dispositional tendencies for particular experiments to give rise to particular results on their singular trial. Popper also states, for example, that

The main point... is that we now take as fundamental *the probability of the result of a single experiment*, with respect to its *conditions*, rather than the frequency of results in a sequence of experiments. Admittedly, if we wish to *test* a probability statement, we have to test an experimental sequence. But now the probability statement (itself) is not a statement *about* this sequence: it is a statement *about* certain properties of the experimental conditions, of the experimental set-up.[3]

In this case, probabilities are properties of singular trials – rather than of long sequences of trials – on experimental set-ups.

Popper's analysis therefore suggests not one but two different constructions of the probability concept:

(1) the *long run* construct, according to which probabilities are dispositional tendencies for experimental set-ups to generate certain outcomes with characteristic relative frequencies during long runs of trials; and,

(2) the *single case* construct, according to which probabilities are dispositional tendencies for particular set-ups to produce particular results on their singular trials.

The difficulty is that, although the significance of probability values interpreted as designating hypothetical results of long run trials is clear, the significance of these same values interpreted as designating features of singular trials is not at all evident. This problem has two aspects: first, the meaning of attributing a probability for a certain result to a single trial on a chance set-up; second, the relations that may obtain between the single case and long run constructs.

Perhaps the most illuminating of Popper's remarks on the question of meaning is the following:

...propensities may be explained as possibilities (or as measures or 'weights' of possibilities) which are endowed with tendencies or dispositions to realize themselves, and which are taken to be responsible for the statistical frequencies with which they will in fact realize themselves in long sequences of repetitions of an experiment.[4]

As a consequence, "The propensity distribution attributes weights to all possible results of the experiment."[5] With respect to the single case construct, therefore, the value of a propensity may be taken to refer to either (i) the weight of the dispositional tendency for a particular outcome, or (ii) the set of all such weighted tendencies, present on a single trial. A fair die and tossing device, for example, possesses not only the propensity to yield an ace with a weight of $\frac{1}{6}$, but also the propensity to yield a duce, a trey, and so on, with weights of $\frac{1}{6}$; and therefore possesses the tendency to yield either an ace or a duce or a trey and so forth with a weight of 1.

These considerations suggest a more precise formulation of the single case construct; however, in view of the fact that the term 'weight' is also employed by the frequency definition and, indeed, within a similar context, it seems preferable to rely upon another term for this purpose. Since the term 'strength' is dispositional in character and available for use, it may suitably be utilized here. Accordingly, we shall henceforth assume that the numerical values of probability statements under the single case interpretation indicate *the strength of the dispositional tendency for an experimental set-up to produce a particular result on its singular trial.*

Popper has also sought to clarify the relationship which may obtain between single case probabilities and long run frequencies. Thus, for example, he asserts that the single case construct

> ... allows us to interpret the probability of a *singular* event as a property of the singular event itself, to be measured by a conjectured *potential or virtual* statistical frequency rather than by an actual one.[6]

Notice that Popper only claims to *measure* probabilities on the basis of hypothetical frequencies; he does not deny that subjecting these statements to direct empirical *test* can only be done on the basis of actual frequencies. In elaboration, Popper explains

> ... we have to visualize the conditions as endowed with a tendency, or disposition, or propensity, to produce sequences whose frequencies are equal to the probabilities; which is precisely what the propensity interpretation asserts.[7]

The strength of the dispositional tendency for a chance set-up to produce a particular result on its singular trial, therefore, may be measured theoretically by means of the hypothetical relative frequencies to which it would characteristically give rise during a long run of trials.

Although these remarks may serve to clarify Popper's conception of the relationship between single case probabilities and long run frequencies, they also tend to obliterate the distinction between the single case and long run propensity constructs. Popper's position here, for example, is in complete accord with the following passage by Ian Hacking, which constitutes an unambiguous endorsement of the long run construct:

It has been convenient,..., to speak of a property and label it with the phrase, "frequency in the long run." But already it is apparent that this term will not do. "Frequency in the long run" is all very well, but it is a property of the coin and tossing device, not only that, in the long run, heads falls more often than tails, but also that this would happen even if in fact the device were dismantled and the coin melted. This is a dispositional property of the coin: what the long run frequency is or would be or would have been.[8]

In order to justifiably claim that the differences between them are significant, after all, it is necessary to identify precisely what those differences are and to explain why they are important.

To recognize a logical difference between them, notice that the hold which single case probabilities have on long run results differs categorically from the grip exercised upon them by long run probabilities. Under the long run construction, the relative frequencies for particular outcomes displayed by an infinite sequence of trials *necessarily* equal their probabilities. This results from the fact that long run probabilities are just what the long run frequencies are or would be or would have been. Under the single case construction, however, the relative frequency for a particular outcome within such an infinite sequence only *probably* equals its probability. This results from the fact that these long run outcomes must be computed on the basis of Bernoulli's theorem – even when the sequence itself consists of an infinite set of trials.

Consequently, although a long run probability permits the prediction of long run results with the force of logical necessity, a single case probability only permits the prediction of long run results with a probability of 1. It would therefore be mistaken to assume that the hypothetical frequencies which would be generated during an infinite set of trials necessarily equal their probabilities; for although this is indeed the case for long run probabilities, it is only overwhelmingly probable for single case probabilities. As a matter of fact, of course, the sequences accessible within our range of experience are always finite; consequently, the logical difference between these two constructs is important in principle but not in practice.

The fundamental theoretical difference between them, naturally, is that one interprets probabilities as long run dispositional properties, while the other interprets them as single case dispositional properties. Interpretations which take long run probabilities as basic, however, may succeed in developing an acceptable long run construct but must still contend with the problem of accounting for the assignment of 'probabilities' to singular events. Reichenbach, for example, resolved this difficulty by introducing the concept of weight as the predictional value of a sentence pertaining to a single case.[9] The meaning of 'weight', however, clearly differs from that of 'probability'; indeed, in Reichenbach's view, its meaning is fictitious but nevertheless justified, "because it serves the purpose of action to deal with such statements as meaningful."[10]

The point is not to appraise the success of Reichenbach's concept of weight for its intended purpose but rather to take note of the fact that, in order to account for single case 'probabilities', the advocate of a long run construct finds it necessary to introduce a further concept for application to singular events. A single case construct, by contrast, not only accounts for the meaning of single case probabilities but also solves the problem of long run probabilities. The fundamental advantage of the single case interpretation, therefore, is that it yields a construct which is theoretically significant for *both* the long run and the single case – and it achieves this effect without introducing any new and difficult problems for the theory of science. Admittedly, an interpretation of probability as a dispositional property does not resolve the thorny issue of explicating the concept of dispositionality itself; but that problem was with us already.

II

Perhaps the most serious reservations about adopting a single case interpretation of probability are epistemological in kind. Consequently, it may be instructive to consider the objections directed against such a concept by Reichenbach himself:

> To state the argument briefly: if probability has something to do with the reliability of predictions, the probability statement must be verifiable in terms of the occurrence of the event predicted; otherwise the statement will be empty so far as predictions are concerned. The frequency interpretation satisfies this condition inasmuch as it verifies a degree of probability through repeated occurrence of the event. If, however, the meaning of the probability statement refers to a single event, it is impossible to verify the statement in terms of the occurrence of the event; and therefore the statement has no predictional value.[11]

This argument, of course, clearly reflects Reichenbach's commitment to the verifiability criterion of meaning. This assumption, moreover, is evendently necessary for the argument's validity; thus, in Reichenbach's view, a probability statement is empirically meaningful if and only if it is completely verifiable, in the sense that its assertive content is theoretically exhaustible by means of predictable occurrences. The assertive content of a single case probability statement cannot be theoretically exhausted by means of predictable occurrences, however, since it refers to a single event which either occurs or fails to occur – neither of which provides a suitable basis for its empirical verification. Consequently, such statements must be regarded as empirically meaningless.

At least four remarks appear to be in order with regard to Reichenbach's argument. The first, of course, is that the verifiability criterion has long been recognized as imposing far too stringent demands upon the language of empirical science. It seems clear that, by any reasonable contemporary standard, the epistemological status of a probability statement is secure when it is physically possible to subject that statement to empirical tests by means of which evidence tending to confirm it or tending to disconfirm it may be obtained. The appropriate criterion here, in other words, is that of confirmability, rather than verifiability, in principle.

The second is that, although a single case probability statement surely does pertain to singular events, it is somewhat misleading to suppose that the events involved only occur a single time. A single case propensity statement, after all, pertains to every event of a certain kind, namely: to all those events characterized by a particular set of conditions. Such a statement asserts of all such events that they possess a specified propensity to generate a certain kind of outcome. Consequently, although the interpretation itself refers to a property of conditions possessed by single events, such statements may actually be tested by reference to the class of all such singular occurrences – which is surely not restricted to a single such event.

In the third place, even if the probability assertion involved were only represented by a single actual occurrence, or perhaps by not even one appropriate event, it would still not be the case that that statement could not be subjected to empirical test. Under such circumstances, of course, the relevant evidence would have to be of an indirect kind. Reichenbach himself, however, was highly resourceful in developing a battery of method for this purpose, including the theory of higher level probabilities and the method of concatenated inductions.[12] So long as these probabilities are displayed by the frequencies to which they characteristically give rise, these methods are applicable for the purpose for which they were originally intended – to provide relevant data for estimating probabilities when the direct evidence is limited or nonexistent.

Finally, even though these probabilities are envisioned as properties of singular events, precisely the same strategy that Reichenbach nurtured with great care to deal with the problem of the single case as it arises within the frequency interpretation applies in equal measure here: the rule of posit and the method of anticipation.[13] The very same kind of long run success guaranteed by the frequency theory is also guaranteed by the single case propensity account (granting, of course, the subtle difference between logical necessity and probability of 1). Consequently, there appears to be no justification for Reichenbach's claim that single case probability statements possess no predictive significance and are therefore incapable of empirical test.

III

It may be illuminating, by way of conclusion, to contrast the single case propensity account which has been presented with another recently advanced by Mellor. Mellor claims that both Popper and Hacking have been misleading in comparing the propensity of an experiment to the fragility of a glass, on the grounds that the results of trials with a chance set-up are not strictly comparable with the results of dropping a fragile glass:

> The main point is that it cannot be the *result* of a chance trial that is analogous to the breaking of a dropped glass. It is true that the breaking may be regarded as the result of dropping the glass, but to warrant the ascription of a disposition, it must be supposed to be the *invariable* result. Other things being equal, if a glass does not break when dropped, that suffices to show that it is not fragile. But if propensity is to be analogous to fragility, the result of a chance trial is clearly *not* the analogue of the glass breaking since, in flat contrast to the latter, it must be supposed *not* to be the invariable result of any trial on the set-up. If it were so, the trial simply would not be of a chance set-up at all. Other things being equal, if a chance trial does not have any given result, that does not suffice to show that the set-up lacks the corresponding propensity.[14]

Thus, in Mellor's view, the results of testing a 'disposition' are always uniform, i.e., invariably the same, necessarily; while the results of trials on a 'chance set-up', by contrast, cannot possibly be invariably the same. If 'propensity' is supposed to be a dispositional property like 'fragility', therefore, then the feature of a propensity which is analogous to the breaking of a glass when it is dropped cannot possibly be the result of a trial on a chance set-up:

> If propensity, then, is a disposition of a chance set-up, its display, analogous to the breaking of a fragile glass, is not the result, or any outcome, of a trial on the set-up. The display is the chance distribution over the results, ...[15]

Thus, by identifying the 'display' of a propensity with its chance distribution, Mellor believes he has succeeded in developing the propensity concept of probability as a dispositional construct perfectly analogous to fragility.

It is evident that Mellor's purpose is the explication of a propensity construct which is 'dispositional' in a strong sense. This requires that the presence of that property be 'displayed' on each and every trial to which it is subjected; otherwise, it would fail to be completely analogous to fragility, which Mellor takes as a dispositional paradigm. By identifying the 'display' of a propensity with its chance distribution, therefore, Mellor has indeed succeeded in obtaining a concept which is 'dispositional' in this strong sense.

It is also evident, however, that Mellor obtains this 'success' at considerable expense. In the first place, the claim that the 'display' of a propensity always accompanies its trial has now become semantically tautologous. For if there were no chance distribution, there would certainly be no propensity: after all, a propensity *is* a chance distribution. In the second place, although Mellor may call its chance distribution the 'display' of a propensity, this claim is obviously empirically vacuous. For while its chance distribution is necessarily present upon each and every trial of an experimental arrangement, this 'display' is not experientially accessible.

The basic difficulty with Mellor's analysis, therefore, is that it relies upon an ambiguous concept of display. Normally, of course, a property is 'displayed' by its observable manifestations. In Mellor's scheme a property is also 'displayed' by its mere theoretical presence. It seems far preferable, after all, to envision the 'display' of a propensity as the relative frequencies it generates during long and short runs of trials. Thus, it is apparently Mellor, rather than Popper and Hacking, who is confused on this point.

These considerations suggest the underlying cause of Mellor's troubles, namely: that he has mistaken an analogy for an isomorphism. These properties, after all, may perfectly well be alike in certain respects without being alike in all. A propensity, for example, is a statistical disposition; fragility, by contrast, is a universal disposition. These properties, therefore, are alike in being dispositions, yet unalike insofar as one is statistical, the other universal. In taking fragility as the paradigm for his analysis, therefore, Mellor placed himself in a conceptual dilemma.

Although Mellor therefore fails to take seriously the concept of a statistical disposition, our examination of his analysis provides an opportunity to distinguish several senses in which propensities may or may not be viewed as 'invariable', i.e., as physical constants. These senses pertain to (a) their strength distribution, (b) their hypothetical long run displays, and (c) their single case results.

(a) Although a propensity may be envisioned as either (i) the strength of the dispositional tendency for a particular outcome, or (ii) the strength distribution over all the possible outcomes, in both cases propensities are regarded as invariable properties of chance set-ups.

(b) These properties, moreover, generate characteristic hypothetical long run relative frequencies for their various outcomes (which may be calculated on the basis of Bernoulli's theorem); and these relative frequencies, like the propensities they display, are also regarded as invariable.

(c) But while the propensity to produce a particular outcome on a single trial with an experimental set-up is a constant property of that set-up, the results of trials with such set-ups are not at all invariable, unless, of course, the propensity involved happens to be a universal, rather than a statistical, disposition.

University of Kentucky

NOTES

[1] K. Popper, 'The Propensity Interpretation of Probability', *British Journal for the Philosophy of Science* **10** (1959) 30.

[2] K. Popper, 'The Propensity Interpretation of the Calculus of Probability, and the Quantum Theory', in *Observation and Interpretation in the Philosophy of Physics* (ed. by S. Körner), Dover Publications, Inc., New York, 1955, p. 67.

[3] Popper, *op. cit.*, p. 68.

[4] Popper, 'The Propensity Interpretation of Probability', p. 30.

[5] Popper, 'The Propensity Interpretation of the Calculus of Probability, and the Quantum Theory', p. 68.

[6] Popper, 'The Propensity Interpretation of Probability', p. 37.

[7] Popper, *op. cit.*, p. 35.

[8] I. Hacking, *Logic of Statistical Inference*, Cambridge University Press, Cambridge, 1965, p. 10.
[9] H. Reichenbach, *Experience and Prediction*, University of Chicago Press, Chicago, 1938, pp. 313–14.
[10] H. Reichenbach, *The Theory of Probability*, University of California Press, Berkeley, 1949, pp. 376–77.
[11] Reichenbach, *op. cit.*, p. 371.
[12] Reichenbach, *Experience and Prediction*, pp. 363–73.
[23] Reichenbach, *op. cit.*, pp. 352–53.
[14] D. H. Mellor, 'Chance', *Proceedings of the Aristotelian Society*, Supplementary Volume, 1969, p. 26.
[15] Mellor, *op. cit.*, p. 26.

BERNARD GENDRON

ON THE RELATION OF NEUROLOGICAL AND PSYCHOLOGICAL THEORIES: A CRITIQUE OF THE HARDWARE THESIS

The Hardware Thesis (HT) is a claim that psychological theories are irreducible to neurological theories; it has its basis in analyses of the distinction between the psychological *functions* postulated to account for the law-like relations between perceptual input and behavioral output, and the neural *structures* (or 'hardware') which *embody* these functions.[1] HT's particular commitment to psychology-neurology irreducibility follows from the conjunction of these three assertions:

(A) The role of psychological theories is to postulate systems of functions which mediate between perceptual input and behavioral output.

(B) The role of neurology is to identify and describe the structures (or 'hardware') which embody the functions postulated by psychological theories.

(C) Theories postulating systems of functions are not reducible to, or eliminable in favor of, the theories which identify the structures performing or embodying such functions. Generally, theories whose role is limited to the identification of the structures embodying the functions postulated by other theories have little or no explanatory value.

The ambiguity of (HT), as presently formulated, is tied to the ambiguity of the structure-function distinction. In this paper, I investigate three relevant variants of the structure-function distinction, and then attempt to show that no variant generates a justifiable version of (HT); that is, I shall argue that for no variant of the structure-function distinction is the conjunction (A)–(C) acceptable.

HT has received significant support in recent philosophical and psychological literature, especially in Jerry Fodor's *Psychologiqal Explanation* and J. A. Deutsch's *The Structural Basis of Behavior*.[2] Both Fodor and Deutsch have ties with the newly emergent information-processing approach to theorizing in psychology.

Boston Studies in the Philosophy of Science, VIII.

I

The structure-function distinction may first be construed as a distinction between parts, or systems of parts, on the one hand, and the events, states, or dispositions ascribable to such parts, or systems of parts, on the other. Broadly speaking, a postulated mechanism may be identified in terms of the parts which constitute it or the whole of which it is a constituent (i.e., in terms of its structure_1), or it may be identified in terms of the sorts of events or results it is disposed to bring about (i.e., in terms of its functions_1). In its crudest form, HT_1 may be formulated as follows:

(A1) Psychological theories postulate the systems of functions_1 in virtue of which, presumably, central mechanisms mediate between perceptual input and behavioral output. Examples of such functions_1 (or mechanisms identified in terms of such functions) are sensory filters, memory traces, trace-interference, trace-consolidation, drives, and drive-reduction.

(B1) The role of neurological theories is to specify and describe the anatomical structures_1 which subserve the functions_1 postulated in psychological theories. For example, the neurologist may attempt to determine the extent to which the hippocampus is implicated in the storage of short-term memories and the frontal lobes involved in the retrieval of memories.

(C1) The specification of the anatomical structures_1 embodying the functions postulated by psychologists has little or no theoretical explanatory value. Specifically, one does not provide a theoretical account of the laws relating the postulated functions_1 with one another, and with their inputs and outputs, by indicating what anatomical structures_1 subserve these functions_1.

So understood, HT_1 appears to be a component in the theories of both Deutsch and Fodor, although other versions of HT are also involved; Deutsch and Fodor explicitly construe neurological taxonomies as anatomical or morphological taxonomies.[3]

The reason initially for subscribing to (C1) may be formulated as follows. In a pure anatomical system, the classification of structures is carried out independently of the functions_1 performed or subserved by the entities described in that system. Anatomical structures_1 are classified typically in terms of their phenomenological and topographical characteristics; for example, the hippocampus is identified as that tubular sea-horsed shaped tissue deep in the cerebrum, extending over the dorsal surface of the corpus callosum, etc. Thus, there is no guarantee that there will be any one-to-one correspondence between the parts of the nervous system identified in anatomical-structural$_1$ terms and those identified in functional$_1$ terms. But, even if there were, the correlation of anatomical characteristics with functional$_1$ characteristics would not have any explanatory value. We do not learn why a certain part performs or exhibits a certain function$_1$ by being given its relative location, color, shape, texture, etc.; for example, one does not account for a mechanism's ability to act in accordance with the laws governing the storage and retrieval of short-term memory traces, by indicating that such a mechanism is a hippocampus (i.e., a tubular sea-horse shaped tissue deep in the cerebrum).

These considerations notwithstanding, HT$_1$ still appears, in its present form, to be implausible, because (B1) and (C1) are still seriously open to question. First, in response to (B1), it seems decidedly wrong to say that all the concepts of neurology are anatomical-structural$_1$ concepts. There are neural functions$_1$ as well as neural structures$_1$. Among these are gross events, such as EEG states and evoked potentials, etc., and unit events, such as synapses and nerve impulses. Secondly, in response to (C1), the claim that anatomical distinctions have no explanatory relevance seems noncontroversial only when restricted to gross anatomical rather than micro-anatomical classifications. The neuron, for example, is an anatomical structural$_1$ unit whose introduction into neurology was based not only on microscopic observations, but also on theoretical needs. Not only are the primary unit events and dispositions of the nervous system predicable of the neuron, but the very possibility of subscribing to the existence of some of these events (e.g., the synapse) presupposes the neuron theory. Also, since the nervous system is essentially a conductive system, the topographical distribution of neurons is of crucial importance in the explanation of the manner in which electrical messages are processed in the nervous system.

The proponent of HT$_1$ may respond to these objections as follows. (i) Though there are classes of neural functions$_1$ as well as anatomical structures$_1$, such functions$_1$ cannot be adequately taxonomized without reference to their anatomical locus (e.g., hippocampal slow waves), if they are to be appropriately correlated with psychological functions$_1$. (ii) Though particular groups of microanatomical structures$_1$ may be correlated with, and may have explanatory import for, particular types of microneural functions$_1$, they still may not be correlatable in any one-to-one fashion with postulated psychological functions$_1$ (or mechanisms identified in terms of such functions$_1$), and, as a consequence, they may not have any important role in the theoretical account for such functions$_1$.

It must be clear, however, that in moves (i) and (ii) the proponent of HT$_1$ is making empirical claims based on available neurological evidence rather than assertions about the intrinsic limits of neurology. Now, as an empirical claim in theoretical neurology, (i) has received substantial support although it has been countered by the Gestaltist claim that neural events correlated with particular psychic states have no fixed anatomical locus and hence are classifiable only in terms of the patterns which they form; (ii) however has received only limited support in certain restricted areas of psychophysiology, the most dramatic being Lashley's work on the physical basis of learning.[4] Traditionally, it had been assumed that the long-term changes in behavioral responses associated with learning were represented by changes in anatomical connections between neurons (e.g., through the growth of new dentritic branches or synaptic nobs); differences between memories presumably were reflected in the location and complexity of corresponding anatomical changes. Lashley and others have argued, however, that there can be no one-to-one correlation between permanent anatomical changes and acquired memories. As a consequence, it was assumed that no neuron or set of neurons has any unique or necessary role in the retrival of any particular memory (e.g., a firing neuron may contribute sometimes to the 'noise' and sometimes to the 'signal' in the retrieval of the same memory). This work has contributed somewhat to the growing tendency to look for the memory-trace on a sub-neuronal level, e.g., in the macromolecules involved directly or indirectly in the manufacture of proteins.) If these speculations are justified, then we must say that there are no explanatorily relevant neural structural$_1$ or functional$_1$ correlates for stored memory traces; proposition (ii), as applied to the domain of learning, would then have empirical warrant. However, though reasonable when restricted to learning, (ii) becomes highly implausible when asserted about every type of postulated psychological functions$_1$. For example, in the past decade, Mountcastle (in the physiology of somatic sensation) and Hubel and Wiesel (in the psysiology of vision) have proposed imaginative descriptions of the way in which different perceptual and sensory states are mapped structurally$_1$ and functionally$_1$ into the nervous system. Even in the domain of learning, (ii) is now being challenged very seriously.[5] Thus, as an empirical thesis. HT$_1$ appears highly implausible and must for the present be rejected. Even in those restricted areas (e.g., learning) where it does have application, HT$_1$ is a relatively weak thesis; for all it claims is that the discovery of some structural$_1$ correlates for psychological functions$_1$ must involve bypassing neurology in favor of some otheı biological discipline; it does not deny (in its present form) that for every psychological function$_1$ there is an explanatorily relevant biological (though non-neurological) structure$_1$ of function$_1$.

Though HT$_1$ seems adhered to by the major proponents of HT, its rejection does not necessarily undermine their position, if, as is the case, they subscribe to other versions of HT as well. I shall now investigate two other versions of HT which, at least in the eyes of their proponents, play a more strategic role.

II

The structure-function contrast generating the second version of HT (HT$_2$) is based on the distinction between a general system of laws and the particular systems of laws which embody or realize it. For example, Newton's laws of mechanics are realized in the system of laws governing planetary motion, Maxwell's in the system of laws governing the behavior of light, while the set of laws specifying the chemical and electrical properties of metals is realized in the system of laws governing the behavior of potassium. Within the specification of appropriate boundary conditions, the particular systems of laws are not only derivable from, but also explained in terms of the general systems which they exemplify or embody; however, the laws of the general systems cannot be explained in terms of the laws of their exemplificatory or realizing systems.

In some cases, a general system of laws corresponds to a general machine design, and the particular systems of laws exemplifying it correspond to the various types of mechanisms which embody or realize the general machine design; these various mechanisms may be construed as the structures$_2$ which embody the functions$_2$ described in the general machine design.

Now, it might be argued that in theorizing the psychologist is in effect putting forth general systems of laws describing the conditions central mechanisms must meet, and the relations they must have, if they are to mediate successfully between perceptual inputs and behavioral outputs; then it would be added that any one of an indefinite number of possible mechanisms could embody such systems of laws. The laws governing the nervous system and the laws governing electronic machines would thus constitute systems exemplifying or realizing the general systems put forth in psychological theories; that is, psychology would specify the general machine design which is realized or embodied in electronic mechanisms as well as in neural mechanisms. For example, it might be suggested that the psychological laws governing the formation of, and mutual interference between, memory traces constitute a general system realized in the systems of laws specifying how the memory trace is formed in biological systems or in electronic systems (perceptrons). HT_2 may thus be stated as follows:

(A2) The systems propounded in theoretical psychology are general systems.

(B2) The systems put forth in neurology are particular embodiments or realizations of the theoretical systems of psychology.

(C2) The laws of general systems cannot be explained or accounted for by the laws of the particular systems which realize them; consequently, theories describing general systems cannot be reduced to, or eliminated in favor of, the theories describing the particular systems embodying such general systems.

A well-known version of HT_2, so formulated, is defended in Deutsch's *Structural Basis of Behavior*.[6]

One can concede that theoretical psychology does delineate general systems which can be embodied in an indefinite number of possible mechanisms of which the nervous system constitutes just one example. The question is whether such general systems are of the same type as the paradigmatic general systems we have taken from physics (e.g., Newton's laws of motion). The answer appears to be negative: unlike the general systems described above, postulated psychological systems cannot be used to explain theoretically the behavior of the systems which realize them (e.g., the nervous system). But, if this is the case, there is no longer any reason for subscribing to (C2).

It seems that HT_2 is rooted in a confusion between the general systems described above and what I shall refer to as *abstract systems*. Consider the following example.[7] Network theory deals with the properties of electrical circuits by interrelating three idealized kinds of electrical structures: resistors, conductors, and capacitors. Now, one can abstract from an interpreted network theory some abstract systems (*S*) which describes not only the behavior of electrical circuits, but also that of vibrating mechanical systems whose components have roles analogous to those of the counterpart electrical components: springs would thus correspond to capacitors, masses to inductors, and dampers to resistors. (*S*) is therefore realized both in electrical circuit systems and in mechanical vibratory systems. However, there seem to be significant differences between this abstract system and general systems like Maxwellian electromagnetic theory. One would not account for the laws of vibratory mechanisms and electrical circuits by appeal to (*S*). There is no stipulation in (*S*) about the classes of entities which will satisfy (*S*). In fact, it would appear that in order to account for the fact that (*S*) has physical application, we would have to appeal to the laws governing electrical circuits and vibratory mechanisms, and the theories underlying them. That is, it is by appeal to the laws and theories governing the particular systems embodying or realizing (*S*) that we can explain theoretically how it is possible for any mechanism to exhibit (*S*). Thus, the abstract system (*S*) is theoretically reducible to, or eliminable in favor of, its embodying systems.

The logical net models of McCulloch-Pitts and von Neumann are good examples of what I have called abstract systems. A logical net is a set of interconnected units whose behavior is governed in accordance with the calculus of Boolean algebra; each of these units is either in an 'on' or 'off' state depending on the level of excitatory or inhibitory inputs from other units, and on the way in which these input units are organized. The system of logical nets has been proposed as a model for such diverse intelligence-exhibiting mechanisms as the nervous system and binary circuit systems. From the fact that the nervous system and binary circuit systems embody the same logical net model, it is sometimes suggested that they belong to the same natural kind, and consequently that the distinction between animate is no longer scientifically relevant. This move, however, is made too quickly. Whether the neuron and the binary circuit belong to the same natural kind depends on whether their respective behavior can be accounted for in terms of the same theoretical mechanisms; for it is in the light of the concepts used in theoretical explanations that scientific taxonomies are ultimately constructed. If logical nets were general systems in the Newtonian or Maxwellian sense, then the nervous system and binary circuit systems would belong to the same natural kind. Logic nets, however, are abstract systems. Logical net systems do not delineate the classes of entities which satisfy them, and consequently cannot be used to explain theoretically the laws of the systems which embody them. Rather, it is by appeal to the laws and theories governing the behavior of neurons and binary circuits that one may explain how it is possible for any mechanism to act in accordance with the laws defining logical nets. Since the physics of the neuron contrasts quite significantly with the physics of the binary circuit, the nervous system and binary circuit systems no more belong to the same natural class than do springs and capacitors.

I suggest that, if the systems postulated by psychologists to mediate between input and output can be realized by an indefinite number of possible mechanisms, they are to be construed as abstract systems rather than general systems. Consider, for example, a psychological theory which specifies the laws governing the growth, possible decay, and mutual interference of memory traces; here we are provided with an abstract machine design which would be satisfied by electronic as well as neuronal mechanisms. However, this theory does not delineate the types of mechanisms which will satisfy the set of abstract laws governing the memory trace; thus, reference to this set of laws is of no use in the theoretical account of the memory storage and retrieval behavior of exemplificatory mechanisms, such as the nervous system or electronic machines (e.g., perceptrons). Rather, the converse is the case. It is only by bringing to bear the laws and theories governing the storage and retrieval capacities of nervous systems and electronic machines that we can explain how it is possible for the postulated memory systems of psychology to work as they do. Thus, psychological theories are reducible to, or eliminable in favor of, neurological theories (or theories about electronic networks), even though the latter are exemplificatory embodiments of the former. HT_2 then must be rejected. The temptation to subscribe to HT_2 is a consequence of running together the concepts of *general system* and *abstract system*, as defined above. If one restricted himself to the former concept, then (A2) should be rejected; if the latter, then (C2) should be rejected.

The admission that abstract systems of laws are theoretically reducible to their exemplificatory embodiments is incompatible with a standard about reduction, namely, the view that a theory (T1) is reducible to some other theory (T2) only on the condition that the laws of (T1) are derivable from the laws of (T2). It is clear that the laws governing abstract systems, having greater extension, cannot be derived from the laws governing their exemplificatory or embodying systems. However, this incompatibility can be overcome if the standard view about reduction is revised to state: the theory (T1) is reducible to the conjunction of theories (T2) only if, for any specific type of embodying mechanism, *M*, one can deduce from the laws of (T2) which laws of (T1) (if any) apply to *M*; thus, one may deduce from theoretical neurology that the laws governing a particular abstract system for information storage and retrieval do apply to the nervous system.

III

HT_1 was premised on a structure-function *category* distinction whereas HT_2 was premised on a structure-function *generality* distinction. HT_3 operates on a structure-function distinction involving both category disparity and generality disparity. The notion of *function* ($function_3$) involved here is somewhat like that used in discourse about teleological explanations. To describe something in $functional_3$ terms is to describe it in terms of the purposes or ends which it subserves (e.g., a mousetrap, waterpump, aversive response, sleep-inducing mechanism) and/or in terms of the rules which govern it (e.g., a chess piece, hypothesis-selection mechanism, transistorized *and* computer circuit). To describe something in $structural_3$ terms is to ascribe to it either properties which are $nonfunctional_3$ or law-like characteristics which are nonteleological. The distinction, in this case, is not between a general machine design and the various kinds of machines which realize it, but between a set of programmed instructions and the various general types of machines which may carry them out. HT_3 may thus be formulated as follows:

(A3) Psychological theories identify internal mechanisms in terms of their $functions_3$, i.e., in terms of the purposes they subserve or the rules which govern them.

(B3) Neurological theories identify internal mechanisms in terms of their $structural_3$ properties or the physico-chemical laws which govern them.

(C3) Explanations involving $functional_3$ concepts are not theoretically reducible to, or eliminable in favor of, explanations involving $structural_3$ concepts.

(A3) would not be accepted by everyone; generally cognitivists would agree and S–R theorists disagree. The crucial claim however is (C3). Now there seem to be two reasons allegedly for justifying (C3). (i) In view of the category disparity between psychological ($functional_3$) concepts and neurological ($structural_3$) concepts, one might argue that states denoted by the former concepts cannot be theoretically identified with, or reduced to, states denoted by the latter; (ii) or one might, in defending (C3), stress the generality contrast between $functional_3$ and $structural_3$ concepts, i.e. the fact that particular $functions_3$ can be achieved by an indefinite number of types of mechanisms. The argument from category disparity seems inadmissible. Cross-category theoretical identification or elimination is not only tolerated but also encouraged in the theoretical sciences, since some categories are oftentimes thought to have greater explanatory power than other categories. For example, water-solubility, which is a *dispositional* state of macro-objects, may be identified theoretically with a *structural* state of molecules, namely, the polar distribution of electrical charges in such molecules. Regarding the argument from generality contrast, we have already demonstrated its unacceptability in our critique of HT_2.

The clearest version of HT_3 is provided by Jerry Fodor in his *Psychological Explanation.* He opposes the standard view that psychological theories are micro-reducible to neurological theories; be suggests, as an alternative, that we interpret "statements that relate psychological and neurological constructs ... as attributing particular psychological functions to corresponding neurological systems," not as articulating microanalyses.[8] Thus, he concludes that "if the doctrine of the unity of science is to be preserved, it requires something else (or other) than reducibility as a relation between the constructs of psychology and neurology."[9]

In defending his position, Fodor stresses the analogies between func-

$tional_3$ concepts such as *valve lifter* and *mousetrap*, on the one hand, and the concepts of theoretical psychology, on the other. In effect, the core of his argument seems to be the thesis of generality disparity between $structures_3$ and $functions_3$.

There is, in particular, no sense to the question "What does a valve lifter consist of," if it is understood as a request for microanalysis... When, in paradigmatic cases, entities in one theory are reduced to entities in another, it is presupposed that both theories have available conceptual mechanisms for saying what the entities have in common... It is patent that functional analysis does not share this property of reductive analysis. When we identify a certain mousetrap with a certain mechanism, we do not commit ourselves to the possibility of saying in mechanistic terms what all members of the set of mousetraps have in common. Because it is roughly a sufficient condition for being a mousetrap that mechanisms be used in a certain way, there is nothing in principle that requires that a pair of mousetraps have any shared mechanical properties.[10]

Fodor assumes that the microreduction of entities of $functional_3$ type K (e.g., being a mousetrap) in terms of, say, physico-chemical constructs presupposes that all members of this class share certain $nonfunctional_3$ properties of relevance to physics and chemistry. But this seems wrong. For, where there are no such shared properties, the option is obviously to redistribute the members of class K into classes each of which is defined in terms of shared mechanical properties (e.g., spring-controlled mousetraps, electronic mousetraps) and then to perform the appropriate microreductions for each of these special classes. If it is objected that such entities have not been theoretically reduced insofar as they belong to class K, the answer is: so much the worse for being a member of class K. For, it might be argued, only those entities which share the same mechanical properties or are microreducible in terms of the same physico-chemical constructs can be said ultimately to belong to the same natural kind. I take it that it is partly in virtue of its role in determining natural kinds that microreduction provides the basis for the unity of science. So, the claim that members of the class of mousetraps do not share any mechanical properties having relevance for microreduction is not an argument against microreduction but against the claim that mousetraps constitute genuine natural kinds.

In conclusion, it appears that no version of the Hardware Thesis is acceptable. Each is generated by a particular interpretation of the structure-function distinction operating in the shared claims (A) that psychological constructs are functional, (B) that neurological constructs are structural, and (C) that functional constructs cannot be theoretically reduced to structural constructs. HT_1 distinguishes between anatomical structures$_1$ and the psychological functions$_1$ which they subserve, HT_2 between abstract machine designs and the various types of mechanisms which embody them, and HT_3 between goals or rules and the mechanisms which subserve them. In this essay, I argued that whenever (A) and (B) are minimally plausible, (C) is not acceptable. Against HT_1, it was shown that anatomical classifications do have explanatory import of a theoretical kind; the claim that there is no correlationship between anatomical structures$_1$ and psychological functions$_1$ was found to be empirically implausible and philosophically uninteresting. It was argued that HT_2 initially appears plausible only because it runs together the concept of general system (Maxwellian electromagnetics) and the concept of abstract system (abstract machine designs), although only the latter may plausibly be applied to psychological theories; unlike the former, however, the latter type of system can be theoretically reduced to the particular systems which embody it. Finally, it was shown that HT_3 is undermined by the same sorts of reasons which worked against HT_1 and HT_2.

University of Texas at Austin

NOTES

[1] Cf. J. Fodor, *Psychological Explanations*, Random House, New York, 1968, p. 109.
[2] Fodor, *op. cit.*; J. A. Deutsch, *The Structural Basis of Behavior*, University of Chicago Press, Chicago, 1960.
[3] "It still remains quite conceivable that identical psychological functions could sometimes be ascribed to anatomically heterogeneous neural mechanisms... It is in short quite conceivable that a parsing of the nervous system by reference to anatomical or morphological similarities may often fail to correspond in any uniform way to its parsing in terms of psychological function." Fodor, p. 116. "For instance, to attempt to guess at the particular change which occurs in the central nervous system during learning ... is not only unnecessary but also purely speculative. ... This can be shown by taking the example of the insightful learning machine. To be told that the semi-permanent change in the machine which occurs when it learns is due to a uniselector arm coming to rest does not help us to understand the behavioral properties of the machine. ... For the change could equally well be due to a self-holding relay, a dekatron selector, or any type of gadget known to technology capable of being turned from one steady state to another. In the same way to speculate about terminal end boutons in the way that Hebb does or about changes of synaptic resistance seems to be trying to answer a question irrelevant, strictly speaking, to the psychological theorist." Deutsch, p. 12.

[4] Beach, Hebb, Morgan, and Nissen (eds.), *The Neuropsychology of Lashley*, McGraw-Hill, New York, 1960, pp. 478–506; see also, E. Roy John, *Mechanisms of Memory*, Academic Press, New York, 1967.
[5] E. R. Kandel and W. A. Spencer, 'Cellular Neurophysiology and Learning', *Physiological Review* **48** (1968) 65–134.
[6] "The precise properties of the parts do not matter; it is only their general relationships to each other which give the machine as a whole its behavioral properties. These general relationships can be described in a highly abstract way, for instance, by the use of Boolean algebra. This highly abstract system, thus derived, can be embodied in a theoretically infinite variety of physical counterparts. Nevertheless, the machines thus. will have the same behavioral properties, given the same sensory and motor side. Therefore, if we wish to explain the behavior of one of these machines, the relevant and enlightening information is about the abstract system and not about its particular embodiment. ... On the other hand, the knowledge that the machine operates chemically, electromechanically, or electronically does not help us very much at all... A complete specification of its embodiment would add very little to the explanatory power of this system." Deutsch, pp. 13–14.
[7] J. R. Peirce, *Symbols, Signals, and Noise*, Harper and Row, New York, 1961, pp. 3–6.
[8] Fodor, p. 114.
[9] Fodor, p. 120.
[10] Fodor, pp. 113, 115–16.

NELSON POLE

'SELF-SUPPORTING' INDUCTIVE ARGUMENTS

Starting with David Hume, most philosophers have maintained that it is illicit to argue that inductive arguments will be successful in the future because they have been successful in the past. This philosophic tradition claims that such an argument is circular since it is an attempt to prove the principle that the future will resemble the past by means of an inductive argument which principle all inductive arguments presuppose. Hans Reichenbach is such a philosopher but he goes on to add that even though induction can not be so justified, our using induction may be justified.[1] Max Black, on the other hand, may be identified as one of the leaders of the movement which accepts what others deem the circular argument.[2] In this paper the position of Reichenbach and Black will be examined. It is my thesis that both of their arguments are vitiated by their failure to employ the distinction between using a logic and talking about a logic, which distinction is the point of the difference between an object language and a meta-language. Once their positions are restated in terms of the distinction, it will be seen that the positions are not very dissimilar after all.

I. BLACK'S DEFENSE OF INDUCTION

Black asks us to consider the following argument which employs the inference rule:

R1: To argue from *Most instances of A's examined under a wide variety of conditions have been B* to (probably) *The next A to be encountered will be B.*

The argument is:

(1) In most instances of the use of R1 in argument with true premisses examined in a wide variety of conditions, R1 has been successful.

Boston Studies in the Philosophy of Science, VIII.

Hence (probably)
In the next instance to be encountered of the use of R1 in an argument with a true premiss, R1 will be successful.

Black claims correctly that the premiss of this argument is not a proposition identical to the conclusion nor does it logically follow from the conclusion. He therefore argues that the argument does not beg the question as the critics maintain. Notice further that the employment of R1 in (1) is not one of the employments of R1 which the premiss of (1) maintains is successful since we do not know, yet, whether the conclusion of (1) is true, nor is the employment of R1 which the conclusion of (1) maintains is successful identical to the employment of R1 in (1) since the employment of R1 in (1) is not the *next* employment of R1. This would be a further consideration of the kind Black claims shows that (1) does not beg the question.

The most interesting criticisms of Black are the kind offered by both Wesley Salmon[3] and Peter Achinstein.[4] What they do is to propose inference rules other than R1 which no one would be willing to accept and they defend them with self-supporting arguments analogous to the one that Black employs. For example Salmon asks us to consider the principle:

R2: To argue from *Most instances of A's examined in a wide variety of conditions have not been B* to (probably) *The next A to be encountered will be B.*

Which is 'self-supported' in argument

(2): In most instances of the use of R2 in arguments with true premises examined in a wide variety of conditions, R2 has been unsuccessful.
Hence (probably)
In the next instance to be encountered of the use of R2 in an argument with a true premiss, R2 will be successful.

R2 is the principle that the future will be unlike the past. The force of argument (2) is that since, up till now, the future has been like the past, by R2 it will cease to be like the past and hence the next application of R2 will be successful. Salmon points out, correctly, that in just those ways in which Black claims that argument (1) is noncircular so is argument (2). Achinstein's argument is similar to Salmon's except that his R3 is a deductive principle.

R3: To argue from *No F is G* and *Some G is H* to *All F is H*.

The 'self-supporting' argument which employs R3 is, of course, an instance of the invalid syllogism IEA-1 which inference R3 permits.

(3) No argument using R3 as its rule of inference is an argument which contains a premiss beginning with the term 'All'.
Some arguments containing a premiss beginning with the term 'All' are valid.
∴ All arguments using R3 as their rule of inference are valid.

So that if the critics are correct, by procedures logically identical to the ones employed by Black to defend induction, an inference rule logically incompatible with induction, Salmon's R2, and an invalid deductive inference, Achinstein's R3, may be defended. If Black is correct in arguing that argument (1) does not beg the question, two specious principles, R2 and R3, will also have to be accepted by philosophers. The critics conclude, then, that Black is wrong.

II. REICHENBACH'S DEFENSE OF INDUCTION

Reichenbach sides with the critics. Salmon's article, I take it, is, in fact, a defense of Reichenbach against Black. Reichenbach who rejects the 'self-supporting' induction defense asks us to justify our using induction on pragmatic grounds instead of justifying the inductive principle itself. The problem, as he sees it, is to justify our use of inductive methods to predict the future instead of non-inductive procedures such as card reading, crystal ball gazing, and, we might add, R2 or R3. He reasons as follows: If there is no way at all of predicting the future, if there are no indicators in the past or the present, by which the events of the future may be predicted, then any method at all, inductive or noninductive may be used with equal success. Pragmatically, it just does not matter what inference rule we use, none will give success except by utter chance. However, if there is a method for predicting the future, if there are marks in the past or present, by which we may come to predict the future then induction, R1, has a pragmatic edge. Either R1 itself will predict the future successfully *or* a noninductive procedure, R4, will be successful. If a noninductive procedure is successful we would still be best off using induction since it will enable us to find a correlation between application of the successful noninductive procedure and the realization of a correct prediction: Suppose some noninductive procedure, R4, is usually successful in a wide variety of applications, then the following argument which employs R1 as its inference rule will defend R4.

(4): In most instances of the use of R4 in arguments with true premisses examined in a wide variety of conditions, R4 has been successful.
Hence (probably)
In the next instance to be encountered of the use of R4 in an argument with a true premiss, R4 will be successful.

Pragmatically we should continue to use inductive as a means of predicting the future since, if there is no way of successfully predicting the future, induction is as good a method as any other, but, if there is a way of predicting the future, induction should be used for it is either that successful method or it will enable us to discover the successful method by picking out a constant conjunction between application of the correct noninductive method and success of the prediction.

It might be objected to Reichenbach that his argument, too, is spurious since an analogous argument will defend our using crystal ball gazing or any other method. If there is no way of predicting the future, crystal ball gazing is as good as any other way, but, if there is a way of predicting the future, it will either be crystal ball gazing or, by looking into the ball, we will find the correct way. However, this objection, like all the other objections and arguments above, ignores the fundamental distinction employed by logicians between reasoning about a logic and reasoning with a logic, the object language/metalanguage distinction.

III. OBJECT LANGUAGE VS. METALANGUAGE

If we examine any contemporary book on the foundations of deductive logic, we find a distinction between the object language which contains valid arguments, rules of inference, and a formal language, and the metalanguage for that object language which contains sentences and arguments about the logic employed in the object language. In the metalanguage we find sentences such as '*Modus Ponens* is a valid argument form' and 'The propositional calculus contains all propositional tautologies as axioms or theorems'. For example, Copi's *Symbolic Logic*, a standard text on the subject, contains an axiom system for the propositional logic and a rule of inference, *modus ponens*, in its object language. In the metalanguage, Copi proves a metatheorem that each axiom of the object language is a tautology and that each theorem derived from the axioms by repeated uses of *modus ponens* is also a tautology.[5] The proof itself seems to employ *modus ponens* as a rule of inference in the metalanguage. (The proof will not be repeated here, since it is highly technical). The logic employed in the metalanguage is presented informally but it is a much more powerful logic than propositional logic. As such, the metalogic seems to employ, or contain as a part, propositional logic. If it actually did this, then the use of the metalogic to reason about the object language logic would beg the question. For this reason it would seem that the principles of inference employed in the metalanguage to reason about a logic in the object language are not identical to the principles of reasoning employed in that object language. If this is questioned, the very foundations of deductive logic are questioned and this will not be done here. However, what is fair for deduction is also fair for induction. There too a metalanguage may be employed to reason about the inductive logic used in the object language and the principles of logic used in the metalanguage to reason about the inductive logic in the object language are not identical to any principle of logic in that object language.

Keeping the distinction in mind will enable us to readily see the mistakes of Black, Reichenbach, and the other critics. Arguments (1), (2) and (3) are all illicit since they are attempts to reason about an inference principle by means of that very inference principle, but all reasoning about an inference principle occurs in a metalanguage for the object language in which the inference principle occurs. In the metalanguage no principle occurs for reasoning about the object language which is identical to a principle of logic in the object language, no matter how similar they may appear. Take argument (1) as an example, the argument is about the employment of R1 and hence is in the metalanguage for the object language in which R1 occurs. The principle of inference upon which argument (1) depends, cannot be R1 itself no matter how much they may look alike. Black's argument, so considered, amounts to this: In the object language we find many arguments which are attempts to predict the future. In the metalanguage we count up such arguments which are about a wide variety of things and we discover that most of such arguments which employ principle R1 are successful, they do correctly predict the future. Hence, we conclude, in the metalanguage, that the next application of R1 will also be successful. The latter sentence, since it occurs in the metalanguage, is not proven by means of an argument which employs R1 as its principle of inference since that rule of inference occurs in the object language. Let us call the principle in the metalanguage which is the principle of inference employed in the metalanguage argument the inductive-like principle. The argument is not circular, but it is reminiscent of Reichenbach's argument once it too is reconstructed in the metalanguage. Reichenbach's pragmatic justification of induction now goes this way: In the object language we find many different principles of logic to predict the future. If we use an inductive-like principle in the metalanguage to pick out which object language principle is successful, then, if there is a method of predicting the future in the object language, the metalinguistic inductive-like principle will find a correlation between application of the object language principle and success of prediction. The object language principle so picked out, is the one which should be continued in use as a means of predicting the future.

Once their arguments are reconstructed using an object language/metalanguage distinction, Black's and Reichenbach's views agree in that an inductive-like principle is employed in the metalanguage to determine which of the available rules of inference in the object language are successful. Their views differ in that Black claims it will pick out the inductive principle in the object language while Reichenbach simply leaves this question open, the rule might pick out a noninductive principle in the object language.

IV. THE DEFENSE OF INDUCTION IN THE METALANGUAGE

The last issue to be faced is the grounds for using an inductive-like principle in the metalanguage for adjudicating the question of which object language principle successfully predicts the future. After all, why not use a principle-like R2 or even crystal ball gazing in the metalanguage? The answer is an old one in part, truth in an empirical context has to do with the way the world goes, truth lies with correspondence. The statement 'this piece of copper when heated, will expand' is true because copper when heated does expand. The preference for induction in the metalanguage, in part, lies in the fact that induction does look directly at the phenomena. If induction is employed in the metalanguage, correlation between application of an object language rule and its successes are examined. The criterion for success is truth, correspondence with the world, not correspondence with a crystal ball. Since induction does look at correspondence with the world, it is preferable to the crystal ball. But such a consideration does not rule out R2. It, like induction, looks for actual success, but it, unlike induction, presupposes that patterns established in the past and present will not occur in the future. Part of what is wrong here is just that if some method which is not counter inductive is successful, R2 will *not* find it. For example, if crystal ball gazing were always successful, R2 could not find it, for it would predict its future lack of success. One of the advantages of an inductive-like rule in the metalanguage, as Reichenbach claimed, is that it will find *any* successful method in the object language and not just ones which are inductive. It is for the last reason, which is pragmatic, that an inductive-like principle is employed in the metalanguage for the object language in which rules for predicting the future occur.

In summary, once the object language/metalanguage distinction is observed when discussing the problem of induction, it is seen that people with such divergent views as Black and Reichenbach have, in fact, considerable agreement. By using an inductive-like principle in the metalanguage we will find the correct method in the object language for predicting the future. Most of us suppose that the method picked out will be inductive and there seems to be no reason to suppose otherwise.

V. POSTSCRIPT

There is one objection to all of the above programs, Black's, Reichenbach's and mine, to which the answer is beyond the scope of this paper and which indicates a new area of philosophic research. All three programs claim that a proper method of predicting the future will be successful more often than not. This implies that we know how to construct instances of the correct method which are not successful. This is, after all, part of what so perplexed Hume about induction. If we can construct instances of the correct method which are not successful, then as a project for our students we can have them construct so many unsuccessful examples of the correct method that it turns out that the correct method is *un*successful more often than not. This possibility, I believe, indicates that we have to reconsider what is meant by a successful method of predicting the future or that we have to reconsider the supposed instances in which the prediction is unsuccessful. Even though such a reconsideration is beyond the scope of this paper, it is important to note that this postscript does not vitiate the earlier thesis.

Cleveland State University

NOTES

[1] *Experience and Prediction*, University of Chicago Press, 1938, pp. 348–63.
[2] 'Self-Supporting Inductive Arguments', *Journal of Philosophy* (1958) 718–25.
[3] 'Should We Attempt to Justify Induction?', *Philosophical Studies* (1957) 37–48.
[4] 'The Circularity of A Self-Supporting Inductive Argument', *Analysis* (1962) 138–41. A fuller bibliography of papers on these issues appears in J. J. C. Smart, *Between Science and Philosophy*, Random House, 1968, p. 204. Many of these papers are reprinted in P. H. Nidditch, *The Philosophy of Science*, Oxford University Press, 1968.
[5] Macmillan and Company, 3rd edition.

VIII. Scientific Theories: Comparison and Change

SCOTT A. KLEINER

ONTOLOGICAL AND TERMINOLOGICAL COMMITMENT AND THE METHODOLOGICAL COMMENSURABILITY OF THEORIES

I

Scientific revolutions have been described as episodes in the history of science in which old empirical research activities, standards of achievement, problem fields and even ontologies have been exchanged for new (Kuhn, 1962; Feyerabend, 1962).

Also it has been claimed that rational canons for the evaluation of theories cannot reach across scientific revolutions. The revolutionary new theory can be defended only by rhetorical techniques, not by hypothetico-deductive or other inductive argumentation to or from empirically ascertained evidence. A revolutionary new theory achieves dominance over its predecessor only after the authority sustaining the latter theory is no longer effective. Thus, succeeding theories in scientific revolutions are held to be *methodologically incommensurable*.[1] That is, logical argumentation cannot be used to assign differential values to succeeding theories.

A superficial examination of what it means for two theories to differ ontologically might lead to the conclusion that theories so differing are methodologically incommensurable. If the ontology of the old theory differs from that of the new, both can be consistently held because their subject matter is different. The theory of evolution cannot compete with or be evaluated relatively to nuclear physics simply because populations of living organisms are not the same things as atomic or sub-atomic particles. On similar grounds it has been claimed (Kuhn, 1962) that succeeding theories in the history of science, such as Galilean and Aristotelian dynamics, are methodologically incommensurable. Aristotle is concerned with heavy, light and celestial bodies whereas Galileo is concerned with pendulums, projectiles and material bodies, including planets, that move horizontally. Both men are confronted by different objects and both of them make true claims, but the claims are true of different objects.

I think this comparison overlooks an important distinction. It is one thing to classify or to describe the same objects differently, and it is quite another to investigate, observe and discuss different objects. In what follows I will attempt to sharpen and to defend this distinction by considering what I think is a case of genuine ontological diversity. I will also specify the components of a theory, i.e., explicit or implicit postulates which are relevant to a theory's ontological commitment. These components are logically independent of the components determining the manner in which the objects are to be classified or described, where the latter components determine what I call a theory's *terminological commitment*. Finally I will examine the ways in which methodological commensurability can arise and will argue that ontologically diverse theories can be methodologically commensurable.

II

In the light of recent developments of fundamental particle theory, we can interpret in two ways the process by which electromagnetic radiation is emitted from the electronic orbits of atoms. The transition of an electron from an orbit of greater to one of lesser energy can be described in the usual way as a change in the energy-state of a single atom accompanied by the emission of a photon. However, this process might also be described as the decay of an excited atom into a photon and something else, *viz*, an unexcited atom. In the first description we contemplate an object whose life-span includes the entire photon-emission process. In the second description the initial atom ceases to exist at the time of decay. In one description two objects are mentioned and in the other three are mentioned. (Pais, 1963, pp. 299f).

These alternative descriptions of electronic radiation correspond to two views of neutron decay. Corresponding to the second of the above descriptions, we might call the neutron (n) and the proton (p) different particles. The decay reaction

$$n \rightarrow p + e + \bar{\nu}_e$$

(e stands for an electron, $\bar{\nu}_e$ an anti-neutrino associated with the electron) accordingly is described as the annihilation of an n and the production of a p, e and $\bar{\nu}_e$. However, in correspondence with the usual description of energy emission from electronic orbital transitions, the reaction might also be described as a change in the stage of a single fundamental particle, *viz*, a *baryon*. On this latter view, baryons have other states, the Λ^0, Σ^+, Σ^0, Σ^-, Ξ^0 and Ξ^- *hyperons*, listed here in order of increasing mass and decreasing stability ('lower' to 'higher' baryon states).

The baryons also appear in families, (n, p), (Λ^0), $(\Sigma^+, \Sigma^0, \Sigma^-)$, (Ξ^0, Ξ^-), each family containing particles of nearly the same mass. These families can be viewed as doublet, singlet, triplet, and doublet states respectively, again analogous to our usual description of the atom's electronic energy-levels. We attribute a *spin* of $\frac{1}{2}$ to electrons, entailing that in a magnetic field (say along some arbitrarily chosen z-axis) the electron in a given atomic orbit can be in two states. One state occurs when the z-component of the spin is $\frac{1}{2}$ ('spin up') and the other when the same component is $-\frac{1}{2}$ ('spin down'). Analogously, the doublets (n, p) and (Ξ^0, Ξ^-) are assigned an *isotopic spin* of $\frac{1}{2}$, meaning that n is a baryon in the 'ground state' with a z-component isotopic spin (I_3) of $-\frac{1}{2}$ and p is a ground-state baryon with $I_3 = +\frac{1}{2}$. Λ^0 is assigned an isotopic spin 0, meaning that there is only one state corresponding to this baryon 'level'. But Σ has an isotopic spin of 1, entailing that there are three possible isotopic spin states for Σ, *viz.* Σ^0 where $I_3 = 0$, Σ^+ where $I_3 = 1$ and Σ^- where $I_3 = -1$.

These states bear one further similarity to the plain spin of a charged particle: The different electronic energy levels in the atom due to spin orientation converge in value (become *degenerate*) if the magnetic field in which the electrons are immersed vanishes. Similarly, in the strong-interaction fields of the atomic nucleus, the differences denoted by the different z-components of the isotopic spin become degenerate. For all practical purposes, $n-n$, $n-p$ and $p-p$ interactions are identical inside the atomic nucleus. Here electromagnetic interactions are negligibly small compared with the strong interactions binding together the nucleus.

Consider now a bubble-chamber track recording the reaction

$$\Xi^- \rightarrow \pi^- + \Lambda^0$$

where π^- stands for a negatively charged π meson. The tracks in the chamber should look something like the diagram (Swartz, 1965, page 72). The solid lines in this diagram represent visible tracks of charged particles, and the dashed lines represent the inferred tracks of neutral particles.[2] If Ξ^-, Λ^0, and p are regarded as distinct particles, the path *ABCD* in the diagram is the path of three distinct particles. The first particle (Ξ^-) lasts

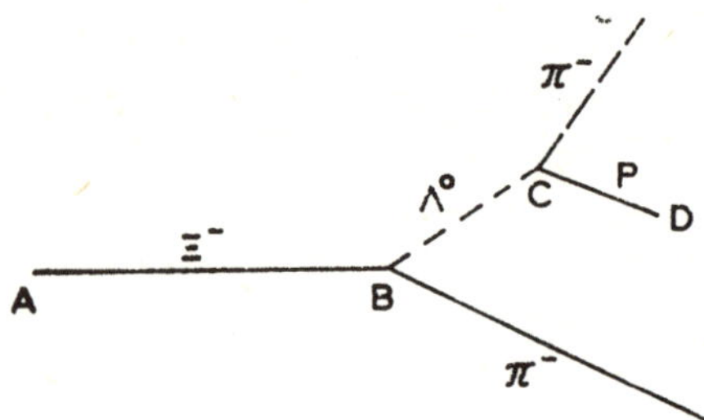

approximately 5×10^{-10} seconds and traverses AB, and the second (Λ^0) lasts approximately 9×10^{-10} seconds over path BC. The third (p) is stable, traverses along CD, and eventually is lost.

On the other hand, if the reaction is viewed as a change in baryon-state, $ABCD$ in the figure is the path of one and only one particle. Under this interpretation, the same particle is observed in CD as was observed in AB and BC.

Consider two observers of the bubble-chamber trace of the reaction depicted above, observers O_1 and O_2. Suppose O_1 believes that three different particles take part in the reaction, a Ξ^- in AB, a Λ^0 in BC and a p in CD. O_2, on the other hand, believes that the same particle appears in all three segments of the path $ABCD$.

O_1 and O_2 can also be assumed to agree regarding the identity of the particle they observe in CD: The particle there is a positively charged material object of rest mass 938.2 MeV and is of the same kind as particles constituting hydrogen 1 nuclei and as particles that are the positively charged components of other atomic nuclei.

Their agreement here is consistent with differences in their views as to the p's age. According to O_1, the p began its existence at the time the Λ^0 was annihilated, say at approximately $t_0 + (5 \times 10^{-10}) + (9 \times 10^{-10})$ sec. (t_0 is the initial moment at which the particle is at A, 5×10^{-10} sec is the lifetime of the Ξ^- and 9×10^{-10} sec the lifetime of the Λ^0.) According to O_2, the p began its existence (if it is not the case that it always existed) sometime before t_0.

If O_1 and O_2 agree about the identity of the particle in CD, they differ about the identity of the particles in AB and BC. O_2 believes that what appears in AB and BC is the same particle as that appearing in CD. By transitivity of material object identity, O_2 should also believe that this particle is identical to that observed or observable in *CD* by O_1 or anyone else. By contrast, O_1 believes that the particles in *AB* and *BC* differ from the particle that he or anyone else, O_2 included, observes or can observe in *CD*. In sum, O_1 and O_2 are committed to the existence of distinct objects in regions *AB* and *BC*. Insofar as 'observing' has cognitive import, O_1 and O_2 observe different objects in these regions. O_1 and O_2 refer to different objects in their reports of what they observe is *AB* and *BC*.

The different ontologies of O_1 and O_2 are consequences of different assumptions regarding the temporal extent or the lifetimes of the objects under consideration. O_1 believes that the Ξ^- and Λ^0 particles are short-lived entities that are annihilated in 5×10^{-10} and 9×10^{-10} sec. after their creation. On the other hand, O_2 believes that the Ξ^- and Λ^0 are merely states of the same object which is initially captured in the bubble chamber and then is lost shortly thereafter.

Assumptions of the same kind are either implicit or explicit in our common sense beliefs as well as our theories in other fields. For example, in astronomy the sun is assumed to be sufficiently long lived that astronomers, from the Ancient Greeks (c. 500 B.C.) to the present, have been able to claim that the same sun can be observed from day to day, year to year, century to century. This assumption is also shared by most plain men. Anyone discussing, e.g., the annual motion of *the* sun through the zodiac is committed to such an assumption, albeit implicitly. By contrast, one could consistently reject this assumption and hold that the sun is created anew in the east at dawn, and annihilated in the west at dusk at regular 24-hour intervals.

III

In an earlier paper (Kleiner, 1970) I argued that if theories such as classical particle mechanics and the special theory of relativity are axiomatized within set theory, certain presuppositions pertaining to the theory's extralogical terminology become explicit assumptions. By elementary principles of *erotetic logic* (the logic of questions), these assumptions can be shown to permit the asking of some and forbid the asking of other questions.

For example, if special relativity is so axiomatized, its axioms will include the claim that mass is a function of velocity, and velocity in turn is a function of the material reference frame chosen for a given problem. On the other hand, a similar axiom for classical particle mechanics claims without further qualification that each material particle has a unique positive real value for its mass. These axioms can be used to generate problems and hence as guides to research. The latter axiom informs the researcher working within classical particle mechanics that a material object's mass is relevant to its mechanical behavior and that its values are restricted to positive real numbers. The former axiom informs an investigator under the special theory that he must specify a material reference frame and determine an object's velocity in that frame before he can ask for its mass.

Now, the terminological axioms just mentioned are logically independent of assumptions regarding the temporal extent of the objects investigated. For example, the electron can be viewed either as a classical or as a relativistic particle. Normally it is considered as a classical particle when its velocity is much less than the velocity of light, *c*. But in cases in which velocity approaches *c*, as in high speed accelerators or in beta rays, the effects of the electron's velocity on its mass have to be taken into consideration. Nevertheless, in both theories the electron is regarded as a stable particle, i.e., a particle whose lifetime can be indefinitely long. Two investigators operating within respectively the relativistic and classical frameworks can be expected to report the same results (within experimental error) in their counts of electrons striking some target.

The independence of terminological and ontological commitment also entails that the logical and pragmatic consequences of these commitments are logically independent. The relativist and the classicist must conduct their research activities differently in the presence of the same objects, such as a shower of bombarding electrons. The relativist's terminological commitment entails that he reject the classical problem of determining an electron's invariant or absolute mass.

There may also be cases in which investigators share problems but differ in ontological commitment. In the case discussed in the preceding section O_1 and O_2 might both seek to evaluate the mass, velocity, or spin of the particles appearing in *AB* and *BC*.

IV

Two propositions (or theories) can be methodologically incommensurable under these circumstances: (1) The propositions are mutually incompatible (contraries or contradictories) but there are (can be) no logical arguments for or against either. One person could not hold both propositions, and he would (could) not have any reason for or against holding either. If each of two persons held one of these propositions, they would be in disagreement, but neither would (could) have any rational means of settling the disagreement. (2) The propositions are mutually compatible (neither contraries nor contradictories, but possibly sub-contraries) and there are (can be) no arguments supporting either. In this case, one person could consistently hold both and two persons could each hold one of the propositions without disagreeing. However, neither would have a reason for or against holding either proposition.

Similarly, two propositions could be methodologically commensurable under these circumstances: (3) There are arguments for one only or there are arguments for one and other arguments against the other, but no further logical relations between them need exist. (4) The propositions are contraries or contradictories, and there is an argument for one of them. In this case it follows that the same argument is against the other proposition.

These cases do not exhaust all circumstances for commensurability and incommensurability, but the list will suffice for my purposes here, *viz.* to show that ontologically divergent theories can still be methodologically commensurable. The fundamental particle theories of O_1 and O_2 are commensurable in at least the first sense (3). There are good reasons for accepting O_2's theory preferentially to O_1's. There are also arguments showing that the usual theory interpreting electronic orbital transitions as changes of state is better than its rival, viz, that these transitions are the annihilation of old and creation of new particles. I'll elaborate these arguments in reverse order:

In quantum mechanics the hydrogen atom is treated as a two particle material system capable of a number of energy states. By a fundamental principle shared by quantum and other branches of mechanics, any material object embedded in a force-field can have either a denumerably or a non-denumerably infinite number of energy-states at a given time. An application of quantum theory, *viz.* a conjunction of the Schrödinger equation with additional assumptions about force-fields in the region of application and perhaps some assumptions about relativistic effects, enables us to deduce exactly the energy states evidenced by the observable spectra of hydrogen. We have no other theory that describes orbital transitions in hydrogen as the anihilation of one and the production of another particle. No theory so describing electronic radiation has been applied to the laws and boundary conditions governing nucleus-orbital electron interaction thereby entailing the existence of the specific and distinct particles possible undeı these circumstances.

Turning now to the theory held by O_2, it is by reference to different values taken by the z-component of isotopic spin that we account for the diffeıent properties of the same objects that appear in the multiplets (n, p), $(\Sigma^-, \Sigma^0, \Sigma^+)$ and (Ξ^-, Ξ^0). Now, are there good reasons for introducing isotopic spin as a state-variable for describing microphysical particles and processes or is this quantity just an arbitrary and unnecessary embellishment in our descriptive apparatus? We should answer this question affirmatively if this quantity appears in one or more well supported laws and if the law(s) cannot be expressed in terms of state variables in use before its introduction.

In the first place, isotopic spin is conserved in strong interactions and strong interactions only. Thus the strong interactions between nucleons and between nucleons and mesons are independent of the charges borne by these objects.

Secondly, the z-component of isotopic spin is conserved in particle reactions governed by the strong and the electromagnetic interaction, but not in reactions governed by the weak interaction. It is by virtue of this conservation law that the 'strange' particles can be produced in time intervals of the order of 10^{-23} sec but continue to exist for periods of the order of 10^{-10} sec. Thus, Λ^0 decays by the weak reaction

$$\Lambda^0 \rightarrow \rho + \pi^-$$

in 9×10^{-10} sec rather than by the strong interaction which would take about 10^{-23} sec. For the above reaction

$$I_3 : 0 \neq -1 + \tfrac{1}{2}$$

By contrast, for a reaction producing Λ^0, *viz.*

$$\pi^- + p \rightarrow \Lambda^0 + K^0$$

note that

$$I_3 : -1 + \tfrac{1}{2} = 0 + (-\tfrac{1}{2})$$

Finally, let us consider whether the theories held by O_1 and O_2 are mutually compatible. That is, should we say that the difference between these two observers is that in addition to being commensurable, their theories are mutually contradictory or contrary, the last of the four possible circumstances mentioned above? More specifically, do these theories each contain at least one statement which is contradictory or contrary to a statement in the other theory?

Remember that O_1 claims that three different particles appear in the bubble-chamber record. O_2 claims that only one such particle appears there. *Prima facie* these claims seem to be contraries.

I believe that the statements are in fact contrary, i.e., that the acceptance of one logically entails that the other be rejected.

I have already noted that criteria for object-identity determines what counts as one object. Specific criteria of this kind are presupposed in any attempt to count these objects (Quine, 1953, pp. 65–79; 1960, pp. 90–5; Feyerabend, 1957). But sometimes we count different objects and the counts of different objects can be expected to produce different enumerative values.

For example, the count of the number of suns appearing in a year's time should produce the value '1'. By contrast, we could count the number of sunrises taking place throughout the year and get an average value '365 ¼'. This latter value coincides with the value we would expect from an observer counting suns that he supposes are annihilated and recreated daily. Now the count offered by the person counting sunrises (365 ¼) is complementary to the usual count of the number of suns appearing in a year's time (1). The person counting sunrises is concerned with objects of a different category from the category of material objects.

The latter category consists of objects that, at least over limited time intervals, exist independently of other objects (observing subjects included). Material objects also have masses and centers of mass. The mass-center points are assumed to possess unique positions and velocities. Material objects can bear a relation of participation to events. Events, like material objects, have duration and a location. One event can involve several material objects, as in a collision between two protons. The same material object can participate in a large number of events; e.g., the sun participates in a large number of sunsets, conjunctions and oppositions with other material bodies, and it could participate in a collision with another star. All of these events are of shorter duration than the sun itself, all have dates but no mass, and all can be assigned a position (occupancy of a region of space) but no velocity.[3] Finally, the existence of events in which material objects participate depends upon the existence of these objects. No sun, no sunrise.

Now to suppose that the sun is annihilated in the evening is to suppose that something has happened to that massive material object whose radiation warms us and whose gravitation sustains the earth's stable elliptical orbit. It is to suppose that the sun has ceased to exist. To say that the sun has ceased to exist patently contradicts the statement that the sun has not ceased to exist.

A similar point can be made about the creation and annihilation of fundamental particles. O_1's theory about the particle event will include the statement that the proton began its life at $t_0+(9\times10^{-10})+(5\times10^{-10})$. O_2's theory claims that the proton did not begin to exist at least in the time interval recorded by the experiment. O_2 simply denies O_1's statement that a proton was created in the course of the experiment.

In early geocentric astronomy, evidence for the supposed continued existence of the sun was indirect. This supposition was made because it is necessary for explaining the regularities appearing in the sun's motion. More direct evidence for the same claim is now available from elementary polar and extraterrestrial observations. Yet, the one and many sun hypothesis remain contraries. Thus, the logical relations between hypotheses are not affected by the kind of evidence available at a given time for either hypothesis. Similarly the logical relations between O_1 and O_2's hypotheses are independent of contingencies that at present make the evidence for both indirect.

V

What I have and what I have not tried to do heie can be summarized as follows:

1. One aspect of the meaning of non-logical expressions in a theory can change without generating incompatibility and *a fortiori* without generating methodological incommensurability between the old and new theories. This aspect has been described (Quine, 1953, IV; 1960) as the manner in which general terms 'is a proton,' 'is a lambda particle,' etc. divide their extension. The same aspect of meaning is reflected in the use of associated

singular terms 'the baryon in path ABCD in chamber ... in lab ...' (where '...' are filled with appropriate referring expressions for bubble chambers and laboratories) as referring expressions.

I have distinguished between this aspect of meaning and another that is reflected in a theory's terminological commitment. An account of this latter aspect is another story that is now in preparation.

2. I've considered only one of a number of possible cases of ontological diversity. Another more obvious kind of case would be between two theories with distinct hypotheses regarding an object's composition. Specifying conditions for complementarity and incompatibility in view of the variety of relations by which parts may compose wholes would carry us far beyond the space allotted here.

3. That O_1's and O_2's theories are incompatible entails that they share capabilities of referring to at least one object, the p in the case studied above. I believe that this entailment holds generally. For example, incompatibles theses about the composition of an object must refer to the same object whose composition is in question.

University of Georgia

BIBLIOGRAPHY

Clagett, M., *Science of Mechanics in the Middle Ages*, University of Wisconsin Press, Madison, 1961.

Cohen, M. R. and Drabkin, I. E. (eds.), *A Source Book in Greek Science*, Harvard University Press, Cambridge, 1958.

Dijksterhuis, E. J., *Mechanization of the World Picture*, Oxford University Press, Oxford, 1961.

Dreyer, J. L. E., *A History of Astronomy from Thales to Kepler*, Harper and Brothers, New York, 1953.

Duhem, P., *The Aim and Structure of Physical Theory*, Princeton University Press, Princeton, 1954.

Feyerabend, P. K., 'Explanation, Deduction and Empiricism', in *Minnesota Studies in the Philosophy of Science*, Vol. 3 (ed. by H. Feigl and G. Maxwell), University of Minnesota Press, Minneapolis, 1962.

Feyerabend, P. K., 'How to Be a Good Empiricist – A Plea for Tolerance in Matters Epistemological', in *Philosophy of Science: The Delaware Seminar* (ed. by B. Baumrin), Interscience Publishers, New York, 1963.

Feyerabend, P. K., 'On the "Meaning" of Scientific Terms', *J. Phil.* **62** (1965) 266.

Feyerabend, P. K., 'An Attempt at a Realistic Interpretation of Experience', *Proceedings of the Aristotelian Society* **58** (1967) 143–70.

Hanson, N. R., *Patterns of Discovery*, Cambridge University Press, Cambridge, 1958.

Kleiner, S. A., 'Erotetic Logic and the Structure of Scientific Revolution', *BJPS* **21**, 1970.
Kuhn, T. S., *The Structure of Scientific Revolutions*, University of Chicago Press, Chicago, 1962.
Magie, W. F., *A Source Book in Physics*, Harvard University Press, Cambridge, 1935.
Pais, A., 'The Structure of Matter', in *Philosophy of Science: The Delaware Seminar* (ed. by B. Baumrin), Interscience, New York, 1963.
Putnam, H., 'The Analytic and the Synthetic', in *Minnesota Studies in the Philosophy of Science*, Vol. 3 (ed. by H. Feigl and G. Maxwell), Minneapolis 1962.
Quine, W. V., *From a Logical Point of View*, Harper and Brothers, New York, 1953.
Quine, W. V., *Word and Object*, M.I.T. Press, Cambridge, 1960.
Rosser, W. K. V., *An Introduction to the Theory of Relativity*, Butterworth's, London, 1964.
Swartz, C. E., *The Fundamental Particles*, Addison-Wesley, Reading, Mass., 1965.
Wheelwright, P. (Ed.), *The Presocratics*, Odyssey Press, New York, 1966.

NOTES

[1] Methodological incommensurability is to be distinguished from Feyerabend's *semantic incommensurability* (1962, 1965), although semantic incommensurability might have methodological incommensurability as a consequence.

[2] Neutral particles leave no visible traces in the usual materials used to record microphysical events.

[3] A wave is best regarded as a sequence of events whose locations are in a one to one correspondence with successive instants. I believe that this analysis is preferable to viewing a wave as a unique event that moves with a determinate velocity.

CARL R. KORDIG

OBJECTIVITY, SCIENTIFIC CHANGE, AND SELF-REFERENCE*

Kuhn, Feyerabend, Polanyi, and Whorf would at times maintain that scientific objectivity is a myth. Kuhn writes:

> We may, to be more precise, have to relinquish the notion, explicit or implicit, that changes of paradigms carry scientists and those who learn from them closer and closer to the truth. ([10], p. 169)

He says that he

> ...would argue, rather, that in these matters neither proof nor error is at issue. ([10], p. 150)

On the basis of his interpretation of the history of science, Polanyi, also, feels that in a strong sense scientific knowledge is not objective ([11], p. 16). On the basis of his study of widely differing languages Whorf makes a similar claim:

> The relativity of all conceptual systems, ours included, and their dependence upon language stand revealed. ([15], pp. 214–215)

Feyerabend, too, perhaps because of a felt lack of success with his previous attempts to objectively compare different theories, has very recently come around to such a view:

> ...an enterprise whose human character can be seen by all is preferable to one that looks "Objective" and impervious to human actions and wishes. The sciences, after all, are our own creation including all the severe standards they seem to impose on us. It is good to be constantly reminded of this fact. And what better reminder is there than the realization that the choice of our basic cosmology may become a matter of taste? ([14])

Such are some consequences of their views. "Adoption of a new scientific theory would be an intuitive, mystical, or merely, aesthetic affair; it would be a matter primarily for psychological description rather than for 'objective' logical and methodological justification or evaluation in the usual senses of these terms." (Scheffler [16], p. 18.)

Such wholesale rejection of objectivity, however, is problematic. "Objectivity is presupposed by any statement which purports to make a cognitive claim. To put forth any such claim in earnest involves a presuppositional commitment to the view that the claim has an objective truth value" [*ibid.*, p. 21]. Consider the particular claim that no claim is objective. It is itself put forth as an objective truth. If, however, it is true then by its very statement it must be false.

Again we follow Scheffler's formulation. There is problematic self-contradictoriness: (a) in the earnest effort to communicate to others that communication is impossible; (b) in appeal to neutral facts about observation in order to deny that neutral facts of observation exist; (c) in arguing from the given realities of the history of science to the conclusion that reality is not discovered but made by the scientist. To accept these claims implies that we are – in the absence of objective standards – free to assert what we wish; hence, in particular, we are free to reject them.

Feyerabend ([2], pp. 29, 59; [3], p. 36), Hanson ([5], p. 7, 54–58, 61, 96–99, 154–156; [6], p. 38), Hesse ([7], p. 102; [8], pp. 51–52), Kuhn ([10], pp. 145, 148–149), Smart ([12], p. 162), and Toulmin ([13], pp. 18, 20–21; [14], p. 57) each maintain that transitions from one theory to another force an incommensurable change in the meanings of the terms employed. But the statement that it is impossible for there to be statements whose meaning or truth is invariant from theory to theory is itself a statement in fact put forward as true and as having the same meaning no matter what different theories we may hold. If it is true we would therefore have it itself as an example of a statement whose truth and meaning stays the same no matter what different theories we may hold; thus, by its very statement, it would be false. Hence, it *cannot* be true.[1]

When Whorf asserts that "the relativity of all conceptual systems, ours included, and their dependence upon language stand revealed," he is making a statement which he considers to be: (a) true; (b) part of his total conceptual system; and (c) determinate in meaning. If, however, this statement is true then, since it is expressed in English, it too must be relative to the English language. If this is so then, this particular meta-claim might, presumably, not hold in a different language and associated conceptual system. If it were false in some conceptual system, by its very statement, it would then be true that there exists at least one conceptual system which is not relative and which is independent of language. But if this is true then Whorf's self-referential meta-claim is false, even when expressed in English. In short, assuming it true, we can deduce that it is false; it is, therefore, untenable.[2] Similarly, the claim that there is no objectivity itself purports to be objective. Hence, if it is true, it must be false. It is, therefore, an untenable claim.

One might object by reinterpreting the above consequences. Roughly, one might want the claim to be only that there can be no objectivity *in science*. There would be no self-referential problem in then, for example, appealing to neutral meta-facts which are *about* observation in order to deny that neutral facts *of* observation could exist.

Let us, therefore, examine the position that there is no objectivity *in science*. Consider the claim that the choice between only scientific theories is "a matter of taste" (Feyerabend) where "neither proof nor error is at issue" (Kuhn). Such a restricted claim although not self-referentially inconsistent like the more general claim, is, nevertheless, problematic. It leads to an unjustified dualism. On the one hand we are supposed to hold (1):

(1) Science is a subjective enterprise (in the sense specified above, cf., e.g., Feyerabend and Kuhn) whose concepts and domain are theory-laden.

But, (1) is indeed put forward as the true, correct, proper, etc., way to view scientific transitions. Thus, on the other hand, in order to claim (1), and on the reinterpretation of these matters that we are considering, we are supposed to also hold (2):

(2) The philosophy of science is an objective enterprise whose concepts and domain are not theory-laden.

People who adhere to (1) would, to avoid the self-referential problem, *have to* maintain that their own views of scientific change are uninfluenced by the fact that they are (say) Kuhnians; that is, they would have to hold (2). Assume they instead held that their own views of scientific change were so influenced, and were, therefore, as with scientific theories, also 'a matter of taste' where 'neither proof nor error is at issue;' in short, assume that they held that their own views of scientific change were subjective. Then they would have to hold that (say) Kuhn's view of scientific change is no more correct than (say) Hempel's 'logical empiricist' account. This, however, neither Kuhn nor Feyerabend would want to do. Each believes 'the logical empiricist' account to be defective and in error. And each believes his own account to be correct and to remedy the objective excesses of logical empiricism. The dualism, therefore, remains. The facts scientists deal with are infected by the particular theory employed. The facts which (say) Kuhnian philosophers of science deal with are unexplainedly uninfected by their meta-theory, even though this theory is far more *conceptual and theoretical* in nature than any scientific theory. What is the difference between these two domains that makes this so? Is it because hard-and-fast and '*non*-theory-laden' facts are relevant to the philosophy of science but not to science? If this were affirmed, radical meaning variance theorists would be dangerously close to a new sort of 'logical empiricism' at a meta-level, one quite similar in structure to the old one they wish to eschew.

My conclusion, therefore, is that the doctrine of radical meaning variance and the denial of scientific objectivity are either demonstrably untenable (for they give rise to consequences which are self-referentially inconsistent) or else lead to an unjustified and problematic dualism with 'neo-positivistic' aspects.

University of Oxford

BIBLIOGRAPHY

[1] Cole, M. *et al.*, 'Linguistic Structure and Transposition', *Science* **164** (1969) 90–91.
[2] Feyerabend, P. K., 'Explanation, Reduction, and Empiricism', in *Minnesota Studies in the Philosophy of Science* (ed. by H. Feigl and G. Maxwell) Scientific Explanation, Space, and Time, Vol. III, pp. 28–97.
[3] Feyerabend, P. K., 'How to Be a Good Empiricist: A Plea for Tolerance in Matters Epistemological', in *The Delawar Seminar in the Philosophy of Science*, Vol. II (ed. by B. Baumrin), Interscience, New York, 1963, pp. 3–39.
[4] Feyerabend, P. K., 'Consolations for the Specialist', in *Criticism and the Growth of Knowledge* (ed. by I. Lakatos and A. Musgrave), The University Press, Cambridge, 1970, pp. 197–230.
[5] Hanson, N. R., *Patterns of Discovery*, The University Press, Cambridge, 1958.
[6] Hanson, N. R., *The Concept of the Positron*, The University Press, Cambridge, 1963.
[7] Hesse, M., 'A New Look at Scientific Explanation', *Review of Metaphysics* **17** (1963) 98–108.
[8] Hesse, M., 'Fine's Criteria of Meaning Change', *The Journal of Philosophy* **65** (1968) 46–52.

[9] Kordig, C. R., *The Justification of Scientific, Change* in Synthese Library, Reidel, Dordrecht, forthcoming.
[10] Kuhn, T. S., *The Structure of Scientific Revolutions*, University of Chicago Press, Chicago, 1962.
[11] Polanyi, M., *Personal Knowledge*, revised ed., University of Chicago Press, Chicago, 1962.
[12] Smart, J. J. C., 'Conflicting Views, about Explanation, in *Boston Studies in the Philosophy of Science*, Vol. PI (ed. by R. S. Cohen and M. W. Wartowsky), Harper and Row, New York, 1953, pp. 157–69.
[13] Toulmin, S., *The Philosophy of Science*, Harper and Row, New York, 1953.
[14] Toulmin, S., *Foresight and Understanding*, Harper and Row, New York, 1961.
[15] Whorf, B. L., 'Science and Linguistics', in *Language Thought and Reality* (ed. by J. B. Carrol), M.I.T. Press, Cambridge, Mass., 1957.
[16] Scheffler, I., *Science and Subjectivity*, Bobbs-Merrill, New York, 1965.

NOTES

* I wish to acknowledge my debt to Professor Richmond H. Thomason for very helpful comments on earlier versions of this paper.

1 I might add here that a positive account of just how and why objectivity and meaning invariance is possible another topic which is beyond the scope of this paper. My own sketch of such an account is to be found in my book, [9].

2 The above points up some conceptual difficulties with Whorf's hypothesis. There is, however, also empirical evidence which disconfirms his hypothesis, cf. M. Cole *et al.* [1].

WARREN D. SIEMENS

A LOGICAL EMPIRICIST THEORY OF SCIENTIFIC CHANGE?

I

There is in the philosophy of science a view of the nature of science which is often referred to as the orthodox view. In my opinion Dudley Shapere has succinctly characterized the main tenets of this view in his article 'Meaning and Scientific Change'.[1] According to this view, let me refer to it as the orthodox Logical Empiricist (LE) view, the problem of developing a theory of scientific change was not considered to be a legitimate problem for the philosophy of science. A primary concern of the Logical Empiricist was to analyze the logical structure which is shared by all scientific theories. From the point of view of the development of science the logical structure of scientific theories is a constant. What varies from one theory to another is their empirical content not their logical form.

> ...since philosophy of science, so conceived, does not deal with particular scientific theories, it is immune to the vicissitudes of science – the coming and going of particular scientific theories, for those changes have to do with the content of science, whereas the philosopher of science is concerned with its structure: not with specific mortal theories, with the characteristics of any possible theory, with the meaning of the word 'theory' itself.[2]

Consequently, scientific change from the orthodox LE point of view is not a problem for the philosophy of science; if it is a problem at all, it is a problem for the history, psychology, or sociology of science.

In recent years the orthodox LE view has been vigorously criticized and what might be called a radical element in the philosophy of science has developed.[3] This radical view holds that history of science is relevant (in a way that is ill-defined as yet) to the philosophy of science. It holds that scientific change is a legitimate problem for philosophy of science, and cannot be so summarily dismissed to other disciplines. What interests me in this paper is a reactionary view that seems to be emerging within the LE tradition. Although the reactionary LE view, if I may call it that, has not yet developed anything like a full-fledged theory of scientific change, there are beginnings which are worthy of attention. I suggest briefly but do not argue here that the reactionary LE view is the result of certain developments within the orthodox LE view and of an effort to meet the radicals' criticisms. The latter raised certain questions about the nature of the development of science which otherwise would have been ignored by the LE. In addition, there were problems which were generated by the LE's interest in theses about the unity of science and physicalism – problems concerning the reduction of one branch of science or theory to another. In order to provide an answer to a question like "Can all of science in principle be reduced to physics?," one needed an answer to the question "What are the conditions, formal and non-formal, for which it can be said that reduction has occurred?" Various types of reductions were distinguished and illustrated with examples from the history of science in what has become the standard account.[4] These illustrations from the history of science of course provided much of the fuel for the radicals' fire. What has emerged, however, within the last several years is the beginnings of something which can be called an LE theory of scientific change. In the next section, I will sketch what I think a full-fledged LE theory of scientific change will look like and try to indicate its potential. I will then outline various points at which such an enlightened LE theory might be vulnerable to criticism and see to what extent the criticism of certain radicals still apply.

The reasons then for developing a sketch of what I think an enlightened LE theory of scientific change will look like are:

(1) to show that a rather elaborate theory of scientific change can be developed within the LE tradition, a view which is more realistic than the orthodox view.

(2) With such a sketch before us, we are less likely to shrug off the LE view as drastically over-simplified and obviously counter to "the actual practice of science" (a favorite phrase of the radical view). Strengthening the LE position in this way, we are able to sharpen the criticism and hopefully gain insight into some of the considerations involved in determining the adequacy of a theory of scientific change – considerations which will specify what a theory of scientific change is a theory of.

II

This paper is part of a larger investigation into the idea that science can be viewed as a system and that scientific change can be viewed as a succession of states of the system. Questions which arise are: for the purpose of providing insight into the development of science (1) how are the states of science to be described? (What are the parameters in terms of which the state-descriptions are given?) and (2) what sort of an account of the transitions from one state to another can and should be given? In this section I am going to sketch one type of answer to these questions which has nowhere been fully developed but which seems to be implicit in recent developments within the LE tradition. Answers of this type will be referred to as LE theories of scientific change. An LE theory of scientific change holds that the state of science at any particular moment in its history is characterized primarily in terms of the theories held at the time. One of the main tenets of the LE tradition is that a scientific theory can be represented as a formal deductive system, generally finitely axiomatizable. Scientific change, then can be described as a succession of interpreted deductive systems (or sets of deductive systems). As was pointed out earlier, problems about intertheoretic reduction did not originally arise within the LE tradition over concerns about an account of scientific change, but not much imagination is required to see how it might be applied. The study of intertheoretic reduction is essentially a study of certain logical interrelations between deductive systems. If what has been learned by these studies is applied to the more general study of various types of transitions from one deductive system to another, then we have a descriptive framework for an important part of a theory of scientific change.

The following sketch borrows heavily from several recent articles.[5] I am not claiming that the sketch is in any way a reconstruction or synthetization of their views. I have used their distinctions and definitions for my own purposes and while I would expect that the sketch is consonant with their general views, I don't expect they will always agree with the particulars. My own interest is not to develop a full-fledged LE theory of scientific change but merely to indicate as clearly as I can what such a theory might look like.

How, then, should the states of science according to the logical empiricist be characterized? What are the parameters, the elements which change in the transition from one state to another?

First, Let L be a formal language (for convenience, we restrict our attention to first-order languages) which we characterize as a set of basic symbols V and a set of well-formed formulae. Since the formation rules will be considered a constant over transitions and the set of formulae is determined by V and the formation rules, we may consider V to be the only part of the formal language to vary from theory to theory. A formal theory F will be characterized as the ordered pair $\langle L, A\rangle$, where A is the set of basic formulae or axioms. The transformation rules are also considered to be constant over theory transitions.

Let M be a model for F; i.e., a structure for L such that A is valid, which we characterized in terms of a universe of individuals U and a set of denotations D for the predicate and function symbols of L.

Let $\text{Cn}(X)$ be the consequence class of a set of formulae X, the we will say that a formal deductive system is a set of formulae such that $X=\text{Cn}(X)$ i.e. closed. A finitely axiomatizable deductive system is one which follows from a finite subset of X called the set of axioms A; i.e., $X=\text{Cn}(A)$.

We might refer to $C=\langle L, M\rangle$ as the conceptual framework and a change in L or M as a change in the conceptual framework. We will say that C is a subpart of C^* when L is a sublanguage of L^* and M is a submodel of M^*.

A scientific theory T will be specified as the ordered triple, $\langle L, A, M\rangle$. We can now consider various types of T-transitions. 'T/T^*' will denote a transition from theory T to theory T^*. A T-transition is defined as a change in any of the parameters – L, A, M.
Consider the following definitions of various types of T-transitions.

1. *T is equivalent to T^** when $L=L^*$, $M=M^*$, $\text{Cn}(A)=\text{Cn}(A)^*$. In the case where $A\neq A^*$ we will call T^* an alternate formulation of T; in the case where $A=A^*$ we will call T^* identical to T.
2. *T is a part of T^** when $L=L^*$, $M=M^*$, and $\text{Cn}(A)$ is a subset of $\text{Cn}(A^*)$. It need not be the case that A is a subset of A^*.
3. *T^* is an expansion of T* when L is a sublanguage of L^*, M is a submodel of M^* and $\text{Cn}(A)$ is a subset of $\text{Cn}(A^*)$. We will say that L is a sub-language of L^* when V is a subset of V^* and that M is a sub-model of M^* when U is a subset of U^* and D is a subset of D^*.
4. *T^* is a U-expansion of T* when $D=D^*$ and U is a subset of U^*, $\text{Cn}(A)$ is a subset of $\text{Cn}(A^*)$. (T^* uses the same concepts to describe an expanded universe.)

5. *T^* is a D-expansion of T* when D is a subset of D^* and $U = U^*$, $\text{Cn}(A)$ is a subset of $\text{Cn}(A^*)$. (T^* adds new concepts to describe the same universe.)

These few definitions illustrate the type of machinery that can be developed. There is no attempt here to be complete. It becomes readily apparent that a variety of types of T-transitions can easily be distinguished given this representation. We might try to represent with this apparatus the types of inter-theoretic reduction specified by Sklar.[6] This requires the addition of the relation '$\approx$', as yet undefined but which reads 'is approximately equal to'.

6. *T/T^* is strict-complete–homogeneous* when C is a subpart of C^* and $\text{Cn}(A)$ is a subset of $\text{Cn}(A)^*$; i.e., T^* is an expansion of T.

7. *T/T^* is approximate-complete-homogeneous* when C is a subpart of C^* and $\text{Cn}(A) \approx \text{Cn}(A')$ is a subset of $\text{Cn}(A^*)$.

8. *T/T^* is strict-partial-homogeneous* when C is a subpart of C^* and $\text{Cn}(A')$ is a subset of $\text{Cn}(A)$ and $\text{Cn}(A')$ is a subset of $\text{Cn}(A^*)$.

9. *T/T^* is approximate-partial-homogeneous* when C is a subpart of C^* and $\text{Cn}(A')$ is a subset of $\text{Cn}(A)$, and $\text{Cn}(A') \approx \text{Cn}(A'')$, and $\text{Cn}(A'')$ is a subset of $\text{Cn}(A^*)$.

10. *T/T^* is strict-complete-inhomogeneous* when $\text{Cn}(A)$ is a subset of $\text{Cn}(\{A^* \cup A_R\})$ where A_R is a complete set of reduction axioms.[7] ($\{A^* \cup A_R\}$ is the set formed by the union of A^* with A_R).

11. *T/T^* is approximate-complete-inhomogeneous* when $\text{Cn}(A') \approx \text{Cn}(A)$ and $\text{Cn}(A')$ is a subset of $\text{Cn}(\{A^* \cup A_R\})$ where A_R is a complete set of reduction axioms.

12. *T/T^* is strict-partial-inhomogeneous* when $\text{Cn}(A')$ is a subset of $\text{Cn}(A)$ and $\text{Cn}(A')$ is a subset of $\text{Cn}(\{A^* \cup A_R\})$ where A_R is either a complete or partial set of reduction axioms.

13. *T/T^* is approximate-partial-inhomogeneous* when $\text{Cn}(A')$ is a subset of $\text{Cn}(A)$, $\text{Cn}(A') \approx \text{Cn}(A'')$, and $\text{Cn}(A'')$ is a subset of $\text{Cn}(\{A^* \cup A_R\})$ where A_R is either a complete or partial set of reduction/axioms.

One of the difficulties with these formulations is the lack of a definition for '$\approx$'. Later in this paper another concept will be introduced to improve upon the one used here, "is approximately equal to."

So far, we have only touched on state-transitions involving the theoretical component of a science. We have, as it were, concentrated on characterizing the developments of science which are a result of theoretical-theoretical interactions. The LE theory can be expanded to include theoretical-empirical interactions as well; that is, state-transitions also involving changes in the empirical base parameters. This, however, requires a more explicit characterization of the state of the empirical base and the nature of its interaction with the theoretical component.

According to the LE view the acceptability of the empirical base of a scientific theory is considered to be independent, perhaps not of all scientific theories, but at least of those under consideration. Alternative theories are to be comparable with respect to a common empirical base and their acceptability is dependent on the relation between the theory and its empirical base – on the degree to which the theory is in concordance with the empirical base. A basic requirement for a theory of concordance is that a theory and its empirical base must have languages such that statements in the theory and statements in the empirical base are comparable with respect to what is expressed. Sometimes this is put as follows; some of the terms of theory must share a common core of meaning with the terms of the empirical base.

According to the orthodox LE view, the empirical base consists of observation statements whose acceptability can readily be decided by straightforward observation procedures. Exactly what these procedures are by which the empirical base is established has generally been left unanalyzed by the LE's. In addition, the orthodox view makes a sharp distinction between observation terms (out of which observation sentences are constructed) and theoretical terms. What allows comparability between a theory and its empirical base is that they share observation terms whose meanings are somehow derived from experience independent of theory.

An enlightened LE view need hold neither of these theses. What is required of an LE view, is (1) that a theory and its empirical base be comparable in terms of what is expressed and (2) that the acceptability of the empirical base be independent of the acceptability of the theory but not vice versa. In other words, an LE requires (1) a theory of concordance between a theory and its empirical base, and (2) a theory of experimental procedures (including observational ones) by means of which an empirical base for a theory can be established independent of the theory.

A full-fledged LE theory of scientific change then will consist of at least three sub-theories:

(1) a theory of intertheoretic reduction
(2) a theory of concordance
(3) a theory of observational and experimental procedures.

Once we have a clear idea what constitutes a state-description and therefore a description of a change of state, we are faced with the problem of specifying what counts as a successor state. The problem is to find some way, if at all, of specifying which state transitions constitute an advance over previous states and which do not. It is a requirement of an enlightened LE theory of scientific change that it specify the conditions under which a transition S/S^* constitutes 'progress', i.e., the conditions under which S^* can be said to be an advance over S.

Within the LE tradition there are two basic ways in which one theory might be said to be an advance over another theory:

(1) if the successor theory is in better agreement than preceding theories with a common empirical base, and (2) if the preceding theories are synthesized or reduced to a more general or fundamental theory; that is, the preceeding theories are special cases or sub-parts of the successor theory.

Often, though not necessarily always, in cases of the first type the predecessor may be found to disagree with the empirical base and therefore be modified or replaced by a 'corrected' theory, though it may, of course, happen that a more accurate theory is discovered without the predecessor actually having been refuted first.

In cases of the second type of advance, it is usually assumed that the predecessor is derivable from the successor; what occurs is a reduction or an integration of one or more empirically adequate theories. What about mixed cases; i.e., cases where we might want to say of a predecessor that it was both 'reduced' and 'corrected' by a successor? Criticism of Nagel's theory of reduction, such as Feyerabend's, usually attempts to point out such cases by showing that strictly speaking the predecessor cannot be logically derived from the successor either with or without the help of reduction axioms, thus violating one of Nagel's formal conditions for reduction. An enlightened LE theory of scientific change such as the one we have been sketching would try to account for mixed cases by introducing some notion of an intermediate theory which is derivable from the successor and which gives approximately the same results as the predecessor.

Before considering these and other points in greater detail and in the light of Feyerabend's criticisms, I want to summarize the LE program and the principles on which it is based.

First of all, I am assuming the following two conditions of adequacy for any theory of scientific change:

(1) it must provide an account of how to describe states of science and therefore how to describe changes of state. A state will be specified when all the state parameters are assigned a value: a change of state is specified whenever a change in value of any state parameter is specified. This first condition amounts to saying that a theory of scientific change must specify what the state parameters are and what counts as a change in value for each parameter.

(2) In addition, it must specify what the conditions on state-transitions are such that if they are satisfied the transition constitutes an advance. Our theory must specify what sort of state changes count as advances.

An LE theory of scientific change specifies a state of science in terms of parameters describing an interpreted formal deductive system and an empirical base. State transitions are specified in terms of the degree of concordance between a theory and its empirical base and certain logical relations between a theory and its successor such as derivability and consistency.

Special conditions on an LE theory of scientific change then are, first, that a theory and its successor must be *comparable* (either in terms of their relative degree of concordance with a common empirical base or in terms of certain logical interrelations). In either case a theory and its empirical base must be *comparable* in terms of their degree of concordance. Finally, if a higher degree of concordance is to count as an advance, the acceptability of the empirical base must be *independent* of the acceptability of the theory. We will refer to these conditions respectively as:

(3) the comparability condition on a theory and its successor

(4) the comparability condition on a theory and its empirical base

(5) the independence condition on an empirical base.

What I have described in this section is a program for developing a theory of scientific change which is compatible with an LE account of science. The problem I have set for myself is not to try to fill in the details of the program but to indicate its potential and to see whether the program is a reasonable one. First, I am interested in whether there are good reasons for thinking that conditions (3)–(5) are not, should not, and/or cannot be satisfied. If I understand him correctly, Feyerabend claims to have arguments against all three of these conditions. In the next section I will be concerned with the question whether Feyerabend's arguments are damaging to the LE theory of scientific change outlined in this section. This paper will be restricted to some aspects of Feyerabend's arguments against the consistency condition and will defend the LE account against these arguments. Elsewehre, I will defend this theory against arguments centering around the meaning invariance condition.

III

In a number of articles Feyerabend has attacked the orthodox theories of reduction (Nagel's) and explanation (Hempel's). These theories, he says, require that two conditions be satisfied, both of which he argues against first by producing examples from the history of science which *do not* satisfy them and second by arguing that for methodological reasons they *should not* be satisfied.

What are the conditions which Feyerabend claims must be satisfied by Nagel's theory of reduction? He formulates them as follows:

> "Only such theories are then admissible in a given domain which either *contain* the theories already used in this domain, or which are at least *consistent* with them inside the domain; and meanings will have to be invariant with respect to scientific progress; that is, all future theories will have to be framed in such a manner that their use in explanations does not affect what is said by the theories, or factual reports to be explained.
>
> 'These two conditions I shall call the *consistency condition* and the *condition of meaning invariance*, respectively."[8]

The first point that ought to be made is that it is a mistake to say that, as they are stated, these are conditions of Nagel's theory of reduction. Feyerabend has said in the papers we are discussing that "my aim has been to present an abstract model for the acquisition of knowledge."[9] Nagel's aim, on the other hand, has been to provide an analysis of the notion of one theory being reduced to another. In the previous section we mentioned that reduction has been considered one of the basic mechanisms of change in science but not the only one.

Nagel's claim is not that in all transitions T/T^* is T reduced by T^*. Yet Feyerabend assumes that Nagel's theory of reduction comprises a full theory of scientific change. The first sort of consideration Feyerabend brings to bear is that *some* T-transitions are not Nagelian reductions. The most Nagel seems to be committed to is that some T-transitions are Nagélian reductions. This thesis is refuted when it can be shown that *no* T-transitions are Nagelian reductions. What Feyerabend does try to show is that the cases which Nagel uses as examples of reduction and certain other interesting candidates do not satisfy the conditions.

Feyerabend is right in the cases he considers about the following point: He says, for example, that strictly speaking the Galilean laws of planetary motion are logically inconsistent with the Newtonian laws of motion (and therefore not derivable from Newton's laws).[10]

Nagel has *prima facie* two options open to him. He may claim (1) that the analysis of reduction is satisfactory and so the examples from the history of science he chose to analyze (and probably most other *interesting* cases) are not cases of reduction or (2) that the examples are cases (paradigm cases if you wish) of reduction and so the analysis of reduction will have to be modified or supplemented in some way. No one seems to have considered the first a live option. The reason, I suppose, is that Nagel's theory of reduction was conceived of as an explication of cases like the transition from the Galilean laws to the Newtonian laws.

Feyerabend says, "What I have shown ... is that some very important cases which have been, or could be used as examples of reduction (and explanation) are not in agreement with the conditions of derivability."[11]

There is little doubt about this conclusion; one may want to quibble over its priority – giving credit perhaps to Popper or to Duhem. To conclude that some cases (important ones at that) of T-transitions are not Nagelian reductions is not to conclude that no cases are Nagelian reductions. Nevertheless Feyerabend's conclusion seems to be just as effective for the reason given above: Nagel's first option is not considered a live option.

Is Nagel then forced to modify his account of reduction in the light of Feyerabend's conclusion? I don't think so. Consider the derivability or consistency condition. Nagel has, it seems to me, a third option open to him. He might, claim that his account of reduction is satisfactory, and that in the important cases of T-transitions the successor theory is in fact inconsistent with the predecessor, but that the reduction process is not the only process at work in such transitions. An adequate analysis of such transitions will also include the other basic process of change mentioned previously; i.e., a T-transition where the successor is in better agreement than the predecessor with a common empirical base. We will call this process the *correction process.* A reconstruction of this important class of T-transitions will, in effect, consist of two transitions: first, a correction transition T/T', and second a reduction transition T'/T^*. What is claimed is that in the T-transition, T/T^*, there is or can be constructed an intermediate theory T' such that it is in better agreement than T with a common empirical base and such that it is derivable from T^*. The claim is not that historically, first the correction transition occurred and then the reduction transition occurred. It may have happened that way, but it need not have. All that is claimed is that for an important class of T-transition T/T^*, a T' is constructible such that T/T' is a correction transition and T'/T^* is a reduction transition.

Nagel's reply to Feyerabend then might be: my account of reduction is satisfactory, at least in terms of the considerations put forward here, but an account of T transitions only in terms of reduction is incomplete. Once we have included the correction process as well, we can see how T^* can be inconsistent with T (since T' can be inconsistent with T) and we have explained what it is to say that T has been 'corrected' and 'reduced' to T^*. There are then three basic classes of T-transitions: (1) T is corrected but not reduced by T^*, (2) T is reduced but not corrected by T^*, and (3) T is both corrected and reduced by T^*. Feyerabend's mistake was to assume that Nagel's account of reduction was a complete account of T-transitions. It was never meant to be that but it has been shown above in what way reduction plays an important role in most T-transitions, how it can play a role in T-transitions even when the successor is inconsistent with the predecessor,[12,13]

ABT Associates, Cambridge, Mass.

NOTES

[1] in *Mind and Cosmos* (ed. by Colodny), Pittsburgh 1966, pp. 41–85.
[2] *op. cit.* p. 43.
[3] I refer to such authors as Hanson, Feyerabend, Kuhn, and Toulmin.
[4] For an orthodox account of reduction, see Nagel's *Structure of Science*, N.Y. 1961, Chapter 11.

[5] L. Sklar, 'Types of Inter-Theoretic Reduction', *British Journal for the Philosophy of Science* **18** (1967) 109–24.
K. Schaffner, 'Approaches to Reduction', *Philosophy of Science* **34** (1967) 137–47.
R. Suszko, 'Formal Logic and the Development of Knowledge', in *Problems in the Phil. of Sci.* (ed. by I. Lakatos and Musgrave), Amsterdam 1968, pp. 210–22.
[6] Sklar, L., *op. cit.*
[7] For a formal specification of reduction axioms see Schaffner's 'Approaches to Reduction', *Philosophy of Science* **34** (1967) 137–47. Roughly, a reduction axiom identifies the concepts in T not shared by T^* with concepts in T^*. A complete set of reduction axioms specifies the relation of all the concepts in T not shared by T^* with concepts in T^*; a partial set relates only some, not all. The usual example of an informal reduction axiom is "Temperature is the mean molecular kinetic energy."
[8] 'Problems of Empiricism', in *Beyond the Edge of Certainty* (ed. by R. Colodny) Englewood Cliffs 1965, p. 164.
[9] 'Reply to Criticism', in *Boston Studies in the Philosophy of Science*, Volume II (ed. by R. S. Cohen and M. Wartofsky), N. Y. 1965, p. 223.
[10] For a good analysis of this case, see Popper's 'The Aim of Science', *Ratio* **1** (1957) pp. 24–35, and Duhem's: *The Aim and Structure of Physical Theory*, Chapters 9 and 10.
[11] 'Explanation, Reduction, and Empiricism' in *Minnesota Studies in the Philosophy of Science*, Volume III (ed. by H. Feigl and G. Maxwell) Minneapolis 1962, p. 60.
[12] This account relates to the Sklar account as follows: A reduction transition might be thought of as, what was earlier called strict-transitions (complete or partial, homogeneous or inhomogeneous). The approximate-transitions referred to earlier would be replaced by a correction transition. Instead of 'A is approximately equal to B,' we have 'A is in better agreement with a common empirical base than B'.
[13] Putnam in 'How not to Talk About Meanings', in *Boston Studies in the Philosophy of Science*, Volume II, p. 206–07, says,
"It is perfectly clear what it means to say that a theory is approximately true, as it is clear what it means to say that an equation is approximately correct: it means that the relationships postulated by the theory hold not exactly, but with a certain specifiable degree of error. In short, it means that the theory is not true, but that a certain *logical consequency* of the theory, obtained, for example, by replacing 'equals' with 'equals plus or minus delta' is true. Such a logical consequency may be called an approximation theory." Putnam's approximation theory as characterized here will not do for an intermediate theory T', as Feyerabend has pointed out in his 'Reply to Criticism' (*op. cit.*, pp. 229–30), for the reason that not just any approximation theory will do. Only those approximation theories which are an advance over the preceding theory; i.e., are in better agreement with a common empirical base, will do.

IX. The Future of Philosophy of Science; Theory in the Social Sciences

RONALD N. GIERE

THE STRUCTURE, GROWTH AND APPLICATION OF SCIENTIFIC KNOWLEDGE: REFLECTIONS ON RELEVANCE AND THE FUTURE OF PHILOSOPHY OF SCIENCE

I. INTRODUCTION

I would like to make it clear at the outset that this paper is not about a problem *in* the philosophy of science; it is *about* the philosophy of science itself. Furthermore, the discussion will not be confined to developments *internal* to philosophy of science but will include considerations of *external* factors and their possible relations to the philosophy of science. The justification for this departure from standard practice is simply that the issues to be discussed seem to be enough on many peoples' minds to justify bringing them out into the public arena.

II. ANALYSIS AND RELEVANCE

There is no need to dwell on the all too familiar *external* factors labeled by well-known catch phrases such as 'over-population', 'pollution', 'violence', 'runaway technology', 'social disintegration', etc., etc. Although one may debate their precise nature and speculate on their eventual relative significance, the seriousness of these problems, especially in conjunction, can hardly be doubted.

For several years now, and for many years to come, the main question concerning relations between an academic field and external problems is this: How *relevant* are these studies to the resolution of the well-known external problems within a reasonable time-span? If one is to obtain a general answer to this question for the philosophy of science, he must have some characterization of what philosophers of science are now doing. The most specific characterization on which there is a wide-spread consensus is that philosophers of science are engaged in the *analysis of scientific concepts.* This includes most Anglo-American philosophers *except* the Popperians. I will return to this latter group below.

Under the umbrella of concept analysis I find two main cross-classifications. One is based on the distinction between *general metascientific concepts* and *substantive concepts*. The chapters of Peter Achinstein's book *Concepts of Science* and the papers in Baruch Brody's recent anthology,[1] for example, all concern completely general metaconcepts like explanation, confirmation, theories and observation, models, etc. But some philosophers of science are primarily interested in concepts pertaining to a particular subject matter, e.g., Adolf Grünbaum and others are concerned with congruence and simultaneity, Henry Margenau and his students investigate various concepts of quantum theory, Patrick Suppes writes on concepts of learning in psychology.

A second main division among concept analysts may be made on the basis on the *type of analysis* being done. Roughly speaking the kinds are these: formal, informal (including ordinary language), and historical. A liberal view of the formal approach includes Grünbaum and Carl Hempel as well as Rudolf Carnap and Suppes. Herbert Feigl, Wilfred Sellars, and Stephan Körner are more informal, but do not lean as strongly on ordinary language as do Michael Scriven and, even more so, Achinstein. N. R. Hanson, Stephen Toulmin, Thomas Kuhn and Dudley Shapere would be representatives of the historical approach.

Not what about the relevance of all this analysis to the eventual solution of external problems? At best it is weak and indirect. The clarification gained by philosophical analysis could perhaps be of some help either to those using various sciences to attack the external problems or to those thinking about the larger questions raised by an increasingly technological society. But anyone *outside* the philosophy of science will have great difficulty applying what he learns by reading philosophers of science either to substantive scientific inquiries, or to relevant applications of scientific knowledge, or to questions concerning the nature of a society highly dependent on science and technology. And there are very few philosophers of science who are themselves interested in filling this gap, i.e., in doing what might be called 'applied philosophy of science."

Postponing the question whether the philosophy of science *should* be more relevant to external problems, there are several ways it *might* be made more relevant without abandoning the concept analysis framework One way is by devoting more attention to concepts in biology, psychology, and the social sciences – the sciences that now seem most directly relevant to current external problems. A second way is by engaging more directly in the application of philosophical analysis in actual scientific concepts and in the further process of applying scientific knowledge. I will not pursue these suggestions, however, because I think there is a better way of viewing the philosophy of science as a whole. In addition to providing a new topography of the field, this rival account provides a clear picture of what philosophers of science should do *if* they want to be more relevant.

III. THE STRUCTURE, GROWTH AND APPLICATION OF SCIENTIFIC KNOWLEDGE

I propose to make the main divisions within the philosophy of science in terms of three distinct aspects of scientific knowledge. (1) The logical-mathematical *structure* of a fixed body of knowledge as represented, for example, in a scientific theory. (2) The *process* by which new knowledge is *acquired*, i.e., the process by which science *grows*. (3) The process by which scientific knowledge is *applied*. I will now briefly indicate what each of the three divisions is intended to include.

Studies of the logical-mathematical structure [2] of scientific knowledge are of two types, 'pure' and 'applied'. Pure studies of the structure of knowledge are concerned, first, with the logical-mathematical structure of *theories in general*, regardless of subject matter, and, secondly, with *meta-questions* about theories, e.g., the nature of explanation, empirical significance, causality, etc. These are all basically questions about theories, e.g. What requirements must be satisfied by a theory and a set of facts so that the theory can be said to *explain* the facts? Those who view the task of philosophy of science as the analysis of concepts tend to view these meta-structural questions as comprising the central core of the field. On the present analysis these questions represent only one of several important kinds of questions for a philosopher of science.

The defining characteristic of *applied structural studies* is that they are concerned with specific scientific theories and therefore with *substantive* scientific concepts. Here also we might distinguish two types of inquiry. One might be primarily interested in drawing out some intrinsically interesting consequences of a *given* theory, e.g., consequences of special relativity for our conception of spatio-temporal relations. On the other hand, one may be primarily concerned to *formulate* a body of knowledge that has not yet been formulated sufficiently rigorously. Such studies are commonly called *foundational* studies, e.g., studies in the foundations of quantum mechanics.

A vast amount of the legitimate professional literature in Anglo-American philosophy of science since 1920 has been concerned with the *structure* of scientific knowledge. According to the present analysis, there is nothing *intrinsically* wrong with this tradition. Indeed, thanks to recent developments in logic and mathematics, e.g., model theory, structural studies are now more interesting than ever. The only trouble is that there are many important questions about scientific knowledge that are not solely about the logical – mathematical structure of a fixed body of knowledge.

As emphasized by Popper and his associates, e.g., Paul Feyerabend and Imre Lakatos, a notable fact about scientific knowledge is that it grows, i.e., we are constantly acquiring new knowledge. I will refer to studies of the process of acquiring knowledge as studies in the *methodology of science*. Here the word 'methodology' is clearly intended in its 19th century 'process' sense, according to which it has something to do with *methods*, and not in its 20th century sense which includes all meta-scientific inquiry. 'Heuristics' is another current name for the intended type of methodology.[3]

Now it seems that methodological studies can also be classified as either pure or applied. Among the tasks for *pure methodology* of science I would include the construction of a general theory of the efficient acquisition of knowledge as well as theories of measurement and inductive reasoning.

Meta-questions about measurement, inductive logic and knowledge acquisition would also be included in pure methodology. The task of *applied methodology* is to determine optimal strategies for the acquisition of desired new knowledge in specific scientific fields (or subfields). It is assumed that such inquiries would require some familiarity with the present state of knowledge in the field in question.

From a pragmatist viewpoint, my distinction between the structure of knowledge and the process of acquiring knowledge may seem unjustifiably sharp. Here I can only state my conviction that it is the pragmatist viewpoint and not the way the distinction is drawn that is at fault. I can see no non-trivial logical connection between the *content* of a theory and the *process* by which it came to be conceived and perhaps even 'accepted'.

What a theory asserts, i.e., what one learns from the relevant textbooks, is one thing; how this theory came to be held, which may be learned from historians of science, is something else.

It may also be objected that a general theory of efficient knowledge acquisition is unattainable. The discovery of important new knowledge depends on the inarticulable intuitions of the trained scientist. But this must be an empirical claim, and at the moment it is far from empirically established. Furthermore, even if scientific intuition must intervene at some point, this does not preclude there being significant parts of the scientific process open to the methodologist. Methodological inquiries may be attempted at many levels, from the daily research strategies of the individual scientist in his own lab, to the decisions of an administrator choosing among the demands of physicists, biologists and sociologists for federal funds. It is hardly to be expected that methodological studies will be equally fruitful at all levels. The most fruitful levels for methodological inquiry by philosophers of science will have to be determined empirically. Finally, that there are no good, detailed methodologies currently available may be due mainly to lack of effort in the past several generations and not to the basic impossibility of the task.

My inclusion of inductive logic in the pure methodology of science may raise a few eyebrows, and, indeed, this classification is not philosophically neutral. In particular, it rules out the Carnapian program for basing inductive logic on a probability measure defined over the language of science. I assume, on the contrary, that an inductive logic tells us which hypotheses may legitimately be added to the body of scientific knowledge, and I would go on to claim that certain parameters in the logic must ultimately be determined by *methodological* factors, e.g., the *rate* of knowledge acquisition. But these are clearly specialized issues which cannot be dealt with here.

Turning finally to the third major division in the philosophy of science, studies concerning the process of *applying knowledge*, which I shall call the *methodology of technology*, seem also to be either pure or applied.[4] The *pure* methodology of technology would include a general theory of the rational application of knowledge. At the moment this would seem to involve first order and meta-questions about parts of decision theory, systems theory and perhaps value theory as well. Attempts to develop a general theory of technology assessment would also be located here, as would general investigations into relations between technology and the rest of society. The *applied* methodology of technology would get one into the study and evaluation of particular applications including such well-known technological ventures as the space program, atomic generators, and *SST*'s.

To date there is very little philosophical literature that could be considered explicit studies of the application of knowledge. A few things have come out of Häken Törnebohm's Institute for the Study of Science. Joseph Agassi and Mario Bunge, both of the Popperian school, have also written briefly on the subject. But, unless one counts a lot of literature on induction and statistical explanation, which is only *implicitly* concerned with the application of knowledge, the methodology of technology is philosophically nearly virgin territory.[5]

The above divisions are summarized, with some examples, in Table I.

TABLE I

	Structure	Growth	Application
Pure Studies	Nature of theories	Inductive logic	Decision theory
Applied:			
Physics	Simultaneity Quantum logic	Role of altern. theories	Atomic generators
Biology		Inference in genetics	Genetic surgery
Psychology	Perception		Teaching machines
Linguistics			
Sociology	Functionalism		
Poly Sci			

It will be noted that there is no place in this topography for distinguishing *kinds* of analysis, e.g., formal, informal, or historical. This seems to me entirely appropriate. The kind of analysis required depends mainly on the problem at hand. No one kind of analysis is necessary or sufficient for the solution of all legitimate problems in the philosophy of science. On the other hand, I am inclined to apply the pattern found in the sciences and to take increasing formalization as an indication of increasing maturity.

Thus the fact that concern with the growth of knowledge is now concentrated among those employing the informal and historical approach seems to me evidence of the relative immaturity of this area.

The recent emphasis on kinds of analysis is not only unnecessary, it is often harmful in that it leads people to neglect the structure – methodology distinction and thus to manufacture unnecessary controversies. Popper's well-known dispute with Carnap over confirmation is a case in point. I agree with Popper's *conclusion* that Carnap's program for confirmation theory is fundamentally mistaken, but most of Popper's *reasons* for this conclusion are made irrelevant by a failure to distinguish clearly enough between the *logical structure* of a language and the *methodology* of science. I suspect a similar failure lies behind much of the current literature on meaning and meaning change.

IV. JUSTIFYING THE NEW TOPOGRAPHY

Those who see philosophy of science as primarily concerned with questions about the structure of knowledge, especially meta-questions, will no doubt object to incorporating the methodology of science and technology into the philosophy of science proper. One traditional reason for this rejection has been the sound Humean point that there can be no Baconian method, i.e., a method guaranteed to yield true theories. But there are at least two conceptions of methodological rules compatible with the Humean insight. One is that the rules of methodology *define* the concept of a *rational method* independently of any reference to success in actually acquiring new knowledge. A second view is that methodological rules are based on *empirical* truths, and generally *probabilistic* truths, concerning the acquisition of knowledge. My own preference is for taking methodological truths to be *empirical*. But this brings up another standard reason for not counting methodology as part of the philosophy of science, namely, that it is *merely* an empirical inquiry, i.e., part of psychology, or sociology, or the history of science. But if methodology is the study of *optimum methods* for acquiring new knowledge, this surely is not *just* part of psychology, sociology, or history, though psychology and sociology may make important contributions to methodology. So methodology truths must be empirical in a broader and more profound sense than that which makes psychology an empirical science. But in a broad sense even pure studies of the structure of scientific knowledge are empirical. This may require further argument.

In the first place, one might argue on generally Quinean grounds that all truths are empirical, including those about the nature of theories, explanation, causation, etc.. But one may not need to go so far. Anyone who holds that meta-structural questions are answered by providing an *explication* of some concept need only be reminded that any explicatum must satisfy a descriptive requirement. It must *resemble* the explicandum. Finally, one might simply ask how it could be acceptable, for example, for one to advance *a priori*, certain requirements on any scientific explanation when in fact no accepted explanations in any specific field could even be viewed as approximating satisfaction of the stated requirements. For these reasons it seems difficult to bar methodological studies from the philosophy of science on the ground that they would be 'merely empirical' rather than properly 'logical'. The same argument applies to the methodology of technology as well.

An immediate corollary to the above conclusion is that some work on applied studies of various kinds is necessary if one is to have well-founded conclusions regarding more general (pure) issues. This should hold in all three of the major areas, i.e., studies of structure, growth and application. Thus in addition to the external reasons to be discussed below, there are *internal* reasons for placing a reasonable emphasis on applied studies in the philosophy of science.

One may object that these latter arguments are unnecessary in that the Popperians, Kuhn and others have made at least the growth of knowledge *de facto* a topic for professional philosophers of science. But it is important to establish that the remaining objections to including genuine methodology within the philosophy of science are not philosophical but are simply a matter of tradition supported by personal interest and inclination. More on this in a moment.

As a final objection to incorporating methodology (heuristics) into the philosophy of science, one should mention the view that a methodology would constitute a 'logic of discovery' which, in spite of Hanson's efforts, many philosophers of science think non-existent. But even if one grants that Hanson failed completely in his attempt to distinguish a logic of discovery from a logic of justification, it does not follow that there are no discoverable truths concerning the efficient acquisition of scientific knowledge. If these truths are empirical, then I would hesitate to consider any rules of strategy based on them as a 'logic' of any kind. If methodological truths were to be *a priori*, perhaps 'logic of discovery' would be a suitable term. But here the issue is mainly over the proper extension of the word 'logic' and thus not worth disputing at great length.

V. RELEVANCE RECONSIDERED

Still postponing the question, *Should* philosophy of science be more relevant to the solution of external problems?, it seems obvious that relevance increases as one moves from structural studies to the methodology of science and technology. Relevance also increases as one moves from pure studies, to applied studies, especially those in the softer sciences. Surely pure structural studies and the foundations of physics are only very weakly and very indirectly relevant to solving today's most pressing external problems.

At this point an example might be helpful, and a nice case is provided by the current state of oceanography.[5] The really big problem in oceanography is to develop an account of the dynamics of ocean circulation, but few oceanographers work on this problem directly because nobody has the capability for gathering sufficiently extensive data to test any really general hypotheses. Recently, however, the threat of pollution has led several national and international agencies to consider setting up a world-wide monitoring system of several hundred deep-sea buoys with a fleet of perhaps 20 ships to tend them. Oceanographers who have been asked what should be measured, where, at what depths, etc. have to reply that they do not really know because they lack a good account of ocean circulation. These scientists are inclined to view the whole project as another expensive boondoggle of little scientific value. The bureaucrats, on the other hand, see the scientists as too interested in their own narrow individual research projects and unresponsive to the general social need.

That this dispute is to a great extent methodological can be seen by considering the question, How might the needs of both groups best be met? The answer to this question will require knowledge of oceanography, and some details of the answer may rest solely on scientific judgment. But detailed methodological principles and a theory of scientific development are also needed. Unfortunately, the existing philosophical accounts are not especially helpful. The rejection of naive inductivism for the method of conjecture and refutation does not get us very far. Lakatos' views on research programs may sound plausible applied to the 'heroic' programs of Newton, Einstein and Bohr, but they are of little help here. Kuhn's view would be that since the forces of scientific development are *internal* to the field, the attempt to force development into socially desirable directions will be scientifically fruitless at best. But this seems plainly wrong. Surely there is some rational way to use all that money in ways that will be both scientifically and socially beneficial.

This example, and methodology in general, need more discussion, but it is time to face the tough question, *Should* philosophers of science put greater emphasis on the more relevant areas of their field? Granting that there are no *philosophical* objections to such a shift in emphasis, the question focuses attention on philosophical tradition as well as on personal interest and inclination.[7]

Turning first to the personal level, many philosophers of science now face at least a potential conflict between their social conscience and their interest, training and experience in dealing with pure or applied structural issues. This conflict may be eliminated, of course, by insisting that, properly judged in terms of intellectual enrichment rather than merely practical benefits, philosophers of science already contribute their fair share to the general welfare. But such traditional justifications of scholarly pursuits are suspiciously self-serving and are beginning to sound hollow to the general public, the federal government, state legislators, alumni, students, and university administrators, in roughly that order. It seems a safe prediction that during the next decade there will be diminishing resources available for theoretical studies of very little relevance to the problems of society as a whole. Unless there are some changes, this means less money for research and new faculty in the philosophy of science. It also means less support for graduate students and fewer jobs for those entering the field. These latter difficulties, however, may be offset by their being fewer good people seeking to enter the field. In this regard, I take the great interest shown by beginning students in a book like Kuhn's *The Structure of Scientific Revolutions*, despite its philosophical shortcomings, as a symptom of a desire for relevance. If the field does not develop in these directions, I fear fewer students will go on in the philosophy of science.

Another way to mitigate the conflict between interests and conscience is to adopt an extremely pessimistic or an extremely optimistic view of the external crises. If nothing anyone can do will ward off the coming disaster, then one may be justified in continuing to do whatever he likes as long as possible. Similarly if one thinks the present crises are due mainly to instant mass communications and a lack of historical perspective. But the available evidence provides little support for either extreme, especially not the latter. The rational position is a qualified pessimism: Things are bad and will probably get worse, but the middle to long run prospects are not yet hopeless. On this assessment, the potential conflict between interest and social conscience remains.

Another resolution may be effected by dividing one's energies. This strategy is exemplified in the extreme by Noam Chomsky. But this course requires a high tolerance for intellectual schizophrenia as the connections between one's professional and non-professional concerns become ever more tenuous. To those who do not like living fragmentary lives, the opportunity to remain a full-time philosopher of science while contributing fairly directly to the social welfare may be quite inviting. The chief remaining obstacle to this solution is philosophical tradition.

The dominant movement in 20-th-century Anglo-American philosophy of science was forged by Carnap and others out of elements of traditional empiricist epistemology, the 'new-physics', i.e., relativity theory and Quantum Mechanics, and Russell's logic together with his vision of philosophy as the analysis of language. Nor was this the first time in the history of philosophy that epistemology, physics, and logic came together to form a powerful movement. Consider Descartes or Kant. Thus it is no surprise that today many philosophers of science tend to see philosophy of science as merely a branch of epistemology, to emphasize questions concerning the structure of knowledge, and to equate the philosophy of science with the philosophy of physics. But there is another tradition, one which emphasizes social and political philosophy as much as epistemology, methods of acquiring and applying knowledge as much as the structure of knowledge, and the social sciences as much as physics. This tradition includes at least Bacon and the 19th century positivists, e.g., Mill. It was even represented within the Vienna Circle in the person of Otto Neurath. My overall recommendation is that we revive this tradition – with appropriate modifications. My hope is that work in this alternative tradition will prove to be both intellectually stimulating and socially relevant. At the moment, of course, this is still more hope than promise.

VI. CONCLUSION

Some will view the above as mildly heretical while probably more will think it too establishmentarian. I myself would have to judge it as being the work of an establishment 'radical-liberal'. But at the moment it is not so important that people agree or disagree as that they seriously consider the nature and relevance of the field as a whole. Philosophy of science cannot be the only field not now suffering the strains related to external concerns, though it might be among the least self-conscious when it comes to dealing with their consequences.

Indiana University

NOTES

[1] P. Achinstein, *Concepts of Science*, Baltimore, Md, Johns Hopkins Press, 1968. B. Brody, *Readings in the Philosophy of Science*, Prentice-Hall, Englewood Cliffs, N.J., 1970.

[2] I assume throughout that the logical-mathematical structure of a set of statements includes *semantical* as well as syntactical relations.

[3] Lakatos would deny that he is contributing to heuristics. Nevertheless, his latest published writings can be viewed as containing general heuristic maxims. See 'Falsification and the Methodology of Scientific Research Programs' in *Criticism and the Growth of Knowledge* (ed. by I. Lakatos and A. Musgrave), Cambridge University Press, Cambridge, 1970.

[4] Note that the term 'application' is used in two distinct ways. One is to distinguish pure (or 'meta') studies from applied (or 'substantive') studies. The other is to distinguish the structure and acquisition of knowledge from the application of knowledge. Thus there can be applied studies of the application of knowledge. One might get by fairly well and perhaps avoid some confusion by relying on the terms 'meta' and 'substantive' to make the former distinction.

[5] H. Törnebohm, *Science of Science*, Institute for the Theory of Science, Gothenburg, 1969. J. Agassi, 'The Confusion Between Science and Technology in the Standard Philosophies of Science', *Technology and Culture* 7 (1966) 348–66. M. Bunge, *Scientific Research*, Springer, New York, 1967. These authors have of course been concerned with the methodology of *science* as well. There are also vast amounts of relevant literature *outside*, or at best at the periphery, of the philosophy of science. Most philosophers of science have at least heard of some of the work of C. West Churchman in Operations Research and related areas. Some of this literature may provide a good starting point for more philosophical studies in the methodology of science and technology.

[6] For an extended discussion and further references see H. Stommel, 'Future Prospects for Physical Oceanography', *Science* **168** (1970) 1531–37. The choice of a 'non-fundamental' science for this example is deliberate. It seems likely that methodological studies will be more fruitful, at least initially, in such sciences than in basic areas of physics.

[7] There are those who would claim that, at the present, anything which contributes to the solution of our many external problems merely prolongs the survival of an evil political system, i.e., the U.S. Government. That major changes in our society are necessary goes without saying, but I do not believe that our present system is *totally* evil and that constructive change is impossible without revolution. Thus in what follows I simply assume that much good can be accomplished within the present political framework. My greatest fear is that those who (irrationally) think otherwise will eventually succeed in making their present claims come true.

STEPHEN TOULMIN

FROM LOGICAL SYSTEMS TO CONCEPTUAL POPULATIONS

I

It is hardly news today to report that philosophers of science share a widespread feeling that methods of analysis which have served them well for half a century are reaching the limits of their usefulness. During much of the twentieth century, they have been concentrating on those aspects of science which lend themselves most readily to analysis using the tools of formal logic; they have given detailed, sophisticated accounts of the ways in which the general principles of a natural science are – or might conceivably be – applied to entail statements about its particular phenomena; and they have done much to make clearer the part played in natural science *by logical systems*. The resulting debate has been complex and on a high level of abstraction: primitive terms, protocol-statements, hypothetico-deductive systems, inductivism ... verification, falsification, confirmation, corroboration, refutation .... From year to year the argument has moved on, and the grounds of debate have shifted. It has all been very exciting.

Very exciting, yes: but also oddly inconclusive – and this is where the current malaise begins. On the one hand, the intellectual program on which 'formal-logic-oriented' philosophers embarked around 1920 – what I shall here call *the logicians' program* for philosophy of science – has been substantially carried through; yet, on the other hand, many urgent questions in the philosophy of science remain entirely unanswered. About the formulation of deductive arguments in science, about the logical relations between singular evidence-reporting propositions and universal hypotheses, and so on and so on, what more is there to be said? By now all the issues have (it seems) been canvassed *ad nauseam*; what remains to be done in this field is largely intellectual filigree. And, though 'a thing of beauty is a joy forever,' this kind of filigree work is – undeniably – yielding diminishing philosophical returns, while leaving many acute and important problems unsolved. Originally, it was hoped to bring all serious

Boston Studies in the Philosophy of Science, VIII.

philosophy of science within the logician's ambit by sufficiently ingenious extensions. Now, this appears increasingly unrealistic, and nobody is entirely clear where to turn next.

My aim here is to diagnose the sources of this uncertainty, and to suggest an alternative direction of advance. Let me first try to pinpoint the group of problems which most stubbornly resist resolution by formal methods of analysis. In a single word, these problems have to do – in one way or another – with the *reasonableness* of science. For some years, logicians invoked a distinction between 'discovery' and 'justification' to sweep these difficulties aside: justification alone was the concern of logic, and therefore of philosophy, while discovery was a matter for psychologists. Then, for a while, there was a counter-attack by Russ Hanson and some others, who pressed the question whether there was not – after all – a 'logic of discovery'. Yet this way of posing the difficulties proved, in turn, to be question-begging: if ever a philosophical issue involved a built-in cross-purpose, that did! So, today, the difficulties remain, and our first step must be to try and repose them more adequately.

Their crux can be stated quite concisely. The logicians' program for philosophy of science deals with the 'structure' of arguments in which *pre-existing* concepts are employed or misemployed: employed (that is) correctly or faultily. This structure is analyzed in a formal manner, so that 'good science' is made to appear a matter of doing one's scientific arguing *impeccably* – i.e. of avoiding inferential blunders, and remaining 'logical'. Inevitably, this approach conceals the distinction between 'logicality' and 'rationality'. For, in practice 'good science' involves not merely doing sums and inferences *right* – i.e. aright, rightly or correctly. It also, more importantly involves doing the *right* sums and inferences – i.e. recognizing what particular concepts and forms of argument are relevant and applicable in any problem-posing situation. Correspondingly, scientific *failure* springs most typically from irrelevance or misjudgment in matching of concepts to situations, rather than from any subsequent logical blundering. I myself doubt, indeed, whether a logical slip or formal error ever amounts by itself, to a failure of scientific 'reasonableness' or 'rationality'.

Instead we need to distinguish sharply between the 'rationality' of of science or scientists and the 'logicality' of scientific arguments. In the actual *doing* of science, logicality represents a kind of intellectual accountancy: concerned with keeping our inferences neat and tidy, avoiding formal blunders, doing our sums aright – and all these competences involve employing already existing theories and concepts in accordance with 'established' procedures. The rationality and reasonableness of scientific thinking, on the other hand, have to do with the ways in which the application of existing concepts is extended, or with the ways in which new concepts are developed, in order to solve outstanding scientific problems. And this is a matter, not of following established procedures rightly (i.e. impeccably) but of recognizing what are the right (i.e. relevant) procedures to employ in a new situation, and of devising new explanatory methods and inferences, of kinds that have previously had no scientific use – and *a fortiori* no 'established' scientific use. In science as elsewhere, that is, man shows that he is acting '*reasonably*' and is 'open to reason,' not by his fixed ideas or by his competence in *standardized, stereotyped* inferences, but by the way in which he *extends and modifies* his ideas and intellectual skills.

To summarize our central problem in a phrase, let me borrow an image from Professor Causey. Analysis in logical terms can give us an instantaneous 'snapshot' of faultless scientific inference, within the scope of existing concepts and methods; but this still leaves us without the 'moving picture' we need if we are to understand the *rational procedures* by which the scope of science is effectively extended. And this is no accident. For the *rationality* of the conceptual changes by which new science is made, and the reasonableness of the scientists who make them, could never have been defined or analyzed as matters of 'faultless inference' alone.

II

This distinction is so elementary that one may want to ask how the problem of rationality or reasonableness, in science ever dropped out of sight in the first place. The prime figure was (I believe) Gottlob Frege. The drive towards mathematical Platonism which has been so powerful in twentieth-century philosophical analysis was triggered by Frege's reaction against the conflation of logic with cognitive psychology. The detailed background to Frege's campaign, in the work of Wundt and the other philosopher-psychologists of the time, would be a first-rate thesis subject for some talented graduate student, but there is room here to deal with only one aspect of the story.

The point I shall pick on is this. The logicians' program for philosophy of science – and for philosophy generally – had a multiple paternity, and its progenitors were not all of one mind. On the contrary, we can order their interests along a spectrum. At one extreme, the program owed a lot to the example of pure mathematicians: especially to Klein, whose techniques for 'mapping' different branches of mathematics on one another inspired Russell's attempts to 'map' arithmetic onto pure logic, and also to Hilbert, whose mathematical program of 'formalization' gave an impulse to the subsequent campaign of 'axiomatization' in philosophy of science. In the center, there was a group of philosophers with a first training in pure mathematics, and mathematicians with philosophical preoccupations, beginning with Frege and Russell, Peano and Vailati. Their concerns were partly internal to mathematics, but they were basically philosophical. In Frege's case, the development of symbolic logic was spurred on by the pursuit of 'pure concepts', untainted by history or psychology, in Russell's by the quest for a genuinely 'perspicuous' language or symbolism, in which the 'true logical forms' of propositions would be immediately evident and explicit.

At the other end of the spectrum there stands a third party, comprising philosophers with a first training in physics, together with physicists interested in philosophical aspects of their own work. Hertz and Boltzmann obviously belong in this group; I myself would also include Wittgenstein; and for good measure one might add their most significant forerunner, Immanuel Kant. In writing and talking about the philosophy of science, all these men focussed on the notion of 'representation': a term that English writers have used to translate both *Vorstellung* – the word employed by Kant, Schopenhauer and Boltzmann – and also *Darstellung* – the variant employed by Hertz, and picked up from him by Wittgenstein. From Kant's *Critique of Pure Reason* right up to the *Tractatus Logico-Philosophicus*, in fact, the central notion in German-language discussions of perception and scientific explanation was this notion of 'representation'. Where ancient Greeks had summarized the task of science in the phrase *sozein ta phainomena*, the corresponding nineteenth-century German thesis was that the task of the human mind is to compose a 'representation' of phenomena.

The multiple paternity of 'logical analysis' had important – and largely overlooked – consequences, which become apparent if one asks (for instance) what purpose was served, for the members of each party, by the procedure of *axiomatization*. For pure mathematicians, of course, axiomatization was its own reward. Integrating a set of previously unrelated theorems into an axiomatic system was for them an *end in itself*, not just a part or means of achieving some further non-mathematical aim. Producing a more comprehensive, powerful and elegant axiom-system was an *intrinsic* mathematical success, quite regardless of its possible applications, since this was an essential element in the pure mathematician's proper intellectual mission. For a physicist like Hertz, the situation was absolutely different. In axiomatizing the science of mechanics, Hertz's aim was not to bring off a feat of formal virtuosity alone: he was not just a pure mathematician who happened to work in the field of 'rational mechanics'. What Hertz stood to gain from a successful axiomatization was someting substantial *within physics*. Axiomatization was to be the means to a larger end; and that end was the more effective conceptual organization of *physical* understanding – or, as he himself put it, a better *Darstellung* of mechanical phenomena.

Even apart from the central philosophical figures, i.e. Frege and Russell, this difference in attitude is already significant. Just because Hertz regarded the axiomatization of mechanics as a scientific means, rather than as an end-in-itself, it was – for him – only one of several distinct alternatives. Far from anticipating Russell's belief that a 'perspicuous' symbolism would make explicit 'the one true logical form' of propositions, Hertz described his own axiom-system as one possible *Bild, Darstellung* or 'model' among others, whose merits must be judged in competition with its rivals in terms of extrinsic explanatory fruits. When so judged (Hertz saw) his axiom-system might prove superior only in certain contexts rather than others, and for certain purposes rather than others, so he did not have to claim that it was the uniquely meritorious representation of mechanical phenomena. Rather, it was *a* way of doing *one* of the jobs of theoretical physics; and for the legitimate purposes of other contexts, it might be proper to employ quite different methods of 'representation' – even Frege's despised 'mental images' or *innere Scheinbilder*.

Hertz was historically influential, in turn, for two reasons. He was the first, the most lasting, and in some ways the most profound philosophical influence on Wittgenstein, in both his contrasted periods; and in addition the example of his *Principles of Mechanics* was regularly appealed to, in the 1920's and 30's as the prime authority for extending the logicians' program out of the philosophy of pure mathematics into the philosophy of natural science. As it happened, however, the philosophers who had the largest stake in the logicians' program saw Hertz's work through the eyes of mathematicians, rather than of working physicists, and so overlooked the deep differences of aim between (say) Hertz and Peano. Meanwhile, that devotee of Hertz, Ludwig Wittgenstein, was grumbling away in the background – Waismann, for instance, reports him criticizing Frege and Russell for being excessively preoccupied with the application of logical symbolism to mathematical examples – but his protests went unheard and disregarded.

The logicians' program for philosophical analysis and philosophy of science accordingly developed in a middle ground, somewhere between the extremes of Klein and Hertz. In both Frege and Russell, the heart of the program was a kind of philosophical reformation. Frege's *Begriffschrift*, like Russell's propositional calculus, was intended to serve as the formal core of a disinfected language, and in this language alone 'pure concepts' or 'true logical forms' could be expressed as unambiguously as that reformation required. So for Russell, in direct contrast to Hertz, the propositional calculus was not just one among several alternative ways of getting an intellectual job done – on a par with (say) using mental images. On the contrary, this was to be the central element in the one truly perspicuous symbolism – the symbolism (to use Frege's revealingly Platonist phrase) of 'concepts in their pure form.' At its most ambitious, the resulting philosophical development culminated in the Unity of Science movement of the 1930's, whose methods and principles combined elements from both extremes. On the one hand, its supporters declared that their aims were *scientific*: the concepts of natural science were to be reorganized on new principles, and integrated into the true *corpus* of positive knowledge. On the other hand, they treated axiomatization in the spirit of pure mathematicians, as self-justifying, and so as an end in itself. Like Frege and Russell, they saw pure mathematics as providing the formal model for *any* perspicuous representation of nature. By adding fresh theoretical terms and definitions (they supposed) we might progressively hope to extend the scope of Russell's logical calculus to embrace not only pure mathematics, but also theoretical physics, genetics, and even – with sufficient ingenuity – physiology and psychology. In this way, the specific virtues and merits of pure mathematics were taken as binding on all intellectual disciplines, and most of all on any natural science worthy of the name. (How Plato and Descartes would have cheered!)

III

So much for the historical background: let me now come to my diagnosis. For too long, I claim, philosophers of science have been blind to the differences between *rationality* and logicality. As a result, they have restricted their attention to those aspects of scientific investigation which can be construed in formal terms, as involving logical systems and strict inferences; and they have been unable to give an account of those other aspects which – while 'rational' enough, in all conscience are non-formalizable. Hence the present impasse in the philosophy of science over the rationality of conceptual change, and also the resulting malaise. This has happened chiefly (I believe) through following too devotedly the philosophical example of Frege and Russell, who have led logic-oriented philosophers to interpret one of the *means* of natural science as its essential *end*. In some cases deliberately, in others unthinkingly, these philosophers have identified the *creation of logical* systems, which is one valid intellectual technique in science among others, with the *conceptual organization of understanding*, which is its more general goal. Yet, while the creation of logical systems has an unquestionable place in the conceptual organization of understanding, these two activities cannot simply be equated. Logical systems can be created for many different purposes, in the context of many different intellectual activities, or *none*: a formal network of theorems would not cease to be 'mathematical' (to cite an example of Wittgenstein's) just because its only application was to determine the patterns on a piece of wallpaper. Logical systems become 'scientific', in short, only through being given a *scientific* application: only through being put to use, that is, within the conceptual organization of *scientific* understanding.

It is one thing then to consider logical systems for themselves, and in their own right: it is another thing to consider them as the intellectual instruments of a natural science, with all its specific concepts and problems. And this brings us to our next important point. As wellas distinguishing between rationality and logicality, we must also take care to distinguish between the intellectual content of an *entire science*, and that of any *single theory* employed in that science. In suitable cases, the content of any one theory or 'conceptual system' can perhaps be represented by a 'logical system': though it would be a mistake – even here – to equate the resulting *logical* system with the *conceptual* system which it is used to represent. (The 'systematicity' of a conceptual system is that of a semantic domain, like that which is established when we segment the color-space into 'red', 'green', 'blue', etc., not that of a formal or syntactic structure.) But the content of an entire 'science' can be cast in a strictly logical form only in quite exceptional circumstances. Normally, any particular science will comprise numerous co-existing, logically-independent theories or conceptual systems, and it will be nonetheless 'scientific' for doing so.

In this respect the dominant influence of Newtonian mechanics, or an example of what a science should become, has been unhappy and misleading. One might even argue that Ernst Mach's book, *The Science of Mechanics*, was ill-named, since mechanics as such is not 'a science', but rather an analytical instrument employed within the science of *physics*. Hertz and Mach were in fact both writing just at the moment when Newton's physical mechanics was in process of crystallizing into a branch of pure mathematics: the branch known today oddly by my standards – as 'rational mechanics'. And the crucial distinction, between mechanics as a branch of pure mathematics and as an instrument of physical science, was made very nearly explicit by Hertz himself, when he set out his argument in two parts – the first concerned with the inner articulation of mechanical theory alone, the second with its application to actual empirical situations.

Granted this distinction between a science and its component theories, we can take one further step away from the present impasse. Any particular science (I have said) normally comprises an aggregate of logically-independent theories and concepts. The scientists working in that field may, in certain special cases, be moving towards a situation in which all its concepts are organized into a single integrated 'conceptual system'. But – pure geometry and rational mechanics apart – that is normally a long-term goal rather than an actuality and, when it is realized, the theoretical system in question is then trembling on the verge of becoming – in turn – a branch of pure mathematics. In the meantime, 'axiomatizing' remains a formal possibility, as an exercise in logic, but there is no scientific occasion – and no scientific profit – in pursuing it.

To go further, there may even be a certain scientific loss incurred as a result of *premature* axiomatization. For there is normally a great deal of logical 'slip', 'gappiness' and even inconsistency, between the constituent concepts and theories of an entire science; and – as Professor Shapere has emphasized elsewhere – this very 'logical gappiness' is one thing that helps to define the characteristic conceptual problems of a science and so powers its rational development. In quantum electrodynamics (to cite Shapere's example) we must at present assume, for certain purposes, that an electron has *zero* radius while, for other equally legitimate purposes within the same science, we must assume that it has *non-zero* radius. For pure mathematicians, this situation would be *catastrophic*. For physicists, it is at worst *unsatisfactory*; a challenge, and a source of conceptual problems. If they could find a way of circumventing this contradiction, their scientific understanding would be improved; but for the moment, if they have to live with it, they know how to do so.

IV

So much by way of criticism of the logicians' systematic ideal; let me now propose a counter-description of science which does better justice to our working conceptions of scientific rationality and reasonableness. Consider a natural science not as a tight and coherent logical system, but rather as a *conceptual aggregate or 'population'*, within which there are – at most – localized pockets of 'logical' systematicity. The whole problem of scientific rationality can then be restated in new terms. And I shall here indicate, as concisely as I can, how this change of standpoint opens up a fresh line of attack on some outstanding problems in the philosophy of science.

We may begin by remarking that, within any particular natural science, numerous different explanatory procedures, concepts and methods of representation are normally current, as means of fulfilling the proper intellectual missions of the science concerned. Between some, but only some, of these concepts and procedures there are formal or 'logical' links – as there are, for instance, between the Newtonian concepts of 'force', 'mass' and 'momentum'. (If there are these formal links, that is because the terms in question were introduced at *one and the same time*, for one and the same purpose, and so *interdefined*.) Meanwhile, alongside these systematically-related concepts and procedures, there will normally be others that are logically independent of, and even logically at variance with, one another. (These co-existing and independent sets of concepts and procedures will commonly have been introduced at *different* stages in the development of the science.) An adequate analysis of 'rationality' and 'reasonableness' in science will then require us to study:

(i) the various possible *non-formal relations* between the co-existing concepts, explanatory procedures and methods of representation circulating in each of the different natural sciences, and

(ii) the ways in which, in each field of science, *conceptual problems* arise and are recognized as such.

As Shapere's example indicates, logical incoherence in the assumptions of a science is *one* source of conceptual problems among others; but scientists most frequently recognize other, nonformal conditions as giving rise to conceptual problems, and these call for a more careful analysis than they have yet had.

Once we have seen how the proper intellectual missions of any science serves to define its outstanding problems, the next task is to study the activity of conceptual innovation by which scientists do their best to meet the intellectual demands of those problems. Often enough, these demands taken as a whole are irreconcilable. Improved understanding in one respect can be obtained, only by paying a price in other respects; and it is a matter for the nonformal theoretical judgment of scientists to decide what intellectual gain justifies what intellectual price. Even where there is no conflict of interests, there will normally be several competing proposals – or 'conceptual variants' – all designed to help us improve our scientific understanding in the field concerned; and it is again a matter of critical judgment to determine which of these variants does most to improve our understanding. So at this point we need to study the manner in which – and the criteria by reference to which – these choices between rival conceptual variants are made. Once more, this is not a formal matter, to be described in terms of the 'logical articulation' of the concepts and propositions in question; rather, it has to do with the ways in which those concepts can be applied, extended and/or modified, in the interest of better scientific understanding – with the ways (that is) in which rival variants can yield more effective procedures of explanation or techniques of representation.

Let me put the issues that arise about the rational development of scientific concepts in a nutshell. (1) Any natural science employs a repertory of concepts and explanatory procedures which, at any stage in its history, goes only part way towards fulfilling its current intellectual missions. (2) The gap between the legitimate goals of the science and its existing explanatory capacities defines a reservoir of outstanding theoretical problems. (3) Alongside the established repertory of concepts and procedures, there is a pool of conceptual variants or 'possibilities' which are under discussion as candidates for incorporation into the repertory of 'established' concepts and procedures. (4) A definitive conceptual change results from the agreed selection of certain theoretical innovations from this pool of variants, as best meeting the intellectual demands of the current theoretical situation. (5) This selection is made with an eye to numerous and partly conflicting criteria of choice. (6) These criteria vary from science to science, and from epoch to epoch within any given science, being determined by developing intellectual missions of the science in question.

About this group of issues, three things must be said in conclusion. In the first place, they all raise matters of substance rather than of form. The *specific* intellectual missions governing the research programs of the various natural sciences are defined in terms that respect the *special* aspects of external nature with which they deal. They cannot be specified in terms of universal formal features, like entailment, inconsistency and elegance. The same is true of the shortcomings that define the outstanding problems in a science, and of the intellectual requirements to be met before conceptual variants can become established. These, in turn, depend on the substantive tasks of the particular science concerned. As for the resulting changes in the theories and explanatory procedures of the science, these generate new populations of concepts and propositions, whose relations to their predecessors are even less formal or 'systematic' than the interrelations between the co-existing sets of concepts and propositions current in a single cross-section of its development. (At most, successive phases in the historical development of a science are linked by partial analogies, like those laid down in the 'correspondence principle', which relates quantum-mechanical and classical interpretations of corresponding physical phenomena.)

In the second place, it is precisely these non-formal questions about conceptual change in the sciences that have most stubbornly resisted answering within the limits of the logicians' program for philosophy of science. If we wish to give an adequate account *either* of the historical development of a science, *or* of the sense in which the intellectual choices involved in this development are 'rational', *or* of the manner in which scientists display their 'reasonableness' in the course of making those choices, our analysis will have to be very different in kind from those that philosophers of science have become accustomed to during the last 50 years. Such an account cannot be given in formal, still less in quasi-mathematical terms. For the steps by which the conceptual and explanatory repertories of the scienoes develop are – by their very nature – *populational* rather than *formal.* Their aim is to devise explanatory techniques which meet the specific intellectual demands of particular scientific problems; and questions about their 'rationality' or 'reasonableness' are concerned with how far and in what respects, they actually succeed in meeting those demands.

Finally, the study of conceptual *populations* is not only a necessary supplement to the study of logical and propositional *systems*, it is the prior and more fundamental task for philosophers of science to undertake. Unless we already have accepted a certain set of concepts as valid, applicable and/or relevant to our scientific inquiries, we are not yet in a position to *formulate* any propositions *at all*, still less to investigate their interrelations. And only in the last few years have we come to realize how far we are, in any field of science, from having any 'uniquely rational' or universally acceptable set of concepts or presuppositions to start with. That is why the analysis of scientific rationality must focus, in our generation, on the problems of *conceptual diversity* and conceptual change.

The main unsolved problems in philosophy of science accordingly raise questions less about 'probability' and 'formal adequacy' than about intellectual ecology, and Professor Causey's 'moving picture' of the historical development of the sciences will have to be an *evolutionary* one. It will involve us in looking at the changing concepts of science as elements not in static 'theorems' but in developing *theories*, with an eye not to their current 'implications' but to their extended *applications*. And it will consider the *rational procedures* adopted by scientists as reflecting the nature of their externally directed tasks: these being, not to formulate a logically-impeccable system of 'unified science' continuous with symbolic logic and pure mathematices, but rather to improve our intellectual grasp over the world of nature in whatever manner, and by whatever means, the actual demands of all their varied problem-situations may suggest or require.

I do not, of course, underestimate either the intrinsic difficulties involved in actually carrying through this new line of philosophical attack, or the temerity involved in encouraging other philosophers of science to move in what must, for them, be a novel and unfamiliar direction. But a moment comes when the only thing left to do is to be blunt and direct. So, as one who has always been more interested in the historical processes and rational procedures of conceptual change than in the logical articulation of static formalisms let me simply present my three central theses.

(a) A man shows his *rationality* not by the ideas he actually holds, but by the manner and the circumstances in which he is prepared to *change* his ideas: this is as true in science as anywhere else.

(b) An adequate account of the rationality of conceptual change, like an adequate account of organic evolution, will avoid any sharp distinctions between different kinds of change (normal, revolutionary etc.), and explain all historical changes in science as the varied outcomes of *one and the same* set of factors working together in different ways.

(c) Talk about *logical systems*, and about the timeless deductive relations within such formal systems, has served its purpose in the philosophy of science. The most serious outstanding problems in the subject – about the rational development of scientific ideas, and about the reasonableness of scientific procedures – require us to think, instead, about *conceptual populations*, and about the temporal sequences of conceptual variation and selective perpetuation by which these populations develop. There may be no such thing as a *logic* of discovery; but conceptual change in science certainly has its characteristic *rationale*. And we shall learn more about the rationality of scientific development from five years' study of 'intellectual ecology' than we shall learn from 'inductive logic' in a century.

Michigan State University

PAUL G. MORRISON

TWO KINDS OF THEORY IN THE SOCIAL SCIENCES

Introduction

In one sense, not every science of social phenomena qualifies as a social science, but only one which investigates the structure of one or more entire societies or cultures. It is within this context, perhaps, that historical linguistics, despite its central concern with an important variety of socially transmitted behavior and its largely deductive character, is less often classified with the social sciences than with the humanities.

From a broader standpoint, however, one may construe as a social science any science which investigates a complex of human behavior, attitude and belief that is transmitted, maintained, modified or discontinued through interpersonal influence. If the social sciences are understood in this broad sense, it is possible to distinguish two kinds of theory at work in them. On one hand, we may discern theories of social phenomena which function in deductive-nomothetic explanation. A typical example is the theory of the evolution of the Indo-European languages. On the other hand, we also come across passages in the works of leading social scientists which combine apparently untestable theories about whole societies, cultures or institutions with methodological prescriptions for investigating them. Examples from social anthropology are Radcliffe-Brown's early functionalism and the organismic theory of Ruth Benedict.

Following a brief account and a preliminary critique of the theories cited, I shall argue that theories of the second sort are not adequately interpreted as bad theories of the first sort, but are better viewed as serving a persistent noncognitive function in fostering indirectly the pursuit of deductive-nomothetic reasoning about social phenomena.

1. A theory of linguistic evolution

Starting from perceived similarities in word form and grammatical composition from one observed language to another, the 'Young Grammarians' of the nineteenth century reconstructed the prehistoric languages of the Indo-European hierachy. They did so by postulating sets of laws which account for pervasive sound shifts distinguishing the word forms of a parent language from those of a tongue immediately derived from it. Since the word forms of the ancestral languages had to be reconstructed right down to their inflections, their postulation together with the sound laws linking them to one another and to the recorded languages comprise a vast testable theoretic edifice.

One way of testing this structure involves projecting the form of an expression in one recorded member language from its cognate in another by recourse to appropriate sound laws from the theoretical superstructure, and then trying to confirm the projection – for example, by looking for the deduced form in a dictionary of the target language. In this way, starting from the East Germanic or Gothic expression '-faths', meaning 'master', one might seek to project its Greek cognate as follows. Applying sound laws which postulate that an initial '*f*' in a Germanic language derives from an initial '*p*' in the hypothetical Indo-European parent language; that a short '*a*' in Germanic roots of the kind in question comes from a short '*o*' in Indo-European; and that Germanic regularly dropped unaccented '*e*' before final '*s*' in nominative singular forms, we reconstruct the parent form of the Gothic word '-faths' as the Indo-European word '-potes'. And since the sound laws connecting the reconstructed parent language with Greek call for no changes in that particular form, we next seek confirmation by looking for a cognate Greek word, having to do with dominion, and containing the form '-potes'. And success is at hand when we hit upon the Greek compound from which our own word 'despot' is borrowed.

Now each such projection qualifies as an empirical inference in Israel Scheffler's sense, since temporally it involves what he has called forward and backward nomic regularities.[1] The uncanny frequency with which such projections can be made shows historical linguistics to be a deductive-nomothetic science of social phenomena.

2. Anthropological Theories of Society and Culture

The functionalist view expressed in Radcliffe-Browns' book *The Andaman Islanders* contrasts sharply with deductive-nomothetic theories. For although he seeks empirical laws, he holds that hypothetical reconstructions of prehistoric cultural practices are not verifiable.[2] The trouble with this is that law-like generalizations in any field refer indifferently to historic and prehistoric events alike. What would become of geology, paleontology or astrophysics, for example, if Radcliffe-Brown's methodological dictum were applied to them? In disciplines in which prehistoric reconstructions are routinely rejected, is not the process of law-like generalization ruled out as well?

Radcliffe-Brown's careful study of the Andamanese surely points up the value of functional studies of stable systems of concurrent practices. On the other hand, much can also be said for studies of major cultural change. The history of the Fiji Islanders provides a case in point. A recent visitor to the Fijis reports that its people are friendly, mild-mannered and happy, in contrast to their anxiety-ridden forebears a hundred years ago, who were "known... as the most blood-thirsty, flesh-craving cannibals in the world."[3] In those days, a Fijian was apprehensive about going outside of his village alone, for fear of being clubbed and eaten by other Fijians at a feast in some neighboring village.

While a functional anthropologist of Radcliffe-Brown's persuasion might prefer to investigate whatever culture was extant in those islands at the time of his study, a devotee of an organismic theory of culture like Ruth Benedict's might prefer to deal with the whole period in which the drastic changes occurred. For on an account like hers, one might hold that the pattern of Fiji culture was discarding cannibalism, and modifying its other practices to support civic, religious and ethical attitudes imported from the West which militate against feasting on human flesh.[4]

From the foregoing, it is evident that there are at least two ways in which one might be tempted to employ the term 'culture' in describing the Fiji experience. One might, in Ruth Benedict's tradition, construe the whole historical spread of changing practices in the Islands as the Fiji culture. Or again, in Radcliffe-Browns' functionalist way, one might want to take some maximum structure of concurrent institutions, customs and beliefs as the Fiji culture of the nineteenth century, or as the Fiji culture of the twentieth century – denying that there is anything spanning both centuries which could properly be called the Fiji culture.

Another feature of the social theories just considered which, like Radcliffe-Brown's injunction against hypothetical reconstructions, would be suspect in the natural sciences, is their anthropomorphism. For Radcliffe-Brown, societies and the ceremonials which keep them in force have needs of their own.[5] For Benedict, the purpose of a culture selects, recasts, or discards various traits. Moreover, that purpose makes demands with which the traits that it uses must conform.[6]

In defense of Ruth Benedict, one might observe that she immediately disavows that any conscious choice or purpose is actually involved in cultural change. It is only that "when we describe the process historically, we inevitably use animistic forms of expression" – as though choice and purpose were involved.[7] On her account, our language initially forces us to claim something that we do not literally mean. Luckily, though, we are subsequently able to make amends for this linguistic compulsion by denying voluntary attributes to a culture pattern. Hers, then, is only a retractile anthropomorphism. Maurice Mandelbaum makes a similar defense of Radcliffe-Brown's position, denying that the latter meant to speak of a "selective power in the society... which accepts or rejects specific beliefs and customs..."[8]

Even if the anthropomorphic overtones of the two theories are not to be taken seriously, however, it is still not clear that either author has a definite concept of what a society or culture is. Of course, those familar with these theories may object that the *holistic* emphasis of each shows that a definite notion of society or culture is employed in it. For Benedict's insistence that a culture is more than the sum of its traits (since it embodies them in a unique arrangement)[9] suggests that she had in mind a criterion for identifying a culture. Moreover, to recommend, as Radcliffe-Brown does, that we investigate the relation between a whole system of ritual and myth and the entire culture within which it operates[10] is, in part, to suggest that we must know how to identify any such whole system and any such entire culture.

The big difficulty, however, is that neither theory reveals how we are to tell when a cultural pattern is finally broken. What is it about these concepts of society or culture which lets us know when the historical complexies to which they refer have come to an end? It is much easier to tell when the social transmission of the Gothic language ceased between native speakers than to tell when the social system of Ancient Greece expired.

Radcliffe-Brown was aware of the problem. For in his essay 'Patrilineal and Matrilineal Succession', he declared that any social system, "to survive, must conform to certain conditions. If we can define adequately one of these universal conditions... we have a sociological law. Thereupon if it can be shown that a particular institution in a particular society is the means by which that society conforms to the law... we may speak of this as the 'sociological origin' of the institution."[11]

The trouble with this position is that we would need to identify all of the institutions essential to the survival of a society, rather than just one or several, in order to know when or under what conditions a particular cociety would become defunct. And the situation is complicated by the fact that two or more competitive institutions for accomplishing the same end are often present in the same populated locality. In Istanbul under the Ottomans, for example, monogamy and polygamy coexisted. But surely these are competitive institutions satisfying the same general needs. How, in such cases, may we determine which of several concurrent alternative institutions with the same general function is 'the' means by which a society conforms to a sociological law? If there can be only one such institutional means, must its competitors not be considered deviant institutions?

From these considerations, it appears that Radcliffe-Brown's criterion for selecting the essential components of a social system is at best a subjective one for backing a personal preference for one competitive institution over its rivals. But this leaves room for an indissoluble disagreement as to which of several largely overlapping institutional complexes comprise the actual social system in a particular locale, since investigators of social phenomena have no special immunity to the random conditioning which brings about virtually ineradicable differences in personal preference.

The same difficulty crops up in Ruth Benedict's theory, according to which a culture selects, modifies or discards social practices. How, for example, would her theory apply to the revolution in the eating habits of the Fiji Islanders? During the transition period, while the eating of human flesh and the Western eating customs which later displaced them were both in vogue, was the culture continuing while discarding cannibalism? Or was it perhaps expiring because cannibalism was dying out? Or had that culture ever selected eating practices of any kind as essential parts of its structure? Benedict's theory provides no means of gauging whether the discontinuance of a practice spells the involuntary death of a culture, or merely shows the culture to be flourishing while discarding that practice. It is noncommittal as to which of several rival kinds of traits or successions of traits are essential features of their cultures and which are only accidental to them.

It would appear, then, that neither Radcliffe-Brown nor Ruth Benedict had a determinate concept of a social system or culture after all. Although each of these authors thought of his field as an empirical science, and although each may have thought that he was developing an empirical concept of society or culture, neither of them, in the last analysis, could bring himself to do so. In my concluding remarks, I shall present a hypothesis to explain why I believe that this was so.

3. The Function of Recent Theories in the Social Sciences

The basic mechanism by which attitudes and kinds of behavior are exemplified and maintained in force is the influence, often unconscious, of one person upon another. Individual human beings almost routinely and automatically conform their attitudes and ways of behaving to those of their neighbors. It is in this way that they learn to express themselves and to deal with their relatives and other associates. When these interpersonally transmitted and reinforced attitudes and ways of behaving are fairly widespread in a community, and when they persist for some time, they may be considered cultural traits.

Time-extended networks of people linked by the transmission and reinforcement of cultural traits stretch from the dawn of the human species down to the present, and show promise of continuing until mankind becomes extinct. If we may refer to these networks of interacting human beings as *social media*, we may distinguish two basic ways in which such media embody cultural traits. First, it is possible to discriminate long periods during which a compatible set of dominant cultural traits holds sway in a social medium. If the traits are mutually supporting, and if they apply to a maximum set of interacting persons, they comprise, roughly speaking, what Radcliffe-Brown called a social system. And it is the presence of such a system of dominant cultural traits that attracts field investigators to make synchronic studies.

In the second place, we find periods during which the members of a social medium let one social system die out in the medium to be replaced by another. In such cases, it seems natural to speak of a social or cultural revolution in the medium. And it is populations undergoing this sort of change to which diachronic studies are customarily directed.

Whether a field investigator is concerned with making a synchronic or a diachronic study, however, he will tend to identify with the social medium whose structure he seeks out. And because the future form of that structure, in the most interesting cases, can never be anticipated completely, he will tend to sympathize with the people who exhibit the attitudes and engage in the practices currently fashionable in the social medium, and to share their untestable belief in an ongoing societal force which sometimes causes a cultural trait to persist, and which sometimes brings about its replacement.

If an investigator is inspired by such a feeling, however, he cannot afford in his theorizing to make precise a set of practices, beliefs and attitudes, however general, which might be construed as structural features of the social complex he is studying. For test results might then awkwardly reveal that the object of his inspiration, the social medium, had outlasted the social structure which was the professed object of his scientific concern. As long as his primary allegiance is to the social medium itself, whatever unanticipated forms it may assume, rather than to a particular set or system of structural relations, any *empirical* notion of a society, culture or complex institution must remain, for him, unspecifiable. To say this is not to condemn the authors of theories which make use of indeterminate notions of society or culture. For these theories are, nevertheless, quite useful, if only indirectly so, in the quest for nomic regularities about societies or cultures. In conclusion, I would like to state my view concerning the positive function of such theories.

To being with, the members of each community investigated tend to share a group rationale – an untestable body of beliefs about the history nature and fate of the community as a whole. After conducting field investigations of several such communities, a veteran investigator like Benedict or Radcliffe-Brown generalizes from the rationales of the various groups, and in this way, arrives at a *pure rationale* which serves, in lieu of a currently unattainable deductive-nomothetic theory, as a theory of society or culture in general.

Such *pro tempore* theories, moreover, serve an important social function within the community of social scientists itself. For they provide an important auxiliary means both of instilling, and of maintaining, a suitable professional attitude in serious students of society or culture who, in anticipation of the discovery of general causal regularities between cultural traits, must meanwhile persist in carefully tracing in further populations the socially transmitted machinery of need satisfaction mentioned earlier, in the hope that some future Kepler or Newton of the social sciences will perceive in it the patterns that sustain the long-awaited law-like connections.

Just as in psychotherapy, the treatment of individual patients must go on even where knowledge of relevant psychological laws is lacking;[12] so in the social sciences, the study of individual social or cultural media must go on even though relevant systematic knowledge is not available. And hopefully, in the latter case as in the former, theoretical knowledge and knowledge gained by piecemeal observation of particular cases will support one another more and more with the passage of time.

State University College,
Brockport, New York

NOTES

[1] Cf. Scheffler, Israel, 'Explanation, Prediction, and Abstraction', *British Journal for the Philosophy of Science* 7 (1957). Reprinted in *Philosophy of Science* (ed. by Danto and Morgenbesser), Meridian, New York, 1960, pp. 274–87. See especially Danto and Morgenbesser, p. 285.

[2] Radcliffe-Brown, A. R., *The Andaman Islanders*, Cambridge University Press, Cambridge, England, 1922, p. 229n.

[3] Lidz, Theodore, *The Person: His Development Throughout the Life Cycle*, Basic Books, New York, 1968, p. 33n.

[4] Cf. Benedict, Ruth, *Patterns of Culture*, Penguin Books, New York, 1946, pp. 42–43.

[5] Radcliffe-Brown, A. R., *op. cit.*, p. 327.

[6] Benedict, Ruth, *op. cit.*, pp. 42–43.

[7] *ibid.*, p. 43.

[8] Mandelbaum, Maurice, 'Functionalism in Social Anthropology', in *Philosophy, Science and Method: Essays in Honor of Ernest Nagel* (ed. by Morgenbesser, Suppes, and White), St. Martin's Press, New York, 1969, p. 319.

[9] Benedict, Ruth, *op. cit.*, p. 43.

[10] Radcliffe-Brown, A. R., *op. cit.*, Preface.

[11] Radcliffe-Brown, A. R., *Structure and Function in Primitive Society*, Free Press, Glencoe, Illinois, 1952, p. 43.

[12] Shoben, Edward J., 'Psychological Theory Construction and the Psychologist', *Journal of General Psychology* 52 (1955) 181–88.

X. Relativity and Congruence

CARLO GIANNONI

EINSTEIN AND THE LORENTZ-POINCARÉ THEORY OF RELATIVITY

The title of this paper has reference to the well known remark which Whittaker has made about Einstein's Special Theory of Relativity in his book *History of the Theories of Aether and Electricity*. In the Chapter entitled 'The Relativity Theory of Poincaré and Lorentz' he writes as follows: "In the autumn of the same year ... Einstein published a paper which set forth the relativity theory of Poincaré and Lorentz with some amplifications, and which attracted much attention ... In this paper Einstein gave the modifications which must now be introduced into the formulae for aberration and the Doppler effect."[1] The question with which I am concerned in this paper is whether from a *strictly empirical* point of view Whittaker's assessment is correct. I shall argue that the physical content of both of Einstein's postulates – the Postulate of Relativity and the Postulate of the Constancy of the Velocity of Light – was in fact already known to Lorentz and Poincaré, and that the significance of Einstein's theory *vis-à-vis* Lorentz's lies purely in its conceptual-epistemological content. Furthermore, we argue that Einstein's theory is derivable from Lorentz's by the addition of these purely conceptual innovations. On the other hand it is, of course, precisely because of its conceptual innovations that Einstein's theory has been the subject of much philosophical interest.

Let us consider first the Postulate of Relativity. One can express this in the words of Einstein as follows: "the phenomena of electrodynamics as well as of mechanics possess no properties corresponding to the idea of absolute rest."[2] That is, on the basis of neither mechanical nor electrodynamical nor optical experiments is it possible to determine whether or not one is at absolute rest in the aether. Therefore, the concept of absolute rest can have no empirical significance. It was already known that mechanical experiments could not differentiate between reference frames at absolute rest and those moving with a constant velocity relative to a frame at absolute rest, but it was believed that by optical experiments one could detect the motion of the Earth through the aether. The Michelson-Morley experiment, which was an attempt to detect this movement through the aether, gave a null result. Now this was in turn explained by the Fitzgerald contraction which postulated that one arm of the Michelson-Morley interferometer was contracted relative to its 'true length'. However, if this latter was the case effects would have shown up in experiments on refraction by Rayleigh and Brace and in experiments on charged condensors by Trouton and Noble. Lorentz had the choice of modifying his theory of electrons to account for the null results in the latter two sets of experiments or to admit that motion through the aether was undetectable. Poincaré urged him to take the latter course and make a fundamental change in his theoretical framework which would explain in all cases at once why motion through the aether was undetectable, and it is on this basis that Lorentz wrote his paper of 1904[3], one year prior to Einstein's paper on Special Relativity. That is, in 1904 Lorentz accepted the Postulate of Relativity when he wrote: "Surely this course of inventing special hypotheses for each new experimental result is somewhat artificial. It would be more satisfactory if it were possible to show by means of certain fundamental assumptions… that many electromagnetic actions are entirely independent of the motion of the system."[4]

A difference between the approach of Einstein and that of Lorentz is that Einstein puts forth the Postulate of Relativity as a postulate, an axiom, and proceeds to deduce the Lorentz transformations, whereas for Lorentz relativity is to be deduced as a consequence and it is the Lorentz transformation equations which are postulated. This is indicative of the different conceptual frameworks from which Lorentz and Einstein were working, but not of a difference in empirical content.

Let us now turn to the postulate of the Constancy of the Velocity of Light. This principle has very important conventional and epistemological ingredients as well as factual ingredients. The issue here is whether the *factual* content of this principle is contained in Lorentz's theory. Since velocity is by definition distance divided by time, it is clear that if two theories differ in their measures of space and time, they will also differ in their measures of the velocity of light and it is also clear that the two theories in question do in fact use different measures of space and time.

Let us first look at spatial measurement in Lorentz's theory. It is useful to distinguish three distinct spatial contractions which we will lable 'the Lorentz-Fitzgerald Contraction Hypothesis', 'the Lorentz contraction', and 'the Einstein contraction'. The Lorentz contraction asserts that a rigid rod when placed in motion is shorter than one would expect it to be on the classical theory. Specifically, let us consider a 'stationary' coordinate system in which is placed a rigid rod of unit length.

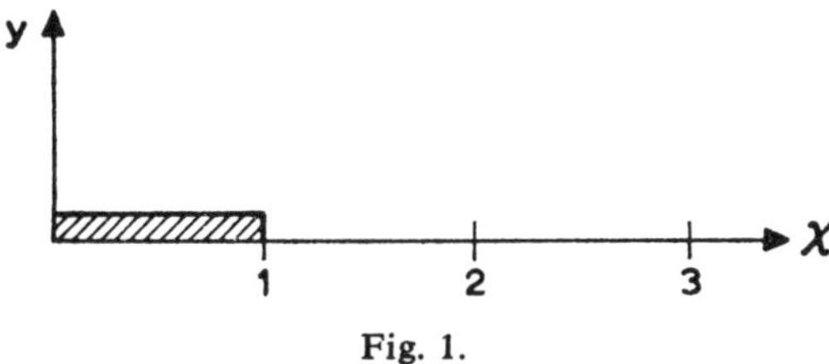

Fig. 1.

Let us now place this same rigid rod in motion with a velocity **v** relative to the 'stationary' system in the direction of the x axis and also place *the same* coordinate system in motion along with the rod. This is the result:

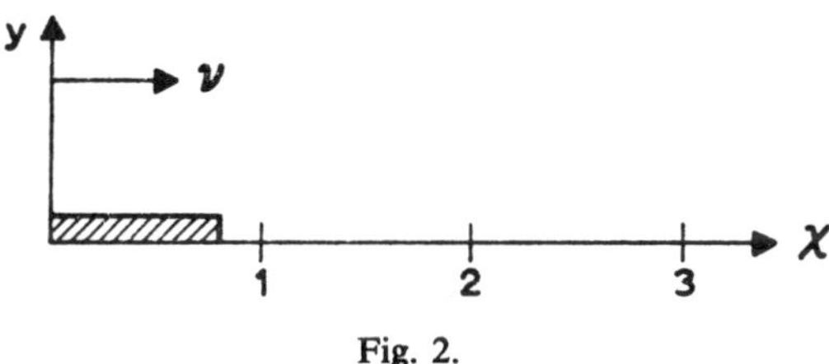

Fig. 2.

(By the two coordinate systems being 'the same' we mean that at the instant at which the two origins coincide all of the other coordinates of the two systems coincide as well.) Now according to classical theory one would expect the coordinate difference of the rod in motion to be one, just as the coordinate difference of the rod at rest is one. The Lorentz contraction asserts that the coordinate difference of the rod in motion is less than one by the factor $\sqrt{(1-v^2/c^2)}$. Now it can be shown mathematically that this Lorentz contraction follows directly from the Michelson-Morley experiment, given that the velocity of light in the stationary system is c on which point both the Special Theory of Relativity and Classical Aether Theory agree (see Appendix A). Therefore, since both STR and Lorentz accepted the result of the Michelson-Morley experiment there is in *both* theories a Lorentz contraction.

In discussing the Lorentz contraction we did *not* talk about the length of the rod in the moving system, but merely about its coordinates in a particular coordinate system. It was one of Einstein's *philosophical* insights to realize that we are free to call the length of the moving rod in the moving coordinate system one if we wish, despite its having a coordinate difference of only $\sqrt{(1 - v^2/c^2)}$. If we make this decision, then while we will say that the rod has a length of one whether at rest or in motion, nevertheless when it is in motion it is shorter than we would expect is to be on the classical theory.

Now this same freedom which permits us to call the length of the moving rod one also gives us the freedom to call its length $\sqrt{(1 - v^2/c^2)}$ if we wish. Whether we call it one or $\sqrt{(1 - v^2/c^2)}$ is a matter of convention. The philosophical mistake in Lorentz's theory is that he insisted that the moving rod *must* have the length $\sqrt{(1 - v^2/c^2)}$. The assertion that the length of the moving rod is *really* $\sqrt{(1 - v^2/c^2)}$ is the Lorentz-Fitzgerald Contraction Hypothesis. The significant point is that it differs from the factually correct Lorentz contraction soley in its philosophical content, just as a version of Newton's theory without a discussion of absolute time would differ from Newton's actual theory solely in its philosophical content. If we grant that Einstein was philosophically correct and Lorentz philosophically mistaken, then we can philosophically correct Lorentz's theory be saying the following: A unit rod when placed in motion has *by convention* the length $\sqrt{(1 - v^2/c^2)}$ and the velocities of light in the moving system are those given by the classical aether theory (i.e., $c - v$ and $c + v$, respectively).

But Lorentz's theory in the form given in his 1904 paper also has an additional postulate, viz., the Postulate of Relativity. Now when we say that we cannot experimentally distinguish between a system in motion with a constant velocity and a system at absolute rest, we mean two things. We mean first of all that the laws of mechanics and electrodynamics and optics will have the same form in a system in motion as in a system at absolute rest. But we also mean that if any intersystemic comparisons between systems moving with respect to each other are made, they will likewise yield no reason for distinguishing one system as being at absolute rest. For example, we say that when the stationary observer placed his coordinate system in motion along with a rigid rod that the rigid rod experienced a Lorentz contraction. But if intersystemic comparisons are to yield no ground for distinguishing one system as at absolute rest, then similarly if the moving coordinate system is considered at rest and if this 'rest' coordinate system is placed in motion with a velocity $-v$ along the x axis and the rigid rod is placed in motion along with it, then the rigid rod will experience a Lorentz contraction relative to that 'rest' coordinate system. But this new 'moving' system is the original rest system. In other words, if the Postulate of Relativity is true, then the Lorentz contraction must be reciprocal. Now it can be shown mathematically that if the Lorentz contraction is to be reciprocal then the following result ensues: the unit of time as indicated by a moving standard clock is longer than would be expected on the classical theory, that is, the unit of time as indicated by a moving standard clock will be longer than the unit temporal coordinate difference of the stationary system (see Appendix B). We refer to this as the Lorentz time dilation.

Again as in the case of length we are free to choose as our unit of time either that indicated by the moving clock or that given by the transported coordinate system. Choosing the latter is tantamount to using the clock readings of the stationary system in both stationary and moving systems. Again Einstein had the philosophical insight to see that it was an arbitrary choice and in the Special Theory of Relativity he chose to accept the readings of standard clocks whether they be at rest or in motion. Lorentz made the philosophical mistake of not realizing that he had a choice and insisted that the readings of the stationary clocks must be accepted and that the readings of the moving clocks must be adjusted. While stationary clocks gave the 'true' time for Lorentz, the moving clocks gave only 'local' time. Again we can philosophically correct Lorentz's theory in the following way: by convention clocks in the rest system are to be used both for the rest system and for the moving systems (see Appendix C).

We now have three elements to Lorentz's theory: there are the two conventions regarding spatial and temporal measurements – a unit rod when placed in motion has the shorter length $\sqrt{(1-v^2/c^2)}$, and the unit indication of the standard clock when placed in motion has the greater length $1/\sqrt{(1-v^2/c^2)}$. Furthermore, the classical aether theoretic velocities are to obtain given the stipulated conventions. Despite the numberically same velocities in Lorentz's theory and the classical aether theory, the two theories are of course factually different for Lorentz's theory entails the Lorentz spatial contraction and time dilation whereas the classical theory does not. Clearly, if one changes his conventions regarding spatial and temporal measurement, one will also change his determination of the velocities of light in the moving system. In particular, if one chooses the conventions adopted by Einstein one arrives at the constancy of the velocity of light, that is, the velocity of light has the value c in all systems. But it is equally true that if one chooses Lorentz's conventions, that the velocities of light will be those given by the classical aether theory (see Appendix D). Therefore, the *factual* content of the Principle of the Constancy of the Velocity of Light is included in Lorentz's theory. But of course not only does Lorentz's theory use different conventions but Lorentz was not even aware of the legitimacy of using different conventions. Therefore, Einstein's addition to the Lorentz theory is conceptual rather than factual.

Similar remarks can be made regarding the status of the aether. In Lorentz's theory the aether no longer plays the experimental role it was expected to play prior to 1900, but it still plays a conceptual role, in the way in which absolute space and time played a conceptual role in Newton's theory. In the Theory of Relativity the aether no longer even plays a conceptual role. But again this is purely a conceptual change between the two theories. I do not mean in any sense to undermine the importance of Einstein's achievements by calling them purely conceptual, for the conceptual changes which Einstein's theory makes in the Poincaré-Lorentz theory are of the greatest significance. One can say that Einstein's conceptual changes in the Lorentz theory have in fact been of greater significance than the changes introduced by Lorentz into classical theory, and moreover it is precisely these conceptual changes which Whittaker fails to appreciate.

We have not yet discussed the Einstein contraction. Whereas the Lorentz contraction is a comparison between the actual extension of a moving rod and the extension expected on the classical theory, the Einstein contraction involves the measurement of the length of the moving rod from the rest system. But the very concept of measuring a moving rod while not moving with it requires conceptual innovation. For example, if we wanted to measure the length of a bus moving in front of us, we would be inclined to run along side the bus with our measuring rod. But then we would not be at rest, but moving with the bus. Einstein introduced the following technique for measuring the length of a moving rod from a rest system. Let two events occur at the end points of the rod which are simultaneous in the rest system. The length of the moving rod as measured from the rest system is the distance between these two simultaneous events as measured in the rest sytem. If a unit rod is placed in motion then using Einstein's technique of measuring its length we would find its length in the rest system to be $\sqrt{(1-v^2/c^2)}$. If we follow Einstein's convention of calling the length of the rod one whether it is at rest or in motion, then the length of the rod as measured in the moving system is one while its length measured in the rest system is only $\sqrt{(1-v^2/c^2)}$. It is interesting to note here, however, that if we use Lorentz's convention regarding length, then the unit rod when placed in motion has a length $\sqrt{(1-v^2/c^2)}$. Therefore, when the moving rod is then measured in the rest system using Einstein's technique it will be found to have the same length, that is, no Einstein contraction occurs if we use Lorentz's convention regarding length. Again we find in Einstein's theory a conceptual innovation, but it is a conceptual innovation whose very significance depends on his having made the prior conventional choice of calling the length of a rod the same whether at rest or in motion, for as we indicated above there is no Einstein contraction if we choose Lorentz's convention, and therefore there would be no point in introducing the new concept of length.

Now we are in a position to evaluate certain comments made by Professor Grünbaum regarding the Lorentz contraction. For example, Grünbaum says the following:

> The Lorentz-Fitzgerald contraction hypothesis asserts a comparison of the actual length of the arm, as measured by the round-trip time of light, to the greater length that the travel-time would have revealed, if the classical aether theory were true.[5]

According to the Lorentz contraction the rod is in fact shorter than it was expected to be on the classical theory, and the Lorentz-Fitzgerald contraction hypothesis goes on to add to this that the rod is *really* shorter than it was when it was at rest. By failing to distinguish between these two contractions. Grünbaum fails to recognize the extent to which the Lorentz-Fitzgerald contraction hypothesis is correct, for the hypothesis asserts the Lorentz contraction and the Lorentz contraction in turn exhausts the empirical content of that hypothesis. That is, Grünbaum fails to realize that in so far as the Lorentz-Fitzgerald Contraction Hypothesis has empirical content it is correct.

Again Grünbaum says the following:

Unlike the L-F contraction... this 'Einstein contraction' is a *symmetrical* or reciprocal relation between the measurements made in any two inertial systems.[6]

Part of the error here is again in not distinguishing between the two contractions, but part of the error also comes from not considering more carefully Lorentz's 1904 paper, and looking primarily at his earlier papers. For as we noted above, Lorentz does use the Principle of Relativity in his 1904 paper, and the Lorentz contraction therefore is reciprocal between any two inertial systems just as the 'Einstein contraction' is reciprocal.

This failure on the part of Grünbaum to give due credit to Lorentz's use of the principle of relativity, also vitiates his discussion of the Kennedy-Thorndike experiment and exhibits a curious confusing of the historical order of events. I will not go into details here except to point out that the Kennedy-Thorndike experiment occurred in 1932. Now Grünbaum quite rightly shows that the Kennedy-Thorndike experiment falsifies a theory which has the Lorentz contraction but no Lorentz time dilation, and he goes on to make the charge that it would be logically ad hoc for Lorentz to introduce the time dilation to save his theory in the face of the Kennedy-Thorndike experiment. But this charge is totally out of place for Lorentz already had the time dilation equation in 1904, 28 years before the Kennedy-Thorndike experiment! Lorentz's time dilation equation is in no sense ad hoc. Again I note that the empirical content of Lorentz's time dilation equation is correct, and it is solely his philosophical addition to it which is mistaken.

One aspect of the Special Theory of Relativity which is often talked about is the Relativity of Simultaneity. If two events are simultaneous in one inertial system they are not simultaneous in another. But this relativity of simultaneity depends on saying that in all inertial systems the velocity of light in every direction is the same constant c. Now if one chooses Lorentz's standards of space and time measurement and one uses the classical aether theoretic velocities of light, which is perfectly acceptable as we have shown above, then the relativity of simultaneity disappears, for in this case we find that if two events are simultaneous in one system they are also simultaneous in the other. This is obvious from the fact that under Lorentz's conventions we are using the rest clocks in all systems. Therefore, we would have a kind of absolute simultaneity in Lorentz's theory. But we also have a relativity of simultaneity in the following sense: since any system can be considered to be the rest system because of the principle of relativity, then depending on which system is considered to be the rest system, different pairs of events will be considered to be simultaneous. There is nevertheless a philosophical difference between the two theories for while Lorentz would say that we cannot determine which system is the rest system, and therefore we cannot determine which pairs of events are absolutely simultaneous, Einstein would say that absolute simultaneity does not exist. Again this is a philosophical difference, but a difference which makes no empirical difference.

One aspect of the Special Theory of Relativity which I have not discussed is the principle that light is the fastest signal chain. This principle has genuine empirical content and does not seem to be included in Lorentz's theory, although it already had been predicted by Poincaré. With the exception of this one principle, I have argued that the empirical content of the Special Theory of Relativity is coextensive with the empirical content of Lorentz's theory and that they differ solely in their conceptual-epistemological content.

APPENDIX A

We wish to show that the Michelson-Morley experiment entails the Lorentz contraction on the assumption that the velocity of light is c in all directions in the 'stationary' system. In the 'stationary' system we have an ordinary set of Cartesian coordinates S. We construct a set of coordinates T' in the moving system which are 'the same' as S in the following respect: when the two origins coincide, all of the spatial and temporal coordinates of both stems coincide. Therefore at $t_S=0$. $(x_S, y_S, z_S, t_S)=(x_{T'}, y_{T'}, z_{T'}, t_{T'})$. At a later time t_S, x_S will be equal to $x_{T'}$ plus the x_S coordinate of the origin of T', i.e., vt_S. Therefore at a time t_S, $(x_S, y_S, z_S, t_S)=(x_{T'}+vt_S, y_{T'}, z_{T'}, t_{T'})$. We have not introduced a metric into T' and therefore one cannot say at this point what length is to be asscoiated with a given coordinate difference $\Delta x_{T'}$, although since S is Cartesian, we can say that the length of *a* coordinate difference Δx_S is Δx_S.

According to the Michelson-Morley experiment we have two rods A and B which when placed along side each other coincide. Now we place these two rods in motion, rod A perpendicular to the direction of motion and rod B parallel to it, rod A lying along the y axis and rod B lying along the x axis. The result of this experiment was that the time which light takes to traverse rod A is equal to the time which light takes to traverse rod B. We will also assume the following experimentally determinable fact: if two rods A and A' are the same length when both are at rest and if rod A' is placed along the y axis of S and A along the y axis of T', then when rod A passes rod A' they coincide. This can be determined independently of any convention.

Let us assume that the lengths of rods A and B in S were one. *We wish to show that the $x_{T'}$ coordinate of rod B is $\sqrt{(1-v^2/c^2)}$.* Let light leave the common origins of S and T' at $t_S=0$. In traveling from the origin to the end of rod B, light travels through a coordinate difference $x_{T'}-0=x_{T'}$ in T'. This is equal to a coordinate difference of $x_S-0=x_S$ in S. But by virtue of the way in which T' is set up we know that $x_{T'}=x_S-vt_{OB}$, where t_{OB} is the time it took light to travel from the common origins of S and T' to x_S. Therefore, $x_S=x_{T'}+vt_{OB}$. Moreover since S is a Cartesian coordinate system, a coordinate difference in S coincides with a length in S, and therefore in traveling through a coordinate difference $x_S=x_{T'}+vt_{OB}$ in S it travels through a distance of exactly the same value. Similarly, in traveling back along rod B to the origin, light travels through a coordinate difference $x_{T'}$ in T' which corresponds to a distance of $x_{T'}-vt_{BO}$ in S. Moreover, by our assumption it traverses each of these distances with a speed of c in S. Therefore, since time in S=distance in S/speed in S,

$$t_{OB} = \frac{x_{T'} + vt_{OB}}{C} \quad t_{OB} = \frac{x_{T'}}{C - v}$$

$$t_{BO} = \frac{x_{T'} - vt_{BO}}{C} \quad t_{BO} = \frac{x_{T'}}{C + v}$$

$$t_{OBO} = t_{OB} + t_{BO} = \frac{x_{T'}}{C - v} + \frac{x_{T'}}{C + v} = \frac{2Cx_{T'}}{C^2 - v^2}.$$

On the other hand in traveling from the origin to the end of rod A, the light travels through a coordinate difference $y_{T'}$ in T', but it travels not only through the same y coordinate difference $y_S=y_{T'}$ in S but also through an x coordinate difference $x_S=vt_{OA}$, for while the light traveled along rod A, the coordinate system T' moved relative to S. Since the x_S and y_S coordinate differences in S are equal to lengths, we can apply the Pythagorem formula to arrive at the distance in S through which the light has traversed in going from the origin to the end of rod A, viz.,

$\sqrt{(y_s^2 + v^2 t_{OA}^2)} = \sqrt{(y_{T'}^2 + v^2 t_{OA}^2)}$. Similarly, in traveling back along rod A to the origin light travels through a distance $\sqrt{(y_{T'}^2 + v^2 t_{AO}^2)}$ in S. Moreover, it traverses each of these distances with a speed of c in S. Therefore,

$$t_{OA} = \frac{\sqrt{y_{T'}^2 + v^2 t_{OA}^2}}{c} \quad t_{OA} = \frac{y_{T'}}{\sqrt{c^2 - v^2}}$$

$$t_{AO} = \frac{\sqrt{y_{T'}^2 + v^2 t_{AO}^2}}{c} \quad t_{AO} = \frac{y_{T'}}{\sqrt{c^2 - v^2}}$$

$$t_{OAO} = t_{OA} + t_{AO} = \frac{2y_{T'}}{\sqrt{c^2 - v^2}}.$$

But by virtue of the Michelson-Morley experiment $t_{OAO} = t_{OBO}$. Therfore,

$$\frac{2y_{T'}}{\sqrt{c^2 - v^2}} = \frac{2cx_{T'}}{c^2 - v^2}$$

$$x_{T'} = y_{T'} \sqrt{1 - \frac{v^2}{c^2}}.$$

But since moving rod A coincides with rest rod A' when they pass each other and the latter has a length of one, $y_{T'} = 1$. Therefore,

$$x_{T'} = \sqrt{1 - \frac{v^2}{c^2}}. \qquad \text{Q.E.D.}$$

APPENDIX B

Given the Lorentz contraction and the principle of relativity, we wish to show that when standard clocks in the moving system indicate a unit of time, standard clocks in the rest system indicate $1/\sqrt{(1 - v^2/c^2)}$ units of time.

By virtue of the principle of relativity we can treat the moving system as if it were at rest. Therefore, we can set up in it a Cartesian coordinate system T using standard rigid rods and standard clocks and the velocities of light in all directions will be c as determined by these rods and clocks. Because of the Lorentz contraction the x coordinates of T and T' will be related as follows: (a) $x_T = x_{T'}/\sqrt{(1 - v^2/c^2)}$. Here we will assume that the clocks of T are synchronized with each other on the assumption that the velocity of light is the same in both directions of T. We also have the following relationship between S and T: (b) $x_{T'} = x_S - vt_S$ (see Appendix A).

Now ,we can set up in the rest system a system of coordinates S' which are 'the same' as T. This means that with regard to the x coordinate the following relationship holds: (c) $x_{S'} = x_T + vt_T$. Now because of the principle of relativity we can also assert a 'reverse' Lorentz contraction and therefore (d) $x_S = x_{S'}/\sqrt{(1 - v^2/c^2)}$.

We wish to show that if two events occur at place x_T of T, then if the time between them as measured by a standard clock at x_T in T is Δt_T, the time between them as measured by standard clocks in S is $\Delta t_S = \Delta t_T \div \sqrt{(1 - v^2/c^2)}$. Let us first derive the general relationship between the standard clocks in T and S.

$$t_S = \frac{x_S - x_{T'}}{v} \qquad \text{(From b)}$$

$$= \frac{\dfrac{x_{S'}}{\sqrt{1 - \dfrac{v^2}{c^2}}} - x_T\sqrt{1 - \dfrac{v^2}{c^2}}}{v} \qquad \text{(From a and d)}$$

$$= \frac{\dfrac{x_T + vt_T}{\sqrt{1 - \dfrac{v^2}{c^2}}} - x_T\sqrt{1 - \dfrac{v^2}{c^2}}}{v} \qquad \text{(From c)}$$

$$= \frac{t_T + x_T v/c^2}{\sqrt{1 - \dfrac{v^2}{c^2}}} \qquad \text{(Simplifying (A))}$$

We notice first that the clock at the origin of T bears the following relationship to the adjacent clocks in S:

$$t_S = \frac{t_T}{\sqrt{1 - \dfrac{v^2}{c^2}}}. \qquad \text{(B)}$$

Secondly, we see that if two events occur at the same point x_T in T at t_{T_1}

and t_{T_2}, respectively, the time between them in S is:

$$\Delta t_S = \frac{\left(t_{T_2} + x_T \frac{v}{c^2}\right) - \left(t_{T_2} + x_T \frac{v}{c^2}\right)}{\sqrt{1 - \frac{v^2}{c^2}}} \tag{C}$$

$$= \frac{\Delta t_T}{\sqrt{1 - \frac{v^2}{c^2}}}. \tag{Q.E.D.}$$

Note: Only the clock at the origin of T bears the simple relationship (B) to the clocks in S, the other clocks bearing the more complicated relationship (A). The reason for this is that the clocks of T are synchronized with the velocities of light being c along both directions of the x axis. In Appendix C we consider a coordinatization $\bar{T}$ in which relationship (A) holds for all of the standard clocks of $\bar{T}$. Thks recoordinatization of T involves only a resetting of the clocks, but not a change in their standard rate, that is, it is a resynchronization of the standard clocks with the clock at the origin. But at this stage in the argument it was merely necessary to deduce relationship (C) from the Lorentz contraction and the Principle of Relativity.

APPENDIX C

When I say that using Lorentz's standard of temporal length is tantamount to using the clocks of the rest system for the moving system, there is an ambiguity for the absolute correctness of this statement depends on how the standard clocks of the moving system happen to have been synchronized. It is correct if the clocks of the moving system were synchronized as follows: when all of the clocks of S read $t_S = 0$, let the clocks of the moving system all likewise read 0, and let the coordinate system $\bar{T}$ be like T except for this synchronization. Therefore, when $t_S = 0$, $t_{\bar{T}} = 0$. This is tantamount to synchronizing the clocks in S and then putting them in motion. Then relation (B) of Appendix B, which holds in T only for $x_T = 0$, will hold for all $x_{\bar{T}}$ and *using Lorentz's standard of temporal equality on the standard clocks of $\bar{T}$ is equivalent to using the clocks of S in the moving system.*

Note that the use of the rest system clocks in a moving system has ample precedence in the General Theory of Relativity. For example, Møller and Grünbaum use these so-called 'coordinate clocks' in describing the situation on the rotating disk. Moreover, they also use spatial coordinates on the rotating disk which are 'the same' as those of the rest system. That is, they use on the rotating disk coordinates which are equivalent to our T' coordinates, and the Lorentz contraction shows up very clearly here as a difference between a unit rod and a unit coordinate difference of T' of the disk. The standard procedure in the Special Theory is, of course, to use the coordinates T in the moving system.

APPENDIX D

We wish to show that using the aether theoretic velocities of light in the moving system together with Lorentz's conventions regarding spatial and temporal measurement is factually equivalent to using Einstein's velocity c for light together with his conventions regarding spatial and temporal measurement. We are concerned here with round trip velocities for light rather than one-way velocities, for again as Einstein has shown, the one-way velocities involve a conventional ingredient. Round trip velocity is defined as follows:

$$rtv = \text{total distance} \,/\, \text{total time}\,.$$

The aether theoretic one way velocities are as follows:

$$\text{away from the origin: } c - v = d_{OA}/t_{OA}$$

$$\text{to the origin: } c + v = d_{AO}/t_{AO}\,.$$

where d and t are measured using Lorentz's conventions. To convert these to Einstein's conventions we divide d by $\sqrt{(1-v^2/c^2)}$ and multiply t by $\sqrt{(1-v^2/c^2)}$. Therefore the aether theoretic velocities measured using Einstein's conventions would be: away from the origin:

$$\frac{d_{OA}\sqrt{1-\dfrac{v^2}{c^2}}}{t_{OA}\sqrt{1-\dfrac{v^2}{c^2}}} = \frac{d_{OA}}{t_{OA}\left(1-\dfrac{v^2}{c^2}\right)} = c^2/c + v \tag{A}$$

to the origin:

$$\frac{d_{AO}\Big/\sqrt{1-\dfrac{v^2}{c^2}}}{t_{AO}\sqrt{1-\dfrac{v^2}{c^2}}} = \frac{d_{AO}}{t_{AO}\left(1-\dfrac{v^2}{c^2}\right)} = c^2/c - v. \qquad \text{(B)}$$

Now to get the round trip velocity we need the total distance and the total time light takes to travel from the origin to some point A and back to the origin:

$$\text{total distance} = 2\, d_{OA}\Big/\sqrt{1-\frac{v^2}{c^2}} = 2k$$

$$\text{total time} = \frac{k}{c^2/c+v} + \frac{k}{c^2/c-v} = 2k/c\,.$$

Therefore,

$$rtv = \frac{2k}{2k/c} = c\,. \qquad \text{(Q.E.D.)}$$

That is, when the aether theoretic one-way velocities are converted into velocities using Einstein's conventions and these are further converted into a round trip velocity, the result is identical with the round trip velocity of the Special Theory of Relativity.

Rice University

NOTES

[1] Whittaker, E., *A History of the Theories of the Aether and Electricity* (1953) II, p. 40.
[2] Einstein, A., 'On the Electrodynamics of Moving Bodies', reprinted in H. A. Lorentz *et al.*, *The Principle of Relativity* (1923), p. 40.
[3] Lorentz, H. A., 'Electromagnetic Phenomena in a System Moving with Any Velocity less than that of Light', reprinted in H. A. Lorentz, *op. cit.*
[4] *Ibid.*, p. 13.
[5] Grünbaum, A., *Philosophical Problems of Space and Time* (1963) p. 402.
[6] *Ibid.*

EDWIN LEVY

COMPETING RADICAL TRANSLATIONS: EXAMPLES, LIMITATIONS AND IMPLICATIONS

INTRODUCTION

In Chapter II of *Word and Object*, Quine presents and analyzes some of the philosophical difficulties found in the following situation: a linguist comes upon some jungle natives whose language is *entirely* unknown to him; the linguist then sets out to provide a translation of the jungle language into English. What the linguist would then be said to produce is a 'radical translation'. Quine's thesis about radical translation is this: while, in a sense, translations can be produced, there are philosophical reasons why there can be no *uniquely* correct equivalence class of radical translations.[1]

Although I have stated Quine's theme with respect to *radical translation*, it must be emphasized that even in this single chapter he is defending a far more general claim, i.e.:

> Thinking in terms of radical translation of exotic languages has helped make factors vivid, but the main lesson to be derived concerns the empirical slack in our own beliefs... To the same degree that the radical translation of sentences is undetermined by the totality of dispositions to verbal behavior, our own theories and beliefs in general are underdetermined by the totality of possible sensory evidence time without end. ([10] p. 78)

In recent works, [3], [6] and [7], Quine has expressed and illustrated his general indeterminancy-of-translation (GIT, hereafter) thesis in terms of natural science. Quine says, for example, that his "argument ... has been and will remain directed to you who already agree that there can be logically incompatible and empirically equivalent physical theories *A* and *B*" ([7] p. 181); i.e. *A* and *B* are and will remain underdetermined by the totality of possible experimental data.

Unpacking the GIT thesis we find that it is composed of the following claims (plus others which will not affect the argument here):[2]

(a) The old indeterminacy between physical theories *A* and *B* recurs in second intension. Insofar as the truth of a physical theory is underdetermined by observables,

Boston Studies in the Philosophy of Science, VIII.

the translation of a physical theory is underdetermined by translation of... observation sentences. ([7] p. 179)

(b) 'Translation' is to be understood in the broadest of senses. A translation from Russian to English is heterophonic. In contrast, a speaker of English usually, but not always, renders another English speaker's discourse homophonically. Thus homophonic rendering is a limiting case of translation and claim (a) applies to the home language situation.

According to Quine, "The indeterminancy of translation (GIT) is not just an instance of the empirically underdetermined character of physics." ([7] p. 180) Roughly speaking, if theories *A* and *B* are underdetermined (and empirically equivalent), GIT proclaims an additional indeterminancy: any *imputation* either of *A* or of *B* to another speaker *will itself be underdetermined*; not only is it impossible to decide empirically between *A* and *B*, but also we cannot experimentally ascertain whether another speaker *embraces A* or *B*. Clearly, then, the GIT thesis will be at least as pervasive as are underdetermined physical theories. Thus, for those, such as Quine, who believe that most discourse is underdetermined, the scope of the GIT thesis is very wide indeed.

Against this background, Quine's radical translation claim is seen to be an attempt to illustrate and to lend plausibility to his general thesis. The difficulty has been, however, that the initial Quinean radical translation example has tended to generate discussion about itself rather than about the GIT thesis.[3] In this paper I attempt to obviate this problem by presenting a radical translation situation which maximizes the connection between radical translation and GIT. After sketching (in Section I) a familiar case involving underdetermined physical theories – namely, physical geometrics – I map these into a radical translation situation in Section II.

In Sections III and IV, respectively, I analyze my radical translation case and then consider a home language 'translation' situation which involves the same physical theories as previously employed. The position I defend in these sections is this: Quine's alleged translation indeterminancy (i.e. GIT) is of a fundamentally different character from the first-order indeterminancy of physical theories. We can gain a glimmer of the distinction by noting that even if there happens to be two totally empirically equivalent theories *A* and *B*, it is entirely reasonable and meaningful to claim, "I believe or embrace *A* (and not *B*)." For this and other reasons, I suspect that there are cases in which empirical evidence, in the form of linguistic and other behavior, will tend to converge to a unique attribution of belief to another speaker (apart from the usual inductive uncertainties). I support this suspicion by showing that whether or not a given case yields to unique translation (within inductive limits) depends crucially on the linguistic and conceptual schemes of the translatee; a sufficiently rich scheme will approach unique translatability (apart from inductive uncertainties). In contrast, the first order indeterminancy is usually considered to be relatively insensitive to the quantity and quality of empirical evidence.

If my analyses are correct, I will have undermined the GIT claim. For, instead of being a necessary concomitant of underdetermined physical theories, the GIT thesis will have been shown to be merely a contingent addition to physical indeterminancy. Section V is devoted to some of the implications of GIT's failure; there will only be time for me to hint at my reasons for believing it to have been an interesting and important mistake.

I

The case from physics I will employ is a standard example used to illustrate a claim which is variously referred to as the 'relativity of geometry' or the 'conventionality of congruence'. The claim is this: since the length attributable to a rod transported through space depends essentially on a *definition* of congruence, the geometry attributable to that space is relative to the particular congruence convention chosen.[4] For example, if we have two congruent solid rods at a given point and we transport one of the rods to a distant point, we cannot attribute a length to the transported rod until we have settled on some congruence definition or metric, such as the customary one (apart from the action of differential forces): a solid rod maintains its length when transported. By recording coincidences with respect to this rod, the geometry of the region can be determined. But under a different congruence convention the transported rod may be assigned a different length and the resultant geometry ascertained by measurements may differ from the first case.

The example I will sketch is one suggested by Poincaré and discussed by many authors ([5] pp. 18–23; [12] and [1]). Imagine a region consisting of a very large blackboard on which is inscribed an ordinary Cartesian coordinate system and the two long lines, G_1 and G_2, such that each has

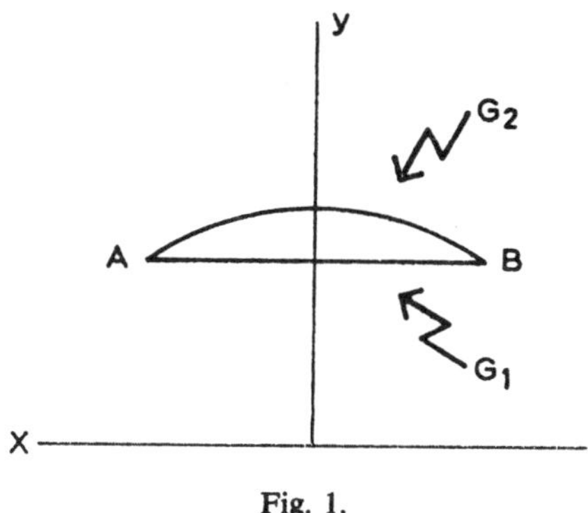

Fig. 1.

terminals at the points *A* and *B* (see Figure 1). Suppose further that two physicists, P_1 and P_2, conduct measurements on this surface; the physicists will use the same invar rod as a measuring stick. Let P_1 adopt the customary definition of congruence given above. After laying the rod along G_1 and G_2, he concludes that G_1 is shorter than G_2. He then inscribes other lines which have terminals at *A* and *B* and he concludes further that, within the usual inductive limits,

(1a) G_1 is the shortest path between *A* and *B*; that is, G_1 is a straight or geodesic.[5]
(1b) Euclidean geometry obtains on this surface.
(1c) G_2 is the arc of a circle whose center is the origin.

Now suppose that P_2 adopts a non-customary congruence convention according to which the length ascribed to the rod decreases in the positive *y* direction; that is, the farther the rod is above the *x* axis, the shorter the length ascribed to it.[6] (The customary metric prevails below the *x* axis). If P_2 carries out the same procedure as P_1 did, with the exception that P_2 'corrects' his 'raw' coincidence results in accordance with the non-customary metric, then P_2 reaches the conclusions:

(2a) G_2 is the shortest path between *A* and *B*; i.e. G_2 is a straight or geodesic.
(2b) A non-Euclidean (hyperbolic) geometry obtains on this surface.
(2c) G_1 is a Jackson-line.

One point about the disparities between the conclusions reached by P_1 and P_2 must be emphasized: the difference between the two are not empirical P_1 and P_2 place the same rod along G_1 and G_2 and it yields the same coincidences. That is, both P_1 and P_2 will find that the rod must be lain down more times to span G_2 than G_1; but while P_1's metric allows him to use his 'raw' data as length, P_2's metric requires him to 'correct' the coincidence data when assigning lengths.[7]

The usual interpretation of this example is succinctly stated by Grünbaum:[8]

> It is apparent that the supplanting of P_1's standard Cartesian metric by that of Poincaré P_2's has had the effect of *renaming* various paths on the semi-blackboard such that the language of hyperbolic geometry describes the very same facts of coincidence of a rod under transport on the semi-blackboard which are customarily rendered in the language of Euclidean geometry... we must conclude that P_2's hyperbolic metrization of the semi-blackboard possesses not only mathematical but also philosophical credentials as good as those of the Euclidean one. ([5] p. 21)

The philosophical credentials of P_2's metrization are further enhanced by considering whether his system could serve as a basis for a physics. Such a system could be developed, though it may be cumbersome, inelegant and uneconomic. For example, even though G_1 is not a geodesic in P_2's scheme, P_2 could identify G_1 as an example of a type of line he calls 'Jackson-lines'. P_2's physics could account for the fact that light follows Jackson-lines (not geodesics) and that although Jackson-lines are not the shortest route, they are the quickest path to traverse on foot (assuming no differences in terrain).

II

To map this example into a radical translation situation, we need only confront P_1 and P_2 with some natives whose language is entirely unknown to them and charge the physicists with the task of translation. Suppose that a native encounters the situation discussed in the above example only the line G_1 has been erased. Figure 2 represents the new example plus the letters G_2, A, B, x, y and the coordinate axes; these additions are included solely to remind us of the relationship to the previous case. Upon confronting the situation the native says, 'Tek'. Our physicists make assumptions and tests a la Quine and conclude that 'Tek' is an occasion sentence

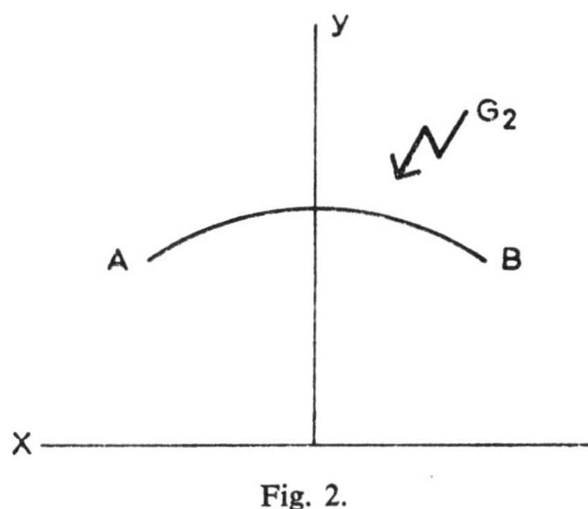

Fig. 2.

but not quite an observation sentence. However, P_1 makes the claim

(3a) 'Tek' is to be translated as 'Arc of a circle', and P_2 concludes

(3b) 'Tek' is to be translated as 'Straight line' or 'Geodesic'.[9]

Furthermore, if the natives are given our old invar rod and are allowed to add inscriptions to the surface and if both P_1 and P_2 agree that the natives are making utterances involving geometry, then their respective translations of these utterances will display the same sort of disparity as the one between (3a) and (3b).[10] It is clear what is happening here: P_1 is *attributing* to the natives the customary metric, while P_2 is *imputing* the non-customary one introduced in Section I. The result is that throughout P_2's translation, any sentence which involves the geometry of the region, for example a statement in optics about the path of a light ray, would assume that the native employs hyperbolic geometry in the region.

Although one could attempt to build in other, non-geometric divergences into the translations, I shall not do so. These then are my sketches of putative competing radical translations: the translations offered by P_1 and P_2 (abbreviated as 'P-translations' or 'P-schemes' hereafter) which differ only in those statements which involve, however indirectly, geometric considerations.

III

As an introduction to the general problem, consider this illustration of how one might set out to attribute a particular metric to the natives:

Under the assumptions of Sections I and II, if a native were to lay a rod successively along G_2, we would observe that approximately 22 coincidences would be required to span the line. Yet P_2 attributes to the native the view that the length of G_2 in 18 units.[11] This difference between 22 coincidences along a line and a length of 18 units assigned to the same line, is the distinction mentioned earlier between 'raw' and 'corrected' data. If the physicists could demonstrate this distinction in native usage, this would reveal that P_2's metric is the one actually employed. And the absence of this distinction would indicate that P_1's metric is the correct one. Now we could test for this distinction if we could induce the native to count aloud each time he lays down the rod. You might find, for example, that the utterance at the end of the line is *the same as the one given in response to a query about the length of* G_2. Then you could conclude that, for the native, length is solely a function of coincidences; i.e., P_2's metric must be rejected.

There are at least two difficulties which undermine the procedure and reasoning in the illustration. (a) Even if you succeed in inducing the native to count, you do not know whether he is counting coincidences or length. Following a tradition passed down in his tribe for millennia, the native might count length directly; i.e. he might automatically compensate for position when he is engaged in geometric measuring. If this were so, his 'raw' data would correspond to our (identifying us with P_1) 'corrected' data and his utterances would be as they are in the illustration. However, P_2's metric would be the right one. (b) In order to query the native about the length of G_2, you must equate some native utterance U to 'length'. If instead, U is translated as 'coincidences', then you would (ignoring difficulty (a) above) reach a conclusion contradictory to the one in the illustration.

Both of these difficulties suggest that there is a close relationship between attributing a metric to the natives and translating some native utterances as 'length' or 'coincidence'. In Quine's analysis the latter relata is described as a choice of 'analytical hypotheses'. According to Quine, translation of observation sentences and truth functions can be accomplished (apart from usual inductive uncertainties). Also, some issues, short of translation, could be settled about other types of sentences. In order to proceed further, however, the translator would have to settle on a set of analytical hypotheses. Roughly, analytical hypotheses are a selected list of native 'words' each of which is hypothetically and approximately equated to an English word or phrase. These hypotheses become the framework for the rest of the translation; other sentences must fit the pattern they outline, though not strictly. Quine's translation thesis (GIT) is now statable in terms of analytical hypotheses: "analytical hypotheses ... are not determinate functions of linguistic behavior" ([10], p. 69); "to project such hypotheses beyond the independently translatable sentences at all is in effect to impute our sence of linguistic analogy unverifiably to the native mind" ([10] p. 72); "... there can be no doubt that rival systems of analytical hypotheses can fit the totality of speech behavior to perfection ... and still specify mutually incompatible translations of countless sentences insusceptible of independent control." ([10] p. 72)

Quine does provide an example of the effects of adopting two rival analytical hypotheses. (Not two full rival sets of hypotheses: he suggests that such an illustration would be extremely laborious)

> ...we saw that stimulus meaning was incapable of deciding among 'rabbit', 'rabbit stage' and various other terms as translations of 'gavagai'. If by analytical hypotheses we take 'are the same' as translation of some construction in the jungle language, we may proceed on that basis to question our informant about sameness of gavagais from occasion to occasion and so conclude that gavagais are rabbits and not rabbit stages. But if instead we take 'are stages of the same animal' as translation of that jungle construction, we will conclude from the subsequent questioning of our informant that gavagais are rabbit stages ([10] pp. 71–72; also see [8] pp. 189–191).

We return to our examination of the *P*-schemes by considering the *P* version of the previous quote:

> ...we saw that we were incapable of deciding between 'arc of circle' and 'straight line' as translations of 'tek'. If by analytical hypotheses we take 'length' as translation of some construction in the jungle language, we might proceed on that basis to question our informant and conclude that teks are arcs of circles. But if instead we take 'coincidences' as translations of that jungle construction, we might conclude from the subsequent questioning of our informant that teks are straight lines.

The question before us now is this: Can we settle on a unique set of geometrical analytical hypotheses? I believe that the determination of a set of analytical hypotheses is a matter of degree. The degree to which we can single out a set of analytical hypotheses depends crucially on the conceptual and linguistic scheme of the natives. That is, it depends on the breadth of the native scheme and the pervasiveness of the particular analytical hypotheses within the account of the scheme. If, for example, the native scheme includes only three utterances, we will be unable to decide between the *P*-translations of 'Tek', one of those utterances. Similarly, we will be unable to choose between *P*-schemes even if the native linguistic apparatus is vast, but we have reason to believe that 'Tek' is the only native utterance having anything directly or indirectly to do with geometric matters.

Certainly these are extreme and uninteresting cases. Let us turn to an intermediate situation like the one in the illustration which began this section. Here the natives not only utter 'Tek', but they also exhibit what we interpret as counting behavior. My remarks above imply that we can get around the difficulties (a) and (b) previously discussed and settle on unique analytical hypotheses.

Before I support my claim, however, let me say in general terms why I believe that the breadth of the native scheme and the pervasiveness of the analytical hypotheses are so significant. Although Quine does not say so,[12] identification of native assent and dissent bears a very strong resemblance to choosing analytical hypotheses. Certainly 'Yes' and 'No' are not observation sentences and, as Quine suggests, we may have to revise our original hypotheses about assent and dissent in light of further evidence. Nevertheless, we can be fairly sure about the identification. Why? I maintain that if the natives have a linguistic scheme of even minimal richness, our decision about yes and no will be tested time and again. In other words, our identification of assent and dissent is tightly woven into our entire translation. We *could* be wrong about our identification, even after several revisions. Native assent behavior could vary as a function of time, of place, of the weather; natives could invert or otherwise vary their assent terminology when they go from gavagai talk to glip talk. But we strongly resist making such hypotheses. The point is this: when you confront a reasonably broad native scheme; and when you make an hypothesized identification, say of assent behavior, which has reverberations throughout a significant portion of your account of the native scheme; and when a conflicting identification could account for empirical evidence only by carrying out a maximum mutilation and complication of your overall description; then you have good, but not indubitable, evidence for claiming that the native is expressing agreement when he displays what you have identified as assent.

I will now attempt to show that geometrical analytical hypotheses share, to a degree, the characteristics of identification of assent. In the illustration at the beginning of this section, we assumed that one could identify counting behavior. Now imagine that you claim to have compiled a great deal of evidence about native counting behavior and you conclude, inductively, that the native utterance, 'Eno', 'Wot', etc., correspond to 'One', 'Two', etc. This would be question-begging if I were implying that we could fully ascertain native individuative mechanisms in all cases or even in many cases. Instead, I am suggesting that we might observe the native uttering the same sequence in a large number of cases. The fine points of native individuation may well remain extremely obscure. When dealing with dishes, days, people, fingers, wheel turns, sandals, huts, spears, trees, and many other things, the natives exhibit what we tentatively identify as a particular kind of counting behavior. Perhaps when dealing with rabbits and robins, the native behavior is anomolous.[13] But no matter, we do have some evidence for claiming that the natives do usually count in the hypothesized manner. Further evidence is easy to imagine, for the natives might manipulate these symbols in an arithmetic scheme. Even this additional evidence fails to confer certainty on our counting hypotheses, but extensive, interlocking evidence increases our confidence in this case just as it does for assent-dissent.

Now when we induce the native to count as he lays the rod down twenty-two times along G_2 and his utterance at the end of the line is the same as the one we had previously identified with 'Twenty-two', we have some evidence that the native is counting coincidences. Further evidence would be forthcoming if his other measurement-utterances showed a similar correspondence. For example, he might inscribe G_1 and his measurement behavior, when interpreted as above, might reflect that it takes the fewest coincidences to span G_1 than any other line.

Could we now query the native to find out if he equates length with coincidences and thus employs the standard metric? Perhaps so, perhaps not. For, the native linguistic scheme may or may not reflect coincidences and length as two separate and distinct concepts. Roughly speaking, all of our linguistic queries to the native might be understood as being about coincidences.[14] Yet the natives might mentally convert the coincidences to length in accordance with P_2's metric. But if the native linguistic scheme is poor, he may not have developed means for expressing this conversion and our attribution is in doubt.[15] However, in a more developed linguistic scheme this difference in belief will be reflected. For example, if the natives appear to make corrections for substance-specific distortions or if the natives appear to utilize analytical geometry, we would have some basis for hypothesizing and testing a distinction between coincidences and length.

In sum, what I claim to have established is this: although P_1 and P_2's physical theories might be permanently underdetermined, their respective translations are not indeterminate in the same sense. While the first order uncertainty will be relatively insensitive to physical evidence, translational indeterminancy will tend to vary inversely with the alleged geometric richness of the native linguistic scheme; of course the translational indeterminancy will always be subject to the usual inductive uncertainties. By establishing a parallel between the identification of assent-dissent and the choice of analytical hypotheses, I claim that the uncertainty between the two P-translations can approach, as richer native schemes are considered, the uncertainty inherent in a single putative translation.

IV

There are two major reasons for sketching a home language situation. In the first place, in home language cases the linguistic and conceptual schemes of the translator and translatee are comparable. Thus, no claim can be established by a disguised appeal to linguistic disparity. In the second place, this case will counter any contention to the effect that the previous analysis involved the assumption of a 'neutral' metalanguage. The setting is this: our old Euclidean physicist, P_1, encounters a new English-speaking physicist, P_3. After a few conversations, an insidious suspicion takes hold of P_1: although P_3 *says* he is a Euclidean and although P_3 has agreed with P_1 on the two occasions they identified geodesics, P_1 suspects that P_3 is a 'secret non Euclidean'.[16] Now, P_1 is going to pay special attention to P_3's speech. Up to now P_1 'translated' P_3's utterances homophonically, e.g. when P_3 said 'G_1 is a geodesic', P_1 translated this as 'G_1 is a geodesic'. P_1 decides to undertake a heterophonic translation of P_3's speech. That is, P_1's new translation will accord with the beliefs he attributes to P_3. A fragment of P_1's effort is contained in Table I.
As is clear from Table I, P_2's procedure has been to interchange 'Lengths' and 'Coincidences'. This method has failed because heterophonic renderings 6b and 7b contradict 2b. (See [9], Chapter 6). Of course, the failure of this method does not prove much; there are lots of other procedures to consider. For example, P_1 could persevere with his original method *and* heterophonically translate the numerals. Or P_1 could try to translate 'Coincidence' for 'Length' but not *vice versa*.

I do not have the space to examine these and other alternatives; I will

TABLE I

P_1 and P_3's statements		P_1's heterophonic translation of P_3's statements	
1a	G_1 is a geodesic	1b	G_1 is a Jackson-line
2a	G_2 is an arc-of a circle	2b	G_2 is a geodesic
3a	Length of $G_1 = 19$ units	3b	Coincidences to span $G_1 = 19$
4a	Length of $G_2 = 22$ units	4b	Coincidences to span $G_2 = 22$
5a	Length of $G_1 <$ length of G_2	5b	Coincidences for G_1 < coincidences for G_2
6a	Coincidences for $G_1 = 19$	6b	Length of $G_1 = 19$ units
7a	Coincidences for $G_2 = 22$	7b	Length of $G_2 = 22$ units

only offer a general conclusion. P_1 can offer many consistent heterophonic translations which will accord with the non-Euclidean beliefs he attributes to P_3. However, the deviations from homophonic translation will be so enormously extensive that to say P_1 and P_2 share a geometric home language will be true only in the most Pickwickean of senses. (At the very least, large scale heterophonic translation will enter into counting behavior, arithmetic, analytical geometry.) To claim that two people have a language in common is not to deny them heterophonic translation; but at the point that heterophonic translation becomes predominant, they no longer share a language. Thus the notion of sharing a language is called into question by the adoption of GIT.

V

I believe that Quine's General Indeterminancy of Translation thesis has been seriously undermined. What survives could be called the Possible Problem in Translation (PPT) thesis:

Given two physical theories, *A* and *B*, which are underdetermined by all physical evidence *obtained thus far*, an attribution of one of these theories to another speaker may suffer indeterminancies above and beyond the usual inductive uncertainties.

Here, I can only outline some reasons in support of PPT and some of its implications.

(1) PPT captures the essential spirit of GIT: both warn us against unjustifiably attributing beliefs and theories to others.

(2) By moving to theories which are underdetermined only by extant

physical evidence, PPT is of significantly wider scope than GIT. The farther one proceeds away from mathematical physics and other formalized disciplines, the more difficult it is to establish total empirical equivalence. And it is precisely in such cases that the stakes are highest. For if we know that *A* and *B* are totally empirically equivalent, then an error in attribution might be regarded simply as an error or uncertainty in record-keeping. In contrast, if *A* and *B* eventually fail to be totally empirically equivalent, an overhasty attribution may have caused us to seriously misconstrue another's beliefs and behavior.

(3) And yet, PPT is applicable to cases in which two theories are totally empirically equivalent. This is a feature well worth retaining, at least for now. For such works as Glymour's suggest that even totally empirically equivalent theories may ascribe different structures to the world and thus provide different explantions.[17]

(4) PPT does not provide a basis for ontological relativity. However, as Quine has recently pointed out, neither does GIT: "... the question raised by deferred ostension, e.g. as between expressions and their Gödel numbers, are strictly a matter of inscrutability of terms. This, not the indeterminancy of translation, is the substance of ontological relativity" ([7] p. 183).

(5) There is another benefit to be derived from PPT's applicability to theories which may not be totally empirically equivalent. As mentioned in point (2), this enables us to consider theories from less formalized disciplines in which decision procedures for total empirical equivalence are not well specified. But it is also in such fields that theoretical networks are less clearly delineated. From my presentation Sections III and IV, it should be clear that the uncertainty of translation will be lower if the theoretical networks is tightly interlocking; this is why we have confidence in the translation of assent-dissent and in some cases of geometrical analytical hypotheses. In contrast, when the theoretical network is loosely knit, as it is in the less formalized disciplines, the uncertainty is higher and there is even greater need to be chary of hasty attribution.

The University of British Columbia

BIBLIOGRAPHY

[1] Barankin, E. W., 'Heat Flow and Non-Euclidean Geometry', *American Mathematical Monthly* **49** (1942) 4–14.
[2] Barrett, R. B., 'On the Conclusive Falsification of Scientific Hypotheses', *Philosophy of Science* **36** (1969) 363–374.
[3] Davidson, D. and Hintikka, J. (eds.), *Words and Objections: Essays on the Work of W. V. Quine*, D. Reidel Publ. Co., Dordrecht-Holland, 1969.
[4] Glymour, C., 'Theoretical Realism and Theoretical Equivalence', this volume.
[5] Grünbaum, A., *Philosophical Problems of Space and Time*, A. A. Knopf, New York, 1963.
[6] Henkin, L., Suppes, P., and Tarski, A. (eds.), *The Axiomatic Method*, North Holland Publ. Co., Amsterdam, 1959.
[7] Quine, W. V. O., 'On the Reasons for Indeterminancy of Translation', *The Journal of Philosophy* **67** (1970) 178–183.
[8] Quine, W. V. O., *Ontological Relativity and Other Essays*, Columbia University Press, New York, 1969.
[9] Quine, W. V. O., *Philosophy of Language*, Prentice Hall, Englewood Cliffs, New Jersey, 1970.
[10] Quine, W. V. O., *Word and Object*, Technology Press, New York, 1960.
[11] Reichenbach, H., *The Philosophy of Space and Time*, transl. by M. Reichenbach and J. Freund, Dover, New York, 1958.
[12] Robertson, H. P., 'The Philosophical Significance of the Theory of Relativity', in *Albert Einstein Philosopher-Scientist* (ed. by P. A. Schilpp), Harper Torchbook, New York, 1959.

NOTES

[1] Since the argument is unaffected by the number of candidate radical translations – as long as that number is greater than one – I shall speak only of two such candidates. Since I am considering only two possibilities, a 'unique' choice of translations means that one of the candidates can be eliminated. For my purposes the members of an equivalence class of translations will differ from one another only trivially; e.g., two members will differ only in one word and the two expressions will be synonymous.
[2] In [7] Quine suggests that one can maximize the scope of GIT by two sorts of arguments: (a) arguments "meant to persuade anyone to recognize the indeterminancy of translation of such portions of natural science as he is willing to regard as underdetermined by all possible observations" (p. 183) and (b) "whatever arguments for indeterminancy of translations can be based on the inscrutability of terms." (p. 183) Although Quine says he prefers not to speculate on which route is preferable, his article rather clearly endorses (a) and has little encouraging to say about (b). In this paper, I attack only (a).
[3] In [7] Quine says, "My *gavagai* example has figures too centrally in discussions of the indeterminancy of translation." (p. 178)
[4] While Carnap and Reichenbach base the relativity of geometry on testability, Grünbaum claims that the basis is the metrical amorphousness of space. (See [5], Chapter III). This disagreement does not affect the argument in this paper. Throughout I shall be assuming that the regions in question have constant curvature and are free

of differential forces. Just how the differential forces can be determined has been the subject of considerable debate recently. (See [2]) The argument in this paper, however, will not be significantly affected by the outcome of that debate.

[5] A geodesic is usually, but not always, the shortest line between two points. I shall ignore the exceptions here (see [5], p. 13, n. 4).

[6] A metric which entails P_2's congruence definition is

$$ds^2 = \frac{dx^2 + dy^2}{y^2}.$$

According to this metric, the geodesics of the surface satisfy the equation

$$(x-h)^2 + y^2 = c^2$$

which, in Euclidean terms represents a family of semi-circles centered on the x axis.

The distance as measured along a geodesic between points $A(x_1, y_1)$ and $B(x_2, y_2)$ is

$$s_g = \tfrac{1}{2} \log e \left\{ \frac{x_2 - (h-c)}{(h+c) - x_2} \cdot \frac{(h+c) - x_1}{x_1 - (h-c)} \right\}$$

(for mathematical details, see [1]).

Applying these equations to the case $A(-1, 1)$ and $B(1, 1)$, P_2 would conclude that G_1 is 2.2 units long and G_2 is approximately 1.8 units. (To facilitate later discussion, suppose that the units are chosen such that the rod is always regarded $\frac{1}{10}$ unit in length.)

It should be mentioned that the original Cartesian coordinate system involving x's and y's above is totally arbitrary; it simply provides labels which imply no commitment.

[7] The terms 'raw' and 'corrected' appear in quotation marks because they are, in a sense, figurative. This matter is discussed in Section III.

[8] A reason for disagreeing with this interpretation is given in note 15 and in the text to which n. 15 refers.

[9] For Quine, observationality is a matter of degree, but observation sentences are "occasion sentences whose stimulus meanings vary none under the influence of collateral information." ([10] p. 42) The sort of collateral information I imagine intruding in the 'Tek' case is this: some natives may well have carried out measurements on G_2 before experiments began and others may not have. Thus, I have not classified this sentence as an observation sentence; similarly, I would consider the sentence 'Line' to be more observational than 'Geodesic' or 'Straight line'.

[10] It is certainly conceivable that P_1 and P_2 might disagree whether some native utterances involve geometry. That disagreement would clearly add to the possible divergences between the translations. However, I will not examine this issue in the present paper.

[11] See note 6.

[12] Professor Quine has pointed out that he did say so in [3] p. 312.

[13] Actually I believe that Quine's 'Gavagai' example tacitly trades on just this sort of anomaly. Consider the long quote on p. 12. If we had taken 'are the same' to be a translation of some native construction U and if we had observed that the natives had used U in all of the situations listed on p. 14 and if we had observed the native using U in the 'Gavagai' case, then we may have been more hesitant about offering an alternative translation of U for the 'Gavagai' case. That an alternative could have been offered and defended, I do not deny, especially since such an introduction would have exceedingly few reverberations within our overall translation.

[14] I believe that Quine is discussing a closely related matter when he says,
"Something of the true situation verges on visibility when the sentences concerned

are extremely theoretical. Thus who would undertake to translate 'Neutrinos lack mass' into the jungle language? If anyone does, we may expect him to plead in extenuation that the natives lack the requisite concepts; also, that they know too little physics. And he is right, except for the hint of there being some free-floating, linguistically neutral meaning which we capture in 'Neutrinos lack mass', and the native cannot" ([10] p. 76).
A necessary condition for translating 'Neutrinos lack mass' into jungle is that the natives' conceptual and linguistic schemes contain a distinction between massive and massless entities; if that distinction is discovered, translation seems almost feasible. There appear to be undeniable parallels between, on the one hand, 'Neutrinos lack mass', massive and massless entities, and, on the other hand, 'Congruence is conventional' lengths and coincidences.
[15] This, incidentally, represents a divergence from the assent-dissent case. Length and coincidences are not contradictories as are yes and no.
[16] This example was suggested in a discussion with Professors Paul and Pat Churchland of the University of Manitoba.
[17] In an unpublished paper, Clark Glymour offers a means of distinguishing between empirically equivalent theories and of blocking the claim that such theories should be considered members of an equivalence class. (This last claim is the position of Reichenbach and other realists.) According to Glymour, two theories might be empirically equivalent and yet offer non-equivalent explanations *because the two theories fail to have a model in common*. Failure to have a model in common, alleges Glymour, entails that the theories ascribe different structures to the world; hence the difference in explanations. One of Glymour's main examples is close to home: citing the works of Robinson, Tarski and Szmillew (in [6]), he claims that complete parts of Euclidean and non-Euclidean geometries fail to have a model in common. Glymour's work will require further elaboration but it might be a sturdy hook on which to hang the claim that two (or more) empirically equivalent theories or translations or conceptual schemes should be considered non-equivalent.

GERALD J. MASSEY

IS 'CONGRUENCE' A PECULIAR PREDICATE?

The logical status of spatial and temporal congruence has been much debated. Recently the topic of debate has shifted from the alleged conventionality of congruence to the alleged eccentricity of congruence predicates. For example, espousing a point of view suggested to him by Nuel Belnap, Adolf Grünbaum has recently remarked that the predicate '*x* is (spatially) congruent to *y*' deviates from the 'classical account' of the interrelations between the intensional and extensional components of the meaning of a predicate. By the *classical account* (hereafter CA), Grünbaum means the rather common view that the extension of a predicate is completely determined by its intension. Contrary to CA, Grünbaum maintains that "the fact that 'being spatially congruent' means sustaining the relation of spatial equality does not suffice at all to determine its extension uniquely in the class of spatial intervals."[1] Rather, he states, the "term 'spatially congruent' has the same (nonclassical) intension and only a different extension in the context of alternative space metrics."[2] (The parenthetical 'nonclassical' is meant only to underscore the allegation that '*x* is congruent to *y*' deviates from CA.)

Several criticisms have been directed against Grünbaum's intension-extension thesis just described, one of them by me. I thought for a time that Grünbaum was confusing the second-order predicate '*x* is a congruence relation' with the first-order predicate '*x* is congruent to *y*', that his thesis came to nothing more than the view that the intension of the second-order predicate does not fully determine the extension of the first-order predicate. Thus I remarked that whereas the congruence axioms determine the intension of the second-order predicate and therewith also its extension, they fail to determine the extension of the first-order congruence predicate precisely because they underdetermine its intension. The intension of the latter, I added, was given fully only by the specification of a member of the extension of the second-order predicate as the extension of '*x* is congruent to *y*'.[3] Behind my charge that Grünbaum was confusing first- and second-order predicates was the conviction, itself a residue of

CA, that the intension of 'x is congruent to y' could not remain invariant when one passed from the context of one metric to that of another metric yielding a different congruence class. I no longer think that Grünbaum was guilty of the aforesaid confusion, and for reasons that will become clear in the sequel I now acknowledge the invariance of the intension of 'x is congruent to y' in the context of alternative space metrics.

It would be unfair to infer from his emphasis on the deviation of 'x is congruent to y' from CA, or from my earlier defense of CA as applied to that same predicate, that either Grünbaum or I deemed CA otherwise adequate, i.e. as though we considered CA unproblematically true apart from the congruence family of predicates. Philosophical perplexities about the nature of intensions aside, as a general thesis about predicates CA has been discredited many times over, for example by the recognition of conceptually inconsistent predicates (such as 'x is heterological') and of referentially opaque predicates (such as 'necessarily, x is greater than seven'). Even when restricted to conceptually consistent, referentially transparent predicates, CA must be refined to take into account such factors as vagueness, the need to restrict discourse to a circumscribed domain, the tensed character of many predicates, and tacit relativities. What makes the alleged deviation of 'x is congruent to y' from a *suitably refined* version of CA worthy of notice are the grounds thereof, being *prima facie* neither conceptual inconsistency nor referential opacity but geochronometric conventionalism, the thesis that physical space and physical time lack nontrivial intrinsic metrics.[4]

Just as Putnam has tried to trivialize Grünbaum's geochronometric conventionalism by reducing it to the latitude to employ a semantically uncommitted noise as one sees fit,[5] so Richard Swinburne has tried to trivialize Grünbaum's intension-extension thesis by reducing it to a platitude that applies indifferently to almost all predicates. Swinburne takes Grünbaum to be affirming that one can divide the rules for the use of 'x is congruent to y' into formal and material rules. The formal rules relate the predicate to other predicates; the material rules tell how to ascertain whether two intervals are congruent. Then, by identifying the intension of a predicate with the formal rules for its use, Grünbaum is able to claim that the intension of 'x is congruent to y' fails to determine its extension. Having attributed the foregoing position to Grünbaum, Swinburne offers three criticisms of it. First, it shows nothing special about

'x is congruent to y', since the rules for the use of almost any predicate can be divided into formal and material rules. Second, with seemingly equal warrant one could say that the formal and material rules together constitute the intension of 'x is congruent to y', and hence that its intension does indeed determine its extension. Third and finally, even if it were more appropriate to regard only the formal rules as constitutive of the intension of 'x is congruent to y', the resulting underdetermination of extension would stem from a semantical fact about the predicate rather than from a structural feature of space such as its intrinsic metrical amorphousness.[6] As criticisms of the position that he attributes to Grünbaum, Swinburne's comments are devastating. But I shall presently offer an alternative interpretation of Grünbaum's intension-extenson thesis, an interpretation that both seems to be more faithful to Grünbaum's intentions and that is immune to Swinburne's criticisms. In short, I question Swinburne's exegesis, not his argumentation.

It will be helpful first to reformulate CA. Rather than say that intension determines extension, let us say instead that the extension of a predicate is a function of two things, its intension and a (global) state of affairs (a possible world). Accordingly, as usually formulated, CA may be construed as making a tacit reference to the actual state of affairs (the actual world). Now Richard Butrick has lately coined the term 'selector predicate' for

> a predicate whose extension is not a function if its intension and a given state of affairs. Eschewing intensions [sic!], P is a selector predicate if its extension under the same state of affairs could be different without making P a homonym, that is, such that P is synonymous with P' and P'' whereas P' and P'' are not synonymous.[7]

Butrick cites 'x is off limits' and 'x is a capital offense' as paradigmatic selector predicates, and contrasts them with observational predicates thus: "the intention [sic!] of (say) an observational predicate such as 'is blue', determines (for a given state of affairs) the extension, namely, those objects that have the property of being blue. Arbitrary variation of the extension of the predicate 'is blue' would deprive the term of its customary meaning."[8] Butrick seems to regard the possibility of varying extension without changing intension as the defining trait of selector predicates. One might be tempted to reply that Butrick ignores the tensed character of most natural language predicates, namely the fact that they do not possess extension *simpliciter* but only relative to a moment.[9] The extension of 'x

is blue' also varies as things change in color, and someone who is handy with a paint brush can modify this variation almost at will. Consequently, the argument runs, '*x* is blue' is no less a selector predicate than '*x* is a capital offense'. Moreover, when 'state of affairs' is construed broadly enough to include legal statutes, the extension of either predicate at a particular moment is a function of its intension and the actual state of affairs.

But it is probably fairer to Butrick to construe his references to states of affairs as references to *purely physical* states of affairs (possible worlds stripped of all deontic features). Assuming that he can make intuitive sense of the concept of a purely physical state of affairs, one might be inclined to say that the extension of '*x* is blue' is indeed a function of its intension and a purely physical state of affairs, whereas the extension of '*x* is a capital offense' is not.

Butrick adverts to a feature that he mistakenly thinks all selector predicates have, namely that they lend themselves to both prescriptive (Butrick's term is 'performatory') and descriptive use.[10] For example, a law-giving body can use '*x* is a capital offense' prescriptively *to make* rape a capital offense; an ordinary citizen can then use it descriptively *to report* that rape is a capital offense. Notice that some selector predicates do not lend themselves to prescriptive use. For example, '*x* is a capital offender' is presumably a selector predicate, but one that has no prescriptive use. Not even an absolute monarch can make a man a capital offender merely by declaring him to be one, though the monarch can do so indirectly by declaring one of the man's actions to be a capital offense. I shall call (intensionally pre-empted) predicates that lend themselves to both prescriptive and descriptive use *preceptive* predicates.

The promised interpretation of Grünbaum's intension-extension thesis can now be given: *Relative to intrinsically metrically amorphous spaces, 'x is congruent to y' is a preceptive predicate.* Let me explain briefly how '*x* is congruent to *y*', regarded as a preceptive predicate, might be thought to deviate from CA. It is obvious that if 'state of affairs' is understood broadly enough, then neither selector predicates nor preceptive predicates deviate from CA, which asserts only that the extension of a predicate is a function of its intension and a state of affairs. (Butrick's claim that selector predicates deviate from CA was warranted only by his narrow physical construal of 'state of affairs'.) But even without narrowing the sense of 'state of affairs', one might plausibly suppose that CA entails, as a corollary, that the extension of an *intensionally pre-empted physical predicate* (a *physical predicate* is a predicate drawn from a theory within physics) does not depend upon the prescriptive use of language. But the extension of a preceptive predicate does depend on the prescriptive use of language. So, if the intensionally pre-empted physical predicate 'x is congruent to y' is a preceptive predicate, then its behavior is at variance with the foregoing putative corollary of CA, and thereby deviates from CA.

Notice that the intension of a preceptive predicate undergirds both its descriptive and prescriptive use. Unless 'x is a capital offense' had a full measure of intension, it could not be used either to make an action a capital offense or to report that some action is or is not a capital offense. Therefore, if 'x is congruent to y' is a preceptive predicate, using it prescriptively to make certain kinds of quasi-disjoint, non-degenerate spatial intervals congruent cannot be regarded as an instance of trivial semantical conventionalism, i.e. as a case of pre-empting a semantically uncommitted expression.[11] So, if the linguistic facet of Grünbaum's geochronometric conventionalism is adequately rendered by the thesis that 'x is congruent to y' is a preceptive predicate relative to spaces that lack intrinsic metrics, then Putnam's attempt to assimilate geochronometric conventionalism to trivial semantical conventionalism must be adjudged futile.

There are interesting differences among preceptive predicates. Some are such that their descriptive use is wholly governed by the prescriptive use of preceptive predicates. For example, whether 'x is a capital offense' is true of some action depends in every case on the prescriptive use of it and related predicates. Such predicates shall be called *pure* preceptive predicates. Unlike 'x is a capital offense', 'x is congruent to y' is an *impure* preceptive predicate. Let I_1 be a closed spatial interval that extends over a closed interval I_2. Then 'x is congruent to y' is true of $\langle I_1, I_1 \rangle$ and of $\langle I_2, I_2 \rangle$, but false of $\langle I_1, I_2 \rangle$ and of $\langle I_2, I_1 \rangle$, independently of prescriptive uses of preceptive predicates. But if I_1 and I_2 are closed quasi-disjoint spatial intervals of an intrinsically metrically amorphous space, then whether 'x is congruent to y' is true of $\langle I_1, I_2 \rangle$ depends on what prescriptive uses have been made of it and related preceptive predicates such as 'x is the same length as y'.

There is another difference worth noting between '*x* is a capital offense' and '*x* is congruent to *y*'. In default of any prescriptive uses of '*x* is a capital offense' or related preceptive predicates, that predicate is false of every action provided only that there is a legal system. But in default of any prescriptive uses of '*x* is congruent to *y*' or related predicates, that predicate is neither true of nor false of a pair of closed quasi-disjoint intervals; its descriptive use within the class of pairs of closed quasi-disjoint intervals presupposes the prescriptive use of it or related preceptive predicates.

Grünbaum holds that discrete space possesses cardinality-based nontrivial metrics intrinsic to the intervals thereof, two intervals being congruent by any of these metrics if and only if they contain the same number of points. Furthermore, in a qualified sense of 'nontrivial', Grünbaum contends that no other nontrivial metric is intrinsic to the intervals of a discrete space.[12] Does '*x* is congruent to *y*' lend itself to prescriptive use relative to such a space? If so, then my interpretation of Grünbaum's intension-extension thesis (as the thesis that '*x* is congruent to *y*' is a preceptive predicate relative to intrinsically metrically amorphous spaces) must be abandoned, for the preceptive character of the congruence predicate would then be independent of the intrinsic nontrivial metrizability of the intervals of a given spatial manifold. Grünbaum acknowledges that, besides the aforementioned cardinality-based metrics, there are other nontrivial metrics extrinsic to the intervals of discrete space that generate different congruence classes. Moreover he acknowledges that in discrete spaces exhibiting certain kinds of singularities there will be nontrivial metrics that generate congruence classes different from the one generated by the cardinality-based metrics but that are nevertheless intrinsic to the space (although not intrinsic to the intervals).[13] So it appears that even in discrete space the descriptive use of '*x* is congruent to *y*' presupposes prescriptive uses of it or related preceptive predicates. That is, it seems that '*x* is congruent to *y*' functions as a preceptive predicate relative even to discrete space – to the utter discredit of my interpretation of Grünbaum's intension-extension thesis.

It seems to me neither in itself implausible nor exegetically farfetched to understand Grünbaum to be according *intensional primacy* to nontrivial congruences that are *intrinsic to the intervals* of a space. By this I mean that, in a space having exactly one congruence class of intervals that is generated by a nontrivial metric intrinsic to those intervals, the intension of '*x* is congruent to *y*' determines without further ado that congruence class as its extension. But if there are several such congruence classes, or none at all, then the intension of '*x* is congruent to *y*' determines a full-blown extension only through the prescriptive use of '*x* is congruent to *y*' or of related preceptive predicates such as '*x* has the same length as *y*'. Accordingly, because the cardinality-based congruence class is the only congruence class generated by a nontrivial (in the qualified sense of 'nontrivial' mentioned in footnote 12) metric intrinsic to the intervals of a discrete space, the intension of '*x* is congruent to *y*' suffices by itself to determine an extension relative to a discrete space. To make some other congruence class the extension of '*x* is congruent to *y*' in a discrete space, therefore, would require a modification of the rudimentary or unadorned intension of this predicate. Contrariwise, because there are no nontrivial intrinsic metrics of any kind in dense or continuous space, no partition of the intervals of such a space is picked out by the unadorned intension of '*x* is congruent to *y*' as the equivalence class generated by that predicate, apart from the prescriptive use of language. Hence in dense and continuous spaces different congruences may correspond to different patterns of prescriptive use of '*x* is congruent to *y*' and related predicates, the intension of the predicate remaining unchanged.

To anticipate a possible misunderstanding, let me say that by the *rudimentary* or *unadorned intension* of '*x* is congruent to *y*', I mean the intension of the congruence predicate that is implicit in the *Scholium* to Newton's *Principia.* When '*x* is congruent to *y*' is thus intensionally pre-empted, physical measurement serves only the epistemic function of enabling one to discover which intervals are congruent to which others. Unlike Newton, Grünbaum maintains that in continuous physical space the rudimentary or unadorned intension of '*x* is congruent to *y*' fails *by itself* to determine a full-blown extension, in the sense that it is neither true nor false (but quite meaningful nevertheless) to say of a pair of quasi-disjoint intervals that they are congruent or that they are incongruent. Now there are two ways to remedy this deficiency of extension: (1) by prescriptive use of '*x* is congruent to *y*' or (2) by augmenting its intension. The first way of effecting a full-blown extension leaves the intension of the predicate undisturbed. The second way, augmentation of intension, yields a family of congruence predicates with different intensions. For example, one might use '*x* is congruent to *y*' in such a way that meter-stick measurement has not only the epistemic function of ascertaining congruences but also enters into the very intension of the predicate. (This second way is favored by many philosophers. For example, it is advocated by operationalists who stress the importance of *defining* scientific concepts in terms of the operations that are to serve as discovery procedures.) If the rudimentary intension of '*x* is congruent to *y*' is augmented in such wise that a full-blown extension is effected for any conceivable state of affairs, then that expression clearly ceases to be a preceptive predicate and even conforms to CA in the strong sense that its extension depends on nothing more than its (augmented) intension and 'purely physical' states of affairs. The two ways just discussed of effecting a full-blown extension for '*x* is congruent to *y*' represent ideal or pure cases. Although most uses of '*x* is congruent to *y*' incorporate some combination of the two, the philosophical significance of the two ways is more readily grasped in the pure cases than in the impure ones. Because the impure cases, wherein prescription and intensional augmentation combine, admit of resolution into pure components, they occasion no novel philosophical observations.

Grünbaum has repeatedly likened the conventionality of congruence ascriptions to that of simultaneity ascriptions in the STR,[14] so it should come as no surprise that '*x* is simultaneous with *y*' also seems to be a preceptive predicate relative to certain kinds of universes. One can be fairly specific about the intension of '*x* is simultaneous with *y*'; it means that *x* and *y* occupy the same position in the system of temporal order. Because there is only one candidate for the role of the system of temporal order in a Newtonian universe, the intension of '*x* is simultaneous with *y*' determines an extension unproblematically. But in an Einsteinian or quasi-Newtonian universe there are infinitely many candidates for the aforementioned role.[15] Hence either '*x* is the system of temporal order' or some related preceptive predicate must be used prescriptively to award this role to one of the candidates in each inertial frame. This is commonly done by a prescriptive use of some such sentence as 'Outgoing and incoming light signals have equal velocities *in vacuo* relative to each inertial frame,' but it could just as well be done through a prescriptive use of '*x* is simultaneous with *y*', e.g. by decreeing that two distant events are simultaneous in a given inertial frame if and only if infinitely slowly transported clocks give the same reading at either event in that frame.[16]

In short, '*x* is simultaneous with *y*' functions as a preceptive predicate in Einsteinian or quasi-Newtonian universes, but has only a descriptive use in Newtonian universes. Whether our universe is Newtonian is a contingent matter to be settled *a posteriori*, so '*x* is simultaneous with *y*' is, like '*x* is congruent to *y*', contingently a preceptive predicate. And like the latter, the former is an impure preceptive predicate; causally connectible events are non-simultaneous quite apart from any prescriptive use of preceptive predicates.

Some philosophers seem to reject geochronometric conventionalism on the ground that the stipulational freedom it allows is tantamount to a linguistic license belied both by commonsensical and scientific fact. The belief that geochronometric conventionalism renders congruence ascriptions wholly arbitrary is a mistaken belief fostered perhaps by talk of 'congruence definitions', 'coordinative definitions', 'congruence conventions' and the like, since over 'definitions' and 'conventions' *non est disputandum*. Much of the illusion of linguistic license is dispelled when '*x* is congruent to *y*' and '*x* is simultaneous with *y*' are viewed as preceptive predicates. For, first of all, not just anyone can use a preceptive predicate prescriptively. For example, the prescriptive use of '*x* is a capital offense' is limited to certain law-giving bodies. Second, the prescriptive use of preceptive predicates can be criticized in view of the purposes that such prescriptions are meant to serve. The freedom of a lawgiving body to make possession of marijuana a capital offense does not render the body immune to criticism when it so decrees. Third, whereas the prescriptive use of preceptive predicates typically creates obligations, the nature and extent of these obligations can vary significantly from case to case. For example, the physicist's congruence prescriptions sometimes impose only self-obligations, e.g. to calculate velocities, forces, etc. in a certain way. And even these prescriptions are not immune to criticism advanced in view of the purposes to be served by theoretical physics. On the other hand, congruence prescriptions of the Bureau of Standards can create obligations which reach even to the neighborhood hardware store.[17]

University of Pittsburgh

NOTES

[1] A. Grünbaum, 'Reply to Hilary Putnam's "An Examination of Grünbaum's Philosophy of Geometry"', *Boston Studies in the Philosophy of Science* Vol. V (1968) p. 45.
[2] *Ibid.*, p. 82.
[3] G. J. Massey, 'Toward a Clarification of Grünbaum's Conception of an Intrinsic Metric', *Philosophy of Science* **36** (1969) 341.
[4] Concerning the thesis of geochronometric conventionalism see Grünbaum, *op. cit.*; Grünbaum, *Philosophical Problems of Space and Time*, Knopf, New York, 1963, Chapters 1–6; Grünbaum, *Geometry and Chronometry in Philosophical Perspective*, University of Minnesota Press, Minneapolis, 1968; Grünbaum, 'Space, Time, and Falsifiability', *Philosophy of Science* **37** (1970), Part A.
[5] See H. Putnam, 'An Examination of Grünbaum's Philosophy of Geometry', in *Philosophy of Science, The Delaware Seminar* **2** (1963) 205–55, and Grünbaum's *Reply* (*op. cit.*).
[6] R. G. Swinburne, Review of Grünbaum's *Geometry and Chronometry in Philosophical Perspective*, forthcoming in *British Journal of the Philosophy of Science*.
[7] R. Butrick, 'Quine on the "is" in "is analytic"', *Mind* **74** (1970) 262.
[8] *Ibid.*
[9] Concerning tensed predicates, see G. J. Massey, *Understanding Symbolic Logic*, Harper and Row, New York, 1970, pp. 265–6 and pp. 404ff.
[10] Butrick, *op. cit.*, p. 263.
[11] Two intervals are quasi-disjoint if and only if neither is included in the other. (This statement is not entirely accurate; for an unexceptionable definition of quasi-disjointedness, see Grünbaum, 'Space, Time, and Falsifiability', *op. cit.*, p. 548.
[12] In his recent 'Space, Time, and Falsifiability', *op. cit.*, pp. 538ff., Grünbaum points out that in the primary sense of 'nontrivial' that he prefers, there are in discrete space (without extreme elements) two nontrivial congruence classes generated by dyadic properties intrinsic to the intervals of the space. One of these is the cardinality-based congruence class; the other is the partition effected by the identity relation among intervals. Any metric such that equality of measure relative to that metric generates either of these congruence classes is nontrivial and intrinsic to the intervals. My 'progression metric' (discussed in Grünbaum, *ibid.*, pp. 539ff) generates the identity-based congruence class, so it is both nontrivial, intrinsic to the intervals, and discrepant with the cardinality-based metrics. But Grünbaum observes that in a stricter sense of 'nontrivial', my progression metric must be judged *trivial* because it renders no two distinct intervals congruent. Because the rudimentary or unadorned intension of 'x is congruent to y' seems to have nontriviality, in the stricter sense, built into it, I will use 'nontrivial' throughout in the stricter of Grünbaum's two senses.
[13] See especially Grünbaum, *ibid.*, pp. 538ff.
[14] See especially Grünbaum, 'Simultaneity by Slow Clock Transport in the Special Theory of Relativity', *Philosophy of Science*, **36** (1969) 5–43.
[15] For details, see *ibid.*
[16] For details, see *ibid.*
[17] In preparing this paper I have profited greatly from discussions with my colleagues Nuel Belnap, Joseph Camp, and Richard Gale. I am especially indebted to Adolf Grünbaum for detailed criticisms and suggestions concerning a number of specific points.

SYNTHESE LIBRARY

Monographs on Epistemology, Logic, Methodology,
Philosophy of Science, Sociology of Science and of Knowledge, and on the
Mathematical Methods of Social and Behavioral Sciences

‡CARL R. KORDIG, *The Justification of Scientific Change*. 1971, XII + 119 pp. Dfl. 33,–

‡YEHOSHUA BAR-HILLEL, *Pragmatics of Natural Languages*. 1971, VII + 231 pp. Dfl. 50,–

‡STEPHEN TOULMIN and HARRY WOOLF (eds.), *Norwood Russell Hanson: What I do not Believe, and other Essays*. 1971, XII + 390 pp. Dfl. 90,–

‡MILIČ ČAPEK, *Bergson and Modern Physics. A Re-interpretation and Re-evaluation. Boston Studies in the Philosophy of Science*. Volume VII (ed. by ROBERT S. COHEN and MARX W. WARTOFSKY). 1971, XV + 414 pp. Dfl. 70,–

‡JOSEPH D. SNEED, *The Logical Structure of Mathematical Physics*. 1971, XV + 311 pp. Dfl. 70,–

‡JEAN-LOUIS KRIVINE, *Introduction to Axiomatic Set Theory*. 1971, VII + 98 pp. Dfl. 28,–

‡RISTO HILPINEN (ed.), *Deontic Logic: Introductory and Systematic Readings*. 1971, VII + 182 pp. Dfl. 45,–

‡EVERT W. BETH, *Aspect of Modern Logic*. 1970, XI + 176 pp. Dfl. 42,–

‡PAUL WEINGARTNER and GERHARD ZECHA, (eds.), *Induction, Physics, and Ethics. Proceedings and Discussions of the 1968 Salzburg Colloquium in the Philosophy of Science*. 1970, X + 382 pp. Dfl. 65,–

‡ROLF A. EBERLE, *Nominalistic System*. 1970, IX + 217 pp. Dfl. 42,–

‡JAAKKO HINTIKKA and PATRICK SUPPES, *Information and Inference*. 1970, X + 336 pp. Dfl. 60,–

‡KAREL LAMBERT, *Philosophical Problems in Logic. Some Recent Developments*. 1970, VII + 176 pp. Dfl. 38,–

‡P. V. TAVANEC (ed.), *Problems of the Logic of Scientific Knowledge*. 1969, XII + 429 pp. Dfl. 95,–

‡ROBERT S. COHEN and RAYMOND J. SEEGER (eds.), *Boston Studies in the Philosophy of Science*. Volume VI: *Ernst Mach: Physicist and Philosopher*. 1970, VIII + 295 pp. Dfl. 38,–

‡MARSHALL SWAIN (ed.), *Induction, Acceptance, and Rational Belief*. 1970, VII + 232 pp. Dfl. 40,–

‡NICHOLAS RESCHER *et al.* (eds.), *Essays in Honor of Carl G. Hempel. A Tribute on the Occasion of his Sixty-Fifth Birthday*. 1969, VII + 272 pp. Dfl. 50,–

‡PATRICK SUPPES, *Studies in the Methodology and Foundations of Science. Selected Papers from 1951 to 1969.* 1969, XII + 473 pp. Dfl. 72,–

‡JAAKKO HINTIKKA, *Models for Modalities. Selected Essays.* 1969, IX + 220 pp. Dfl. 34,–

‡D. DAVIDSON and J. HINTIKKA (eds.), *Words and Objections: Essays on the Work of W. V. Quine.* 1969, VIII + 366 pp. Dfl. 48,–

‡J. W. DAVIS, D. J. HOCKNEY and W. K. WILSON (eds.), *Philosophical Logic.* 1969, VIII + 277 pp. Dfl. 45,–

‡ROBERT S. COHEN and MARX W. WARTOFSKY (eds.), *Boston Studies in the Philosophy of Science.* Volume V: *Proceedings of the Boston Colloquium for the Philosophy of Science 1966/1968.* 1969, VIII + 482 pp. Dfl. 60,–

‡ROBERT S. COHEN and MARX W. WARTOFSKY (eds.), *Boston Studies in the Philosophy of Science.* Volume IV: *Proceedings of the Boston Colloquium for the Philosophy of Science 1966/1968.* 1969, VIII + 537 pp. Dfl. 72,–

‡NICHOLAS RESCHER, *Topics in Philosophical Logic.* 1968, XIV + 347 pp. Dfl. 70,–

‡GÜNTHER PATZIG, *Aristotle's Theory of the Syllogism. A Logical-Philological Study of Book A of the Prior Analytics.* 1968, XVII + 215 pp. Dfl. 48,–

‡C. D. BROAD, *Induction, Probability, and Causation. Selected Papers.* 1968, XI + 296 pp. Dfl. 54,–

‡ROBERT S. COHEN and MARX W. WARTOFSKY (eds.), *Boston Studies in the Philosophy of Science.* Volume III: *Proceedings of the Boston Colloquium for the Philosophy of Science 1964/1966.* 1967, XLIX + 489 pp. Dfl. 70,–

‡GUIDO KÜNG, *Ontology and the Logistic Analysis of Language. An Enquiry into the Contemporary Views on Universals.* 1967, XI + 210 pp. Dfl. 41,–

*EVERT W. BETH and JEAN PIAGET, *Mathematical Epistemology and Psychology.* 1966, XXII + 326 pp. Dfl. 63,–

*EVERT W. BETH, *Mathematical Thought. An Introduction to the Philosophy of Mathematics.* 1965, XII + 208 pp. Dfl. 37,–

‡PAUL LORENZEN, *Formal Logic.* 1965, VIII + 123 pp. Dfl. 26,–

‡GEORGES GURVITCH, *The Spectrum of Social Time.* 1964, XXVI + 152 pp. Dfl. 25,–

‡A. A. ZINOV'EV, *Philosophical Problems of Many-Valued Logic.* 1963, XIV + 155 pp. Dfl. 32,–

‡MARX W. WARTOFSKY (ed.), *Boston Studies in the Philosophy of Science.* Volume I: *Proceedings of the Boston Colloquium for the Philosophy of Science, 1961–1962.* 1963, VIII + 212 pp. Dfl. 26,50

‡B. H. KAZEMIER and D. VUYSJE (eds.), *Logic and Language. Studies dedicated to Professor Rudolf Carnap on the Occasion of his Seventieth Birthday.* 1962, VI + 246 pp. Dfl. 35,–

*EVERT W. BETH, *Formal Methods. An Introduction to Symbolic Logic and to the Study of Effective Operations in Arithmetic and Logic.* 1962, XIV + 170 pp. Dfl. 35,–

*HANS FREUDENTHAL (ed.), *The Concept and the Role of the Model in Mathematics and Natural and Social Sciences. Proceedings of a Colloquium held at Utrecht, The Netherlands, January 1960.* 1961, VI + 194 pp. Dfl. 34,–

‡P. L. R. GUIRAUD, *Problèmes et méthodes de la statistique linguistique.* 1960, VI + 146 pp. Dfl. 28,–

*J. M. BOCHEŃSKI, *A Precis of Mathematical Logic.* 1959, X + 100 pp. Dfl. 23,–

SYNTHESE HISTORICAL LIBRARY

Texts and Studies
in the History of Logic and Philosophy

Editors:

‡Karl Wolf and Paul Weingartner (eds.), *Ernst Mally: Logische Schriften.* 1971, X + 340 pp. Dfl. 80,–

‡Leroy E. Loemker (ed.), *Gottfried Wilhelm Leibnitz: Philosophical Papers and Letters.* A Selection Translated and Edited, with an Introduction. 1969, XII + 736 pp. Dfl. 125,–

‡M. T. Beonio-Brocchieri Fumagalli, *The Logic of Abelard.* Translated from the Italian. 1969, IX + 101 pp. Dfl. 27,–

Sole Distributors in the U.S.A. and Canada:
*GORDON & BREACH, INC., 440 Park Avenue South, New York, N.Y. 10016
‡HUMANITIES PRESS, INC., 303 Park Avenue South, New York, N.Y. 10010